Student Solutions Manual

Prealgebra and Introductory Algebra
An Applied Approach

THIRD EDITION

AF812763

Richard N. Aufmann
Palomar College

Joanne S. Lockwood
Nashua Community College

Prepared by

Jeremiah Gilbert
San Bernardino Valley College

CENGAGE Learning

Australia • Brazil • Mexico • Singapore • United Kingdom • United States

CENGAGE Learning

ISBN-13: 978-1-285-42038-7
ISBN-10: 1-285-42038-1

Cengage Learning
200 First Stamford Place, 4th Floor
Stamford, CT 06902
USA

Cengage Learning is a leading provider of customized learning solutions with office locations around the globe, including Singapore, the United Kingdom, Australia, Mexico, Brazil, and Japan. Locate your local office at: **www.cengage.com/global**.

Cengage Learning products are represented in Canada by Nelson Education, Ltd.

To learn more about Cengage Learning Solutions, visit **www.cengage.com**.

Purchase any of our products at your local college store or at our preferred online store **www.cengagebrain.com**.

Printed in the United States of America
1 2 3 4 5 6 7 17 16 15 14 13

Contents

Chapter 1: Introduction to Whole Numbers

Prep Test

1. 8
2. 1 2 3 4 5 6 7 8 9 10
3. a and D; b and E; c and A; d and B; e and F; f and C
4. 0
5. fifty

Section 1.1

Concept Check

1. is greater than
3. thousand
5. less

Objective A Exercises

7.
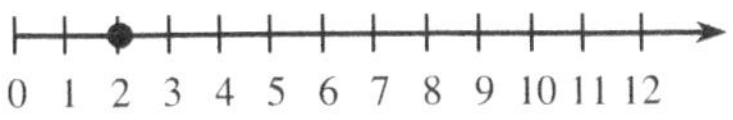

9.
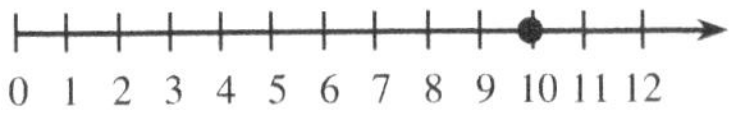

11.
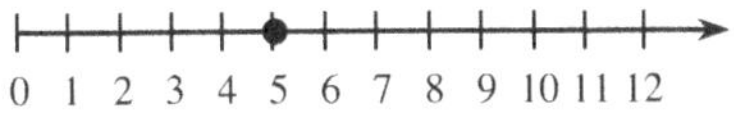

13.
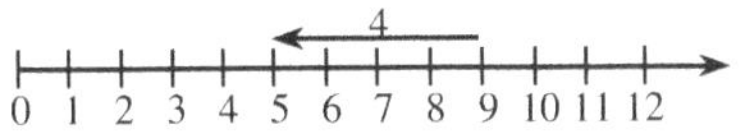

5 is 4 units to the left of 9.

15.
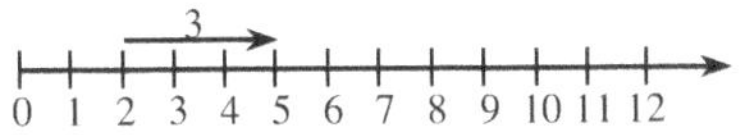

5 is 3 units to the right of 2.

17.
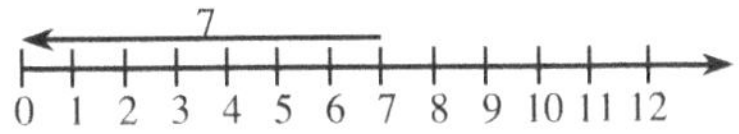

0 is 7 units to the left of 7.

19. $27 < 39$
21. $0 < 52$
23. $273 > 194$
25. $2761 < 3857$
27. $4610 > 4061$
29. $8005 < 8050$
31. Yes
33. 11, 14, 16, 21, 32
35. 13, 48, 72, 84, 93
37. 26, 49, 77, 90, 106
39. 204, 399, 662, 736, 981
41. 307, 370, 377, 3077, 3700

Objective B Exercises

43. five hundred eight
45. six hundred thirty-five
47. four thousand seven hundred ninety
49. fifty-three thousand six hundred fourteen
51. two hundred forty-six thousand fifty-three
53. three million eight hundred forty-two thousand nine hundred five
55. 496
57. 53,340
59. 502,140
61. 9706
63. 5,012,907
65. 8,005,010
67. 7000 + 200 + 40 + 5
69. 500,000 + 30,000 + 2000 + 700 + 90 + 1
71. 5000 + 60 + 4
73. 20,000 + 300 + 90 + 7
75. 400,000 + 2000 + 700 + 8
77. 8,000,000 + 300 + 10 + 6
79. Tens

Objective C Exercises

81. 7108 — Given place value; $8 > 5$

7108 rounded to the nearest ten is 7110.

83. 4962 — Given place value; $6 > 5$

4962 rounded to the nearest hundred is 5000.

85. 28,551 — Given place value
5 = 5

28,551 rounded to the nearest hundred is 28,600.

87. 6808 — Given place value
8 > 5

6809 rounded to the nearest thousand is 7000.

89. 93,825 — Given place value
8 > 5

93,825 rounded to the nearest thousand is 94,000.

91. 629,513 — Given place value
5 = 5

629,513 rounded to the nearest thousand is 630,000.

93. 352,876 — Given place value
2 < 5

352,876 rounded to the nearest ten-thousand is 350,000.

95. Sometimes true

97. Never true

Objective D Exercises

99. Strategy:
To find the person with the greater number of stolen bases, compare the numbers 892 and 937.

Solution:
892 < 937
Billy Hamilton had the greater number of stolen bases.

101. Strategy:
To find the greater number of performances, compare the numbers 2844 and 3242.

Solution:
2844 < 3242
"Fiddler on the Roof" was performed the greater number of times.

103. Strategy:
To find the food which contains more calories, compare the numbers 190 and 114.

Solution:
190 > 114
Two tablespoons of peanut butter contain more calories.

105. Strategy:
a. To determine during which school year enrollment was lowest, use the line graph to find the school year corresponding to the lowest point on the line.

b. To determine if enrollment increased or decreased between 1990 and 2000, read the graph at 1990 and at 2000.

Solution:
a. Student enrollment was lowest during the 1980 school year.

b. Enrollment was greater in 2000 than in 1990. Enrollment increased between 1990 and 2000.

107. Strategy:
a. To determine the most often mentioned complaint, find the complaint with the largest number of responses.

b. To determine the least often mentioned complaint, find the complaint with the smallest number of responses.

Solution:
a. The complaint with the largest number of responses is people talking.
The most often mentioned complaint is people talking.

b. The complaint with the smallest number of responses is uncomfortable seats.
The least often mentioned complaint was uncomfortable seats.

109. Strategy:
To find which chain had more restaurants, compare the two numbers 33,749 and 32,737.

Solution:
33,749 > 32,737
There were more Subway restaurants at the end of 2010.

111. Strategy:
To find the land area to the nearest ten-thousand acres, round 161,546 to the nearest ten-thousand.

Solution:
161,546 rounded to the nearest ten-thousand is 160,000.
To the nearest ten-thousand acres, the land area of the Appalachian Trail is 160,000 acres.

113. Strategy:
To find the U.S. population to the nearest million, round 308,745,538 to the nearest million.

Solution:
308,745,538 rounded to the nearest million is 309,000,000.
To the nearest million, the population of the United States on April 10, 2010 was 309,000,000.

Critical Thinking 1.1

115. The largest three-digit number is 999. (The next largest whole number is 1000, which has four digits.) The smallest five-digit number is 10,000. (The next smallest whole number if 9999, which has 4 digits.)

Projects or Group Activities 1.1

117. Asia (17,000,000 mi^2)
Africa (12,000,000 mi^2)
North America (9,000,000 mi^2)
South America (7,000,000 mi^2)
Antarctica (6,000,000 mi^2)
Europe (4,000,000 mi^2)
Australia (3,000,000 mi^2)

119. Answers to the question of the total enrollment of your school will vary. Generally, the population of a small school (less than 5000 students) is rounded to the nearest hundred; the population of a medium-size school (5000-20,000 students) or a large school (more than 20,000 students) is rounded to the nearest thousand.
The population of your state can be found in most annual almanacs. The place value to which the population is rounded will vary. For example, the population of New York might be rounded to the nearest million (18,000,000) while the population of Wyoming might be rounded to the nearest hundred thousand (500,000).
Generally the population of the United States is rounded to the nearest ten million (310,000,000).

Section 1.2

Concept Check

1. 24 ; 15 ; 39

3. thousand ; ten-thousand

5. addition

Objective A Exercises

7.
$$\begin{array}{r} 732{,}453 \\ +\,651{,}206 \\ \hline 1{,}383{,}659 \end{array}$$

9.
$$\begin{array}{r} \scriptstyle 1\ 1\ 1 \\ 2{,}879 \\ +\,3{,}164 \\ \hline 6{,}043 \end{array}$$

11.
$$\begin{array}{r} \scriptstyle 1\ 1 \\ 4{,}037 \\ 3{,}342 \\ +\,5{,}169 \\ \hline 12{,}548 \end{array}$$

13.
$$\begin{array}{r} \scriptstyle 1\ 1\ 1 \\ 67{,}390 \\ 42{,}761 \\ +\,89{,}405 \\ \hline 199{,}556 \end{array}$$

15.
$$\begin{array}{r} \scriptstyle 2\ 1\ 1\ 2 \\ 54{,}097 \\ 33{,}432 \\ 97{,}126 \\ 64{,}508 \\ 78{,}310 \\ \hline 327{,}473 \end{array}$$

17.
$$\begin{array}{r} 88{,}123 \\ +\ 80{,}451 \\ \hline 168{,}574 \end{array}$$

19.
$$\begin{array}{r} ^{1} \\ 7{,}293 \\ +654 \\ \hline 7{,}947 \end{array}$$

21.
$$\begin{array}{r} ^{1\ 1} \\ 216 \\ 8{,}707 \\ +\ 90{,}714 \\ \hline 99{,}637 \end{array}$$

23.
$$\begin{array}{r} 585 \\ 497 \\ 412 \\ +\ 378 \\ \hline 1{,}872 \end{array}$$

The total number of undergraduates enrolled at the college in 2015 was 1872.

25.
$$\begin{array}{rcr} 6{,}742 & \rightarrow & 7{,}000 \\ +\ 8{,}298 & \rightarrow & +8{,}000 \\ \hline 15{,}040 & & 15{,}000 \end{array}$$

27.
$$\begin{array}{rcr} 972{,}085 & \rightarrow & 1{,}000{,}000 \\ +\ 416{,}832 & \rightarrow & +400{,}000 \\ \hline 1{,}388{,}917 & & 1{,}400{,}000 \end{array}$$

29.
$$\begin{array}{rcr} 387 & \rightarrow & 400 \\ 295 & \rightarrow & 300 \\ 614 & \rightarrow & 600 \\ +\ 702 & \rightarrow & +\ 700 \\ \hline 1{,}998 & & 2{,}000 \end{array}$$

31.
$$\begin{array}{rcr} 224{,}196 & \rightarrow & 200{,}000 \\ 7{,}074 & \rightarrow & 7{,}000 \\ +\ 98{,}531 & \rightarrow & +\ 100{,}000 \\ \hline 329{,}801 & & 307{,}000 \end{array}$$

33. $x + y$

$574 + 698$

$$\begin{array}{r} ^{1\ 1} \\ 574 \\ +\ 698 \\ \hline 1{,}272 \end{array}$$

35. $x + y$

$4{,}752 + 7{,}398$

$$\begin{array}{r} ^{1\ 1\ 1} \\ 4{,}752 \\ +\ 7{,}398 \\ \hline 12{,}150 \end{array}$$

37. $x + y$

$38{,}229 + 51{,}671$

$$\begin{array}{r} ^{1\ 1} \\ 38{,}229 \\ +\ 51{,}671 \\ \hline 89{,}900 \end{array}$$

39. $a + b + c$

$693 + 508 + 371$

$$\begin{array}{r} ^{1\ 1} \\ 693 \\ 508 \\ +\ 371 \\ \hline 1{,}572 \end{array}$$

41. $a + b + c$

$4{,}938 + 2{,}615 + 7{,}038$

$$\begin{array}{r} ^{1\ \ 2} \\ 4{,}938 \\ 2{,}615 \\ +\ 7{,}038 \\ \hline 14{,}591 \end{array}$$

43. $a + b + c$

$12{,}897 + 36{,}075 + 7{,}038$

$$\begin{array}{r} ^{1\ 1\ \ 2\ 2} \\ 12{,}897 \\ 36{,}075 \\ +\ 7{,}038 \\ \hline 56{,}010 \end{array}$$

45. The Commutative Property of Addition

47. The Associative Property of Addition

49. The Addition Property of Zero

51. $28 + 0 = 28$

53. $9 + (4 + 17) = (9 + 4) + 17$

55. $15 + 34 = 34 + 15$

57. Commutative Property of Addition

59. $m + 6 = 13$

$$\begin{array}{c|c} 17+6 & 13 \end{array}$$

$23 \neq 13$

No, 17 is not a solution of the equation $m + 6 = 13$.

61. $n = 17 + 24$

$$\begin{array}{c|c} 41 & 17+24 \end{array}$$

$41 = 41$

Yes, 41 is a solution of the equation $n = 17 + 24$.

63. $38 = 11 + z$

$$\begin{array}{c|c} 38 & 11+29 \end{array}$$

$38 \neq 40$

No, 29 is not a solution of the equation $38 = 11 + z$.

Objective B Exercises

65. $\begin{array}{r} \scriptstyle 8\,11 \\ 591 \\ -238 \\ \hline 353 \end{array}$

67. $\begin{array}{r} \scriptstyle 8\,14\,10 \\ 950 \\ -483 \\ \hline 467 \end{array}$

69. $\begin{array}{r} \scriptstyle 5\,12 \\ 762 \\ -659 \\ \hline 103 \end{array}$

71. $\begin{array}{r} \scriptstyle 7\,9\,15 \\ 805 \\ -147 \\ \hline 658 \end{array}$

73. $\begin{array}{r} \scriptstyle 6\,12\,15\,11 \\ 7{,}361 \\ -4{,}575 \\ \hline 2{,}786 \end{array}$

75. $\begin{array}{r} \scriptstyle 3\,9\,9\,10 \\ 4{,}000 \\ -1{,}873 \\ \hline 2{,}127 \end{array}$

77. $\begin{array}{r} \scriptstyle 6\,12\,9\,10 \\ 7{,}300 \\ -2{,}562 \\ \hline 4{,}738 \end{array}$

79. $\begin{array}{r} \scriptstyle 6\,9\,9\,9\,13 \\ 70{,}003 \\ -8{,}246 \\ \hline 61{,}757 \end{array}$

81. $\begin{array}{r} \scriptstyle 5\,11\,13 \\ 1{,}623 \\ -287 \\ \hline 1{,}336 \end{array}$

83. $\begin{array}{r} \scriptstyle 7\,9\,11 \\ 14{,}801 \\ -3{,}522 \\ \hline 11{,}279 \end{array}$

85. $x - y$

87. $\begin{array}{r} \scriptstyle 1\,9\,10 \\ 200 \\ -175 \\ \hline 25 \end{array}$

The eruption of the Giant is 25 ft higher than the eruption of Old Faithful.

89. $\begin{array}{rcr} 8{,}953 & \to & 9{,}000 \\ -2{,}217 & \to & \underline{2{,}000} \\ \hline 6{,}736 & & 7{,}000 \end{array}$

91. $\begin{array}{rcr} 63{,}051 & \to & 60{,}000 \\ 29{,}478 & \to & -30{,}000 \\ \hline 33{,}573 & & 30{,}000 \end{array}$

93. $\begin{array}{rcr} 58{,}316 & \to & 60{,}000 \\ -19{,}072 & \to & -20{,}000 \\ \hline 39{,}244 & & 40{,}000 \end{array}$

95. $\begin{array}{rcr} 873{,}925 & \to & 900{,}000 \\ -28{,}744 & \to & \underline{30{,}000} \\ \hline 845{,}181 & & 870{,}000 \end{array}$

97. $x - y$

$80 - 33$

$\begin{array}{r} \scriptstyle 7\,10 \\ 80 \\ -33 \\ \hline 47 \end{array}$

99. $x - y$

$623 - 197$

$\begin{array}{r} \scriptstyle 5\,11\,13 \\ 623 \\ -197 \\ \hline 426 \end{array}$

101. $x - y$

$870 - 243$

$\begin{array}{r} \scriptstyle 6\,10 \\ 870 \\ -243 \\ \hline 627 \end{array}$

103. $x - y$

$7{,}814 - 3{,}512$

$\begin{array}{r} 7{,}814 \\ -3{,}512 \\ \hline 4{,}302 \end{array}$

105. $x - y$

$1{,}406 - 968$

$\begin{array}{r} \scriptstyle 0\,13\,9\,16 \\ 1{,}406 \\ -968 \\ \hline 438 \end{array}$

107. $x - y$

56,397 − 8,249

$$\begin{array}{r} \overset{4\,16}{\not5\not6},\overset{8\,17}{39\not7} \\ -\ 8,249 \\ \hline 48,148 \end{array}$$

109. $48 - p = 17$

$$\begin{array}{c|c} 48-31 & 17 \end{array}$$

$17 = 17$

Yes, 31 is a solution of the equation $48 - p = 17$.

111. $34 = x - 9$

$$\begin{array}{c|c} 34 & 25-9 \end{array}$$

$34 \neq 16$

No, 25 is not a solution of the equation $34 = x - 9$.

113. $72 = 100 - d$

$$\begin{array}{c|c} 72 & 100-28 \end{array}$$

$72 = 72$

Yes, 28 is a solution of the equation $72 = 100 - d$.

Objective C Exercises

115. Strategy:
To find the difference, subtract 99 (the largest two-digit number) from 1,000 (the smallest four-digit number).

Solution:

$$\begin{array}{r} 1,000 \\ -\ \ 99 \\ \hline 901 \end{array}$$

The difference is 901.

117. Strategy:
To find the number of calories, add the number of calories in one apple (80), one cup of cornflakes (95), one tablespoon of sugar (45), and one cup of milk (150).

Solution:
$80 + 95 + 45 + 150 = 370$
The breakfast contained 370 calories.

119. Strategy:
To find the perimeter of a rectangle, replace L with 24 and W with 15 in the given formula and solve for P.

Solution:
$P = L + W + L + W$
$P = 24 + 15 + 24 + 15$
$P = 78$
The perimeter is 78 m.

121. Strategy:
To find the perimeter of a triangle, replace a, b, and c with 16, 12, and 15 in the given formula and solve for P.

Solution:
$P = a + b + c$
$P = 16 + 12 + 15$
$P = 43$
The perimeter is 43 in.

123. Strategy:
To find the amount, add the deposit (870) to the balance in the checking account (1,054).

Solution:

$$\begin{array}{r} 870 \\ +\ 1,054 \\ \hline 1,924 \end{array}$$

You have $1,924 in your checking account.

125. Strategy:
To find the total repair bill, add the cost of parts (358), labor (156) and tax (30).

Solution:

$$\begin{array}{r} 358 \\ 156 \\ +\ 30 \\ \hline 544 \end{array}$$

The total repair bill is $544.

127. Strategy:
To find the estimate of the number of miles driven:
→Round the two readings of the odometer.
→Subtract the rounded reading from the beginning of the year from the rounded reading at the end of the year.

Solution:

$$\begin{array}{lcr} 77,912 & \rightarrow & 80,000 \\ 58,376 & \rightarrow & -\ 60,000 \\ \hline & & 20,000 \end{array}$$

The car was driven approximately 20,000 mi during the year.

129. Strategy:
To determine between which two months car sales increased the most, find the difference, if there was an increase, between sales for January and February, February and March, and March and April for 2005.

Solution:
Between January and February, 2005: 132 – 108 = 24
Between February and March, 2005: 152 – 132 = 20
Between March and April, 2005: not an increase
Car sales increased the most between January and February of 2005. The increase was 24 car.

131. Strategy:
To find the number of U.S. households, add the number of households with broadband Internet access to the number of households without broadband Internet access.

Solution:
37+78=115
There are 115 million U.S. households.

133. Strategy:
To determine between which two years electric car sales are projected to increase the most, find the difference between projected sales for each pair of consecutive years.

Solution:
Between 2015 and 2016:
230,000 – 188,000 = 42,000
Between 2016 and 2017:
312,000 – 230,000 = 82,000
Between 2017 and 2018:
359,000 – 312,000 = 47,000
Between 2018 and 2019:
406,000 – 359,000 = 47,000
Between 2019 and 2020:
414,000 – 406,000 = 8,000
Car sales are projected to increase the most between 2016 and 2017. The increase is projected to be 82,000 cars.

135. Strategy:
To find the value of the investment, replace P by 12,500 and I by 775 in the given formula and solve for A.

Solution:
$A = P + I$
$A = 12{,}500 + 775$
$A = 13{,}275$
The value of the investment is \$13,275.

137. Strategy:
To find the mortgage loan, replace S by 290,000 and D by 29,000 in the given formula and solve for M.

Solution:
$M = S - D$
$M = 290{,}000 - 29{,}000$
$M = 261{,}000$
The mortgage loan on the home is \$261,000.

139. Strategy:
To find the ground speed, replace a by 375 and h by 25 in the given formula and solve for g.

Solution:
$g = a - h$
$g = 375 - 25$
$g = 350$
The ground speed of the airplane is 350 mph.

141. It is not possible to tell how many motorists were driving at 70 mph. The data shows that 3,717 drivers were traveling at 66–70 mph. The number 3,717 includes the motorists driving at 66–69 mph as well as those driving at 70 mph. We cannot separate those driving at 70 mph from the others.

143. $b - a$; $b - a$ represents how much longer the side of length b is than the side of length a.

Critical Thinking 1.2

145. The two-digit numbers are the numbers 10 through 99.
99 – 9 = 90.
There are 90 two-digit numbers.
The three-digit numbers are the numbers 100 through 999. 999 – 99 = 900.
There are 900 three-digit numbers.

Projects or Group Activities 1.2

147.

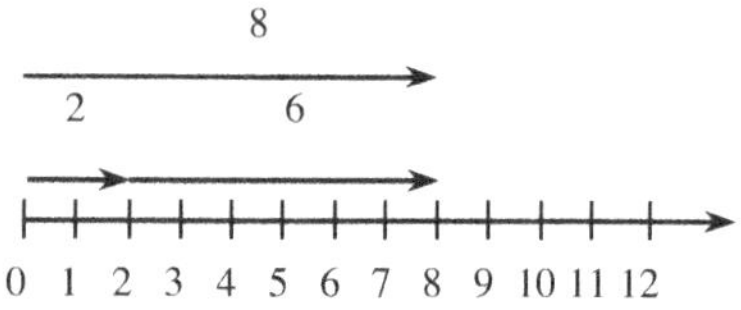

Check Your Progress: Chapter 1

1.

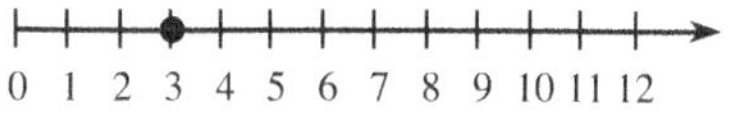

2. 199, 247, 462, 506, 831

3. 397>246; 898<1594

4.

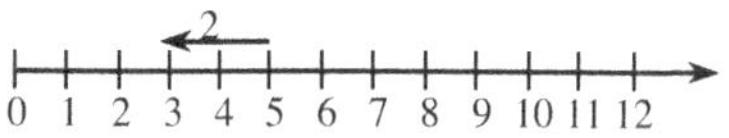

3 is 2 units to the left of 5.

5. six thousand seven hundred two

6. 32,518

7. 900,000 + 3000 + 400 + 80 + 7

8. 16,000

9. 433,000

10. 172,621

11. 1683

12.
$$\begin{array}{rcr} 473 & \to & 500 \\ 879 & \to & 900 \\ 215 & \to & 200 \\ +306 & \to & +300 \\ \hline 1{,}873 & & 1{,}900 \end{array}$$

13. The Associative Property of Addition

14. $x + y + z$
$16 + 42 + 39$
$$\begin{array}{r} {}^{1}16 \\ 42 \\ +39 \\ \hline 97 \end{array}$$

15. $24 = 18 + n$
$$\begin{array}{c|c} 24 & 18+42 \end{array}$$
$24 \neq 60$
No, 42 is not a solution of the equation $24 = 18 + n$.

16. 4600 – 2781 = 1819

17. 13,904 – 8655 = 5249

18.
$$\begin{array}{rcr} 57{,}293 & \to & 60{,}000 \\ -46{,}018 & \to & 50{,}000 \\ \hline 11{,}275 & & 10{,}000 \end{array}$$

19. $x - y$
$704 - 279$
$$\begin{array}{r} 704 \\ -279 \\ \hline 425 \end{array}$$

20. $82 = 143 - x$
$$\begin{array}{c|c} 82 & 143-61 \end{array}$$
$82 \quad = 82$
Yes, 61 is a solution of the equation $82 = 143 - 61$.

21. Strategy:
To find the difference, subtract 6530 from 7460.

Solution:
$$\begin{array}{r} 7460 \\ -\ 6530 \\ \hline 930 \end{array}$$
The difference is $930.

22. Strategy:
To find the total amount of charitable contributions, add each of her contributions for the six-month period.

Solution:
25 + 30 + 13 + 15 + 20 + 27 = 130
The total amount of charitable contributions is $130.

23. Strategy:
To find which planet is smaller, compare the diameter of Uranus (32,200) with the diameter of Neptune (30,800).

Solution:
32,200 > 30,800
Neptune is smaller than Uranus.

24. Strategy:
To find the perimeter of a triangle, replace a, b, and c with 6, 9, and 12 in the given formula and solve for P.

Solution:
P = a + b + c
P = 6 + 9 + 12
P = 27
The perimeter is 27 m.

25. Strategy:
To find the price, replace *C* by 119 and *M* by 79 in the given formula and solve for *P*.

Solution:
$P = C + M$
$P = 119 + 79$
$P = 198$
The price is $198.

Section 1.3

Concept Check

1. Students should note that there are five 6's. Therefore, the addition $6 + 6 + 6 + 6 + 6$ can be written as 5 times 6: 5×6.

3. a. 3; 4

b. 3; 3; 3; 3; 81

5. 495; 6

7. factors; 1; 3; 5; 15

9. division

Objective A Exercises

11.
$$\begin{array}{r} \scriptstyle 2\,6 \\ 127 \\ \times\ \ 9 \\ \hline 1{,}143 \end{array}$$

13.
$$\begin{array}{r} \scriptstyle 4\ \ 6 \\ 6{,}709 \\ \times\ \ \ \ 7 \\ \hline 46{,}963 \end{array}$$

15.
$$\begin{array}{r} \scriptstyle 7\,6\ 5\,7 \\ 58{,}769 \\ \times\ \ \ \ \ 8 \\ \hline 470{,}152 \end{array}$$

17.
$$\begin{array}{r} 683 \\ \times 71 \\ \hline 683 \\ 47\,81 \\ \hline 48{,}493 \end{array}$$

19.
$$\begin{array}{r} 7{,}053 \\ \times 46 \\ \hline 42\,318 \\ 282\ 12 \\ \hline 324{,}438 \end{array}$$

21.
$$\begin{array}{r} 3{,}285 \\ \times 976 \\ \hline 19\,710 \\ 229\,95 \\ 2\,956\,5 \\ \hline 3{,}206{,}160 \end{array}$$

23. $500 \cdot 3 = 1{,}500$

25. $40 \cdot 50 = 2{,}000$

27. $400 \cdot 3 \cdot 20 \cdot 0 = 1{,}200 \cdot 20 \cdot 0$
$= 24{,}000 \cdot 0$
$= 0$

29. a.
$$\begin{array}{r} 373 \\ \times 6 \\ \hline 2{,}238 \end{array}$$
You would burn 2,238 calories by working out vigorously on the stair climber for 6 h.

b.
$$\begin{array}{r} 353 \\ \times\ \ 12 \\ \hline 706 \\ 3\,53 \\ \hline 4{,}236 \end{array}$$

You would burn 4,236 calories by working out moderately on a treadmill for 12 h.

31. $8{,}745 \rightarrow 9{,}000$
$63 \rightarrow 60$
$8{,}745 \cdot 63 = 550{,}935$
$9{,}000 \cdot 60 = 540{,}000$

33. $64{,}409 \rightarrow 60{,}000$
$67 \rightarrow 70$
$64{,}409 \cdot 67 = 4{,}315{,}403$
$60{,}000 \cdot 70 = 4{,}200{,}000$

35. $432 \rightarrow 400$
$91 \rightarrow 90$
$432 \cdot 91 = 39{,}312$
$400 \cdot 90 = 36{,}000$

37. $2{,}837 \rightarrow 3{,}000$
$216 \rightarrow 200$
$2{,}837 \cdot 216 = 612{,}792$
$3{,}000 \cdot 200 = 600{,}000$

39. cd
$381 \cdot 25 = 9{,}525$

41. $6n$
$6 \cdot 382 = 2{,}292$

43. abc
$4 \cdot 20 \cdot 50 = 80 \cdot 50$
$= 4{,}000$

45. $4ab$
$4 \cdot 95 \cdot 33 = 380 \cdot 33$
$= 12{,}540$

47. For example, 20 and 500

49. The Associative Property of Multiplication

51. The Multiplication Property of Zero

53. $(5 \cdot 6)100 = 5(6 \cdot 100)$

55. $1 \cdot 77 = 77$

57. $4 = 4n$
$4 \mid 4(0)$
$4 \neq 0$
No, 0 is not a solution of the equation $4 = 4n$.

59. $56 = 4c$
$56 \mid 4(14)$
$56 = 56$
Yes, 14 is a solution of the equation $56 = 4c$.

61. $44 = 3a$
$44 \mid 3(11)$
$44 \neq 33$
No, 11 is not a solution of the equation $44 = 3a$.

Objective B Exercises

63. $3 \cdot 3 \cdot 3 \cdot 3 \cdot 3 \cdot 3 \cdot 5 \cdot 5 \cdot 5 = 3^6 \cdot 5^3$

65. $7 \cdot 7 \cdot 11 \cdot 11 \cdot 11 \cdot 19 \cdot 19 \cdot 19 \cdot 19 = 7^2 \cdot 11^3 \cdot 19^4$

67. $d \cdot d \cdot d = d^3$

69. $a \cdot a \cdot b \cdot b \cdot b \cdot b = a^2 b^4$

71. $2^6 = 2 \cdot 2 \cdot 2 \cdot 2 \cdot 2 \cdot 2 = 64$

73. $10^9 = 1{,}000{,}000{,}000$

75. $2^4 \cdot 3^2 = (2 \cdot 2 \cdot 2 \cdot 2) \cdot (3 \cdot 3) = 16 \cdot 9 = 144$

77. $2^4 \cdot 10^2 = (2 \cdot 2 \cdot 2 \cdot 2) \cdot (10 \cdot 10)$
$= 16 \cdot 100 = 1{,}600$

79. $4^3 \cdot 0^3 = (4 \cdot 4 \cdot 4) \cdot (0 \cdot 0 \cdot 0) = 64 \cdot 0 = 0$

81. $5^2 \cdot 2 \cdot 3^4 = (5 \cdot 5) \cdot 2 \cdot (3 \cdot 3 \cdot 3 \cdot 3)$
$= 25 \cdot 2 \cdot 81 = 50 \cdot 81 = 4{,}050$

83. $6^3 = 6 \cdot 6 \cdot 6 = 36 \cdot 6 = 216$

85. $11^2 = 11 \cdot 11 = 121$

87. t^5

89. $x^2 y$
$3^2 \cdot 4 = (3 \cdot 3) \cdot 4$
$= 9 \cdot 4$
$= 36$

91. ab^3
$7 \cdot 4^3 = 7 \cdot (4 \cdot 4 \cdot 4)$
$= 7 \cdot 64$
$= 448$

93. $m^3 n^3$
$5^3 \cdot 10^3 = (5 \cdot 5 \cdot 5)(10 \cdot 10 \cdot 10)$
$= 125 \cdot 1{,}000$
$= 125{,}000$

Objective C Exercises

95.
$$\begin{array}{r} 540 \\ 4 \overline{)\, 2{,}160} \\ \underline{-20} \\ 16 \\ \underline{-16} \\ 0 \\ \underline{-0} \\ 0 \end{array}$$

97.
$$\begin{array}{r} 204 \text{ r } 4 \\ 8 \overline{)\, 1{,}636} \\ \underline{-16} \\ 3 \\ \underline{-0} \\ 36 \\ \underline{-32} \\ 4 \end{array}$$

99.
$$\begin{array}{r} 8{,}700 \\ 5\overline{)\,43{,}500} \\ \underline{-40} \\ 35 \\ \underline{-35} \\ 0 \\ \underline{-0} \\ 0 \\ \underline{-0} \\ 0 \end{array}$$

101.
$$\begin{array}{r} 21 \text{ r } 18 \\ 41\overline{)\,879} \\ \underline{-82} \\ 59 \\ \underline{-41} \\ 18 \end{array}$$

103.
$$\begin{array}{r} 180 \text{ r } 21 \\ 23\overline{)\,4{,}161} \\ \underline{-23} \\ 186 \\ \underline{-184} \\ 21 \\ \underline{-0} \\ 21 \end{array}$$

105.
$$\begin{array}{r} 200 \text{ r } 7 \\ 26\overline{)\,5{,}207} \\ \underline{-52} \\ 0 \\ \underline{-0} \\ 7 \\ \underline{-0} \\ 7 \end{array}$$

107.
$$\begin{array}{r} 609 \\ 64\overline{)\,38{,}976} \\ \underline{-384} \\ 57 \\ \underline{-0} \\ 576 \\ \underline{-576} \\ 0 \end{array}$$

109.
$$\begin{array}{r} 40 \text{ r } 27 \\ 223\overline{)\,8{,}947} \\ \underline{-8\,92} \\ 27 \\ \underline{-0} \\ 27 \end{array}$$

111.
$$\begin{array}{r} 908 \\ 9\overline{)\,8{,}172} \\ \underline{-81} \\ 7 \\ \underline{-0} \\ 72 \\ \underline{-72} \\ 0 \end{array}$$

113.
$$\begin{array}{r} 461 \text{ r } 4 \\ 9\overline{)\,4{,}153} \\ \underline{-36} \\ 55 \\ \underline{-54} \\ 13 \\ \underline{-9} \\ 4 \end{array}$$

115. **a.** $300{,}000 \div 12 = 25{,}000$
The average monthly claim for theft would be $25,000.

b. $(560{,}000 + 300{,}000 + 80{,}000 + 50{,}000 + 20{,}000 + 20{,}000 + 110{,}000) \div 12 = 95{,}000$
The average claims per month for all sources combined would be $95,000.

117.
$$\begin{array}{rcr} 62{,}176 & \rightarrow & 60{,}000 \\ 58 & \rightarrow & 60 \end{array}$$
$62{,}176 \div 58 = 1{,}072$
$60{,}000 \div 60 = 1{,}000$

119.
$$\begin{array}{rcr} 637{,}072 & \rightarrow & 600{,}000 \\ 29 & \rightarrow & 30 \end{array}$$
$637{,}072 \div 29 = 21{,}968$
$600{,}000 \div 30 = 20{,}000$

121.
$$\begin{array}{rcr} 11{,}792 & \rightarrow & 10{,}000 \\ 53 & \rightarrow & 50 \end{array}$$
$11{,}792 \div 53 = 222 \text{ r}26$
$10{,}000 \div 50 = 200$

123.
$$\begin{array}{rcr} 632{,}124 & \rightarrow & 600{,}000 \\ 324 & \rightarrow & 300 \end{array}$$
$632{,}124 \div 324 = 1{,}951$
$600{,}000 \div 300 = 2{,}000$

125. $\dfrac{x}{y}$

$\dfrac{56}{56} = 1$

127. $\dfrac{x}{y}$

$\dfrac{0}{23} = 0$

129. $\frac{x}{y}$

$\frac{16,200}{3}$

$$\begin{array}{r} 5,400 \\ 3\overline{)\,16,200} \\ \underline{-15} \\ 12 \\ \underline{-12} \\ 0 \\ \underline{-0} \\ 0 \\ \underline{-0} \\ 0 \end{array}$$

131. $\frac{n}{12} = 5$

$\frac{60}{12} \;\Big|\; 5$

$5 = 5$

Yes, 60 is a solution of the equation.

133. $6 = \frac{48}{y}$

$4 \;\Big|\; \frac{48}{16}$

$6 \neq 3$

No, 16 is not a solution of the equation.

Objective D Exercises

135. $20 \div 1 = 20$
$20 \div 2 = 10$
$20 \div 4 = 5$
$20 \div 5 = 4$
The factors of 20 are 1, 2, 4, 5, 10, and 20.

137. $9 \div 1 = 9$
$9 \div 3 = 3$
The factors of 9 are 1, 3, and 9.

139. $16 \div 1 = 16$
$16 \div 2 = 8$
$16 \div 4 = 4$
The factors of 16 are 1, 2, 4, 8, and 16.

141. 17 is a prime number.
The factors of 17 are 1 and 17.

143. $24 \div 1 = 24$
$24 \div 2 = 12$
$24 \div 3 = 8$
$24 \div 4 = 6$
$24 \div 6 = 4$
The factors of 24 are 1, 2, 3, 4, 6, 8, 12, and 24.

145. $36 \div 1 = 36$
$36 \div 2 = 18$
$36 \div 3 = 12$
$36 \div 4 = 9$
$36 \div 6 = 6$
The factors of 36 are 1, 2, 3, 4, 6, 9, 12, 18, and 36.

147. $45 \div 1 = 45$
$45 \div 3 = 15$
$45 \div 5 = 9$
$45 \div 9 = 5$
The factors of 45 are 1, 3, 5, 9, 15, and 45.

149. $32 \div 1 = 32$
$32 \div 2 = 16$
$32 \div 4 = 8$
$32 \div 8 = 4$
The factors of 32 are 1, 2, 4, 8, 16, and 32.

151. $64 \div 1 = 64$
$64 \div 2 = 32$
$64 \div 4 = 16$
$64 \div 8 = 8$
The factors of 64 are 1, 2, 4, 8, 16, 32, and 64.

153. $75 \div 1 = 75$
$75 \div 3 = 25$
$75 \div 5 = 15$
$75 \div 15 = 5$
The factors of 75 are 1, 3, 5, 15, 25, and 75.

155.
$$\begin{array}{l} 3 \\ 2\overline{)6} \\ 2\overline{)12} \\ 2\overline{)24} \end{array}$$
$24 = 2 \cdot 2 \cdot 2 \cdot 3 = 2^3 \cdot 3$

157.
$$\begin{array}{l} 3 \\ 3\overline{)9} \\ 3\overline{)27} \end{array}$$
$27 = 3 \cdot 3 \cdot 3 = 3^3$

159.
$$\begin{array}{l} 3 \\ 3\overline{)9} \\ 2\overline{)18} \\ 2\overline{)36} \end{array}$$
$36 = 2 \cdot 2 \cdot 3 \cdot 3 = 2^2 \cdot 3^2$

161. $5\overline{)25}$ quotient 5

$2\overline{)50}$

$50 = 2 \cdot 5 \cdot 5 = 2 \cdot 5^2$

163. 83 is a prime number.

165. $2\overline{)10}$ quotient 5

$2\overline{)20}$

$2\overline{)40}$

$2\overline{)80}$

$80 = 2 \cdot 2 \cdot 2 \cdot 2 \cdot 5 = 2^4 \cdot 5$

167. $7\overline{)49}$ quotient 7

$49 = 7 \cdot 7 = 7^2$

169. $3\overline{)9}$ quotient 3

$3\overline{)27}$

$3\overline{)81}$

$81 = 3 \cdot 3 \cdot 3 \cdot 3 = 3^4$

171. 89 is a prime number.

173. $3\overline{)15}$ quotient 5

$2\overline{)30}$

$2\overline{)60}$

$2\overline{)120}$

$120 = 2 \cdot 2 \cdot 2 \cdot 3 \cdot 5 = 2^3 \cdot 3 \cdot 5$

Objective E Exercises

175. Strategy:
To find the average number of funerals per day, divide the number of funerals per year (10,200) by the number of days in a year (365). Then round to the nearest whole number.

Solution:
$\frac{10{,}200}{365} \approx 27.945$
There are approximately 28 funerals per day.

177. Strategy:
a. To determine the number of marriages per week, multiply the number of marriages per day (542) by the number of days in a week (7).

b. To determine the number of marriages per year, multiply the number of marriages per day (542) by the number of days in a year (365).

Solution:
a. $542 \cdot 7 = 3{,}794$
There are 3,794 marriages per week.

b. $542 \cdot 365 = 197{,}830$
There are 197,830 marriages per year.

179. Strategy:
To find the length of fencing needed to surround a square corral, substitute 55 for s in the perimeter formula below.

Solution:
$P = 4 \cdot s = 4 \cdot 55 = 220$
The length of fencing needed to surround a square corral is 220 ft.

181. Strategy:
To find the area of the patio, substitute 9 for s in the area formula below.

Solution:
$A = s^2 = 9^2 = 81$
The area of the patio is 81 ft^2.

183. Strategy:
To find the estimate of the total cost:
→Round the number of suits to hundreds and the cost of each suit to tens.
→Multiply the rounded numbers.

Solution:
215 → 200
83 → 80
$200 \cdot 80 = 16{,}000$
The total cost of the suits is approximately $16,000.

185. Strategy:
To find Melissa's monthly starting salary, divide her yearly salary ($69,048) by the number of months in a year (12).

Solution:
$\frac{69{,}048}{12} = 5{,}754$
Melissa's monthly starting salary is $5,754.

187. Strategy:
To find the total wages paid, multiply the number of plumbers (4), the number of hours each will work (23), and the wage per hour for each ($30).

Solution:
$4 \cdot 23 \cdot 30 = 2{,}760$
The total wages paid are $2,760.

189. Strategy:
To find the total amount paid, replace M by 285 and N by 24 in the given formula and solve for A.

Solution:
$A = MN$
$A = 285 \cdot 24$
$A = 6{,}840$
The total amount paid on the loan is $6,840.

191. Strategy:
To find the time to drive the distance, replace d by 513 and r by 57 in the given formula and solve for t.

Solution:

$t = \frac{d}{r}$
$t = \frac{513}{57}$
$t = 9$
It takes 9 h to drive 513 mi.

193. Strategy:
To find the value per share, replace C by 10,500,000 and S by 500,000 in the given formula and solve for V.

Solution:
$V = \frac{C}{S}$
$V = \frac{10{,}500{,}000}{500{,}000}$
$V = 21$
The stock has a value of $21 per share.

Critical Thinking 1.3

195. The largest possible number that can be written using the digits 1, 3, 8, 2, and 7 is 87,321.
87,321 is not divisible by 4.
The next largest possible number that can be written using these digits is 87,312.
87,312 is divisible by 4.
87,312 is the largest possible number that can be written using the digits 1, 3, 8, 2, 7 and that is divisible by 4.

197. **a**. We need to consider each of the digits from 0 through 9. We know that any number ending with an even digit is not prime as it is divisible by 2. Also, any number ending with 5 is not prime as it is divisible by 5. That leaves the four digits 1,3,7, and 9.

b. There are many possible answers. For example, 21, 33, 27, and 39 are not prime numbers.

Projects or Group Activities 1.3

199. To use the Sieve of Eratosthenes to find all prime numbers less than 100, write out all numbers from 1 to 100 in order. Cross out 1, since 1 is not prime. Mark 2 as prime and cross out all multiples of 2. Proceed in order by marking the next crossed out number as prime, and crossing out each of its multiples. Using this method, we see that the prime numbers less than 100 are 2, 3, 5, 7, 11, 13, 17, 19, 23, 29, 31, 37, 41, 43, 47, 53, 59, 61, 67, 71, 73, 79, 83, 89, and 97.

Section 1.4

Concept Check

1. to multiply 7 times 2

Objective A Exercises

3. $8 \div 4 + 2 = 2 + 2$
$= 4$

5. $6 \cdot 4 + 5 = 24 + 5$
$= 29$

7. $4^2 - 3 = 16 - 3$
$= 13$

9. $5 \cdot (6 - 3) + 4 = 5 \cdot 3 + 4$
$= 15 \div 4$
$= 19$

11. $9 + (7 + 5) \div 6 = 9 + 12 \div 6$
$= 9 \div 2$
$= 11$

13. $13 \cdot (1 + 5) \div 13 = 13 \cdot 6 \div 13$
$= 78 \div 13$
$= 6$

15. $6 \cdot 3^2 + 7 = 6 \cdot 9 + 7$
$= 54 + 7$
$= 61$

17. $14 + 5 \cdot 2^3 = 14 + 5 \cdot 8$
$= 14 + 40$
$= 54$

19. $10 + (8 - 5) \cdot 3 = 10 + 3 \cdot 3$
$= 10 + 9$
$= 19$

21. $2^3 + 4(10 - 6) = 2^3 + 4 \cdot 4$
$= 8 + 4 \cdot 4$
$= 8 + 16$
$= 2$

23. $6(7) + 4^2 \cdot 3^2 = 6(7) + 16 \cdot 9$
$= 42 + 16 \cdot 9$
$= 42 + 144$
$= 186$

25. $18 + 3(7) = 18 + 21$
$= 39$

27. $6(8 - 3) - 12 = 6(5) - 12$
$= 30 - 12$
$= 18$

29. $16 - (13 - 5) \div 4 = 16 - 8 \div 4$
$= 16 - 2$
$= 14$

31. $17 + 1 - 8 \cdot 2 \div 4 = 17 + 1 - 16 \div 4$
$= 17 + 1 - 4$
$= 18 - 4$
$= 14$

33. $x - 2y$
$8 - 2 \cdot 3 = 8 - 6$
$= 2$

35. $x^2 + 3y$
$6^2 + 3 \cdot 7 = 36 + 3 \cdot 7$
$= 36 + 21$
$= 57$

37. $x^2 + y \div x$
$2^2 + 8 \div 2 = 4 + 8 \div 2$
$= 4 + 4$
$= 8$

39. $4x + (x - y)^2$
$4 \cdot 8 + (8 - 2)^2 = 4 \cdot 8 + 6^2$
$= 4 \cdot 8 + 36$
$= 32 + 36$
$= 68$

41. $x^2 + 3(x - y) + z^2$
$2^2 + 3(2 - 1) + 3^2 = 2^2 + 3 \cdot 1 + 3^2$
$= 4 + 3 \cdot 1 + 9$
$= 4 + 3 + 9$
$= 7 + 9$
$= 16$

43. $11 + (8 + 4) \div 6 = 11 + 12 \div 6$
$= 11 + 2$
$= 13$
$12 + (9 - 5) \cdot 3 = 12 + 4 \cdot 3$
$= 12 + 12$
$= 24$
Since $24 > 13$,
$12 + (9 - 5) \cdot 3 > 11 + (8 + 4) \div 6$

45. $5 + 7 \cdot (3 - 1)$

47. $5 + (7 \cdot 3) - 1$

Critical Thinking 1.4

49. $15 + (8 - 3)(2^4) = 15 + (5)(2^4)$
$= 15 + (5)(16) = 15 + 80 = 95$
We are looking for the smallest prime number greater than 95.
96 is not prime (for example, it is divisible by 2).
97 is prime.
97 is the smallest prime number greater than $15 + (8 - 3)(2^4)$

Projects and Group Activities 1.4

51. Using 3 as the digit:
$3 + 3^3 = 3 + 27 = 30$

Chapter Review

1.

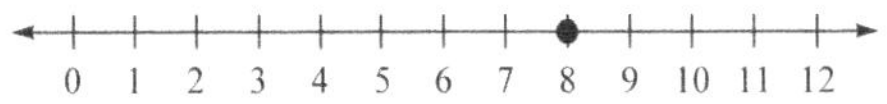

2. $10^4 = 10{,}000$

3. $$\begin{array}{r} 4,207 \\ -\,1,624 \\ \hline 2,583 \end{array}$$

4. $3 \cdot 3 \cdot 5 \cdot 5 \cdot 5 \cdot 5 = 3^2 \cdot 5^4$

5. $$\begin{array}{r} 319 \\ 358 \\ +\,712 \\ \hline 1,389 \end{array}$$

6. 38,729 (Given place value: 7; $2 < 5$)

38,729 rounded to the nearest hundred is 38,700.

7. $247 > 163$

8. 32,509

9. $2xy$

$2 \cdot 50 \cdot 7 = 100 \cdot 7$

$= 700$

10. $$\begin{array}{r} 2,607 \\ 6\overline{)15,642} \\ -12 \\ \hline 36 \\ -36 \\ \hline 4 \\ -0 \\ \hline 42 \\ -42 \\ \hline 0 \end{array}$$

11. $$\begin{array}{r} 6,407 \\ -\,2,359 \\ \hline 4,048 \end{array}$$

12. $$\begin{array}{rcr} 482 & \rightarrow & 500 \\ 319 & \rightarrow & 300 \\ 570 & \rightarrow & 600 \\ 146 & \rightarrow & +\,100 \\ \hline & & 1,500 \end{array}$$

13. $50 \div 1 = 50$

$50 \div 2 = 25$

$50 \div 5 = 10$

$50 \div 10 = 5$

The factors of 50 are 1, 2, 5, 10, 25, and 50.

14. $$\begin{array}{c|c} 24 - y & 17 \\ \hline 24 - 7 & 17 \end{array}$$

$17 = 17$

Yes, 7 is a solution of the equation.

15. $16 + 4(7-5)^2 \div 8 = 16 + 4 \cdot 2^2 \div 8$

$= 16 + 4 \cdot 4 \div 8$

$= 16 + 16 \div 8$

$= 16 + 2$

$= 18$

16. The Commutative Property of Addition

17. four million nine hundred twenty-seven thousand thirty-six

18. $x^3 y^2$

$3^3 \cdot 5^2 = (3 \cdot 3 \cdot 3) \cdot (5 \cdot 5)$

$= 27 \cdot 25$

$= 675$

19. Strategy:

a. To determine by many times more PG-13 films were released than NC-17 films:

→Find the number of PG-13 films (112) and the number of NC-17 films (7) from the pie chart.

→Divide the number of PG-13 films by the number of NC-17 films.

b. To determine by many times more R rated films were released than NC-17 films:

→Find the number of R rated films (427) and the number of NC-17 films (7) from the pie chart.

→Divide the number of R rated films by the number of NC-17 films.

Solution:

a. $\dfrac{112}{7} = 16$

There were 16 times more PG-13 films released than NC-17 films.

b. $\dfrac{427}{7} = 61$

There were 61 times more R rated films released than NC-17 films.

20.
$$\begin{array}{r} 67 \text{ r } 70 \\ 92\overline{)\ 6,234} \\ \underline{-552} \\ 714 \\ \underline{-644} \\ 70 \end{array}$$

21.
$$\begin{array}{r} 659 \\ \underline{\times\ 4} \\ 2,636 \end{array}$$

22. $x - y$
$270 - 133 = 137$

23.
$$\begin{array}{r} 5 \\ 3\overline{)15} \\ 3\overline{)45} \\ 2\overline{)90} \end{array}$$
$90 = 2 \cdot 3 \cdot 3 \cdot 5 = 2 \cdot 3^2 \cdot 5$

24. $\frac{x}{y}$
$\frac{480}{6} = 80$

25. $1 \cdot 82 = 82$

26. $58 - 3 \cdot 4^2$
$= 58 - 3 \cdot 16$
$= 58 - 48$
$= 10$

27. $x + y$
$683 + 249 = 932$

28.
$$\begin{array}{r} 18 \\ \underline{\times 24} \\ 72 \\ \underline{36\ \ } \\ 432 \end{array}$$

29. $(a + b)^2 - 2c$
$(5+3)^2 - 2 \cdot 4 = 8^2 - 2 \cdot 4$
$= 64 - 2 \cdot 4$
$= 64 - 8$
$= 56$

30. Strategy:
To find the person with the greater number of rebounds, compare the numbers 17,440 and 16,279.

Solution:
$17,440 > 16,279$
Kareem Abdul-Jabbar had more rebounds.

31. Strategy:
To find the total cost, multiply the number of square feet of floor space (2,800) by the cost per square feet (65).

Solution:
$2,800 \cdot 85 = 238,000$
The total cost of the contractor's work will be $238,000.

32. Strategy:
a. To find the perimeter of the rectangle, substitute 25 for L and 12 for W in the formula below.

b. To find the area of the rectangle, substitute 25 for L and 12 for W in the formula below.

Solution:
a. $P = 2L + 2W$
$P = 2 \cdot 25 + 2 \cdot 12$
$P = 50 + 24$
$P = 74$
The perimeter is 74 m.

b. $A = LW$
$A = 25 \cdot 12$
$A = 300$
The area is 300 m^2.

33. Strategy:
To determine the projected growth from 2016 to 2019, subtract the number of adults projected to be enrolled in 2016 (3,621,000) from the number of adults projected to be enrolled in 2019 (3,904,000).

Solution:
$3,904,000 - 3,621,000 = 283,000$
The enrollment is projected to increase by 283,000 adults.

34. Strategy:
To find the distance traveled, substitute 3 for t and 14 for r in the given formula and solve for d.

Solution:
$d = rt$
$d = 14 \cdot 3$
$d = 42$
The cyclist traveled 42 mi.

35. Strategy:
To find the markup, substitute 2,224 for S and 1,775 for C in the given formula and solve for M.

Solution:
$M = S - C$
$M = 2{,}224 - 1{,}775$
$M = 449$
The markup on the copy machine is $449.

Chapter Test

1. $3{,}297 \cdot 100 = 329{,}700$

2. $2 \cdot 2 \cdot 2 \cdot 2 \cdot 10 \cdot 10 \cdot 10 = 16{,}000$

3. $\begin{array}{r} 4{,}902 \\ -873 \\ \hline 4{,}029 \end{array}$

4. $x \cdot x \cdot x \cdot x \cdot y \cdot y \cdot y = x^4y^3$

5. $23 = p + 16$
$23 = 7 + 16$
$23 = 23$
Yes

6. 2,961 — Given place value: 9; $6 > 5$
2,961 rounded to the nearest hundred is 3,000.

7. $7{,}177 < 7{,}717$

8. 8,490

9. three hundred eighty-two thousand nine hundred four.

10. $\begin{array}{rcr} 392 & \rightarrow & 400 \\ 477 & \rightarrow & 500 \\ 519 & \rightarrow & 500 \\ +648 & \rightarrow & 600 \\ \hline 2{,}036 & & 2{,}000 \end{array}$

11. $\begin{array}{r} 1{,}376 \\ \times\ 8 \\ \hline 11{,}008 \end{array}$

12. $36{,}479 \rightarrow 40{,}000$
$50 \rightarrow 60$
$40{,}000 \cdot 60 = 2{,}400{,}000$

13. $92 \div 1 = 92$
$92 \div 2 = 46$
$92 \div 4 = 23$
$92 \div 23 = 4$
The factors of 92 are 1, 2, 4, 23, 46, 92.

14. $3\overline{)15}$ = 5
$2\overline{)30}$
$2\overline{)60}$
$2\overline{)120}$
$2\overline{)240}$
$240 = 2 \cdot 2 \cdot 2 \cdot 2 \cdot 3 \cdot 5 = 2^4 \cdot 3 \cdot 5$

15. $x - y$
$39{,}241 - 8{,}375$
$= 30{,}866$

16. The Commutative Property of Addition

17. $\frac{x}{y}$
$\frac{3{,}588}{4}$
$= 897$

18. $27 - (12 - 3) \div 9$
$= 27 - 9 \div 9$
$= 27 - 1$
$= 26$

19. Strategy:
To determine between which two years girls grow the most, find the difference in median height for each pair of consecutive years.

Solution:
Birth to age 1: 74 – 49 = 25
Age 1 to age 2: 84 – 74 = 10
Age 2 to age 3: 95 – 84 = 11
Age 3 to age 4: 100 – 95 = 5
Age 4 to age 5: 108 – 100 = 8
Girls grow the most between birth and age 1.

20. $5+2(4-3)^6$
$=5+2(1)^6$
$=5+2(1)$
$=5+2$
$=7$

21. 3000 + 900 + 70 + 2

22. $5x+(x-y)^2$
$5\cdot 8+(8-4)^2$
$=40+4^2$
$=40+16$
$=56$
The solution is 56.

23. 7

24. $6\cdot 5\cdot 4\cdot 3\cdot 2\cdot 1=720$

25. Strategy:
First, to find the total cost of the computer system, add the prices of the components. Then, to find the balance of the checking account, subtract the total cost of the computer system from $2276.

Solution:
850 + 270 + 175 + 425 = 1720
2276 – 720 = 556
The balance is $556.

26. Strategy:
a. To find the perimeter of the square, substitute 24 for *s* in the formula below.

b. To find the area of the square, substitute 24 for *s* in the formula below.

Solution:
a. $P=4s$
$P=4\cdot 24$
$P=96$
The perimeter is 96 cm.

b. $A=s^2$
$A=(24)^2$
$A=576$
The area is 576 cm^2.

27. Strategy:
To find the data processor's take home pay, subtract the sum of his deductions, taxes (854), retirement (272), and insurance (108), from the total salary (5,690).

Solution:
5,690 – (854 + 272 + 108)
= 5,690 – 1,234
= 4,456
The data processor's take home pay is $4,456.

28. Strategy:
To determine the how many more registered users are on Facebook than Twitter, subtract the number users on Twitter (175 million) from the number of users of Facebook (640 million).

Solution:
640 – 175 = 465
There are 465 million more registered users on Facebook than on Twitter.

29. Strategy:
To find the commission earned, substitute 3 for R and 480 for U in the given formula and solve for C.

Solution:
$C=U\cdot R$
$=480\cdot 3$
$=1440$
The commission earned is $1440.

30. Strategy:
To find the value per share of the fund, substitute 5,500,000 for C and 500,000 for S in the given formula and solve for V.

Solution:

$$V = \frac{C}{S} = \frac{5{,}5000{,}000}{500{,}000} = 11$$

The value per share is \$11.

Chapter 2: Fractions and Decimals

Prep Test

1. 20

2. 120

3. 9

4. 10

5. 7

6.
$$30\overline{)63}\quad 2\text{ r}3$$
$$\begin{array}{r} 63 \\ -60 \\ \hline 3 \end{array}$$

7. 1, 2, 3, 4, 6, 12

8. $8 \cdot 7 + 3 = 56 + 3 = 59$

9. 7

10. <

Section 2.1

Concept Check

1. 5, 10, 15, 20

3. 10, 20, 30, 40

5. Multiples of 6:
6, 12, 18, 24, 30, 36, 42, 48, 54, 60
Multiples of 8:
8, 16, 24, 32, 40, 48, 56, 64, 72, 80
Common multiples: 24, 48
Least common multiple: 24

7. 1, 2, 4, 5, 10, 20

9. 1, 2, 4, 7, 14, 28

Objective A Exercises

11.

	2	5
5 =		5
8 =	2 · 2 · 2	

LCM = $2 \cdot 2 \cdot 2 \cdot 5 = 40$

13.

	2	3
3 =		3
8 =	2 · 2 · 2	

LCM = $2 \cdot 2 \cdot 2 \cdot 3 = 24$

15.

	2	3
4 =	2 · 2	
6 =	2	3

LCM = $2 \cdot 2 \cdot 3 = 12$

17.

	2	3
8 =	2 · 2 · 2	
12 =	2 · 2	3

LCM = $2 \cdot 2 \cdot 2 \cdot 3 = 24$

19.

	2	3	5
5 =			5
12 =	2 · 2	3	

LCM = $2 \cdot 2 \cdot 3 \cdot 5 = 60$

21.

	2	7
8 =	2 · 2 · 2	
14 =	2	7

LCM = $2 \cdot 2 \cdot 2 \cdot 7 = 56$

23.

	2
8 =	2 · 2 · 2
32 =	2 · 2 · 2 · 2 · 2

LCM = $2 \cdot 2 \cdot 2 \cdot 2 \cdot 2 = 32$

25.

	2	3
9 =		3 · 3
36 =	2 · 2	3 · 3

LCM = $2 \cdot 2 \cdot 3 \cdot 3 = 36$

27.

	2	3	5	11
44 =	2 · 2			11
60 =	2 · 2	3	5	

LCM = $2 \cdot 3 \cdot 5 \cdot 11 = 660$

29.

	2	3	17	23
102 =	2	3	17	
184 =	2 · 2 · 2			23

LCM = $2 \cdot 2 \cdot 2 \cdot 3 \cdot 17 \cdot 23 = 9384$

31.

	2	3
4 =	2 · 2	
8 =	2 · 2 · 2	
12 =	2 · 2	3

LCM = 2 · 2 · 2 · 3 = 24

33.

	2	3	5
3 =		3	
5 =			5
10 =	2		5

LCM = 2 · 3 · 5 = 30

35.

	2	3
3 =		3
8 =	2 · 2 · 2	
12 =	2 · 2	3

LCM = 2 · 2 · 2 · 3 = 24

37.

	2	3
9 =		3 · 3
36 =	2 · 2	3 · 3
64 =	2 · 2 · 2 · 2 · 2 · 2	

LCM = 2 · 2 · 2 · 2 · 2 · 2 · 3 · 3 = 576

39.

	2	3	5	7
3 =		3		
7 =				7
20 =	2 · 2		5	

LCM = 2 · 2 · 3 · 5 · 7 = 420

41. True

Objective B Exercises

43.

	3	5
3 =	3	
5 =		5

GCF = 1

45.

	2	3
6 =	2	3
9 =		3 · 3

GCF = 3

47.

	3	5
15 =	3	5
25 =		5 · 5

GCF = 5

49.

	2	5
25 =		5 · 5
100 =	2 · 2	5 · 5

GCF = 5 · 5 = 25

51.

	2	3	17
32 =	2 · 2 · 2 · 2 · 2		
51 =		3	17

GCF = 1

53.

	2	3	5
12 =	2 · 2	3	
80 =	2 · 2 · 2 · 2		5

GCF = 2 · 2 = 4

55.

	2	5	7
16 =	2 · 2 · 2 · 2		
140 =	2 · 2	5	7

GCF = 2 · 2 = 4

57.

	2	3	11
44 =	2 · 2		11
96 =	2 · 2 · 2 · 2 · 2	3	

GCF = 2 · 2 = 4

59.

	3	5	11
3 =	3		
5 =		5	
11 =			11

GCF = 1

61.

	2	7
7 =		7
14 =	2	7
49 =		7 · 7

GCF = 7

63.

	2	3	5
10 =	2		5
15 =		3	5
20 =	2 · 2		5

GCF = 5

65.

	2	3	5
24 =	2 · 2 · 2	3	
40 =	2 · 2 · 2		5
72 =	2 · 2 · 2	3 · 3	

GCF = 2 · 2 · 2 = 8

67.

	3	17	31
17 =		17	
31 =			31
81 =	3 · 3 · 3 · 3		

GCF = 1

69.

	5
25 =	(5 · 5)
125 =	5 · 5 · 5
625 =	5 · 5 · 5 · 5

GCF = 25

71.

	2	3	7
32 =	2 · 2 · 2 · 2 · 2		
56 =	2 · 2 · 2		7
72 =	(2 · 2 · 2)	3 · 3	

GCF = 2 · 2 · 2 = 8

73. True

Critical Thinking 2.1

75. Joe has a 4-day cycle (3 workdays + 1 day off). Raya has a 6-day cycle (5 workdays + 1 day off). The least common multiple of 4 and 6 is 12. After Joe and Raya have a day off together, they will have another day off together in 12 days.

77. The GCF of 3 and 5 is 1. The GCF of 7 and 11 is 1. The GCF of 29 and 43 is 1. Because two prime numbers do not have a common factor other than 1, the GCF of two prime numbers is 1. Because three prime numbers do not have a common factor other than 1, the GCF of three prime numbers is 1.

Projects or Group Activities 2.1

79 **a.** No; the GCF of 48 and 50 is 2. 48 and 50 are not coprime.

b. Yes; $25 = 5 \cdot 5$ and $36 = 2 \cdot 2 \cdot 3 \cdot 3$, so their GCF is 1.

c. Yes; $22 = 2 \cdot 11$ and $27 = 3 \cdot 3 \cdot 3$, so their GCF is 1.

d. Yes; 71 and 73 are both prime numbers, so their GCF is 1.

Section 2.2

Concept Check

1. **a.** 9; 4

b. improper

c. divided by; 9; 4

3. 3; 2; 17; $\frac{17}{5}$

5. 1; 2

Objective A Exercises

7. $\frac{4}{5}$

9. $\frac{1}{4}$

11. $\frac{4}{3}; 1\frac{1}{3}$

13. $\frac{13}{5}; 2\frac{3}{5}$

15. $\begin{array}{r} 3 \\ 4\overline{)\,13} \\ \underline{-12} \\ 1 \end{array}$ $\quad \frac{13}{4} = 3\frac{1}{4}$

17. $\begin{array}{r} 4 \\ 5\overline{)\,20} \\ \underline{-20} \\ 0 \end{array}$ $\quad \frac{20}{5} = 4$

19. $\begin{array}{r} 2 \\ 10\overline{)\,27} \\ \underline{-20} \\ 7 \end{array}$ $\quad \frac{27}{10} = 2\frac{7}{10}$

21. $\begin{array}{r} 7 \\ 8\overline{)\,56} \\ \underline{-56} \\ 0 \end{array}$ $\quad \frac{56}{8} = 7$

23. $\begin{array}{r} 1 \\ 9\overline{)\,17} \\ \underline{-9} \\ 8 \end{array}$ $\quad \frac{17}{9} = 1\frac{8}{9}$

25. $5\overline{)12}$ = 2, -10, remainder 2; $\frac{12}{5}=2\frac{2}{5}$

27. $1\overline{)18}$ = 18, -1, 8, -8, 0; $\frac{18}{1}=18$

29. $15\overline{)32}$ = 2, -30, remainder 2; $\frac{32}{15}=2\frac{2}{15}$

31. $\frac{8}{8}=1$

33. $3\overline{)28}$ = 9, -27, remainder 1; $\frac{28}{3}=9\frac{1}{3}$

35. $2\frac{1}{4}=\frac{(4\cdot 2)+1}{4}=\frac{8+1}{4}=\frac{9}{4}$

37. $5\frac{1}{2}=\frac{(2\cdot 5)+1}{2}=\frac{10+1}{2}=\frac{11}{2}$

39. $2\frac{4}{5}=\frac{(5\cdot 2)+4}{5}=\frac{10+4}{5}=\frac{14}{5}$

41. $7\frac{5}{6}=\frac{(6\cdot 7)+5}{6}=\frac{42+5}{6}=\frac{47}{6}$

43. $7=\frac{7}{1}$

45. $8\frac{1}{4}=\frac{(4\cdot 8)+1}{4}=\frac{32+1}{4}=\frac{33}{4}$

47. $10\frac{1}{3}=\frac{(3\cdot 10)+1}{3}=\frac{30+1}{3}=\frac{31}{3}$

49. $4\frac{7}{12}=\frac{(12\cdot 4)+7}{12}=\frac{48+7}{12}=\frac{55}{12}$

51. $8=\frac{8}{1}$

53. $12\frac{4}{5}=\frac{(5\cdot 12)+4}{5}=\frac{60+4}{5}=\frac{64}{5}$

55. $a>b$

Objective B Exercises

57. $12\div 2=6$
$\frac{1}{2}=\frac{1\cdot 6}{2\cdot 6}=\frac{6}{12}$
$\frac{6}{12}$ is equivalent to $\frac{1}{2}$.

59. $24\div 8=3$
$\frac{3}{8}=\frac{3\cdot 3}{8\cdot 3}=\frac{9}{24}$
$\frac{9}{24}$ is equivalent to $\frac{3}{8}$.

61. $51\div 17=3$
$\frac{2}{17}=\frac{2\cdot 3}{17\cdot 3}=\frac{6}{51}$
$\frac{6}{51}$ is equivalent to $\frac{2}{17}$.

63. $32\div 4=8$
$\frac{3}{4}=\frac{3\cdot 8}{4\cdot 8}=\frac{24}{32}$
$\frac{24}{32}$ is equivalent to $\frac{3}{4}$.

65. $18\div 1=18$
$6=\frac{6}{1}=\frac{6\cdot 18}{1\cdot 18}=\frac{108}{18}$
$\frac{108}{18}$ is equivalent to 6.

67. $90\div 3=30$
$\frac{1}{3}=\frac{1\cdot 30}{3\cdot 30}=\frac{30}{90}$
$\frac{30}{90}$ is equivalent to $\frac{1}{3}$.

69. $21\div 3=7$
$\frac{2}{3}=\frac{2\cdot 7}{3\cdot 7}=\frac{14}{21}$
$\frac{14}{21}$ is equivalent to $\frac{2}{3}$.

71. $49\div 7=7$
$\frac{6}{7}=\frac{6\cdot 7}{7\cdot 7}=\frac{42}{49}$
$\frac{42}{49}$ is equivalent to $\frac{6}{7}$.

73. $18\div 9=2$
$\frac{4}{9}=\frac{4\cdot 2}{9\cdot 2}=\frac{8}{18}$
$\frac{8}{18}$ is equivalent to $\frac{4}{9}$.

75. $4 \div 1 = 4$

$7 = \frac{7}{1} = \frac{7 \cdot 4}{1 \cdot 4} = \frac{28}{4}$

$\frac{28}{4}$ is equivalent to 7.

77. $\frac{3}{12} = \frac{3}{2 \cdot 2 \cdot 3} = \frac{1}{4}$

79. $\frac{33}{44} = \frac{3 \cdot 11}{2 \cdot 2 \cdot 11} = \frac{3}{4}$

81. $\frac{4}{24} = \frac{2 \cdot 2}{2 \cdot 2 \cdot 2 \cdot 3} = \frac{1}{6}$

83. $\frac{8}{33} = \frac{2 \cdot 2 \cdot 2}{3 \cdot 11} = \frac{8}{33}$

85. $\frac{0}{8} = 0$

87. $\frac{42}{36} = \frac{2 \cdot 3 \cdot 7}{2 \cdot 2 \cdot 3 \cdot 3} = \frac{7}{6}$

89. $\frac{16}{16} = 1$

91. $\frac{21}{35} = \frac{3 \cdot 7}{5 \cdot 7} = \frac{3}{5}$

93. $\frac{16}{60} = \frac{2 \cdot 2 \cdot 2 \cdot 2}{2 \cdot 2 \cdot 3 \cdot 5} = \frac{4}{15}$

95. $\frac{12}{20} = \frac{2 \cdot 2 \cdot 3}{2 \cdot 2 \cdot 5} = \frac{3}{5}$

97. $\frac{12m}{18} = \frac{\overset{1}{\cancel{2}} \cdot 2 \cdot \overset{1}{\cancel{3}} \cdot m}{\underset{1}{\cancel{2}} \cdot \underset{1}{\cancel{3}} \cdot 3} = \frac{2m}{3}$

99. $\frac{4y}{8} = \frac{\overset{1}{\cancel{2}} \cdot \overset{1}{\cancel{2}} \cdot y}{\underset{1}{\cancel{2}} \cdot \underset{1}{\cancel{2}} \cdot 2} = \frac{y}{2}$

101. $\frac{24a}{36} = \frac{\overset{1}{\cancel{2}} \cdot \overset{1}{\cancel{2}} \cdot 2 \overset{1}{\cancel{3}} \cdot a}{\underset{1}{\cancel{2}} \cdot \underset{1}{\cancel{2}} \cdot \underset{1}{\cancel{3}} \cdot 3} = \frac{2a}{3}$

103. $\frac{8c}{8} = \frac{\overset{1}{\cancel{2}} \cdot \overset{1}{\cancel{2}} \cdot \overset{1}{\cancel{2}} \cdot c}{\underset{1}{\cancel{2}} \cdot \underset{1}{\cancel{2}} \cdot \underset{1}{\cancel{2}}} = c$

105. $\frac{18k}{3} = \frac{2 \cdot 3 \cdot \overset{1}{\cancel{3}} \cdot k}{\underset{1}{\cancel{3}}} = 6k$

107. ii

109. iii; One example is $\frac{5}{6}$ and $\frac{15}{18}$.

Objective C Exercises

111. $\frac{5}{7} = \frac{15}{21}$ $\quad \frac{2}{3} = \frac{14}{21}$

$\frac{15}{21} > \frac{14}{21}$

$\frac{5}{7} > \frac{2}{3}$

113. $\frac{7}{12} = \frac{14}{24}$ $\quad \frac{5}{8} = \frac{15}{24}$

$\frac{14}{24} < \frac{15}{24}$

$\frac{7}{12} < \frac{5}{8}$

115. $\frac{11}{14} = \frac{22}{28}$ $\quad \frac{3}{4} = \frac{21}{28}$

$\frac{22}{28} > \frac{21}{28}$

$\frac{11}{14} > \frac{3}{4}$

117. $\frac{11}{12} = \frac{33}{36}$ $\quad \frac{7}{9} = \frac{28}{36}$

$\frac{33}{36} > \frac{28}{36}$

$\frac{11}{12} > \frac{7}{9}$

119. $\frac{5}{8} = \frac{35}{56}$ $\quad \frac{4}{7} = \frac{32}{56}$

$\frac{35}{56} > \frac{32}{56}$

$\frac{5}{8} > \frac{4}{7}$

121. $\frac{11}{30} = \frac{44}{120}$ $\quad \frac{7}{24} = \frac{35}{120}$

$\frac{44}{120} > \frac{35}{120}$

$\frac{11}{30} > \frac{7}{24}$

123. $\frac{9}{11} = \frac{72}{88}$ $\quad \frac{7}{8} = \frac{77}{88}$

$\frac{72}{88} < \frac{77}{88}$

$\frac{9}{11} < \frac{7}{8}$

125. $\frac{3}{4} = \frac{39}{52}$ $\quad \frac{11}{13} = \frac{44}{52}$

$\frac{39}{52} < \frac{44}{52}$

$\frac{3}{4} < \frac{11}{13}$

127. $\frac{2}{3}=\frac{20}{30}$ $\frac{7}{10}=\frac{21}{30}$

$\frac{20}{30}<\frac{21}{30}$

$\frac{2}{3}<\frac{7}{10}$

129. $\frac{3}{10}=\frac{15}{50}$ $\frac{7}{25}=\frac{14}{50}$

$\frac{15}{50}>\frac{14}{50}$

$\frac{3}{10}>\frac{7}{25}$

131. One example is $\frac{4}{7}$ and $\frac{3}{5}$.

Critical Thinking 2.2

133. Strategy:

To find which service is more important,

→Rewrite the food quality fraction $\frac{1}{4}$ as a fraction with a denominator of 100.

→Rewrite location fraction $\frac{13}{50}$ as a fraction with a denominator of 100.

→Compare the two fractions.

Solution:

$\frac{1}{4}=\frac{1\cdot 25}{4\cdot 25}=\frac{25}{100}$

$\frac{13}{50}=\frac{13\cdot 2}{50\cdot 2}=\frac{26}{100}$

$\frac{26}{100}>\frac{25}{100}$

The location was more important than the food quality.

135. Strategy:

To determine if you answered more or less than $\frac{8}{10}$ of the questions correctly:

→Write a fraction with the number of questions answered correctly (42) in the numerator and the total number of questions (50) in the denominator.

→Rewrite $\frac{8}{10}$ as a fraction with a denominator of 50.

→Compare the two fractions.

Solution:

$\frac{42}{50}$

$\frac{8}{10}=\frac{8\cdot 5}{10\cdot 5}=\frac{40}{50}$

$\frac{42}{50}>\frac{40}{50}$

You answered more than $\frac{8}{10}$ of the questions correctly.

137. a. The midpoint or halfway point between $\frac{2}{a}$ and $\frac{4}{a}$ is one-half the sum of $\frac{2}{a}+\frac{4}{a}$.

$$\frac{1}{2}\left(\frac{2}{a}+\frac{4}{a}\right)=\frac{1}{2}\left(\frac{6}{a}\right)=\frac{3}{a}$$

The diagram below also indicates that the midpoint is $\frac{3}{a}$.

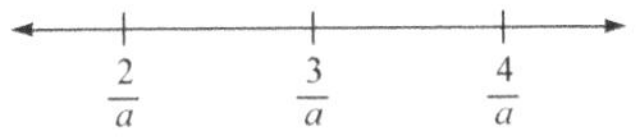

b. The diagram below shows p and q evenly spaced between A and B. The point p is $\frac{1}{3}$ of the distance between A and B.

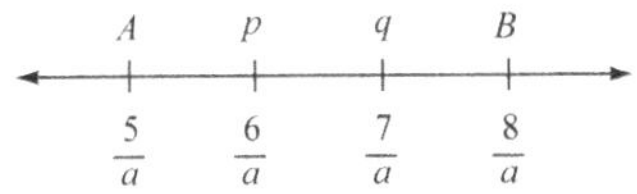

$$\frac{1}{3}\left(\frac{8}{b}-\frac{5}{b}\right)=\frac{1}{3}\left(\frac{3}{b}\right)=\frac{1}{b}$$

Add $\frac{1}{b}$ to $\frac{5}{b}$ to find p. $\frac{1}{b}+\frac{5}{b}=\frac{6}{b}$

Add $\frac{1}{b}$ to $\frac{6}{b}$ to find q. $\frac{1}{b}+\frac{6}{b}=\frac{7}{b}$

Thus the two fractions $\frac{6}{b}$ and $\frac{7}{b}$ are evenly spaced between $\frac{5}{b}$ and $\frac{8}{b}$

Projects or Group Activities 2.2

139. Strategy:

a. To find the fraction, write a fraction with the number of spades (13) in the numerator and the total number of cards (52) in the denominator. Write the fraction in simplest form.

b. To find the fraction, write a fraction with the number of aces (4) in the numerator and the total number of cards (52) in the denominator. Write the fraction simplest form.

Solution:

a. $\frac{13}{52}=\frac{13}{4\cdot 13}=\frac{1}{4}$

Spades are $\frac{1}{4}$ of a standard deck of cards.

b. $\frac{4}{52}=\frac{2\cdot 2}{2\cdot 2\cdot 13}=\frac{1}{13}$

Aces are $\frac{1}{13}$ of a standard deck of cards.

Section 2.3

Concept Check

1. denominators

3. **a.** $\frac{1}{2}$; $-\frac{3}{7}$

   b. $\frac{1}{2}$; $\frac{3}{7}$

5. subtraction

Objective A Exercises

7. $\frac{4}{11}+\frac{5}{11}=\frac{4+5}{11}=\frac{9}{11}$

9. $\frac{2}{3}+\frac{1}{3}=\frac{2+1}{3}$
$=\frac{3}{3}=1$

11. $\frac{5}{6}+\frac{5}{6}=\frac{5+5}{6}$
$=\frac{10}{6}=\frac{5}{3}=1\frac{2}{3}$

13. $\frac{7}{18}+\frac{13}{18}+\frac{1}{18}=\frac{7+13+1}{18}$
$=\frac{21}{18}=\frac{7}{6}=1\frac{1}{6}$

15. $\frac{7}{b}+\frac{9}{b}=\frac{7+9}{b}$
$=\frac{16}{b}$

17. $\frac{5}{c}+\frac{4}{c}=\frac{5+4}{c}$
$=\frac{9}{c}$

19. $\frac{1}{x}+\frac{4}{x}+\frac{6}{x}=\frac{1+4+6}{x}$
$=\frac{11}{x}$

21. $\frac{1}{4}+\frac{2}{3}=\frac{3}{12}+\frac{8}{12}$
$=\frac{3+8}{12}=\frac{11}{12}$

23. $\frac{7}{15}+\frac{9}{20}=\frac{28}{60}+\frac{27}{60}$
$=\frac{28+27}{60}=\frac{55}{60}=\frac{11}{12}$

25. $\frac{2}{3}+\frac{1}{12}+\frac{5}{6}=\frac{8}{12}+\frac{1}{12}+\frac{10}{12}$
$=\frac{8+1+10}{12}$
$=\frac{19}{12}=1\frac{7}{12}$

27. $\frac{7}{12}+\frac{3}{4}+\frac{4}{5}=\frac{35}{60}+\frac{45}{60}+\frac{48}{60}$
$=\frac{35+45+48}{60}$
$=\frac{128}{60}=\frac{32}{15}=2\frac{2}{15}$

29. $8+7\frac{2}{3}=15\frac{2}{3}$

31. $2\frac{1}{6}+3\frac{1}{2}=2\frac{1}{6}+3\frac{3}{6}$
$=5\frac{4}{6}=5\frac{2}{3}$

33. $8\frac{3}{5}+6\frac{9}{20}=8\frac{12}{20}+6\frac{9}{20}$
$=14\frac{21}{20}=15\frac{1}{20}$

35. $5\frac{5}{12}+4\frac{7}{9}=5\frac{15}{36}+4\frac{28}{36}$
$=9\frac{43}{36}=10\frac{7}{36}$

37. $2\frac{1}{4}+3\frac{1}{2}+1\frac{2}{3}$
$=2\frac{3}{12}+3\frac{6}{12}+1\frac{8}{12}$
$=6\frac{17}{12}=7\frac{5}{12}$

39. $\frac{2}{7}+\frac{3}{14}+\frac{1}{4}=\frac{8}{28}+\frac{6}{28}+\frac{7}{28}$
$=\frac{8+6+7}{28}$
$=\frac{21}{28}=\frac{3}{4}$

41. $5\frac{4}{9}+6\frac{5}{6}=5\frac{8}{18}+6\frac{15}{18}$
$=11\frac{23}{18}=12\frac{5}{18}$

43. $3\frac{7}{12}+2\frac{5}{8}=3\frac{14}{24}+2\frac{15}{24}$
$=5\frac{29}{24}=6\frac{5}{24}$

45. $x+y$

$\frac{3}{5}+\frac{4}{5}=\frac{3+4}{5}$

$=\frac{7}{5}=1\frac{2}{5}$

47. $x+y$

$\frac{5}{6}+\frac{8}{9}=\frac{15}{18}+\frac{16}{18}$

$=\frac{15+16}{18}=\frac{31}{18}=1\frac{13}{18}$

49. $x+y+z$

$\frac{3}{8}+\frac{1}{4}+\frac{7}{12}=\frac{9}{24}+\frac{6}{24}+\frac{14}{24}$

$=\frac{9+6+14}{24}$

$=\frac{29}{24}=1\frac{5}{24}$

51. $x+y+z$

$1\frac{1}{2}+3\frac{3}{4}+6\frac{5}{12}=1\frac{6}{12}+3\frac{9}{12}+6\frac{5}{12}$

$=10\frac{20}{12}=10\frac{5}{3}=11\frac{2}{3}$

53. $x+y+z$

$4\frac{3}{5}+8\frac{7}{10}+1\frac{9}{20}$

$=4\frac{12}{20}+8\frac{14}{20}+1\frac{9}{20}$

$=13\frac{35}{20}=13\frac{7}{4}=14\frac{3}{4}$

55. $z+\frac{1}{4}=\frac{7}{20}$

$\frac{3}{5}+\frac{1}{4}$	$\frac{7}{20}$
$\frac{12}{20}+\frac{5}{20}$	$\frac{7}{20}$
$\frac{17}{20}$	$\frac{17}{20}$

$\frac{7}{20}=\frac{7}{20}$

Yes, $\frac{3}{5}$ is a solution of the equation.

57. $\frac{19}{50}+\frac{6}{25}=\frac{19}{50}+\frac{12}{50}=\frac{31}{50}$

$\frac{31}{50}$ of money borrowed on home-equity loans are spent on debt consolidation and home improvement.

59. iii

Objective B Exercises

61. $\frac{7}{12}-\frac{5}{12}=\frac{7-5}{12}=\frac{2}{12}=\frac{1}{6}$

63. $\frac{11}{24}-\frac{7}{24}=\frac{11-7}{24}=\frac{4}{24}=\frac{1}{6}$

65. $\frac{8}{d}-\frac{3}{d}=\frac{8-3}{d}=\frac{5}{d}$

67. $\frac{10}{n}-\frac{5}{n}=\frac{10-5}{n}=\frac{5}{n}$

69. $\frac{3}{7}-\frac{5}{14}=\frac{6}{14}-\frac{5}{14}$

$=\frac{6-5}{14}=\frac{1}{14}$

71. $\frac{2}{3}-\frac{1}{6}=\frac{4}{6}-\frac{1}{6}$

$=\frac{4-1}{6}=\frac{3}{6}=\frac{1}{2}$

73. $\frac{11}{12}-\frac{2}{3}=\frac{11}{12}-\frac{8}{12}$

$=\frac{11-8}{12}=\frac{3}{12}=\frac{1}{4}$

75. $4\frac{11}{18}-2\frac{5}{18}=2\frac{6}{18}=2\frac{1}{3}$

77. $8\frac{3}{4}-2=6\frac{3}{4}$

79. $8\frac{5}{6}-7\frac{3}{4}=8\frac{10}{12}-7\frac{9}{12}$

$=1\frac{1}{12}$

81. $7-3\frac{5}{8}=6\frac{8}{8}-3\frac{5}{8}$

$=3\frac{3}{8}$

83. $10-4\frac{8}{9}=9\frac{9}{9}-4\frac{8}{9}$

$=5\frac{1}{9}$

85. $7\frac{3}{8}-4\frac{5}{8}=6\frac{11}{8}-4\frac{5}{8}$

$=2\frac{6}{8}=2\frac{3}{4}$

87. $12\frac{5}{12}-10\frac{17}{24}=12\frac{10}{24}-10\frac{17}{24}$

$=11\frac{34}{24}-10\frac{17}{24}$

$=1\frac{17}{24}$

89. $6\frac{2}{3}-1\frac{7}{8}=6\frac{16}{24}-1\frac{21}{24}$
$=5\frac{40}{24}-1\frac{21}{24}$
$=4\frac{19}{24}$

91. $10\frac{2}{5}-8\frac{7}{10}=10\frac{4}{10}-8\frac{7}{10}$
$=9\frac{14}{10}-8\frac{7}{10}$
$=1\frac{7}{10}$

93. $\frac{7}{8}-\frac{2}{3}=\frac{21}{24}-\frac{16}{24}=\frac{5}{24}$

95. $8-1\frac{7}{12}=7\frac{12}{12}-1\frac{7}{12}$
$=6\frac{5}{12}$

97. $x-y$
$\frac{8}{9}-\frac{5}{9}=\frac{8-5}{9}=\frac{3}{9}=\frac{1}{3}$

99. $x-y$
$\frac{7}{15}-\frac{3}{10}=\frac{14}{30}-\frac{9}{30}=\frac{5}{30}=\frac{1}{6}$

101. $x-y$
$5\frac{7}{9}-4\frac{2}{3}=5\frac{7}{9}-4\frac{6}{9}$
$=1\frac{1}{9}$

103. $x-y$
$7\frac{9}{10}-3\frac{1}{2}=7\frac{9}{10}-3\frac{5}{10}$
$=4\frac{4}{10}=4\frac{2}{5}$

105. $x-y$
$5-2\frac{7}{9}=4\frac{9}{9}-2\frac{7}{9}$
$=2\frac{2}{9}$

107. $x-y$
$10\frac{1}{2}-5\frac{7}{12}=10\frac{6}{12}-5\frac{7}{12}$
$=9\frac{18}{12}-5\frac{7}{12}=4\frac{11}{12}$

109. $\frac{4}{5}=\frac{31}{20}-y$

$\frac{4}{5}$	$\frac{31}{20}-\frac{3}{4}$
$\frac{4}{5}$	$\frac{31}{20}-\frac{15}{20}$
$\frac{4}{5}$	$\frac{31-15}{20}$
$\frac{4}{5}$	$\frac{16}{20}$
$\frac{4}{5}$	$\frac{4}{5}$

$\frac{4}{5}=\frac{4}{5}$

Yes, $\frac{4}{5}$ is a solution of the equation.

111. i

Objective C Exercises

113. Strategy:
To find the amount of property owned, subtract the amount sold $\left(1\frac{1}{2}\right)$ from the amount originally owned $\left(3\frac{1}{4}\right)$.

Solution:
$3\frac{1}{4}-1\frac{1}{2}=3\frac{1}{4}-1\frac{2}{4}$
$=2\frac{5}{4}-1\frac{2}{4}$
$=1\frac{3}{4}$
You now own $1\frac{3}{4}$ acres.

115. Strategy:
To find the number of hours still required, subtract the number of hours already contributed $\left(12\frac{1}{4}\right)$ from the original amount required (20).

Solution:
$20-12\frac{1}{4}=19\frac{4}{4}-12\frac{1}{4}$
$=7\frac{3}{4}$
$7\frac{3}{4}$ h of community service are still required.

117. Strategy:
To find the amount of roofing that remains:
→Add the amounts already done $\left(\frac{1}{3}+\frac{1}{4}\right)$.
→Subtract the amount already done from the total job (1).
To determine if the roofers can finish in another day, compare the amount already done with $\frac{1}{2}$.

Solution:
$\frac{1}{3}+\frac{1}{4}=\frac{4}{12}+\frac{3}{12}=\frac{7}{12}$
$1-\frac{7}{12}=\frac{12}{12}-\frac{7}{12}=\frac{5}{12}$
$\frac{5}{12}$ of the roofing job remains to be done.
Since $\frac{5}{12}$ is less than $\frac{1}{2}$, the roofing can be finished in another day.

119. Strategy:
To find the feet of fencing, replace a, b, and c with $6\frac{1}{4}$, $10\frac{3}{4}$, and $12\frac{1}{2}$ in the given formula and solve for P.

Solution:
$P = a + b + c$
$P=6\frac{1}{4}+10\frac{3}{4}+12\frac{1}{2}=6\frac{1}{4}+10\frac{3}{4}+12\frac{2}{4}=29\frac{1}{2}$
The number of feet of fencing needed is $29\frac{1}{2}$ ft.

121. Strategy:

a. To find miles for the first two days, add the distance for the first day $\left(7\frac{3}{8}\text{ mi}\right)$ to the distance for the second day $\left(10\frac{1}{3}\text{ mi}\right)$.

b. To find the distance for the third day, subtract the distance for the first two days from the total distance $\left(27\frac{1}{2}\text{ mi}\right)$.

Solution:

a. $7\frac{3}{8}+10\frac{1}{3}=7\frac{9}{24}+10\frac{8}{24}=17\frac{17}{24}$

The distance for the first two days is $17\frac{17}{24}$ mi..

b. $27\frac{1}{2}-17\frac{17}{24}=27\frac{12}{24}-17\frac{17}{24}$

$=26\frac{36}{24}-17\frac{17}{24}=9\frac{19}{24}$

The distance to be traveled the third day is $9\frac{19}{24}$ mi.

123. Strategy:

To find the difference in heights, subtract the low 1970s $\left(\frac{3}{16}\right)$ from the high in 1960s $\left(\frac{7}{32}\right)$.

Solution:

$\frac{7}{32}-\frac{3}{16}=\frac{7}{32}-\frac{6}{32}=\frac{1}{32}$

The difference in heights between 1970s and 1960s was $\frac{1}{32}$ in.

125. Strategy:

To find the amount of weight to gain:

→Add the amounts already gained $\left(4\frac{1}{2}+3\frac{3}{4}\right)$.

→Subtract the amount gained from the goal amount (15).

Solution:

$4\frac{1}{2}+3\frac{3}{4}=4\frac{2}{4}+3\frac{3}{4}=7\frac{5}{4}=8\frac{1}{4}$

$15-8\frac{1}{4}=14\frac{4}{4}-8\frac{1}{4}=6\frac{3}{4}$

The boxer has $6\frac{3}{4}$ lb left to gain.

Critical Thinking 2.3

127. Answers will vary. This is an example. Fractions with the same denominator are added by adding the numerators and placing the sum over the common denominator.

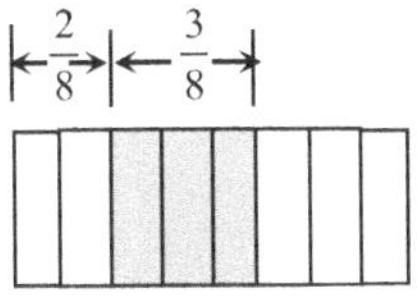

$$\frac{2}{8}+\frac{3}{8}=\frac{2+3}{8}=\frac{5}{8}$$

Projects or Group Activities 2.3

129. Not possible.

131. Answers will vary. One example is: $\frac{4}{6}+\frac{4}{3}=\frac{4}{6}+\frac{8}{6}=\frac{12}{6}=2$

133. Not possible.

Section 2.4

Concept Check

1. **a.** can

   b. product; product

3. three

5. division

Objective A Exercises

7. $\frac{2}{3}\cdot\frac{9}{10}=\frac{2\cdot 9}{3\cdot 10}$

$=\frac{2\cdot 3\cdot 3}{3\cdot 2\cdot 5}$

$=\frac{3}{5}$

9. $\frac{14}{15}\cdot\frac{6}{7}=\frac{14\cdot 6}{15\cdot 7}$
$=\frac{2\cdot 7\cdot 2\cdot 3}{3\cdot 5\cdot 7}$
$=\frac{4}{5}$

11. $\frac{6}{7}\cdot\frac{0}{10}=\frac{6\cdot 0}{7\cdot 10}=\frac{0}{70}=0$

13. $\frac{2}{3}\cdot\frac{3}{8}\cdot\frac{4}{9}=\frac{2\cdot 3\cdot 4}{3\cdot 8\cdot 9}$
$=\frac{2\cdot 3\cdot 2\cdot 2}{3\cdot 2\cdot 2\cdot 2\cdot 3\cdot 3}=\frac{1}{9}$

15. $\frac{9}{x}\cdot\frac{7}{y}=\frac{9\cdot 7}{x\cdot y}$
$=\frac{63}{xy}$

17. $\frac{y}{5}\cdot\frac{z}{6}=\frac{y\cdot z}{5\cdot 6}=\frac{yz}{30}$

19. $6\cdot\frac{1}{6}=\frac{6}{1}\cdot\frac{1}{6}$
$=\frac{6\cdot 1}{1\cdot 6}$
$=\frac{6}{6}=1$

21. $\frac{3}{4}\cdot 8=\frac{3}{4}\cdot\frac{8}{1}$
$=\frac{3\cdot 8}{4\cdot 1}$
$=\frac{3\cdot 2\cdot 2\cdot 2}{2\cdot 2\cdot 1}=\frac{6}{1}=6$

23. $\frac{6}{7}\cdot 0=0$

25. $3\frac{1}{2}\cdot 5\frac{3}{7}=\frac{7}{2}\cdot\frac{38}{7}$
$=\frac{7\cdot 38}{2\cdot 7}$
$=\frac{7\cdot 2\cdot 19}{2\cdot 7}$
$=\frac{19}{1}=19$

27. $8\cdot 5\frac{1}{4}=\frac{8}{1}\cdot\frac{21}{4}$
$=\frac{8\cdot 21}{1\cdot 4}$
$=\frac{2\cdot 2\cdot 2\cdot 3\cdot 7}{1\cdot 2\cdot 2}$
$=\frac{42}{1}=42$

29. $3\frac{1}{2}\cdot 1\frac{5}{7}\cdot\frac{11}{12}=\frac{7}{2}\cdot\frac{12}{7}\cdot\frac{11}{12}$
$=\frac{7\cdot 12\cdot 11}{2\cdot 7\cdot 12}$
$=\frac{7\cdot 2\cdot 2\cdot 3\cdot 11}{2\cdot 7\cdot 2\cdot 2\cdot 3}$
$=\frac{11}{2}=5\frac{1}{2}$

31. $\frac{3}{4}\cdot\frac{14}{15}=\frac{3\cdot 14}{4\cdot 15}$
$=\frac{3\cdot 2\cdot 7}{2\cdot 2\cdot 3\cdot 5}$
$=\frac{7}{10}$

33. $4\frac{4}{5}\cdot\frac{3}{8}=\frac{24}{5}\cdot\frac{3}{8}$
$=\frac{24\cdot 3}{5\cdot 8}$
$\frac{2\cdot 2\cdot 2\cdot 3\cdot 3}{5\cdot 2\cdot 2\cdot 2}$
$=\frac{9}{5}=1\frac{4}{5}$

35. Strategy:
To find the amount spent on housing, multiply the housing fraction $\left(\frac{13}{45}\right)$ by the income (45,000).

Solution:

$$45{,}000\cdot\frac{13}{45}=13{,}000$$

The typical household spends \$13,000 per year on housing.

37. xy

$$\frac{5}{16}\cdot\frac{7}{15}=\frac{5\cdot 7}{16\cdot 15}$$
$$=\frac{5\cdot 7}{2\cdot 2\cdot 2\cdot 2\cdot 3\cdot 5}$$
$$=\frac{7}{48}$$

39. xy

$$\frac{4}{7}\cdot 6\frac{1}{8}=\frac{4}{7}\cdot\frac{49}{8}$$
$$=\frac{4\cdot 49}{7\cdot 8}$$
$$=\frac{2\cdot 2\cdot 7\cdot 7}{7\cdot 2\cdot 2\cdot 2}$$
$$=\frac{7}{2}=3\frac{1}{2}$$

41. xyz

$$\frac{3}{8}\cdot\frac{2}{3}\cdot\frac{4}{5}=\frac{3\cdot 2\cdot 4}{8\cdot 3\cdot 5}$$
$$=\frac{3\cdot 2\cdot 2\cdot 2}{2\cdot 2\cdot 2\cdot 3\cdot 5}$$
$$=\frac{1}{5}$$

43. xyz

$$2\frac{3}{8}\cdot\frac{3}{19}\cdot\frac{4}{9}=\frac{19}{8}\cdot\frac{3}{19}\cdot\frac{4}{9}$$
$$=\frac{19\cdot 3\cdot 4}{8\cdot 19\cdot 9}$$
$$=\frac{19\cdot 3\cdot 2\cdot 2}{2\cdot 2\cdot 2\cdot 19\cdot 3\cdot 3}$$
$$=\frac{1}{6}$$

45. xyz

$$\frac{5}{6}\cdot 3\cdot 1\frac{7}{15}=\frac{5}{6}\cdot\frac{3}{1}\cdot\frac{22}{15}$$
$$=\frac{5\cdot 3\cdot 22}{6\cdot 1\cdot 15}$$
$$=\frac{5\cdot 3\cdot 2\cdot 11}{2\cdot 3\cdot 1\cdot 3\cdot 5}$$
$$=\frac{11}{3}=3\frac{2}{3}$$

47.

$$\frac{4}{5}x=\frac{5}{3}$$

$\frac{4}{5}x$	$\frac{5}{3}$
$\frac{4}{5}\cdot\frac{3}{4}$	$\frac{5}{3}$
$\frac{4\cdot 3}{5\cdot 4}$	$\frac{5}{3}$
$\frac{2\cdot 2\cdot 3}{5\cdot 2\cdot 2}$	$\frac{5}{3}$

$$\frac{3}{5}\neq\frac{5}{3}$$

No, $\frac{3}{4}$ is not a solution of the equation.

49. Less than

51.

$$\left(\frac{3}{4}\right)^2=\frac{3}{4}\cdot\frac{3}{4}$$
$$=\frac{3\cdot 3}{4\cdot 4}=\frac{9}{16}$$

53.

$$\left(\frac{5}{8}\right)^3\cdot\left(\frac{2}{5}\right)^2=\frac{5}{8}\cdot\frac{5}{8}\cdot\frac{5}{8}\cdot\frac{2}{5}\cdot\frac{2}{5}$$
$$=\frac{5\cdot 5\cdot 5\cdot 2\cdot 2}{8\cdot 8\cdot 8\cdot 5\cdot 5}$$
$$=\frac{5}{128}$$

55.

$$\left(\frac{18}{25}\right)^2\cdot\left(\frac{5}{9}\right)^3=\frac{18}{25}\cdot\frac{18}{25}\cdot\frac{5}{9}\cdot\frac{5}{9}\cdot\frac{5}{9}$$
$$=\frac{18\cdot 18\cdot 5\cdot 5\cdot 5}{25\cdot 25\cdot 9\cdot 9\cdot 9}$$
$$=\frac{4}{45}$$

57.

$$7^2\cdot\left(\frac{2}{7}\right)^3=\frac{7}{1}\cdot\frac{7}{1}\cdot\frac{2}{7}\cdot\frac{2}{7}\cdot\frac{2}{7}$$
$$=\frac{7\cdot 7\cdot 2\cdot 2\cdot 2}{1\cdot 1\cdot 7\cdot 7\cdot 7}$$
$$=\frac{8}{7}=1\frac{1}{7}$$

59. x^4

$\left(\frac{2}{3}\right)^4 = \frac{2}{3}\cdot\frac{2}{3}\cdot\frac{2}{3}\cdot\frac{2}{3}$

$= \frac{2\cdot2\cdot2\cdot2}{3\cdot3\cdot3\cdot3}$

$= \frac{16}{81}$

61. x^3y^2

$\left(\frac{2}{3}\right)^3\cdot\left(1\frac{1}{2}\right)^2 = \frac{2}{3}\cdot\frac{2}{3}\cdot\frac{2}{3}\cdot\frac{3}{2}\cdot\frac{3}{2}$

$= \frac{2\cdot2\cdot2\cdot3\cdot3}{3\cdot3\cdot3\cdot2\cdot2}$

$= \frac{2}{3}$

Objective B Exercises

63. $\frac{5}{7}\div\frac{2}{5} = \frac{5}{7}\cdot\frac{5}{2}$

$= \frac{5\cdot5}{7\cdot2}$

$= \frac{25}{14} = 1\frac{11}{14}$

65. $0\div\frac{7}{9} = 0$

Zero divided by a non-zero number is 0.

67. $6\div\frac{3}{4} = \frac{6}{1}\cdot\frac{4}{3}$

$= \frac{6\cdot4}{1\cdot3}$

$= \frac{2\cdot3\cdot2\cdot2}{1\cdot3}$

$= \frac{8}{1} = 8$

69. $\frac{3}{4}\div6 = \frac{3}{4}\div\frac{6}{1}$

$= \frac{3}{4}\cdot\frac{1}{6}$

$= \frac{3\cdot1}{4\cdot6}$

$= \frac{3\cdot1}{2\cdot2\cdot2\cdot3}$

$= \frac{1}{8}$

71. $\frac{9}{10}\div0$

Division by zero is undefined.

73. $\frac{5}{12}\div\frac{15}{32} = \frac{5}{12}\cdot\frac{32}{15}$

$= \frac{5\cdot32}{12\cdot15}$

$= \frac{5\cdot2\cdot2\cdot2\cdot2\cdot2}{2\cdot2\cdot3\cdot3\cdot5}$

$= \frac{8}{9}$

75. $\frac{8}{x}\div\frac{y}{4} = \frac{8}{x}\cdot\frac{4}{y}$

$= \frac{8\cdot4}{x\cdot y}$

$= \frac{32}{xy}$

77. $\frac{b}{6}\div\frac{5}{d} = \frac{b}{6}\cdot\frac{d}{5}$

$= \frac{b\cdot d}{6\cdot5}$

$= \frac{bd}{30}$

79. $3\frac{1}{3}\div\frac{5}{8} = \frac{10}{3}\cdot\frac{8}{5} = \frac{10\cdot8}{3\cdot5}$

$= \frac{2\cdot5\cdot2\cdot2\cdot2}{3\cdot5}$

$= \frac{16}{3} = 5\frac{1}{3}$

81. $5\frac{1}{2}\div11 = \frac{11}{2}\div\frac{11}{1}$

$= \frac{11}{2}\cdot\frac{1}{11}$

$= \frac{11\cdot1}{2\cdot11}$

$= \frac{1}{2}$

83. $5\frac{2}{7}\div1 = \frac{37}{7}\div\frac{1}{1}$

$= \frac{37}{7}\cdot\frac{1}{1}$

$= \frac{37\cdot1}{7\cdot1}$

$= \frac{37}{7} = 5\frac{2}{7}$

85. $2\frac{4}{13} \div 1\frac{5}{26} = \frac{30}{13} \div \frac{31}{26}$
$= \frac{30}{13} \cdot \frac{26}{31}$
$= \frac{30 \cdot 26}{13 \cdot 31}$
$= \frac{2 \cdot 3 \cdot 5 \cdot 2 \cdot 13}{13 \cdot 31}$
$= \frac{60}{31} = 1\frac{29}{31}$

87. Strategy:
To find the number of servings, divide the net weight of Kellogg Honey Crunch Corn Flakes (24) by $\frac{3}{4}$.

Solution:
$24 \div \frac{3}{4} = \frac{24}{1} \cdot \frac{4}{3} = \frac{24 \cdot 4}{1 \cdot 3} = 32$
There are 32 servings in a box of Kellogg Honey Crunch Corn Flakes.

89. $\frac{9}{10} \div \frac{3}{4} = \frac{9}{10} \cdot \frac{4}{3}$
$= \frac{9 \cdot 4}{10 \cdot 3}$
$= \frac{3 \cdot 3 \cdot 2 \cdot 2}{2 \cdot 5 \cdot 3}$
$= \frac{6}{5} = 1\frac{1}{5}$

91. $\frac{7}{8} \div 3\frac{1}{4} = \frac{7}{8} \div \frac{13}{4}$
$= \frac{7}{8} \cdot \frac{4}{13}$
$= \frac{7 \cdot 4}{8 \cdot 13}$
$= \frac{7 \cdot 2 \cdot 2}{2 \cdot 2 \cdot 2 \cdot 13}$
$= \frac{7}{26}$

93. $x \div y$
$-\frac{5}{8} \div \left(-\frac{15}{2}\right) = \frac{5}{8} \div \frac{15}{2}$
$= \frac{5}{8} \cdot \frac{2}{15}$
$= \frac{5 \cdot 2}{8 \cdot 15}$
$= \frac{5 \cdot 2}{2 \cdot 2 \cdot 2 \cdot 3 \cdot 5}$
$= \frac{1}{12}$

95. $x \div y$
$18 \div \frac{3}{8} = \frac{18}{1} \cdot \frac{8}{3}$
$= \frac{18 \cdot 8}{1 \cdot 3}$
$= \frac{2 \cdot 3 \cdot 3 \cdot 2 \cdot 2 \cdot 2}{1 \cdot 3}$
$= \frac{48}{1} = 48$

97. greater than

Objective B Exercises

99. $\frac{\frac{9}{16}}{\frac{3}{4}} = \frac{9}{16} \div \frac{3}{4}$
$= \frac{9}{16} \cdot \frac{4}{3} = \frac{3}{4}$

101. $\frac{\frac{2}{3} + \frac{1}{2}}{7} = \frac{\frac{7}{6}}{\frac{7}{1}}$
$= \frac{7}{6} \div \frac{7}{1}$
$= \frac{7}{6} \cdot \frac{1}{7} = \frac{1}{6}$

103. $\frac{2 + \frac{1}{4}}{\frac{3}{8}} = \frac{\frac{9}{4}}{\frac{3}{8}}$
$= \frac{9}{4} \div \frac{3}{8}$
$= \frac{9}{4} \cdot \frac{8}{3} = 6$

105. $\frac{\frac{9}{25}}{\frac{4}{5} - \frac{1}{10}} = \frac{\frac{9}{25}}{\frac{7}{10}}$
$= \frac{9}{25} \div \frac{7}{10}$
$= \frac{9}{25} \cdot \frac{10}{7} = \frac{18}{35}$

107. $\frac{3 + 2\frac{1}{3}}{5\frac{1}{6} - 1} = \frac{5\frac{1}{3}}{4\frac{1}{6}}$
$= 5\frac{1}{3} \div 4\frac{1}{6}$
$= \frac{16}{3} \div \frac{25}{6}$
$= \frac{16}{3} \cdot \frac{6}{25} = \frac{32}{25} = 1\frac{7}{25}$

109. $\dfrac{5\frac{2}{3}-1\frac{1}{6}}{3\frac{5}{8}-2\frac{1}{4}}=\dfrac{4\frac{1}{2}}{1\frac{3}{8}}$

$=4\frac{1}{2}\div 1\frac{3}{8}$

$=\frac{9}{2}\div\frac{11}{8}$

$=\frac{9}{2}\cdot\frac{8}{11}=\frac{36}{11}=3\frac{3}{11}$

111. $\dfrac{x+y}{z}$

$\dfrac{\frac{2}{3}+\frac{3}{4}}{\frac{1}{12}}=\dfrac{\frac{17}{12}}{\frac{1}{12}}$

$=\frac{17}{12}\div\frac{1}{12}$

$=\frac{17}{12}\cdot\frac{12}{1}=17$

113. $\dfrac{x-y}{z}$

$\dfrac{2\frac{5}{8}-1\frac{1}{4}}{1\frac{3}{8}}=\dfrac{1\frac{3}{8}}{1\frac{3}{8}}$

$=1\frac{3}{8}\div 1\frac{3}{8}=\frac{11}{8}\div\frac{11}{8}$

$=\frac{11}{8}\cdot\frac{8}{11}=1$

115. 1

117. 0

Objective D Exercises

119. Strategy:
To find the length of time, multiply the length of time for one chukker $\left(7\frac{1}{2}\right)$ by 4.

Solution:
$7\frac{1}{2}\cdot 4=\frac{15}{2}\cdot\frac{4}{1}=\frac{15\cdot 4}{2\cdot 1}=30$
Four periods of play takes 30 min.

121. Strategy:
→To find the number of feet in one rod, multiply the number of yards in one rod $\left(5\frac{1}{2}\right)$ by 3.
→To find the number of inches in one rod, multiply the number of yards in one rod $\left(5\frac{1}{2}\right)$ by 36.

Solution:
$5\frac{1}{2}\cdot 3=\frac{11}{2}\cdot\frac{3}{1}=\frac{11\cdot 3}{2\cdot 1}=16\frac{1}{2}$
$5\frac{1}{2}\cdot 36=\frac{11}{2}\cdot\frac{36}{1}$
$=\frac{11\cdot 36}{2\cdot 1}=198$
One rod is equivalent to $16\frac{1}{2}$ ft.
One rod is equivalent to 198 in.

123. Strategy:
To find the amount of time, multiply the number of hours cleaning per week $\left(4\frac{1}{2}\right)$ by the number of weeks (52).

Solution:
$4\frac{1}{2}\cdot 52=\frac{9}{2}\cdot\frac{52}{1}=\frac{9\cdot 52}{2\cdot 1}=234$
The average couple spends 234 h a year cleaning house.

125. Strategy:
To find the number of house plots:
→Subtract the number of acres set aside (3) from the total number of acres $\left(25\frac{1}{2}\right)$ to find the number of acres to be sold in parcels.
→Divide the number of acres available by the number of acres in one parcel $\left(\frac{3}{4}\right)$.

Solution:
$25\frac{1}{2}-3=22\frac{1}{2}$
$22\frac{1}{2}\div\frac{3}{4}=\frac{45}{2}\cdot\frac{4}{3}=\frac{45\cdot 4}{2\cdot 3}=30$
The developer plans to build 30 houses.

127. Strategy:
To approximate the asteroid's distance from Earth at its closest point, multiply the distance the earth is from the moon in miles (250,000) by the fraction of that distance which the asteroid is from the earth $\left(\frac{9}{10}\right)$.

Solution:
$250{,}000\cdot\frac{9}{10}=\frac{250{,}000}{1}\cdot\frac{9}{10}$
$=\frac{2{,}250{,}000}{10}=225{,}000$
At its closest point, the asteroid is approximately 225,000 miles from Earth.

129. Strategy:
To find the area, use the formula below, substitute 21 for b, 13 for h, and solve for A.

Solution:
$A=\frac{1}{2}bh$
$A=\frac{1}{2}\cdot 21\cdot 13=\frac{1}{2}\cdot\frac{21}{1}\cdot\frac{13}{1}$
$=\frac{651}{2}=136\frac{1}{2}$
The area of the vegetable garden is $136\frac{1}{2}$ ft^2.

131. Strategy:
To find the number of bags of seed:
→Find the area of the triangle, use the formula below, substitute 20 for b, 24 for h, and solve for A.
→Divide the area by 120.

Solution:
$A=\frac{1}{2}bh$
$A=\frac{1}{2}\cdot 20\cdot 24=\frac{1}{2}\cdot\frac{20}{1}\cdot\frac{24}{1}=240$
$240\div 120=2$
Two bags of grass seed should be purchased.

133. Strategy:
To find the pressure, substitute $12\frac{1}{2}$ for D in the given formula and solve for P.

Solution:
$P=15+\frac{1}{2}D$
$P=15+\frac{1}{2}\left(12\frac{1}{2}\right)$
$P=15+\frac{1}{2}\left(\frac{25}{2}\right)$
$P=15+6\frac{1}{4}$
$P=21\frac{1}{4}$
The pressure is $21\frac{1}{4}$ pounds per square inch.

Critical Thinking 2.4

135. Divide $3\frac{1}{8}$ by $\frac{1}{8}$ to find the number of 50-mile segments between the two cities.
$3\frac{1}{8}\div\frac{1}{8}=\frac{25}{8}\div\frac{1}{8}=\frac{25}{8}\cdot\frac{8}{1}=25$
Multiply 25 times 50 to find the number of miles.
$25\cdot 50=1{,}250$
The distance between the two cities is 1,250 mi.

Projects or Group Activities 2.4

137. Our calendar year is 365 days. A complete orbit around the sun takes $365\frac{1}{4}$ days, so one solar year is $365\frac{1}{4}$ days. Therefore, after 365 days, the Earth is $\frac{1}{4}$ day short of a complete orbit around the sun. After two years, the earth is $\frac{1}{4}$ day $+\frac{1}{4}$ day $=\frac{1}{2}$ day short of a complete orbit around the sun. After 4 years, the earth is $\frac{1}{4}$ day $+\frac{1}{4}$ day $+\frac{1}{4}$ day $+\frac{1}{4}$ day = 1 day short of a complete orbit around the sun. Therefore, every four years, our calendar incorporates a leap year, which is a 366-day year, in order to make up for the one extra day in the four solar years.

Check Your Progress: Chapter 2

1. $5=5$
$12=2^2\cdot 3$
$15=3\cdot 5$
The LCM $=2^2\cdot 3\cdot 5=60$.

2. $26=2\cdot 13$
$52=2^2\cdot 13$
The GCF $=2\cdot 13=26$.

3. $3\overline{)\,10}$ quotient 3; -9; remainder 1
$\frac{10}{3}=3\frac{1}{3}$

4. $9\overline{)\,81}$ quotient 9; -81; remainder 0
$\frac{81}{9}=9$

5. $3\frac{1}{4}=\frac{(4\cdot 3)+1}{4}=\frac{12+1}{4}=\frac{13}{4}$

6. $17=\frac{17}{1}$

7. $12\div 3=4$
$\frac{2}{3}=\frac{2\cdot 4}{3\cdot 4}=\frac{8}{12}$
$\frac{8}{12}$ is equivalent to $\frac{2}{3}$.

8. $\frac{24}{64}=\frac{2\cdot 2\cdot 2\cdot 3}{2\cdot 2\cdot 2\cdot 2\cdot 2\cdot 2}=\frac{3}{8}$

9. $\frac{2}{3}=\frac{16}{24}$ $\quad\frac{5}{8}=\frac{15}{24}$
$\frac{16}{24}>\frac{15}{24}$
$\frac{2}{3}>\frac{5}{8}$

10. $\frac{4}{7}+\frac{2}{21}=\frac{12}{21}+\frac{2}{21}=\frac{14}{21}=\frac{2}{3}$

11. $4\frac{5}{6}+\frac{3}{4}=\frac{29}{6}+\frac{3}{4}=\frac{58}{12}+\frac{9}{12}=\frac{67}{12}=5\frac{7}{12}$

12. $\frac{2}{3}-\frac{1}{15}=\frac{10}{15}-\frac{1}{15}=\frac{9}{15}=\frac{3}{5}$

13. $1\frac{3}{8}-\frac{7}{10}=\frac{11}{8}-\frac{7}{10}=\frac{55}{40}-\frac{28}{40}=\frac{27}{40}$

14. $2\frac{1}{4}\cdot 1\frac{2}{9}=\frac{9}{4}\cdot\frac{11}{9}$
$=\frac{9\cdot 11}{4\cdot 9}$
$=\frac{3\cdot 3\cdot 11}{2\cdot 2\cdot 3\cdot 3}$
$=\frac{11}{4}=2\frac{3}{4}$

15. $\frac{3}{4}\cdot\left(\frac{2}{9}\right)\cdot\left(\frac{2}{5}\right)=\frac{3\ 2\ 2}{4\cdot 9\cdot 5}$
$=\frac{3\cdot 2\cdot 2}{2\cdot 2\cdot 3\cdot 3\cdot 5}$
$=\frac{1}{15}$

16. $\frac{6}{m} \div \frac{n}{5} = \frac{6}{m} \cdot \frac{5}{n}$

$= \frac{6 \cdot 5}{m \cdot n}$

$= \frac{30}{mn}$

17. $8\frac{1}{3} \div 25 = \frac{25}{3} \cdot \frac{1}{25} = \frac{25 \cdot 1}{3 \cdot 25}$

$= \frac{1}{3}$

18. $\frac{\frac{3}{4} - \frac{1}{3}}{\frac{2}{3} + \frac{1}{6}} = \frac{\frac{9}{12} - \frac{4}{12}}{\frac{4}{6} + \frac{1}{6}} = \frac{\frac{5}{12}}{\frac{5}{6}}$

$= \frac{5}{12} \div \frac{5}{6}$

$= \frac{5}{12} \cdot \frac{6}{5} = \frac{6}{12} = \frac{1}{2}$

19. $x - y$

$4 - 2\frac{4}{11} = \frac{4}{1} - \frac{26}{11} = \frac{44}{11} - \frac{26}{11} = \frac{18}{11} = 1\frac{7}{11}$

20. $\frac{3}{5}x = \frac{1}{5}$

$$\begin{array}{c|c} \frac{3}{5} \cdot \frac{1}{3} & \frac{1}{5} \\ \frac{3 \cdot 1}{5 \cdot 3} & \frac{1}{5} \end{array}$$

$\frac{1}{5} = \frac{1}{5}$

Yes, $\frac{1}{3}$ is a solution of the equation.

21. $2\frac{5}{8} \div 5\frac{1}{4} = \frac{21}{8} \div \frac{21}{4}$

$= \frac{21}{8} \cdot \frac{4}{21}$

$= \frac{21 \cdot 4}{8 \cdot 21}$

$= \frac{3 \cdot 7 \cdot 2 \cdot 2}{2\ 2\ 2 \cdot 3\ 7}$

$= \frac{1}{2}$

22. $\frac{x}{y}$

$\frac{\frac{2}{3}}{\frac{7}{9}} = \frac{2}{3} \div \frac{7}{9}$

$= \frac{2}{3} \cdot \frac{9}{7}$

$= \frac{2 \cdot 3\ 3}{3 \cdot 7}$

$= \frac{6}{7}$

23. Strategy:

To find course length, replace a, b, and c with $12\frac{1}{12}$, $29\frac{1}{3}$, and $26\frac{3}{4}$ in the given formula and solve for P.

Solution:

$P = a + b + c$

$P = 12\frac{1}{12} + 29\frac{1}{3} + 26\frac{3}{4}$

$= 12\frac{1}{12} + 29\frac{4}{12} + 26\frac{9}{12}$

$= 67\frac{14}{12} = 68\frac{2}{12}$

$= 68\frac{1}{6}$

The total length of the course is $68\frac{1}{6}$ yd.

24. Strategy:
To find the dimensions of the board when folded, divide the length (14) by 2 and multiply the height $\left(\frac{7}{8}\right)$ by 2.

Solution:
$14 \div 2$

$\frac{7}{8} \cdot 2 = \frac{7}{8} \cdot \frac{2}{1} = \frac{7 \cdot 2}{8 \cdot 1}$

$= \frac{7 \cdot 2}{2 \cdot 2 \cdot 2 \cdot 1} = \frac{7}{4} = 1\frac{3}{4}$

The dimensions of the board are 14 in. by 7 in. by $1\frac{3}{4}$ in.

Section 2.5

Concept Check

1. hundredths; thousandths; hundred-thousandths; millionths

3. and; thousandths

5. <; <

Objective A Exercises

7. The digit 5 is in the thousandths' place.

9. The digit 5 is in the ten-thousandths' place.

11. The digit 5 is in the hundredths' place.

13. $\frac{3}{10} = 0.3$ [three tenths]

15. $\frac{21}{100} = 0.21$ [twenty-one hundredths]

17. $\frac{461}{1{,}000} = 0.461$ [four hundred sixty-one thousandths]

19. $\frac{93}{1{,}000} = 0.093$ [ninety-three thousandths]

21. $0.1 = \frac{1}{10}$ [one tenth]

23. $0.47 = \frac{47}{100}$ [forty-seven hundredths]

25. $0.289 = \frac{289}{1{,}000}$ [two hundred eighty-nine thousandths]

27. $0.09 = \frac{9}{100}$ [nine hundredths]

29. thirty-seven hundredths

31. nine and four tenths

33. fifty-three ten-thousandths

35. forty-five thousandths

37. twenty-six and four hundredths

39. 3.0806

41. 407.03

43. 246.024

45. 73.02684

Objective B Exercises

47. $0.7 = 0.70$
$0.70 > 0.56$
$0.7 > 0.56$

49. $3.605 > 3.065$

51. $9.04 = 9.040$
$9.004 < 9.040$
$9.004 < 9.04$

53. $9.31 = 9.310$
$9.310 > 9.031$
$9.31 > 9.031$

55. $4.6 < 40.6$

57. $0.07046 > 0.07036$

59. 0.609, 0.660, 0.696, 0.699
0.609, 0.66, 0.696, 0.699

61. 1.237, 1.327, 1.372, 1.732

63. 21.780, 21.805, 21,870, 21.875
21.78, 21.805, 21.87, 21.875

65. $0.62 > 0.062$

Objective C Exercises

67. 5.398 — Given place value: 3; $9 > 5$

5.398 rounded to the nearest tenth is 5.4.

69. 30.0092 — Given place value: 0; $0 < 5$

30.0092 rounded to the nearest tenth is 30.0.

71. 413.5972 — Given place value: 9; $7 > 5$

413.5972 rounded to the nearest hundredth is 413.60.

73. 6.061745 — Given place value: 1; $7 > 5$

6.061745 rounded to the nearest thousandth is 6.062.

75. 96.8027 — Given place value: 6; $8 > 5$

96.8027 rounded to the nearest whole number is 97.

77. 5,439.83 — Given place value: 9; $8 > 5$

5,439.83 rounded to the nearest whole number is 5,440.

79. 0.023591 — Given place value: 5; $9 > 5$

0.023591 rounded to the nearest ten-thousandth is 0.0236.

Objective D Exercises

81. Strategy:

a. To find if the cost to mint a penny is greater or less than the face value of a penny, compare the cost to mint a penny (1.67 cents) with the value of a penny (1 cent).

b. To find the cost to mint a penny to the nearest cent, round the cost to mint a penny (1.67 cents) to the nearest integer.

c. To find if the cost to mint a nickel is greater or less than the face value of a nickel, compare the cost to mint a nickel (7.7 cents) with the value of a nickel (5 cents).

d. To find the cost to mint a nickel to the nearest cent, round the cost to mint a nickel (7.7 cents) to the nearest integer.

Solution:

a. $1.67 > 1$
The cost to mint a penny is greater than the face value of a penny.

b. 1.67 rounded to the nearest integer is 2.
The cost to mint a penny to the nearest cent is 2 cents.

c. $7.7 > 5$
The cost to mint a nickel is greater than the face value of a nickel.

d. 7.7 rounded to the nearest integer is 8.
The cost to mint a nickel to the nearest cent is 8 cents.

83. Strategy:
To find the distance, round 26.21875 to the nearest tenth.

Solution:
26.21875 rounded to the nearest tenth is 26.2.
To the nearest tenth, the Boston Marathon is 26.2 mi.

85. Strategy:
To find the shipping and handling charges for each order, compare each amount ordered with the amounts in the table and read the corresponding shipping and handling charges from the table.

Solution:

a. $\$10.01 < \$12.42 < \$20.00$
The shipping and handling charge is $2.40.

b. $\$20.01 < \$23.56 < \$30.00$
The shipping and handling charge is $3.60.

c. $\$40.01 < \$47.80 < \$50.00$
The shipping and handling charge is $6.00.

d. $\$66.91 > \$50.01$
The shipping and handling charge is $7.00.

e. $\$30.01 < \$35.75 < \$40.00$
The shipping and handling charge is $4.70.

f. $\$20.00 = \$20.00$
The shipping and handling charge is $2.40.

g. $\$10.01 < \$18.25 < \$20.00$
The shipping and handling charge is $2.40.

Critical Thinking 2.5

87. a. Answers will vary.
For example, 0.11, 0.12, 0.13, 0.14, 0.15, 0.16, 0.17, 0.18, and 0.19 are numbers between 0.1 and 0.2. But any number of digits can be attached to 0.1, and the number will be between 0.1 and 0.2. For example, 0.123456789 is a number between 0.1 and 0.2.

b. Answers will vary.
For example, 1.01, 1.02, 1.03, 1.04, 1.05, 1.06, 1.07, 1.08, and 1.09 are numbers between 1 and 1.1. But any number of digits can be attached to 1.0, and the number will be between 1 and 1.1. For example, 1.0123456789 is a number between 1 and 1.1.

c. Answers will vary.
For example, 0.001, 0.002, 0.003, and 0.004 are numbers between 0 and 0.005. But any number of digits can be attached to 0.001, 0.002, 0.003, or 0.004, and the number will be between 0 and 0.005. For example, 0.00123456789 is a number between 0 and 0.005.

Section 2.6

Concept Check

1.
$$\begin{array}{r} {}^{1\ 1} \\ 2.391 \\ 45 \\ +13.0784 \\ \hline 60.4694 \end{array}$$

3. 1; 2; 3; 1.113

5. two; whole; $304\overline{)364.8}$ with quotient 1.2

7. 5; 4; terminating

Objective A Exercises

9.
$$\begin{array}{r} {}^{1\ \ 1\,1} \\ 1.864 \\ 39 \\ +\ 25.0781 \\ \hline 65.9421 \end{array}$$

11.
$$\begin{array}{r} {}^{2\,1} \\ 35.9 \\ 8.217 \\ +146.74 \\ \hline 190.857 \end{array}$$

13.
$$\begin{array}{r} 36.47 \\ -15.21 \\ \hline 21.26 \end{array}$$

15.
$$\begin{array}{r} {}^{7\ 9\ 10} \\ 28.00 \\ -\ 6.74 \\ \hline 21.26 \end{array}$$

17.
$$\begin{array}{r} {}^{5\ \ 9\,11\,10} \\ 6.020 \\ -3.252 \\ \hline 2.768 \end{array}$$

19. $6.24 + 8.573 + 19.06 + 22.488$
$= 14.813 + 19.06 + 22.488$
$= 33.873 + 22.488$
$= 56.361$

21. $62.57 - 8.9 = 62.57 + (-8.9)$
$= 53.67$

23.
$$\begin{array}{r} 6.408 \\ +\,5.917 \\ \hline 12.325 \end{array} \quad \begin{array}{c} \to \\ \to \\ \end{array} \quad \begin{array}{r} 6 \\ +\,6 \\ \hline 12 \end{array}$$

25.
$$\begin{array}{r} 56.87 \\ -\,23.24 \\ \hline 33.63 \end{array} \quad \begin{array}{c} \to \\ \to \\ \end{array} \quad \begin{array}{r} 60 \\ -\,20 \\ \hline 40 \end{array}$$

27.
$$\begin{array}{r} 0.931 \\ -\,0.628 \\ \hline 0.303 \end{array} \quad \begin{array}{c} \to \\ \to \\ \end{array} \quad \begin{array}{r} 0.9 \\ -\,0.6 \\ \hline 0.3 \end{array}$$

29.
$$\begin{array}{r} 87.65 \\ -\,49.032 \\ \hline 38.618 \end{array} \quad \begin{array}{c} \to \\ \to \\ \end{array} \quad \begin{array}{r} 90 \\ -\,50 \\ \hline 40 \end{array}$$

31. a. $49.3 + 5.9 + 1.5 = 56.7$
The total children is 56.7 million.

b. $49.3 - 5.9 = 43.4$
There are 43.4 million more children in public school.

33. $x + y + z$
$6.059 + 3.884 + 15.71$
$= 9.943 + 15.71$
$= 25.653$

35. $x - y$
$6.029 - 4.708$
$= 1.321$

37. $7.01 - 2.325$

39. $19.35 - 8.967$

Objective B Exercises

41.
$$\begin{array}{r} 3.4 \\ \times\,0.5 \\ \hline 1.70 \end{array}$$

43.
$$\begin{array}{r} 8.29 \\ \times\,0.004 \\ \hline 0.03316 \end{array}$$

45. $1,000 \cdot 4.25 = 4,250$

47. $6.71 \cdot 10^4 = 67,100$

49.
$$\begin{array}{r} 9.81 \\ 0.77 \end{array} \quad \begin{array}{c} \to \\ \to \end{array} \quad \begin{array}{r} 10 \\ \times\,0.8 \\ \hline 8.0 \end{array}$$
$9.81 \cdot 0.77 = 7.5537$

51.
$$\begin{array}{r} 6.88 \\ 9.97 \end{array} \quad \begin{array}{c} \to \\ \to \end{array} \quad \begin{array}{r} 7 \\ \times\,10 \\ \hline 70 \end{array}$$
$6.88 \cdot 9.97 = 68.5936$

53.
$$\begin{array}{r} 28.45 \\ 1.13 \end{array} \quad \begin{array}{c} \to \\ \to \end{array} \quad \begin{array}{r} 30 \\ \times\,1 \\ \hline 30 \end{array}$$
$28.45 \cdot 1.13 = 32.1485$

55. $20,000(0.6267) = 12,534$
12,534 British pounds would be exchanged for 20,000 U.S. dollars.

57. ab
$6.27 \cdot 8 = 50.16$

59. $1.6 = 0.2z$
$$\begin{array}{c|c} 1.6 & 0.2(8) \end{array}$$
$1.6 = 1.6$
Yes, 8 is a solution of the equation.

61. two

Objective C Exercises

63.
$$\begin{array}{r} 32.3 \\ 0.5\overline{)\,16.15} \\ \underline{-15} \\ 11 \\ \underline{-10} \\ 15 \\ \underline{-15} \\ 0 \end{array}$$

65. $27.08 \div 0.4 = 67.7$

67. $55.63 \div 8.8 \approx 6.3$

69. $52.8 \div 9.1 \approx 5.8$

71. $6.457 \div 8 \approx 0.81$

73. $0.0416 \div 0.53 \approx 0.08$

75. $52.78 \div 10 = 5.278$

77. $48.05 \div 10^2 = 0.4805$

79. $42.43 \rightarrow 40$
$3.8 \rightarrow 4$
$42.43 \div 3.8 \approx 11.17$
$40 \div 4 = 10$

81. $6.398 \rightarrow 6$
$5.5 \rightarrow 6$
$6.398 \div 5.5 \approx 1.16$
$6 \div 6 = 1$

83. $1.237 \rightarrow 1$
$0.021 \rightarrow 0.02$
$1.237 \div 0.021 \approx 58.90$
$1 \div 0.02 = 50$

85. $33.14 \rightarrow 30$
$4.6 \rightarrow 5$
$33.14 \div 4.6 \approx 7.20$
$30 \div 5 = 6$

87. Strategy:
To determine how many times greater sales were in 2010 than in 1997, divide the amount of sales in 2010 (28.6) by the amount of sales in 1997 (3.6).

Solution:
$28.6 \div 3.6 \approx 8$
Sales in 2010 were approximately 8 times greater than in 1997.

89. $\frac{x}{y}$
$$\frac{3.542}{0.7} = 3.542 \div 0.7 = 5.06$$

91. $\frac{q}{8} = 3.1$

$\frac{24.8}{8}$	3.1

$3.1 = 3.1$
Yes, 24.8 is a solution of the equation.

93. $21 = \frac{t}{0.4}$

21	$\frac{0.84}{0.4}$

$21 \neq 2.1$
No, 0.84 is not a solution of the equation.

95. three

Objective D Exercises

97. $8\overline{)3.000}$ quotient 0.375
$\frac{3}{8} = 0.375$

99. $11\overline{)8.0000}$ quotient 0.7272
$\frac{8}{11} = 0.\overline{72}$

101. $12\overline{)7.00000}$ quotient 0.58333
$\frac{7}{12} = 0.58\overline{3}$

103. $4\overline{)7.00}$ quotient 1.75
$\frac{7}{4} = 1.75$

105. Write $\frac{1}{2}$ as a decimal.
$2\overline{)1.0}$ quotient 0.5
$1\frac{1}{2} = 1.5$

107. Write $\frac{1}{6}$ as a decimal.
$6\overline{)1.0000}$ quotient 0.1666 $= 0.1\overline{6}$
$4\frac{1}{6} = 4.1\overline{6}$

109. Write $\frac{1}{4}$ as a decimal.
$4\overline{)1.00}$ quotient 0.25
$2\frac{1}{4} = 2.25$

111. Write $\frac{8}{9}$ as a decimal.
$9\overline{)8.000}$ quotient 0.888 $= 0.\overline{8}$
$3\frac{8}{9} = 3.\overline{8}$

113. $0.2 = \frac{2}{10} = \frac{1}{5}$

115. $0.75 = \frac{75}{100} = \frac{3}{4}$

117. $0.125 = \frac{125}{1{,}000} = \frac{1}{8}$

119. $2.5 = 2\frac{5}{10} = 2\frac{1}{2}$

121. $4.55 = 4\frac{55}{100} = 4\frac{11}{20}$

123. $1.72 = 1\frac{72}{100} = 1\frac{18}{25}$

125. $0.045 = \frac{45}{1,000} = \frac{9}{200}$

127. $\frac{9}{10} = 0.90$
$0.90 > 0.89$
$\frac{9}{10} > 0.89$

129. $\frac{4}{5} = 0.800$
$0.800 < 0.803$
$\frac{4}{5} < 0.803$

131. $\frac{4}{9} = 0.\overline{444}$
$0.444 < 0.\overline{444}$
$0.444 < \frac{4}{9}$

133. $\frac{3}{25} = 0.12$
$0.13 > 0.12$
$0.13 > \frac{3}{25}$

135. $\frac{5}{16} = 0.3125$
$0.3125 > 0.3120$
$\frac{5}{16} > 0.312$

137. $\frac{10}{11} = 0.9\overline{09}$
$0.9\overline{09} > 0.909$
$\frac{10}{11} > 0.909$

139. $\frac{3}{5}$

Objective E Exercises

141. (iii) and (iv)

143. Strategy:
To find the new balance:
→Add the amount of the deposit (189.53) to the old balance (347.80).
→Subtract the amount of the check (62.89) from the sum

Solution:
347.08 + 189.53 = 536.61
536.61 − 62.89 = 473.72
The new balance is $473.72

145. Strategy:
To find the average annual cost of electricity, multiply the number of months in a year (12) times the average monthly bill (95.66).

Solution:
12 · 95.66 = 1147.92
The average annual cost of electricity $1,147.92.

147. Strategy:
a. To determine the amount of the payments over 36 months, multiply the amount of each payment (499.50) by the number of payments (36).

b. To determine the total cost of the car, add the amount of the payments to the down payment (5000).

Solution:
a. 499.50 · 36 = 17,982
The amount of the payments over 36 months is $17,982.

b. 17,982 + 5000 = 22,982
The total cost of the car is $17,982.

149. Strategy:

a. To determine the difference in average age at first marriage from 1970 to 2010 for women, subtract the average age at first marriage for women in 1970 (20.8) from the average age at first marriage for women in 2010 (26.1).

b. To determine the difference in average age at first marriage from 1970 to 2010 for men, subtract the average age at first marriage for men in 1970 (23.2) from the average age at first marriage for men in 2010 (28.2).

Solution:

a. $26.1 - 20.8 = 5.3$
The difference in average age at first marriage for women from 1970 to 2010 is 5.3 years.

b. $28.2 - 23.2 = 5.0$
The difference in average age at first marriage for men from 1970 to 2010 is 5.0 years.

151. Strategy:
To find the amount received for the 400 cans, multiply the weight of the cans (18.75) by the amount paid per pound (0.75). Round to the nearest cent.

Solution:
$18.75 \cdot 0.75 \approx 14.06$
The amount received for the 400 cans is $14.06.

153. Strategy:
To find the amount of taxes paid, multiply the tax per gallon (0.466) by the number of gallons (12.5).

Solution:
$0.466 \cdot 12.5 \approx 5.83$
The amount of taxes paid is approximately $5.83.

155. Strategy:
To find the amount of coal you will use in one year, divide the number of kilowatt-hours used per month (25) by the number of kilowatt-hours produced by one ton of coal (3000). Then multiply that amount by the number of months in a year (12).

Solution:
$25 \div 3000 = 0.0083333...$
$0.008333 \cdot 12 = 0.1$
You will use 0.1 ton of coal in one year.

157. Strategy:

a. To find the reduction in solid waste per month, divide the reduction in solid waste per year (1.6 billion) by the number of months in a year (12). Round to the nearest whole number.

b. To find the reduction in greenhouse gas emissions per month, divide the reduction in greenhouse gas emissions per year (2.1 million) by the number of months in a year (12). Round to the nearest whole number.

Solution:

a. $1{,}600{,}000{,}000 \div 12 \approx 133{,}333{,}333$
The reduction in solid waste per month is approximately 133,333,333 tons.

b. $2{,}100{,}000 \div 12 = 175{,}000$
The reduction in greenhouse gas emissions per month is 175,000 tons.

159. Strategy:
To find the profit:
→Convert 5 L to milliliters.
→Find the number of bottles by dividing the number of milliliters by 250.
→Find the total cost by multiplying the number of bottles by 0.75 and adding to 175.
→Find the income by multiplying the number of bottles by 15.89.
→Find the profit by subtracting the cost from the income.

Solution:
5 L = 5,000 ml
$5{,}000 \div 250 = 20$
$20 \cdot 0.75 + 175 = 15 + 175 = 190$
$20 \cdot 15.89 = 317.80$
$317.80 - 190 = 127.80$
The profit for the cough syrup was $127.80.

161. Strategy:
To find the perimeter, substitute 4.5 for L and 3.25 for W in the formula below and solve for P.

Solution:
$P = 2L + 2W$
$P = 2 \cdot 4.5 + 2 \cdot 3.25$
$= 9 + 6.5 = 15.5$
The perimeter is 15.5 in.

163. Strategy:
To find the area, substitute 4.5 for L and 3.25 for W in the formula below and solve for A.

Solution:
$A = LW$
$A = 4.5 \cdot 3.25 = 14.625$
The area is 14.625 in^2.

165. Strategy:
To find the perimeter, substitute 2.8, 4.75, and 6.4 for a, b, and c in the formula below and solve for P.

Solution:
$P = a + b + c$
$P = 2.8 + 4.75 + 6.4 = 13.95$
The perimeter is 13.95 m.

167. Strategy:
To find the employee's federal earnings, substitute 694.89 for E and 132.69 for W in the given formula and solve for F.

Solution:
$F = E - W$
$F = 694.89 - 132.69$
$F = 562.20$
The federal earnings are $562.20.

169. Strategy:
To find the force on the falling object, substitute 4.25 for m and 9.80 for g in the given formula and solve for F.

Solution:
$F = ma$
$F = 4.25(9.80)$
$F = 41.65$
The force on the object is 41.65 newtons.

Critical Thinking 2.6

171. $1.3 \times 2.31 = \frac{13}{10} \times \frac{231}{100} = \frac{3003}{1000} = 3.003$

173. $3.46 \times 0.24 = 0.8304$

175. $0.064 \times 1.6 = 0.1024$

177. $3.0381 \div 1.23 = 2.47$

Projects or Group Activities 2.6

179. A batting average in baseball is calculated by dividing the number of hits by the number of times at bat. The quotient is rounded to the nearest thousandth. Some plate appearances are not included in the number of times at bat; for example, a walk is not counted as an "at bat."

Section 2.7

Concept Check

1. addition, division, subtraction

Objective A Exercises

3. $\frac{3}{7}\cdot\frac{14}{15}+\frac{4}{5}=\frac{2}{5}+\frac{4}{5}=\frac{6}{5}=1\frac{1}{5}$

5. $\left(\frac{5}{6}\right)^2-\frac{5}{9}=\frac{25}{36}-\frac{5}{9}$
$=\frac{5}{36}$

7. $\frac{3}{4}\cdot\left(\frac{11}{12}-\frac{7}{8}\right)+\frac{5}{16}=\frac{3}{4}\cdot\frac{1}{24}+\frac{5}{16}$
$=\frac{1}{32}+\frac{5}{16}$
$=\frac{11}{32}$

9. $\frac{11}{16}-\left(\frac{3}{4}\right)^2+\frac{7}{8}=\frac{11}{16}-\frac{9}{16}+\frac{7}{8}$
$=\frac{1}{8}+\frac{7}{8}$
$=1$

11. $\left(1\frac{1}{3}-\frac{5}{6}\right)+\frac{7}{8}\div\left(\frac{1}{2}\right)^2$
$=\frac{1}{2}+\frac{7}{8}\div\left(\frac{1}{2}\right)^2$
$=\frac{1}{2}+\frac{7}{8}\div\frac{1}{4}$
$=\frac{1}{2}+\frac{7}{8}\cdot\frac{4}{1}$
$=\frac{1}{2}+\frac{7}{2}=4$

13. $\left(\frac{2}{3}\right)^2+\frac{8-7}{9-3}\div\frac{3}{8}=\left(\frac{2}{3}\right)^2+\frac{1}{6}\div\frac{3}{8}$
$=\frac{4}{9}+\frac{1}{6}\div\frac{3}{8}$
$=\frac{4}{9}+\frac{1}{6}\cdot\frac{8}{3}$
$=\frac{4}{9}+\frac{4}{9}=\frac{8}{9}$

15. $(0.5)(0.2)^2+1.7=(0.5)(0.04)+1.7$
$=0.02+1.7=1.72$

17. $(1.8)^2-2.52\div1.8=3.24-1.4=1.84$

19. $0.4(3-1.5)+(1.2)^2=0.4(1.5)+1.44$
$=0.6+1.44$
$=2.04$

21. $\frac{1}{2}+\frac{\frac{13}{25}}{4-\frac{3}{4}}\div\frac{1}{5}=\frac{1}{2}+\frac{\frac{13}{25}}{\frac{13}{4}}\div\frac{1}{5}$
$=\frac{1}{2}+\frac{13}{25}\div\frac{13}{4}\div\frac{1}{5}$
$=\frac{1}{2}+\frac{13}{25}\cdot\frac{4}{13}\div\frac{1}{5}$
$=\frac{1}{2}+\frac{4}{25}\div\frac{1}{5}=\frac{1}{2}+\frac{4}{25}\cdot\frac{5}{1}$
$=\frac{1}{2}+\frac{4}{5}=\frac{5}{10}+\frac{8}{10}=\frac{13}{10}=1\frac{3}{10}$

23. $\left(\frac{2}{3}\right)^2+\frac{\frac{5}{8}-\frac{1}{4}}{\frac{2}{3}-\frac{1}{6}}\cdot\frac{8}{9}=\left(\frac{2}{3}\right)^2+\frac{\frac{3}{8}}{\frac{1}{2}}\cdot\frac{8}{9}$
$=\left(\frac{2}{3}\right)^2+\frac{3}{8}\div\frac{1}{2}\cdot\frac{8}{9}$
$=\frac{4}{9}+\frac{3}{8}\div\frac{1}{2}\cdot\frac{8}{9}$
$=\frac{4}{9}+\frac{3}{8}\cdot\frac{2}{1}\cdot\frac{8}{9}$
$=\frac{4}{9}+\frac{3}{4}\cdot\frac{8}{9}$
$=\frac{4}{9}+\frac{2}{3}=\frac{10}{9}=1\frac{1}{9}$

25. $\frac{x}{y}-z^2$
$\frac{\frac{5}{6}}{\frac{1}{3}}-\left(\frac{3}{4}\right)^2=\frac{5}{6}\div\frac{1}{3}-\left(\frac{3}{4}\right)^2$
$=\frac{5}{6}\div\frac{1}{3}-\frac{9}{16}$
$=\frac{5}{6}\cdot\frac{3}{1}-\frac{9}{16}$
$=\frac{5}{2}-\frac{9}{16}$
$=\frac{31}{16}=1\frac{15}{16}$

27. $xy^3 + z$

$\frac{9}{10}\cdot\left(\frac{1}{3}\right)^3 + \frac{7}{15} = \frac{9}{10}\cdot\frac{1}{27} + \frac{7}{15}$

$= \frac{1}{30} + \frac{7}{15}$

$= \frac{1}{2}$

29. $\frac{w}{xy} - z$

$\frac{2\frac{1}{2}}{4\cdot\frac{3}{8}} - \frac{2}{3} = \frac{\frac{5}{2}}{\frac{3}{2}} - \frac{2}{3}$

$= \frac{5}{2} \div \frac{3}{2} - \frac{2}{3}$

$= \frac{5}{2}\cdot\frac{2}{3} - \frac{2}{3}$

$\frac{5}{3} - \frac{2}{3} = \frac{3}{3} = 1$

31. $(a+b)^2 - c$

$(2.5+1.8)^2 - 0.4 = (4.3)^2 - 0.4$

$= 18.49 - 0.4 = 18.09$

33. $\frac{x}{y^2} + 3z$

$\frac{7.2}{(0.6)^2} + 3(3.5) = \frac{7.2}{0.36} + 10.5$

$= 20 + 10.5 = 30.5$

Critical Thinking 2.7

35. $\left(1-\frac{1}{2^2}\right)\left(1-\frac{1}{3^2}\right)\left(1-\frac{1}{4^2}\right)\cdots\left(1-\frac{1}{9^2}\right)\left(1-\frac{1}{10^2}\right)$

$= \left(1-\frac{1}{4}\right)\left(1-\frac{1}{9}\right)\left(1-\frac{1}{16}\right)\cdots\left(1-\frac{1}{81}\right)\left(1-\frac{1}{10^2}\right)$

$= \left(\frac{3}{4}\right)\left(\frac{8}{9}\right)\left(\frac{15}{16}\right)\left(\frac{24}{25}\right)\left(\frac{35}{36}\right)\left(\frac{48}{49}\right)\left(\frac{63}{64}\right)\left(\frac{80}{81}\right)\left(\frac{99}{100}\right)$

$= \left(\frac{3}{1}\right)\left(\frac{2}{9}\right)\left(\frac{15}{2}\right)\left(\frac{3}{25}\right)\left(\frac{35}{3}\right)\left(\frac{4}{49}\right)\left(\frac{63}{4}\right)\left(\frac{5}{81}\right)\left(\frac{99}{100}\right)$

$= \left(\frac{1}{1}\right)\left(\frac{2}{3}\right)\left(\frac{3}{2}\right)\left(\frac{3}{5}\right)\left(\frac{5}{3}\right)\left(\frac{4}{7}\right)\left(\frac{7}{4}\right)\left(\frac{5}{9}\right)\left(\frac{99}{100}\right)$

$= \frac{11}{20}$

Chapter Review Exercises

1. $2\overline{)19}$ quotient 9; -18; remainder 1 $\frac{19}{2} = 9\frac{1}{2}$

2. $6\frac{2}{9} - 3\frac{7}{18} = 6\frac{4}{18} - 3\frac{7}{18}$

$= 5\frac{22}{18} - 3\frac{7}{18}$

$= 2\frac{15}{18} = 2\frac{5}{6}$

3. $x \div y$

$2\frac{5}{8} \div 1\frac{3}{4} = \frac{21}{8} \div \frac{7}{4}$

$= \frac{21}{8}\cdot\frac{4}{7}$

$= \frac{3}{2} = 1\frac{1}{2}$

4. 5.034

5. $0.28 = \frac{28}{100} = \frac{7}{25}$

6. $3\cdot\frac{8}{9} = \frac{3}{1}\cdot\frac{8}{9}$

$= \frac{3\cdot 8}{1\cdot 9}$

$= \frac{3\cdot 2\cdot 2\cdot 2}{1\cdot 3\cdot 3}$

$= \frac{8}{3} = 2\frac{2}{3}$

7. $8.31 = 8.310$

$8.039 < 8.310$

$8.039 < 8.31$

8. $\frac{3}{5} = \frac{9}{15}$ $\frac{7}{15} = \frac{7}{15}$

$\frac{9}{15} > \frac{7}{15}$

$\frac{3}{5} > \frac{7}{15}$

9. $50 = 2\cdot 5^2$

$75 = 3\cdot 5^2$

$\text{LCM} = 2\cdot 3\cdot 5^2 = 150$

10. $0.918 \cdot 10^5 = 91{,}800$

11. xy

$8 \cdot \frac{5}{12} = \frac{8}{1} \cdot \frac{5}{12}$
$= \frac{8 \cdot 5}{1 \cdot 12}$
$= \frac{2 \cdot 2 \cdot 2 \cdot 5}{1 \cdot 2 \cdot 2 \cdot 3}$
$= \frac{10}{3} = 3\frac{1}{3}$

12. $\frac{10}{7}; 1\frac{3}{7}$

13. $\frac{3}{7} \approx 0.4286$ $\quad$ $0.429 = 0.4290$

$0.4286 < 0.4290$
$\frac{3}{7} < 0.429$

14. $\frac{\frac{5}{8} - \frac{1}{4}}{\frac{1}{2} + \frac{1}{8}} = \frac{\frac{3}{8}}{\frac{5}{8}}$

$= \frac{3}{8} \div \frac{5}{8}$
$= \frac{3}{8} \cdot \frac{8}{5}$
$= \frac{3 \cdot 8}{8 \cdot 5} = \frac{3}{5}$

15. $72 \div 9 = 8$

$\frac{4}{9} = \frac{4 \cdot 8}{9 \cdot 8} = \frac{32}{72}$
$\frac{32}{72}$ is equivalent to $\frac{4}{9}$.

16. $x^2 y^3$

$\left(\frac{8}{9}\right)^2 \cdot \left(-\frac{3}{4}\right)^3$
$= \frac{8}{9} \cdot \frac{8}{9} \cdot \left(-\frac{3}{4}\right) \cdot \left(-\frac{3}{4}\right) \cdot \left(-\frac{3}{4}\right)$
$= -\left(\frac{8}{9} \cdot \frac{8}{9} \cdot \frac{3}{4} \cdot \frac{3}{4} \cdot \frac{3}{4}\right)$
$= -\frac{8 \cdot 8 \cdot 3 \cdot 3 \cdot 3}{9 \cdot 9 \cdot 4 \cdot 4 \cdot 4} = -\frac{1}{3}$

17. $ab^2 - c$

$4 \cdot \left(\frac{1}{2}\right)^2 - \frac{5}{7} = 4 \cdot \frac{1}{4} - \frac{5}{7}$
$= \frac{4}{1} \cdot \frac{1}{4} - \frac{5}{7} = 1 - \frac{5}{7}$
$= \frac{7}{7} - \frac{5}{7}$
$= \frac{2}{7}$

18. $42 = 2 \cdot 3 \cdot 7$
$63 = 3 \cdot 3 \cdot 7$
$\text{GCF} = 3 \cdot 7 = 21$

19. $14.2 \div 10^3 = 0.0142$

20. $6.8 \div 47.92 \approx 0.1$

21. $\frac{5}{9} \div \frac{2}{3} = \frac{5}{9} \cdot \frac{3}{2}$

$= \frac{5 \cdot 3}{9 \cdot 2}$
$= \frac{5 \cdot 3}{3 \cdot 3 \cdot 2} = \frac{5}{6}$

22. $\frac{x}{y}$

$\frac{0.396}{3.6} = 0.11$

23.
$$\begin{array}{rcr} 506.81 & \rightarrow & 500 \\ 64.1 & \rightarrow & \underline{60} \\ & & 440 \end{array}$$

24.
$$\begin{array}{r} 9.47 \\ \underline{\times 0.26} \\ 5682 \\ \underline{1894} \\ 2.4622 \end{array}$$

25. $a - b$

$80.32 - 29.577$
$= 50.743$

26. $\left(\frac{3}{8}\right)^2 \cdot 4^2 = \frac{9}{64} \cdot 16$

$= \frac{9}{64} \cdot \frac{16}{1}$
$= \frac{9}{4} = 2\frac{1}{4}$

27. $3\frac{7}{12} + 5\frac{1}{2} = 3\frac{7}{12} + 5\frac{6}{12}$

$= 8\frac{13}{12} = 9\frac{1}{12}$

28. $\frac{30}{105} = \frac{2 \cdot 3 \cdot 5}{3 \cdot 5 \cdot 7} = \frac{2}{7}$

29. $a - b$

$7 - 2\frac{3}{10} = 6\frac{10}{10} - 2\frac{3}{10}$
$= 4\frac{7}{10}$

30. $614.3 \div 100 = 6.143$

31. Strategy:
To find the cost per ounce, divide 11.78 by 7.

Solution:
$\frac{11.78}{7} \approx 1.683$
To the nearest cent, the cost of instant coffee is \$1.68 per ounce.

32. Strategy:
To find the price of the treadmill, substitute 1,124.75 for C and 374.75 for M in the given formula and solve for P.

Solution:
$P = C + M$
$P = 1{,}124.75 + 374.75$
$P = 1{,}499.50$
The price of the treadmill is \$1,499.50.

33. Strategy:
To find the amount of weight to gain:
→Add the amounts already gained
$\left(3\frac{1}{2} + 2\frac{1}{4}\right)$
→Subtract the amount gained from the goal amount (12).

Solution:
$3\frac{1}{2} + 2\frac{1}{4} = 3\frac{2}{4} + 2\frac{1}{4}$
$= 5\frac{3}{4}$
$12 - 5\frac{3}{4} = 11\frac{4}{4} - 5\frac{3}{4} = 6\frac{1}{4}$
The wrestler has $6\frac{1}{4}$ lb left to gain.

34. Strategy:
a. To find the difference, subtract the cost of World War I (0.38) from the cost of World War II (3.1).

b. To find how many times greater the monetary cost of the Vietnam War was, divide the cost of the Vietnam War by the cost of World War I.

Solution:
a. $3.1 - 0.38 = 2.72$
The difference between the monetary costs of the two World Wars was \$2.72 trillion.

b. $0.57 \div 0.38 = 1.5$
The cost of the Vietnam War was 1.5 times the cost of World War I.

35. Strategy:
To find the overtime pay, substitute $6\frac{1}{4}$ for H and 24 for R in the given formula and solve for P.

Solution:
$P = RH$
$P = 24 \cdot 6\frac{1}{4}$
$P = 24 \cdot \frac{25}{4}$
$P = \frac{24}{1} \cdot \frac{25}{4} = 150$
The employee is due \$150 in overtime pay.

36. Strategy:
To find the final velocity, substitute 0 for S and $15\frac{1}{2}$ for t in the given formula and solve for t.

Solution:
$V = S + 32t$
$V = 0 + 32 \cdot 15\frac{1}{2}$
$V = 32 \cdot \frac{31}{2}$
$V = \frac{32}{1} \cdot \frac{31}{2} = 496$
The final velocity is 496 ft/s.

Chapter Test

1. $7\overline{)\,18}$, quotient 2; -14; remainder 4. $\frac{18}{7}=2\frac{4}{7}$

2. $7\frac{3}{4}-3\frac{5}{6}=7\frac{9}{12}-3\frac{10}{12}$
$=6\frac{21}{12}-3\frac{10}{12}$
$=3\frac{11}{12}$

3. xy
$6\frac{3}{7}\cdot 3\frac{1}{2}=\frac{45}{7}\cdot\frac{7}{2}$
$=\frac{45}{2}$
$=22\frac{1}{2}$

4. $\frac{2}{3}\cdot\frac{7}{8}=\frac{2\cdot 7}{3\cdot 8}=\frac{7}{12}$

5. $30=2\cdot 3\cdot 5$
$45=3\cdot 3\cdot 5$
$\text{LCM}=2\cdot 3^2\cdot 5=90$

6. 9.033

7. x^3y^2
$\left(1\frac{1}{2}\right)^3\cdot\left(\frac{5}{6}\right)^2=\left(\frac{3}{2}\right)^3\cdot\left(\frac{5}{6}\right)^2$
$=\frac{3}{2}\cdot\frac{3}{2}\cdot\frac{3}{2}\cdot\frac{5}{6}\cdot\frac{5}{6}$
$=\frac{3\cdot 3\cdot 3\cdot 5\cdot 5}{2\cdot 2\cdot 2\cdot 6\cdot 6}$
$=\frac{75}{32}=2\frac{11}{32}$

8. $3\frac{4}{5}=\frac{(5\cdot 3)+4}{5}$
$=\frac{15+4}{5}=\frac{19}{5}$

9. $\frac{7}{12}\div\frac{3}{4}=\frac{7}{12}\cdot\frac{4}{3}$
$=\frac{7\cdot 4}{12\cdot 3}=\frac{7}{9}$

10. $4.003<4.009$

11. $\frac{x}{yz}$
$\frac{\frac{7}{20}}{\frac{2}{15}\cdot\frac{3}{8}}=\frac{\frac{7}{20}}{\frac{1}{20}}$
$=\frac{7}{20}\div\frac{1}{20}$
$=\frac{7}{20}\cdot\frac{20}{1}=7$

12. $18=2\cdot 3^2$
$54=2\cdot 3^3$
$\text{GCF}=2\cdot 3^2=18$

13. $\frac{13}{14}-\frac{16}{21}=\frac{39}{42}-\frac{32}{42}$
$=\frac{7}{42}=\frac{1}{6}$

14. $\frac{60}{75}=\frac{2\cdot 2\cdot 3\cdot 5}{3\cdot 5\cdot 5}=\frac{4}{5}$

15. $x+y+z$
$1\frac{3}{8}+\frac{1}{2}+\frac{5}{6}=1\frac{9}{24}+\frac{12}{24}+\frac{20}{24}$
$=2\frac{17}{24}$

16. $\frac{5}{6}=\frac{25}{30}$ $\quad\frac{11}{15}=\frac{22}{30}$
$\frac{25}{30}>\frac{22}{30}$
$\frac{5}{6}>\frac{11}{15}$

17. a^2b-c^2
$\left(\frac{2}{3}\right)^2\cdot 9-\left(\frac{3}{5}\right)^2=\frac{2}{3}\cdot\frac{2}{3}\cdot 9-\frac{3}{5}\cdot\frac{3}{5}$
$=4-\frac{9}{25}$
$=\frac{100}{25}-\frac{9}{25}$
$=\frac{91}{25}=3\frac{16}{25}$

18. $0.22 = 0.2200 \qquad \frac{2}{9} \approx 0.2222$

$0.2200 < 0.2222$

$0.22 < \frac{2}{9}$

19. $6.051\overline{3}67$ *Given place value*

$3 < 5$

6.051367 rounded to the nearest thousandth is 6.051.

20. $x \div y$

$\frac{8}{9} \div \frac{16}{27} = \frac{8}{9} \cdot \frac{27}{16}$

$= \frac{3}{2} = 1\frac{1}{2}$

21. $30 - 7.247 = 2.753$

22.

92.34 →	90
−17.95 →	−20
74.39	70

23. $4.58 + 3.9 + 6.017$

$= 8.48 + 6.017$

$= 14.497$

24. $20cd$

$20 \cdot 0.5 \cdot 6.4 = 10 \cdot 6.4$

$= 64$

25. $28 \div 7 = 4$

$\frac{3}{7} = \frac{3 \cdot 4}{7 \cdot 4} = \frac{12}{28}$

$\frac{12}{28}$ is equivalent to $\frac{3}{7}$.

26. $84.96 \div 100 = 0.8496$

27. $63.6 - 22.8 = 40.8$

The gross from *Thunderball* was \$40.8 million more than from *On Her Majesty's Secret Service*.

28. Strategy:

To find the amount of weight to lose:

→Add the amounts already lost $\left(11\frac{1}{6} + 8\frac{5}{8}\right)$.

→Subtract the amount lost from the goal amount (30).

Solution:

$11\frac{1}{6} + 8\frac{5}{8} = 11\frac{4}{24} + 8\frac{15}{24}$

$= 19\frac{19}{24}$

$30 - 19\frac{19}{24} = 29\frac{24}{24} - 19\frac{19}{24} = 10\frac{5}{24}$

The patient has $10\frac{5}{24}$ lbs left to lose.

29. Strategy:

To find the amount of hamburger meat, multiply the number of hamburgers per person (2) by the weight of each hamburger $\left(\frac{1}{4}\right)$ by the number of guests (35).

Solution:

$2 \cdot \frac{1}{4} \cdot 35 = \frac{2}{1} \cdot \frac{1}{4} \cdot \frac{35}{1} = 17\frac{1}{2}$

You should buy $17\frac{1}{2}$ lbs of hamburger.

30. Strategy:

To find the amount of felt needed, substitute 20 for b and 12 for h in the given formula and solve for A.

Solution:

$A = \frac{1}{2}bh$

$A = \frac{1}{2} \cdot 20 \cdot 12 = 120$

The amount of felt needed is 120 in^2.

31. Strategy:
To find the amount of hours still required:
→Add the hours already contributed
$\left(7\frac{1}{4}+2\frac{3}{4}\right)$.
→Subtract the amount already contributed from the total hours required (20).

Solution:
$7\frac{1}{4}+2\frac{3}{4}=10$
$20-10=10$
You must still contribute 10 h of community service.

32. Strategy:
To find the perimeter, substitute 8.75, 5.25, and 4.5 for *a*, *b*, and *c* in the formula below and solve for *P*.

Solution:
$P=a+b+c$
$P=8.75+5.25+4.5=18.5$
The perimeter is 18.5 m.

Cumulative Review Exercises

1. $3a+(a-b)^3$
$3\cdot 4+(4-1)^3=3\cdot 4+3^3$
$=3\cdot 4+27$
$=12+27$
$=39$

2. $4\cdot\frac{7}{8}=\frac{4}{1}\cdot\frac{7}{8}$
$=\frac{4\cdot 7}{1\cdot 8}$
$=\frac{2\cdot 2\cdot 7}{1\cdot 2\cdot 2\cdot 2}$
$=\frac{7}{2}=3\frac{1}{2}$

3. $4\frac{7}{9}+3\frac{5}{6}=4\frac{14}{18}+3\frac{15}{18}$
$=7\frac{29}{18}$
$=8\frac{11}{18}$

4. $387.9\div 10^4=0.03879$

5. $72=2\cdot 2\cdot 2\cdot 3\cdot 3$
$108=2\cdot 2\cdot 3\cdot 3\cdot 3$
$GCF=2\cdot 2\cdot 3\cdot 3=36$

6. $3\frac{1}{13}\cdot 5\frac{1}{5}=\frac{40}{13}\cdot\frac{26}{5}=\frac{40\cdot 26}{13\cdot 5}$
$=\frac{2\cdot 2\cdot 2\cdot 5\cdot 2\cdot 13}{13\cdot 5}$
$=16$

7. $\frac{8}{9}\div\frac{4}{5}=\frac{8}{9}\cdot\frac{5}{4}$
$=\frac{8\cdot 5}{9\cdot 4}$
$=\frac{2\cdot 2\cdot 2\cdot 5}{3\cdot 3\cdot 2\cdot 2}$
$=\frac{10}{9}=1\frac{1}{9}$

8. $25\overline{)19.00}$ = 0.76
$\frac{19}{25}=0.76$

9. $\dfrac{\frac{1}{5}+\frac{1}{4}}{\frac{1}{4}-\frac{1}{5}}=\dfrac{\frac{9}{20}}{\frac{1}{20}}$
$=\frac{9}{20}\div\frac{1}{20}$
$=\frac{9}{20}\cdot\frac{20}{1}=\frac{9\cdot 20}{20\cdot 1}$
$=9$

10. $\frac{7}{11}=\frac{35}{55}$ $\quad\frac{4}{5}=\frac{44}{55}$
$\frac{35}{55}<\frac{44}{55}$
$\frac{7}{11}<\frac{4}{5}$

11. $2\frac{1}{3}\div 1\frac{2}{7}=\frac{7}{3}\div\frac{9}{7}$
$=\frac{7}{3}\cdot\frac{7}{9}$
$=\frac{7\cdot 7}{3\cdot 9}$
$=\frac{7\cdot 7}{3\cdot 3\cdot 3}=\frac{49}{27}$
$=1\frac{22}{27}=-1\frac{22}{27}$

12. $\frac{3}{8}\cdot\frac{2}{5}\cdot\frac{4}{9}=\frac{3\cdot2\cdot4}{8\cdot5\cdot9}$
$=\frac{3\cdot2\cdot2\cdot2}{2\cdot2\cdot2\cdot5\cdot3\cdot3}$
$=\frac{1}{15}$

13. abc
$\frac{4}{7}\cdot1\frac{1}{6}\cdot3=\frac{4}{7}\cdot\frac{7}{6}\cdot\frac{3}{1}$
$=\frac{4\cdot7\cdot3}{7\cdot6\cdot1}$
$=\frac{2\cdot2\cdot7\cdot3}{7\cdot2\cdot3\cdot1}=2$

14. $8\frac{3}{4}-1\frac{5}{7}=8\frac{21}{28}-1\frac{20}{28}$
$=7\frac{1}{28}$

15. $36.92+18.5=55.42$

16. $\frac{2}{5}\div\frac{9-6}{3+7}+\left(\frac{1}{2}\right)^2=\frac{2}{5}\div\frac{3}{10}+\left(\frac{1}{2}\right)^2$
$=\frac{2}{5}\div\frac{3}{10}+\frac{1}{4}$
$=\frac{2}{5}\cdot\frac{10}{3}+\frac{1}{4}$
$=\frac{4}{3}+\frac{1}{4}=\frac{19}{12}$
$=1\frac{7}{12}$

17. $62.8\ \rightarrow\ 60$
$0.47\ \rightarrow\ \times 0.5$
30

18. $1\frac{9}{16}+4\frac{5}{8}=1\frac{9}{16}+4\frac{10}{16}$
$=5\frac{19}{16}=6\frac{3}{16}$

19. 8,072,092

20. $9\overline{)41}$ quotient 4, -36, remainder 5; $\frac{41}{9}=4\frac{5}{9}$

21. $\frac{5}{14}-\frac{9}{42}=\frac{15}{42}-\frac{9}{42}$
$=\frac{6}{42}=\frac{1}{7}$

22. x^3y^4
$\left(\frac{7}{12}\right)^3\left(\frac{6}{7}\right)^4=\frac{7}{12}\cdot\frac{7}{12}\cdot\frac{7}{12}\cdot\frac{6}{7}\cdot\frac{6}{7}\cdot\frac{6}{7}\cdot\frac{6}{7}$
$=\frac{7\cdot7\cdot7\cdot6\cdot6\cdot6\cdot6}{12\cdot12\cdot12\cdot7\cdot7\cdot7\cdot7}$
$=\frac{3}{28}$

23. $2.617\div0.93\approx2.8$

24.
$$\begin{array}{r}6{,}847\\3{,}501\\+924\\\hline 11{,}272\end{array}$$

25. $(x-y)^3+5x$
$(8-6)^3+5\cdot8=2^3+5\cdot8$
$=8+5\cdot8$
$=8+40$
$=48$

23. $\frac{a}{b+c}$
$\frac{\frac{3}{8}}{\frac{1}{2}+\frac{3}{4}}=\frac{\frac{3}{8}}{\frac{5}{4}}$
$=\frac{3}{8}\div\frac{5}{4}$
$=\frac{3}{8}\cdot\frac{4}{5}$
$=\frac{3\cdot4}{8\cdot5}$
$=\frac{3\cdot2\cdot2}{2\cdot2\cdot2\cdot5}=\frac{3}{10}$

27. $89{,}357\ \rightarrow\ 90{,}000$
$66{,}042\ \rightarrow\ -70{,}000$
$\phantom{66{,}042\ \rightarrow\ }20{,}000$

28. $x\div y$
$3\frac{2}{3}\div2\frac{4}{9}=\frac{11}{3}\div\frac{22}{9}$
$=\frac{11}{3}\cdot\frac{9}{22}$
$=\frac{11\cdot9}{3\cdot22}$
$=\frac{11\cdot3\cdot3}{3\cdot2\cdot11}$
$=\frac{3}{2}=1\frac{1}{2}$

29. $7\frac{3}{4}=\frac{(4\cdot7)+3}{4}$
$=\frac{28+3}{4}$
$=\frac{31}{4}$

30.
$$\begin{array}{r} 7 \\ 5\overline{)35} \\ 2\overline{)70} \\ 2\overline{)140} \end{array}$$
$$140 = 2 \cdot 2 \cdot 5 \cdot 7 = 2^2 \cdot 5 \cdot 7$$

31. Strategy:
To determine how many more calories you would burn:
→Calculate the number of calories burned by bicycling at 12 mph for 4 h.
→Calculate the number of calories burned by walking at a rate of 3 mph for 5 h.
→Find the difference between the two calculations.

Solution:
Bicycling: $4 \cdot 410 = 1{,}640$
Walking: $320 \cdot 5 = 1{,}600$
$1{,}640 - 1{,}600 = 40$
You would burn 40 more calories by bicycling at 12 mph for 4 h.

32. Strategy:
To find the population increase, subtract the population in 2000 (13,922,517) from the population in 2010 (14,444,865).

Solution:
$14{,}444{,}865 - 13{,}922{,}517 = 522{,}348$
The projected increase in population is 522,348 people.

33. Strategy:
To calculate the copier service bill:
→Calculate the number of pages the service was used after the first 50 pages.
→Add the standard charge (6.95) to the product of 0.12 and the number of pages the service was used after the first 50 pages.

Solution:
$78 - 50 = 28$
The service was used for 28 pages after the first 50 pages.
$0.12(28) + 6.95$
$= 3.36 + 6.95$
$= 10.31$
The pager service bill is $10.31.

34. Strategy:
To find the length of fencing, substitute $16\frac{1}{2}$ for s in the given formula and solve for P.

Solution:
$P = 4s$
$$P = 4 \cdot 16\frac{1}{2} = \frac{4}{1} \cdot \frac{33}{2}$$
$$P = \frac{132}{2} = 66$$
The length of fencing needed is 66 ft.

35. Strategy:
To find the distance traveled, substitute $5\frac{1}{2}$ for r and $\frac{3}{4}$ for t in the given formula and solve for d.

Solution:
$d = rt$
$$d = 5\frac{1}{2} \cdot \frac{3}{4}$$
$$d = \frac{11}{2} \cdot \frac{3}{4}$$
$$d = \frac{33}{8} = 4\frac{1}{8}$$
The distance traveled is $4\frac{1}{8}$ miles.

36. Strategy:

To find the pressure, substitute $14\frac{3}{4}$ for D in the given formula and solve for P.

Solution:

$$P = 15 + \frac{1}{2}D$$
$$P = 15 + \frac{1}{2} \cdot 14\frac{3}{4}$$
$$P = 15 + \frac{1}{2} \cdot \frac{59}{4}$$
$$P = 15 + 7\frac{3}{8} = 22\frac{3}{8}$$

The pressure is $22\frac{3}{8}$ pounds per square inch.

Chapter 3: Rational Numbers

Prep Test

1. $54 > 45$
2. 4 units
3. $7{,}654 + 8{,}193 = 15{,}847$
4. $6{,}097 - 2{,}318 = 3{,}779$
5. $472 \times 56 = 26{,}432$
6. $144 \div 24 = 6$
7. $\frac{2}{3}+\frac{3}{5}=\frac{10}{15}+\frac{9}{15}=\frac{10+9}{15}=\frac{19}{15}=1\frac{4}{15}$
8. $\frac{3}{4}-\frac{5}{16}=\frac{12}{16}-\frac{5}{16}=\frac{12-5}{16}=\frac{7}{16}$
9. $0.75+3.9+6.408=4.65+6.408$
$=11.058$
10. $5.4-1.619=3.781$
11. $\frac{3}{4}\times\frac{8}{15}=\frac{3\cdot 8}{4\cdot 15}=\frac{3\cdot 2\cdot 2\cdot 2}{2\cdot 2\cdot 3\cdot 5}=\frac{2}{5}$
12. $\frac{5}{12}\div\frac{3}{4}=\frac{5}{12}\times\frac{4}{3}=\frac{5\cdot 4}{12\cdot 3}=\frac{5\cdot 2\cdot 2}{2\cdot 2\cdot 3\cdot 3}=\frac{5}{9}$
13. $23.5\times 0.4=9.4$
14. $0.96\div 2.4=0.4$
15. $(8-6)^2+12\div 4\cdot 3^2$
$=2^2+12\div 4\cdot 3^2$
$=4+12\div 4\cdot 9$
$=4+3\cdot 9$
$=4+27$
$=31$

Section 3.1

Concept Check

1. **a.** left

   b. right

3. negative; positive

5. absolute value

7. 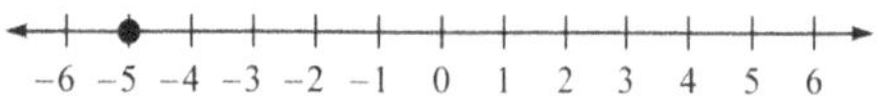

9. 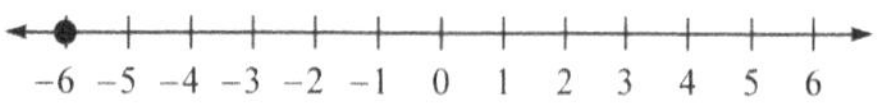

11. 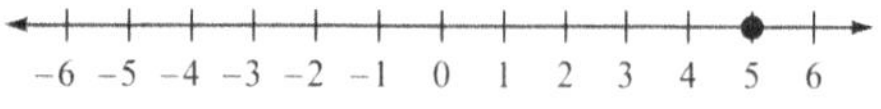

13. 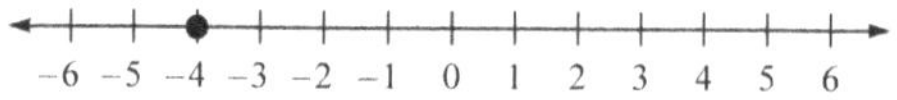

15. 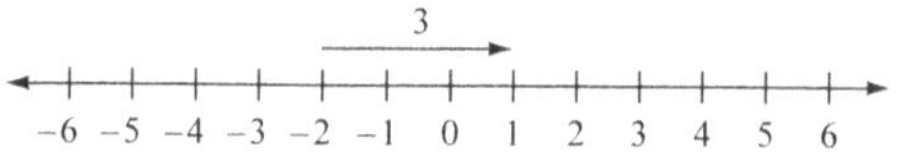

1 is 3 units to the right of –2.

17. 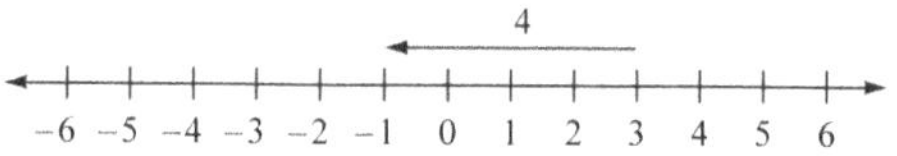

–1 is 4 units to the left of 3.

19. 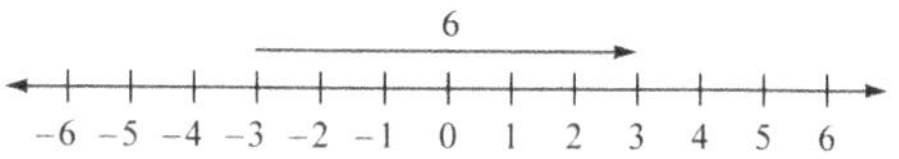

3 is 6 units to the right of –3.

21. 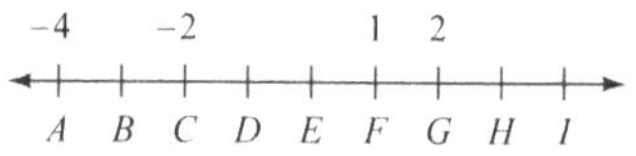

A is –4, and *C* is –2.

23. 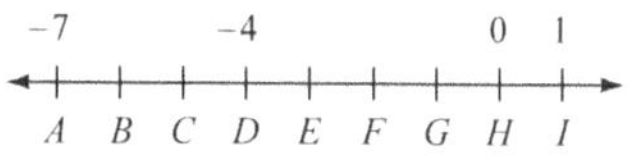

A is –7, and *D* is –4.

25. $-2 > -5$

27. $3 > -7$

29. $-42 < 27$

31. $53 > -46$

33. $-51 < -20$

35. $-131 < 101$

37. $-7, -2, 0, 3$

39. $-5, -3, 1, 4$

41. $-4, 0, 5, 9$

43. $-10, -7, -5, 4, 12$

45. **a.** never true

b. sometimes true

c. sometimes true

d. always true

Objective B Exercises

47. –45

49. 88

51. $-n$

53. d

55. the opposite of negative thirteen

57. the opposite of negative p

59. five plus negative ten

61. negative fourteen minus negative three

63. negative thirteen minus eight

65. m plus negative n

67. $-(-7) = 7$

69. $-(46) = -46$

71. $-(-73) = 73$

73. $-(-z) = z$

75. $-(p) = -p$

77. negative

Objective C Exercises

79. $|-4| = 4$

81. $|9| = 9$

83. $|-11| = 11$

85. $|-12| = 12$

87. $|-23| = 23$

89. $-|27| = -27$

91. $|25| = 25$

93. $-|-41| = -41$

95. $-|-93| = -93$

97. $|x|$
$|-10| = 10$

99. $|-x|$
$|-8| = 8$

101. $|-y|$
$|-(-6)| = |6| = 6$

103. $|-12| = 12, |8| = 8$
$12 > 8$
$|-12| > |8|$

105. $|6| = 6, |13| = 13$
$6 < 13$
$|6| < |13|$

107. $|-1| = 1, |-17| = 17$
$1 < 17$
$|-1| < |-17|$

109. $|x| = |-x|$

111. $-|6| = -6, -(4) = -4, |-7| = 7, -(-9) = 9$
$-|6|, -(4), |-7|, -(-9)$

113. $-|-7| = -7, -9 = -9, -(5) = -5, |4| = 4$
$-9, -|-7|, -(5), |4|$

115. $-(-3) = 3, -|-8| = -8, |5| = 5,$
$-|10| = -10, -(-2) = 2$
$-|10|, -|-8|, -(-2), -(-3), |5|$

Objective D Exercises

117. –329,000,000

119. Strategy:
To determine which quarter had the greater loss, compare the absolute values of the numbers –12,575 and –11,350. The larger number corresponds to the quarter with the greater loss.

Solution:
$|-12,575| = 12,575$
$|-11,350| = 11,350$
$12,575 > 11,350$
The loss was greater during the first quarter.

121. Strategy:
a. To find the earnings per share in 2015, read the number in the bar graph below the bar corresponding to 2015.

b. To find the earnings per share in 2017, read the number in the bar graph below the bar corresponding to 2017.

Solution:
a. The earnings per share for 2015 were –27¢.

b. The earnings per share for 2017 were –40¢.

123. Strategy:
To find a year in which Mycopen had a profit, use the bar graph to find a year in which earnings per share were positive.

Solution:
The only recorded positive earnings per share were 11¢. Earnings per share were 11¢ in 2018.
Mycopen did earn a profit during the years shown.
Mycopen earned a profit in 2018.

125. Strategy:
To determine which stock showed the least net change, compare the absolute values of the numbers –1 and –2. The smaller number represents the least net change.

Solution:
$|-1| = 1, |-2| = 2$
$1 < 2$
Stock B showed the least net change.

127. Strategy:
To find the wind chill factor, use the given table.

Solution:
Find the number where the column with 10 and the row with 20 cross. Read the number –9.
The wind chill factor is –9°F.

129. Strategy:
To find the cooling power, use the given table.

Solution:
Find the number where the column with –15 and the row with 10 cross. Read the number –35.
The cooling power is –35°F.

131. Strategy:
To find the situation which feels colder:
→From the given table, find the wind chill factor with a temperature of –30°F with a 5-mph wind and the wind chill factor with a temperature of –20°F with a 10-mph wind.
→Compare the wind chill factors.

Solution:
The wind chill factor with a temperature of –30°F and 5-mph wind is –46°F.
The wind chill factor with a temperature of –20°F and 10-mph wind is –41°F.
$-46 < -41$
–30°F with a 5-mph wind feels colder.

Critical Thinking 3.1

133. The absolute value of a number is the distance from zero to the number on the number line. If $|y| = 11$, then y must be a number that is 11 units from 0 on the number line. Therefore, y is –11 or 11.

135. x must be less than 7 and greater than –7.
–6, –5, –4, –3, –2, –1, 0, 1, 2, 3, 4, 5, 6

137. **a.** Two numbers that are 4 units from 2 on the number line are –2 and 6.

b. Two numbers that are 5 units from 3 on the number line are –2 and 8.

Projects or Group Activities 3.1

139. Answers will vary. One example is:
Point A is a point on the number line halfway between –3 and 5. Point B is a point halfway between A and the graph of 3 on the number line.
Solution: If A is a point on the number line halfway between –3 and 5, then A is the graph of 1.
B is a point halfway between A and the graph of 3 on the number line.
Therefore, B is a point halfway between 1 and 3.
B is the graph of 2.

Section 3.2

Concept Check

1. the same; negative

3. negative; negative

5. negative; minus

7. (–5)

Objective A Exercises

9. $-3 + (-8) = -11$

11. $-8 + 3 = -5$

13. $-5 + 13 = 8$

15. $6 + (-10) = -4$

17. $3 + (-5) = -2$

19. $-4 + (-5) = -9$

21. $-6 + 7 = 1$

23. $(-5) + (-10) = -15$

25. $-7 + 7 = 0$

27. $(-15) + (-6) = -21$

29. $0 + (-14) = -14$

31. $73 + (-54) = 19$

33. $2 + (-3) + (-4) = -1 + (-4)$
$= -5$

35. $-3 + (-12) + (-15) = -15 + (-15)$
$= -30$

37. $-17 + (-3) + 29 = -20 + 29$
$= 9$

39. $11 + (-22) + 4 + (-5)$
$= -11 + 4 + (-5)$
$= -7 + (-5)$
$= -12$

41. $-22 + 10 + 2 + (-18)$
$= -12 + 2 + (-18)$
$= -10 + (-18)$
$= -28$

43. $-25 + (-31) + 24 + 19$
$= -56 + 24 + 19$
$= -32 + 19$
$= -13$

45. $3 + (-21) = -18$

47. $-5 + 16 = 11$

49. $(-3) + (-8) + 12 = -11 + 12$
$= 1$

51. $x + (-7)$

53. a. $-60{,}100{,}000{,}000 + (-50{,}100{,}000{,}000)$
$= -110{,}200{,}000{,}000$
The total of the U.S. balance of trade in Japan and Mexico is –$110,200,000,000.

b. $-28{,}500{,}000{,}000 + (-50{,}100{,}000{,}000)$
$= -78{,}600{,}000{,}000$
The total of the U.S. balance of trade with Canada and Mexico is –$78,600,000,000.

c. $-60{,}100{,}000{,}000 + (-273{,}100{,}000{,}000)$
$= -333{,}200{,}000{,}000$
The total of the U.S. balance of trade with Japan and China is –$333,200,000,000.

55. $-a + b$
$-(-8) + (-3) = 8 + (-3)$
$= 5$

57. $-x + y$
$-(-5) + (-7) = 5 + (-7)$
$= -2$

59. $a + b + c$
$-10 + (-6) + 5 = -16 + 5$
$= -11$

61. $-x + (-y) + z$
$-(-2) + (-8) + (-11) = 2 + (-8) + (-11)$
$= -6 + (-11)$
$= -17$

63. The Addition Property of Zero

65. The Associative Property of Addition

67. $-13 + 0 = -13$

69. $18 + (-18) = 0$

71. $6 = -3 + z$
$6 \mid -3 + (-8)$
$6 \neq -11$
No, -8 is not a solution of the equation $6 = -3 + z$.

73. $-7 + m = -15$
$-7 + (-8) \mid -15$
$-15 = -15$
Yes, -8 is a solution of the equation $-7 + m = -15$.

75. $1 + z = z + 2$
$1 + (-4) \mid -4 + 2$
$-3 \neq -2$
No, -4 is not a solution of the equation $1 + z = z + 2$.

77. sometimes true

79. always true

Objective B Exercises

81. $6 - 9 = 6 + (-9)$
$= -3$

83. $-9 - 4 = -9 + (-4)$
$= -13$

85. $3 - (-4) = 3 + 4$
$= 7$

87. $-4 - (-4) = -4 + 4$
$= 0$

89. $-10 - 7 = -10 + (-7)$
$= -17$

91. $(-7) - (-4) = -7 + 4$
$= -3$

93. $-4 - (-16) = -4 + 16$
$= 12$

95. $3 - (-24) = 3 + 24$
$= 27$

97. $(-41) - 65 = -41 + (-65)$
$= -106$

99. $-95 - (-28) = -95 + 28$
$= -67$

101. $-10 - (-4) = -10 + 4$
$= -6$

103. $-9 - 6 = -9 + (-6)$
$= -15$

105. $49 - (-33) = 49 + 33$
$= 82$
The difference between the highest and lowest temperatures ever recorded in South America is 82°C.

107. $-4 - 3 - 2 = -4 + (-3) + (-2)$
$= -7 + (-2)$
$= -9$

109. $12 - (-7) - 8 = 12 + 7 + (-8)$
$= 19 + (-8)$
$= 11$

111. $4 - 12 - (-8) = 4 + (-12) + 8$
$= -8 + 8$
$= 0$

113. $-16 - 47 - 63 - 12$
$= -16 + (-47) + (-63) + (-12)$
$= -63 + (-63) + (-12)$
$= -126 + (-12)$
$= -138$

115. $12 - (-6) + 8 = 12 + 6 + 8$
$= 18 + 8$
$= 26$

117. $-8 - (-14) + 7 = -8 + 14 + 7$
$= 6 + 7$
$= 13$

119. $9 - 12 + 0 - 5 = 9 + (-12) + 0 + (-5)$
$= -3 + 0 + (-5)$
$= -3 + (-5)$
$= -8$

121. $5 + 4 - (-3) - 7 = 5 + 4 + 3 + (-7)$
$= 9 + 3 + (-7)$
$= 12 + (-7)$
$= 5$

123. $-13 + 9 - (-10) - 4 = -13 + 9 + 10 + (-4)$
$= -4 + 10 + (-4)$
$= 6 + (-4)$
$= 2$

125. $-x - y$
$-(-3) - 9 = 3 + (-9)$
$= -6$

127. $-x - (-y)$
$-(-3) - (-9) = 3 + 9$
$= 12$

129. $a - b - c$
$4 - (-2) - 9 = 4 + 2 + (-9)$
$= 6 + (-9)$
$= -3$

131. $x - y - (-z)$
$-9 - 3 - (-30) = -9 + (-3) + 30$
$= -12 + 30$
$= 18$

133. $x - 7 = -10$

$-3-7$	-10
$-3+(-7)$	-10

$-10 = -10$
Yes, –3 is a solution of the equation $x - 7 = -10$.

135. $-5 - w = 7$

$-5-(-2)$	7
$-5+2$	7

$-3 \neq 7$
No, –2 is not a solution of the equation $-5 - w = 7$.

137. $-t - 5 = 7 + t$

$-(-6)-5$	$7+(-6)$
$6-5$	$7+(-6)$
$6+(-5)$	$7+(-6)$

$1 = 1$
Yes, –6 is a solution of the equation $-t - 5 = 7 + t$.

139. sometimes true

Objective C Exercises

141. Strategy:
a. To find the difference, subtract the elevation of Death Valley (–86) from the elevation of Mt. Aconcagua (6,960).

b. To find the difference, subtract the elevation of the Lake Assal (–156) from the elevation of Mt. Kilimanjaro (5,895).

Solution:
a. $6{,}960 - (-86) = 6{,}960 + 86$
$= 7{,}046$
The difference in elevation is 7,046 m.

b. $5{,}895 - (-156) = 5{,}895 + 156$
$= 6{,}051$
The difference in elevation is 6,051 m.

143. Strategy:
To determine for which continent the difference between the highest and lowest elevations is smallest:
→Find the difference between the highest and lowest elevation for each continent.
→Compare the differences.

Solution:
Africa: $5{,}895 - (-156) = 5{,}895 + 156 = 6{,}051$
Asia: $8{,}850 - (-411) = 8{,}850 + 411 = 9{,}261$
Europe: $5{,}642 - (-28) = 5{,}642 + 28 = 5{,}670$
America: $6{,}960 - (-86) = 6{,}960 + 86 = 7{,}046$
$5{,}670 < 6{,}051 < 7{,}046 < 9{,}261$
The difference between the highest and lowest elevations is smallest in Europe.

145. Strategy:
To find the difference, subtract the average temperature at 40,000 ft (–70) from the average temperature at 12,000 ft (16).

Solution:
$16-(-70)=16+70=86$
The difference in temperature is 86°.

147. Strategy:
To find the golfer's score, substitute 49 for N and 52 for P in the given formula and solve for S.

Solution:
$S=N-P$
$S=49-52$
$S=49+(-52)$
$S=-3$
The golfer's score is –3.

149. Strategy:

a. To find the difference, subtract the low temperature in the U.S. (–14) from the high temperature in the U.S. (93).

b. To find the difference, subtract the low temperature in the contiguous U.S. (–7) from the high temperature in the U.S. (93).

Solution:
a. $93-(-14)=93+14=107$
The difference in temperature is 107°F.

b. $93-(-7)=93+7=100$
The difference in temperature is 100°F.

151. Strategy:
To find d, replace a by 7 and b by –12 in the given formula and solve for d.

Solution:
$d=|a-b|$
$d=|7-(-12)|$
$d=|7+12|$
$d=|19|$
$d=19$
The distance between the two points is 19 units.

Critical Thinking 3.2

153. Answers will vary. Possible answers include –6 and –1, –5 and –2, –4 and –3.

Projects or Group Activities 3.2

155. Possible answers include:
$-11-(-3)=-11+3=-8$;
$-10-(-2)=-10+2=-8$;
$-9-(-1)=-9+1=8$.
Choose a negative number which is less than –8 as the first number. Subtract 8 from the absolute value of that number, and then use the opposite of that result to get the second number.

Check Your Progress: Chapter 3

1. 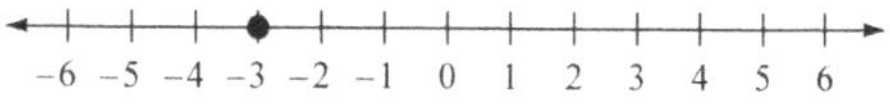

2. –3 is 5 units to the left of 2

3. $-12>-16$

4. –19, –8, 4, 7

5. **a.** 11

b. –13

c. m

6. negative five minus negative seven

7. **a.** $-(-42)=42$

b. $-(t)=-t$

8. **a.** 18

b. 37

9. **a.** $|-51|=51$

b. $-|67|=-67$

10. $-|x|$

$-|2|=-(2)=-2$

11. $|-19|=19, |7|=7$
$19>7$
$|-19|>|7|$

12. $|-5|=5, -(-6)=6, |3|=3,$
$-|-8|=-8, -|12|=-12$
$-|12|, -|-8|, |3|, |-5|, -(-6)$

13. $-8+(-12)=-20$

14. $20+(-3)+(-7)=17+(-7)=10$

15. $5 - 40 = 5 + (-40)$
$= -35$

16. $-32 - (-16) = -32 + 16$
$= -16$

17. $4 - (-15) - 3 + 7$
$= 4 + 15 + (-3) + 7$
$= 19 + (-3) + 7$
$= 16 + 7$
$= 23$

18. $-11 - 16 = -11 + (-16) = -27$

19. $-6 + (-9) + 14 = -15 + 14 = -1$

20. $-x + y$
$-(-6) + (-2) = 6 + (-2) = 4$

21. $a - (-b)$
$-5 - (-7) = -5 + 7 = 2$

22. $-3 = y - 4$

-3	$-7-4$

$-3 \neq -11$
No, –8 is not a solution of the equation $6 = -3 + z$.

23. Strategy:
To determine which temperature is colder, compare the absolute values of the numbers –16 and –4. The larger number is the colder temperature since both temperatures are negative.

Solution:
$|-16| = 16, |-4| = 4$
$16 > 14$
–16 is the colder temperature.

24. Strategy:
To find the temperature, add the increase (8) to the previous temperature (–3).

Solution:
$-3 + 8 = 5$
The temperature is 5°C.

25. Strategy:
To find d, replace a by 9 and b by –5 in the given formula and solve for d.

Solution:
$d = |a - b|$
$d = |9 - (-5)|$
$d = |9 + 5|$
$d = |14|$
$d = 14$
The distance between the two points is 21 units.

Section 3.3

Concept Check

1. a. different; negative

b. the same; positive

3. $\frac{-63}{9}$

Objective A Exercises

5. $-4 \cdot 6 = -24$

7. $-2(-3) = 6$

9. $(9)(2) = 18$

11. $5(-4) = -20$

13. $-8(2) = -16$

15. $(-5)(-5) = 25$

17. $(-7)(0) = 0$

19. $14(3) = 42$

21. $-32(4) = -128$

23. $(-8)(-26) = 208$

25. $9(-27) = -243$

27. $-5 \cdot (23) = -115$

29. $-7(-34) = 238$

31. $4 \cdot (-8) \cdot 3 = -32 \cdot 3$
$= 96$

33. $(-6)(5)(7) = -30(7)$
$= -210$

35. $-8(-7)(-4) = 56(-4)$
$= -224$

37. $2(-20) = -40$

39. $-30(-6) = 180$

41. $-q(r) = -qr$

43. a. $-170{,}000{,}000(4) = -680{,}000{,}000$
The annual net income for Sears Holdings would be –\$680,000,000.

b. $-74{,}000{,}000(4) = -296{,}000{,}000$
The annual net income for Rite Aid would be –\$296,000,000.

45. The Multiplication Property of One

47. The Associative Property of Multiplication

49. $-6 \cdot (5 \cdot 10) = (-6 \cdot 5) \cdot 10$

51. $1(-14) = -14$

53. $-xy$
$-(-3)(-8) = 3(-8)$
$= -24$

55. $-xyz$
$-(-6)(2)(-5) = 6(2)(-5)$
$= 12(-5)$
$= -60$

57. $-7n$
$-7(-51) = 357$

59. $8ab$
$8(7)(-1) = 56(-1)$
$= -56$

61. $-5st$
$-5(-40)(-8) = 200(-8)$
$= -1{,}600$

63. $-5x = -15$

$-5(-3)$	-15

$15 \neq -15$
No, –3 is not a solution of the equation $-5x = 15$.

65. $-8 = -8a$

-8	$-8(0)$

$-8 \neq 0$
No, 0 is not a solution of the equation $-8 = -8a$.

67. $-27 = -3c$

-27	$-3(9)$

$-27 = -27$
Yes, 9 is a solution of the equation $-27 = -3c$.

69. positive

Objective B Exercises

71. $18 \div (-3) = -6$

73. $(-64) \div (-8) = 8$

75. $-49 \div 1 = -49$

77. $-40 \div (-5) = 8$

79. $\frac{44}{-4} = -11$

81. $\frac{-98}{-7} = 14$

83. $-91 \div (-7) = 13$

85. $(-162) \div (-162) = 1$

87. $-130 \div (-5) = 26$

89. $(-92) \div (-4) = 23$

91. $\frac{550}{-5} = -110$

93. $\frac{-333}{-3} = 111$

95. $\frac{-9}{x}$

97. $-318{,}000{,}000 \div 3 = -106{,}000{,}000$
The average monthly net income for Delta Airlines was –\$106,000,000.

99. $-a \div b$
$-(-36) \div (-4) = 36 \div (-4)$
$= -9$

101. $(-a) \div (-b)$
$-(-36) \div (-(-4)) = 36 \div 4$
$= 9$

103. $\frac{-x}{y}$
$\frac{-(-42)}{-7} = \frac{42}{-7}$
$= -6$

105. $\frac{-x}{-y}$

$\frac{-(-42)}{-(-7)} = \frac{42}{7}$

$= 6$

107. $6 = \frac{-c}{-3}$

$6 \;\Big|\; \frac{-18}{-3}$

$6 = 6$

Yes, 18 is a solution of the equation.

109. $\frac{21}{n} = 7$

$\frac{21}{-3} \;\Big|\; 7$

$-7 \neq 7$

No, –3 is not a solution of the equation.

111. $\frac{m}{-4} = \frac{-16}{m}$

$\frac{8}{-4} \;\Big|\; \frac{-16}{8}$

$-2 = -2$

Yes, 8 is a solution of the equation.

113. $\frac{a}{b}$

115. $-\frac{a}{b}$

Objective C Exercises

117. Strategy:
To find the average score, divide the combined scores (–12) by the number of golfers (4).

Solution:
$-12 \div 4 = 3$
The average golf score was –3.

119. Strategy:
To find the average record low temperature for the first three months of the year:
→Add the average temperatures for January (–70), February (–66), and March (–50).
→Divide the sum by the number of months (3).

Solution:
$-70 + (-69) + (-50)$
$= -139 + (-50) = -189$
$-189 \div 3 = -63$
The average record low temperature for the first three months of the year is –63°F.

121. Strategy:
To find the average U.S. trade deficit for March, April, and May of 2011:
→Add the three trade deficits for March (–47 billion), April (–44 billion), and May (–50 billion).
→Divide the sum by the number of months (3).

Solution:
$-47 + (-44) + (-50)$
$= -91 + (-50) = -141$
$-141 \div 3 = -47$
The average U.S. trade deficit for March, April, and May of 2011 is –$47 billion.

123. Strategy:
To find the average daily low temperature for the week:
→Add the seven temperature readings.
→Divide by 7.

Solution:
$4 + (-5) + 8 + (-1) + (-12) + (-14) + (-8) = -28$
$-28 \div 7 = -4$
The average daily low temperature for the week was –4°.

Critical Thinking 3.3

125. Answers will vary. For example:
5 and –2 have different signs.
The product of 5 and –2 is $5(-2) = (-2) + (-2) + (-2) + (-2) + (-2) = -10$, which is a negative number.

Projects or Group Activities 3.3

127. Strategy:
To find the next four numbers in the sequence:
→Find the multiplier by dividing the second number in the sequence (–4) by the first number (2).
→Use the multiplier to find the successive numbers in the sequence.

Solution:
$\frac{-4}{2} = -2$
$8 \cdot (-2) = -16$
$-16 \cdot (-2) = 32$
$32 \cdot (-2) = -64$
$-64 \cdot (-2) = 128$
The next four number-s in the sequence are –16, 32, –64, and 128.

129. Strategy:
To find the next four numbers in the sequence:
→Find the multiplier by dividing the second number in the sequence (–5) by the first number (–1).
→Use the multiplier to find the successive numbers in the sequence.

Solution:
$\frac{-5}{-1} = 5$
$-25 \cdot 5 = -125$
$-125 \cdot 5 = -625$
$-625 \cdot 5 = -3{,}125$
$-3{,}125 \cdot 5 = -15{,}625$
The next four numbers in the sequence are –125, –625, –3125, and –15,625.

Section 3.4

Concept Check

1. **a.** 0, –3

b. $-\frac{15}{2},\ 0,\ -3,\ 2.\overline{33}$

c. $\pi,\ 4.232232223\ldots,\ \frac{\sqrt{5}}{4},\ \sqrt{7}$

d. all

Objective A Exercises

3. $\frac{5}{8} - \frac{5}{6} = \frac{15}{24} - \frac{20}{24}$
$= \frac{15}{24} + \frac{(-20)}{24}$
$= \frac{15 + (-20)}{24}$
$= -\frac{5}{24}$

5. $-\frac{5}{12} - \frac{3}{8} = \frac{-10}{24} - \frac{9}{24}$
$= \frac{-10}{24} + \frac{(-9)}{24}$
$= \frac{-10 + (-9)}{24} = \frac{-19}{24} = -\frac{19}{24}$

7. $-\frac{6}{13} + \frac{17}{26} = \frac{-12}{26} + \frac{17}{26}$
$= \frac{-12 + 17}{26} = \frac{5}{26}$

9. $-\frac{5}{8} - \left(-\frac{11}{12}\right) = \frac{-15}{24} - \left(\frac{-22}{24}\right)$
$= \frac{-15}{24} + \frac{22}{24} = \frac{-15 + 22}{24} = \frac{7}{24}$

11. $\frac{5}{12} - \frac{11}{15} = \frac{25}{60} - \frac{44}{60}$
$= \frac{25}{60} + \frac{(-44)}{60}$
$= \frac{25 + (-44)}{60}$
$= \frac{-19}{60} = -\frac{19}{60}$

13. $-\frac{3}{4} - \frac{5}{8} = \frac{-6}{8} - \frac{5}{8}$
$= \frac{-6}{8} + \frac{(-5)}{8}$
$= \frac{-6 + (-5)}{8} = \frac{-11}{8} = -\frac{11}{8}$

15. $-\frac{5}{2}-\left(-\frac{13}{4}\right)=\frac{-10}{4}-\left(\frac{-13}{4}\right)$
$=\frac{-10}{4}+\frac{13}{4}=\frac{-10+13}{4}=\frac{3}{4}$

17. $-\frac{3}{8}-\frac{5}{12}-\frac{3}{16}=\frac{-18}{48}-\frac{20}{48}-\frac{9}{48}$
$=\frac{-18}{48}+\frac{(-20)}{48}+\frac{(-9)}{48}$
$=\frac{-18+(-20)+(-9)}{48}$
$=\frac{-47}{48}=-\frac{47}{48}$

19. $\frac{1}{2}-\frac{3}{8}-\left(-\frac{1}{4}\right)=\frac{4}{8}-\frac{3}{8}-\left(\frac{-2}{8}\right)$
$=\frac{4}{8}+\frac{(-3)}{8}+\frac{2}{8}$
$=\frac{4+(-3)+2}{8}=\frac{3}{8}$

21. $\frac{1}{3}-\frac{1}{4}-\frac{1}{5}=\frac{20}{60}-\frac{15}{60}-\frac{12}{60}$
$=\frac{20}{60}+\frac{(-15)}{60}+\frac{(-12)}{60}$
$=\frac{20+(-15)+(-12)}{60}$
$=\frac{-7}{60}=-\frac{7}{60}$

23. $\frac{1}{2}+\left(-\frac{3}{8}\right)+\frac{5}{12}=\frac{12}{24}+\frac{(-9)}{24}+\frac{10}{24}$
$=\frac{12+(-9)+10}{24}=\frac{13}{24}$

25. $3.4+(-6.8)=-3.4$

27. $-8.32+(-0.57)=-8.89$

29. $-4.8+(-3.2)=-8.0$

31. $-4.6+3.92=-0.68$

33. $-4.2-6.83=-4.2+(-6.83)=-11.03$

35. $18.4-25.2=18.4+(-25.2)=-6.8$

37. $-4.5+3.2+(-19.4)=-1.3+(-19.4)=-20.7$

39. $-18.39+4.9-23.7$
$=-18.39+4.9+(-23.7)$
$=-13.49+(-23.7)$
$=-37.19$

41. $-3.09-4.6-27.3$
$=-3.09+(-4.6)+(-27.3)$
$=-7.69+(-27.3)=-34.99$

43. $\frac{7}{12}+\left(-\frac{11}{16}\right)=\frac{7}{12}+\frac{-11}{16}$
$=\frac{28}{48}+\frac{-33}{48}$
$=\frac{28+(-33)}{48}$
$=\frac{-5}{48}=-\frac{5}{48}$

45. $-\frac{7}{12}+\left(-\frac{5}{9}\right)=\frac{-7}{12}+\frac{-5}{9}$
$=\frac{-21}{36}+\frac{-20}{36}$
$=\frac{-21+(-20)}{36}$
$=\frac{-41}{36}=-1\frac{5}{36}$

47. $\frac{3}{5}-\left(-\frac{7}{10}\right)=\frac{3}{5}-\frac{-7}{10}$
$=\frac{6}{10}-\frac{-7}{10}$
$=\frac{6-(-7)}{10}$
$=\frac{6+7}{10}=\frac{13}{10}=1\frac{3}{10}$

49. $-138.72-510.64$
$=-138.72+(-510.64)=-649.36$

51. $x+y$
$\frac{3}{10}+\left(-\frac{7}{15}\right)=\frac{3}{10}+\frac{-7}{15}$
$=\frac{9}{30}+\frac{-14}{30}$
$=\frac{9+(-14)}{30}=\frac{-5}{30}=-\frac{1}{6}$

53. $x + y$

$$-\frac{3}{8}+\left(-\frac{5}{6}\right)=\frac{-3}{8}+\frac{-5}{6}$$
$$=\frac{-9}{24}+\frac{-20}{24}$$
$$=\frac{-9+(-20)}{24}$$
$$=\frac{-29}{24}=-1\frac{5}{24}$$

55. $x + y$

$-6.175 + (-19.49) = -25.665$

57. $x - y$

$$-\frac{11}{12}-\frac{5}{12}=\frac{-11}{12}-\frac{5}{12}$$
$$=\frac{-11-5}{12}$$
$$=\frac{-16}{12}=\frac{-4}{3}=-1\frac{1}{3}$$

59. $x - y$

$$-\frac{2}{3}-\left(-\frac{3}{4}\right)=\frac{-2}{3}-\frac{-3}{4}$$
$$=\frac{-8}{12}-\frac{-9}{12}$$
$$=\frac{-8-(-9)}{12}$$
$$=\frac{-8+9}{12}=\frac{1}{12}$$

61. $x - y$

$-21.073 - 6.48 = -21.073 + (-6.48)$
$= -27.553$

63. $x - y$

$-8.21 - (-6.798) = -8.21 + 6.798 = -1.412$

65. $-\frac{1}{4}=x-\frac{7}{8}$

$-\frac{1}{4}$	$\frac{5}{8}-\frac{7}{8}$
$-\frac{1}{4}$	$\frac{5-7}{8}$
$-\frac{1}{4}$	$\frac{-2}{8}$

$-\frac{1}{4}=-\frac{1}{4}$

Yes, $\frac{5}{8}$ is a solution of the equation.

67. $0.8 - p = 3.6$

$0.8-(-2.8)$	3.6
$0.8+2.8$	3.6

$3.6 = 3.6$

Yes, -2.8 is a solution of the equation.

69. Negative

Objective B Exercises

71. $-\frac{2}{9}\times\left(-\frac{3}{14}\right)=\frac{2\cdot 3}{9\cdot 14}=\frac{1}{21}$

73. $\left(-\frac{3}{4}\right)\left(-\frac{8}{27}\right)=\frac{3\cdot 8}{4\cdot 27}=\frac{2}{9}$

75. $\frac{5}{12}\left(-\frac{8}{15}\right)=-\left(\frac{5\cdot 8}{12\cdot 15}\right)=-\frac{2}{9}$

77. $\left(\frac{3}{8}\right)\left(-\frac{15}{41}\right)=-\left(\frac{3\cdot 15}{8\cdot 41}\right)=-\frac{45}{328}$

79. $\left(-\frac{5}{7}\right)\left(-\frac{14}{15}\right)=\frac{5\cdot 14}{7\cdot 15}=\frac{2}{3}$

81. $\left(\frac{1}{2}\right)\left(-\frac{3}{4}\right)\left(-\frac{5}{8}\right)=\frac{1\cdot 3\cdot 5}{2\cdot 4\cdot 8}=\frac{15}{64}$

83. $-\frac{3}{8}\div\frac{7}{8}=-\frac{3}{8}\cdot\frac{8}{7}=-\frac{3}{7}$

85. $\frac{5}{6}\div\left(-\frac{3}{4}\right)=\frac{5}{6}\left(-\frac{4}{3}\right)$
$$=-\left(\frac{5\cdot 4}{6\cdot 3}\right)=-\frac{10}{9}$$

87. $-\frac{5}{16}\div\left(-\frac{3}{8}\right)=-\frac{5}{16}\left(-\frac{8}{3}\right)=\frac{5\cdot 8}{16\cdot 3}=\frac{5}{6}$

89. $-\frac{8}{19}\div\frac{7}{38}=-\frac{8}{19}\left(\frac{38}{7}\right)$
$$=-\left(\frac{8\cdot 38}{19\cdot 7}\right)=-\frac{16}{7}$$

91. $-6\div\frac{4}{9}=-\frac{6}{1}\cdot\frac{9}{4}$
$$=-\left(\frac{6\cdot 9}{1\cdot 4}\right)=-\frac{27}{2}$$

93.
$$\begin{array}{r} 8.9 \\ \times\ 3.5 \\ \hline 445 \\ 267\ \ \\ \hline 31.15 \end{array}$$
$-8.9\times(-3.5)=8.9\times 3.5=31.15$

95.
$$\begin{array}{r} 14.3 \\ \times\ 7.9 \\ \hline 1287 \\ 1001\ \ \\ \hline 112.97 \end{array}$$
$-14.3\times 7.9=-(14.3\times 7.9)=-112.97$

97.
$$\begin{array}{r} 1.21 \\ \times\ 0.03 \\ \hline 0.0363 \end{array}$$
$(-1.21)(-0.03)=(1.21)(0.03)=0.0363$

99.
$$\begin{array}{r} 97. \\ 0.8\overline{)77.6.} \\ \underline{72} \\ 56 \\ \underline{-56} \\ 0 \end{array}$$
$-77.6 \div (-0.8) = 77.6 \div 0.8 = 97$

101.
$$\begin{array}{r} 2.2 \\ 3.2\overline{)7.0.4} \\ \underline{-64} \\ 64 \\ \underline{-64} \\ 0 \end{array}$$
$(-7.04) \div (-3.2) = 7.04 \div 3.2 = 2.2$

103. $-\frac{9}{16}\cdot\frac{4}{27}=-\left(\frac{9\cdot 4}{16\cdot 27}\right)$
$=-\frac{3\cdot 3\cdot 2\cdot 2}{2\cdot 2\cdot 2\cdot 2\cdot 3\cdot 3\cdot 3}$
$=-\frac{1}{12}$

105. $-\frac{7}{24}\cdot\frac{8}{21}\cdot\frac{3}{7}=-\left(\frac{7}{24}\cdot\frac{8}{21}\cdot\frac{3}{7}\right)$
$-\frac{7\cdot 8\cdot 3}{24\cdot 21\cdot 7}$
$=-\frac{7\cdot 2\cdot 2\cdot 2\cdot 3}{2\cdot 2\cdot 2\cdot 3\cdot 3\cdot 7\cdot 7}=-\frac{1}{21}$

107. $\left(-\frac{15}{24}\right)\div\frac{3}{5}=-\left(\frac{15}{24}\cdot\frac{5}{3}\right)$
$=-\frac{15\cdot 5}{24\cdot 3}=-\frac{3\cdot 5\cdot 5}{2\cdot 2\cdot 2\cdot 3\cdot 3}$
$=-\frac{25}{24}=-1\frac{1}{24}$

109. (2.7) (–16)(3.04) = (–43.2)(3.04) = –131.328

111. $-19.04 \div 0.75 \approx -25.4$

113. xy
$-49\cdot\frac{5}{14}=-\left(\frac{49}{1}\cdot\frac{5}{14}\right)$
$=-\frac{49\cdot 5}{1\cdot 14}=-\frac{7\cdot 7\cdot 5}{1\cdot 2\cdot 7}$
$=-\frac{35}{2}=-17\frac{1}{2}$

115. xy
$1\frac{3}{13}\cdot\left(-6\frac{1}{2}\right)=-\left(\frac{16}{13}\cdot\frac{13}{2}\right)$
$=-\frac{2\cdot 2\cdot 2\cdot 2\cdot 13}{13\cdot 2}$
$=-8$

117. $10t$
$10(-4.8) = -48$

119. $x \div y$
$6\frac{2}{5}\div(-4)=-\left(\frac{32}{5}\div\frac{4}{1}\right)$
$=-\left(\frac{32}{5}\cdot\frac{1}{4}\right)=-\left(\frac{32\cdot 1}{5\cdot 4}\right)$
$=-\frac{2\cdot 2\cdot 2\cdot 2\cdot 2\cdot 1}{5\cdot 2\cdot 2}=-\frac{8}{5}=-1\frac{3}{5}$

121. $x \div y$
$-2\frac{5}{8}\div 1\frac{3}{4}=-\left(\frac{21}{8}\div\frac{7}{4}\right)$
$=-\frac{21\cdot 4}{8\cdot 7}=-\frac{3\cdot 7\cdot 2\cdot 2}{2\cdot 2\cdot 2\cdot 7}$
$=-\frac{3}{2}=-1\frac{1}{2}$

123. $x \div y$
$(-64.05)\div(-6.1)-10.5$

125. $x \div y$
$1.173\div(-0.69)=-1.7$

127. $\frac{5}{4}n = -1$

$$\begin{array}{c|c} \frac{5}{4}\left(-\frac{4}{5}\right) & -1 \\ -\frac{5\cdot 4}{4\cdot 5} & -1 \\ -\frac{5\cdot 2\cdot 2}{2\cdot 2\cdot 5} & -1 \end{array}$$

$-1 = -1$

Yes, $-\frac{4}{5}$ is a solution of the equation.

129. $-7.9c = -7.9$

$$\begin{array}{c|c} -7.9(-1) & -7.9 \end{array}$$

$7.9 \neq -7.9$

No, -1 is not a solution of the equation.

131. False

Objective C Exercises

133. Strategy:
To find the amount the temperature fell from 9:00 A.M., subtract the temperature at 9:27 A.M. from the temperature at 9:00 A.M.

Solution:
$12.22 - (-20) = 32.22$
The temperature fell 32.22°C in 27 min.

135. Greater than

137. Strategy:
To find the difference, subtract the melting point of nitrogen (–209.86°C) from its boiling point (–195.8°C).

Solution:
$-195.8 - (-209.86)$
$= -195.8 + 209.86$
$= 14.06$
The difference between the boiling point of nitrogen and its melting point is 14.06°C.

Critical Thinking 3.4

139. **a.** True

b. True

c. False

d. False

141. $-\frac{7}{8} < -\frac{5}{6}$

143. $-\frac{3}{4} < -0.7$

Section 3.5

Concept Check

1. division

Objective A Exercises

3. $2(3-5) - 2 = 2(-2) - 2$
$= -4 - 2$
$= -4 + (-2)$
$= -6$

5. $4 - (-3)^2 = 4 - 9$
$= 4 + (-9)$
$= -5$

7. $3 \cdot (6-2) \div 6 = 3 \cdot 4 \div 6$
$= 12 \div 6$
$= 2$

9. $2^3 - (-3)^2 + 2 = 8 - 9 + 2$
$= 8 + (-9) + 2$
$= -1 + 2$
$= 1$

11. $6 - 2(1-5) = 6 - 2(-4)$
$= 6 - (-8)$
$= 6 + 8$
$= 14$

13. $16 \div 2 - 9 \div 3 = 8 - 9 \div 3$
$= 8 - 3$
$= 8 + (-3)$
$= 5$

15. $(-1)\cdot(4-7)^2 \div 9+6-3-4(2)$
$=(-1)\cdot(-3)^2 \div 9+6-3-4(2)$
$=(-1)\cdot 9 \div 9+6-3-4(2)$
$=-9 \div 9+6-3-4(2)$
$=-1+6-3-4(2)$
$=-1+6-3-8$
$=-1+6+(-3)+(-8)$
$=5+(-3)+(-8)$
$=2+(-8)$
$=-6$

17. $(1.2)^2-4.1\times 0.3=1.44-4.1\times 0.3$
$=1.44-1.23$
$=1.44+(-1.23)=0.21$

19. $(4.1-3.9)-0.7^2=[4.1+(-3.9)]-0.7^2$
$=0.2-0.7^2$
$=0.2-0.49$
$=0.2+(-0.49)=-0.29$

21. $-\frac{1}{2}+\frac{3}{8}\div\left(-\frac{3}{4}\right)=-\frac{1}{2}+\left[\frac{3}{8}\left(-\frac{4}{3}\right)\right]$
$=-\frac{1}{2}+\left(-\frac{12}{24}\right)$
$=-\frac{1}{2}+\left(-\frac{1}{2}\right)=-\frac{2}{2}=-1$

23. $\frac{1}{2}-\left(\frac{3}{4}-\frac{3}{8}\right)\div\frac{1}{3}=\frac{1}{2}-\left(\frac{6}{8}-\frac{3}{8}\right)\div\frac{1}{3}$
$=\frac{1}{2}-\frac{3}{8}\div\frac{1}{3}=\frac{1}{2}-\frac{3}{8}\cdot\frac{3}{1}$
$=\frac{1}{2}-\frac{9}{8}=\frac{4}{8}+\left(-\frac{9}{8}\right)=-\frac{5}{8}$

25. $5-|2-8|=5-|-6|=5-6=-1$

27. $-|-7|+9^2=-7+9^2=-7+81=74$

29. $2|2-6|+4=2|-4|+4$
$=2(4)+4=8+4=12$

31. $||-7|-|-3||=|7-3|=|4|=4$

33. $|3.4(-2)|-9.8=|-6.8|-9.8$
$=6.8-9.8=-3$

35. $|0.9-(1.5)^2|=|0.9-2.25|$
$=|-1.35|=1.35$

37. $\left|\left|-\frac{4}{9}\right|-\left|-\frac{5}{9}\right|\right|=\left|\frac{4}{9}-\frac{5}{9}\right|=\left|-\frac{1}{9}\right|=\frac{1}{9}$

39. $\left(-\frac{1}{4}\right)^2+\left|\frac{3}{8}\left(-\frac{5}{6}\right)\right|=\left(-\frac{1}{4}\right)^2+\left|-\frac{5}{16}\right|$
$=\frac{1}{16}+\frac{5}{16}=\frac{6}{16}=\frac{3}{8}$

41. $3a+2b$
$3(-2)+2(4)=-6+2(4)$
$=-6+8$
$=2$

43. $bc\div(2a)$
$4(-1)\div((2)(-2))=4(-1)\div(-4)$
$=(-4)\div(-4)$
$=1$

45. b^2-c^2
$4^2-(-1)^2=16-1$
$=16+(-1)$
$=15$

47. $(b-a)^2+4c$
$(4-(-2))^2+4(-1)=(4+2)^2+4(-1)$
$=6^2+4(-1)$
$=36+4(-1)$
$=36+(-4)$
$=32$

49. $\frac{d-b}{c}$
$\frac{3-4}{-1}=\frac{3+(-4)}{-1}$
$=\frac{-1}{-1}$
$=1$

51. $(d-a)^2-3c$
$(3-(-2))^2-3(-1)$
$=(3+2)^2-3(-1)$
$=5^2-3(-1)$
$=25-3(-1)$
$=25-(-3)$
$=25+3$
$=28$

53. $\left|\frac{a-b}{c-d}\right|$

$\left|\frac{-2-3}{-1-4}\right| = \left|\frac{-5}{-5}\right| = |1| = 1$

55. $|a+d|^2 + |c-b|^2$

$|-2+4|^2 + |-1-3|^2$

$= |2|^2 + |-4|^2$

$= 2^2 + 4^2$

$= 4 + 16$

$= 20$

57. Any odd power of a negative number is negative.

Critical Thinking 3.5

59. $-2^2 - (-3)^2 + 5(4) \div 10 - (-6)$

$= -4 - 9 + 5(4) \div 10 - (-6)$

$= -4 - 9 + 20 \div 10 - (-6)$

$= -4 - 9 + 2 - (-6)$

$= -4 + (-9) + 2 + 6$

$= -13 + 2 + 6$

$= -11 + 6 = -5$

We are looking for the smallest integer greater than –5.

The smallest integer greater than –5 is –4.

The smallest integer greater than $-2^2 - (-3)^2 + 5(4) \div 10 - (-6)$ is –4.

61. a.

$x^2 - 2x - 8 = 0$	
$(-4)^2 - 2(-4) - 8$	0
$16 - 2(-4) - 8$	0
$16 + 8 - 8$	0
$24 - 8$	0

$16 \neq 0$

No, –4 is not a solution of the equation.

b.

$x^3 + 3x^2 - 5x - 15 = 0$	
$(-3)^3 + 3(-3)^2 - 5(-3) - 15$	0
$-27 + 3(9) - 5(-3) - 15$	0
$-27 + 27 + 15 - 15$	0
$0 + 15 - 15$	0
$15 - 15$	0

$0 = 0$

Yes, –3 is a solution of the equation.

63. $6 - 12 \div 2 \cdot (3-5)^2$

Chapter Review Exercises

1. eight minus negative one

2. $-|-36| = -36$

3. $(-40)(-5) = 200$

4. $-a \div b$

$-(-27) \div (-3) = 27 \div (-3)$

$= -9$

5. $-28 + 14 = -14$

6. $-(-13) = 13$

7.

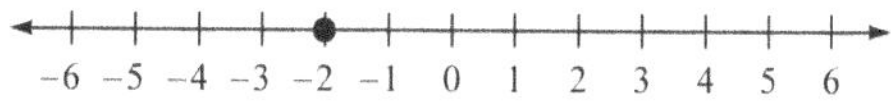

8. $-65.47 + (-32.91) = -98.38$

9. $-51 \div (-3) = 17$

10. $\frac{840}{-4} = -210$

11. $-6 - (-7) - 15 - (-12)$

$= -6 + 7 + (-15) + 12$

$= 1 + (-15) + 12$

$= -14 + 12$

$= -2$

12. $-ab$

$-(-2)(-9) = 2(-9)$

$= -18$

13. $18 + (-13) + (-6) = 5 + (-6)$

$= -1$

14. $-18(4) = -72$

15. $(-2)^2 - (-3)^2 \div (1-4)^2 \cdot 2 - 6$

$= (-2)^2 - (-3)^2 \div (-3)^2 \cdot 2 - 6$

$= 4 - 9 \div 9 \cdot 2 - 6$

$= 4 - 1 \cdot 2 - 6$

$= 4 - 2 - 6$

$= 4 + (-2) + (-6)$

$= 2 + (-6)$

$= -4$

Projects or Group Activities 3.5

16. $-x-y$

$-(-1)-3=1-3$

$=1+(-3)$

$=-2$

17. $4-(-14)=4+14$

$=18$

The difference between Tom Lehman's score and Mark Calcavecchia's score is 18 strokes.

18. $-15-(-28)=-15+28$

$=13$

19. $\left(-\frac{1}{5}\right)\div\left(-\frac{7}{10}\right)=\frac{1}{5}\cdot\frac{10}{7}=\frac{2}{7}$

20.

$$\begin{array}{r|l} \multicolumn{2}{c}{-6-t=3} \\ \hline -6-(-9) & 3 \\ -6+9 & 3 \end{array}$$

$3=3$

Yes, -9 is a solution of the equation $-6-t=3$.

21. $-9+16-(-7)=-9+16+7$

$=7+7$

$=14$

22. $\frac{0}{-17}=0$

23. $-5(2)(-6)(-1)=-10(-6)(-1)$

$=60(-1)$

$=-60$

24. $3+(-9)+4+(-10)=-6+4+(-10)$

$=-2+(-10)$

$=-12$

25. $(a-b)^2-2a$

$(-2-(-3))^2-2(-2)$

$=(-2+3)^2-2(-2)$

$=1^2-2(-2)$

$=1-2(-2)$

$=1-(-4)$

$=1+4$

$=5$

26. $-8>-10$

27. $-21+21=0$

28. $|-27|=27$

29. $-0.8(3.5)=-2.8$

30. $-\frac{11}{16}+\frac{7}{12}=-\frac{33}{48}+\frac{28}{48}=-\frac{5}{48}$

31. $3\div\left(\frac{1}{2}-\frac{1}{4}\right)-3=3\div\left(\frac{2}{4}-\frac{1}{4}\right)-3$

$=3\div\frac{1}{4}-3=\frac{3}{1}\div\frac{1}{4}-3$

$=\frac{3}{1}\cdot\frac{4}{1}-3=\frac{12}{1}-3$

$=12-3=9$

32. Strategy:

To find the colder temperature, compare the numbers -4 and -12. The smaller number represents the colder temperature.

Solution:

$-4>-12$

The colder temperature is -12°C.

33. Strategy:

To find the boiling point of neon:

→Find the highest boiling point shown in the table.

→Multiply the highest boiling point by 7.

Solution:

The highest boiling point shown in the table is -34°.

$-34(7)=-238$

The boiling point of neon is -238°C.

34. Strategy:

To find the temperature, add the increase (5) to the previous temperature (-8).

Solution:

$-8+5=-3$

The temperature is -3°C.

35. Strategy:
To find d, replace a by 7 and b by –5 in the given formula and solve for d.

Solution:
$d = |a - b|$
$d = |7 - (-5)|$
$d = |7 + 5|$
$d = |12|$
$d = 12$
The distance between the two points is 12 units.

Chapter Test

1. negative three plus negative five

2. $-|-34| = -34$

3. $3 - (-15) = 3 + 15$
$= 18$

4. $a + b$
$(-11) + (-9) = -20$

5. $(-x)(-y)$
$(-(-4)) \cdot (-(-6)) = 4 \cdot 6$
$= 24$

6. $\frac{4}{9} + \left(-\frac{5}{6}\right) = \frac{8}{18} + \left(-\frac{15}{18}\right) = -\frac{7}{18}$

7. $-360 \div -30 = 12$

8. $-3 + -6 + 11 = -9 + 11$
$= 2$

9. $16 > -19$

10. $7 - (-3) - 12 = 7 + 3 - 12$
$= 10 - 12$
$= -2$

11. $a - b - c$
$= 6 - (-2) - 11 = 6 + 2 - 11$
$= 8 - 11$
$= -3$

12. $-(-49) = 49$

13. $50 \cdot (-5) = -250$

14. $-|5|, -(3), |-9|, -(-11)$

15.
$$\begin{array}{r|l} \multicolumn{2}{c}{17 - x = 8} \\ \hline 17 - (-9) & 8 \\ 17 + 9 & 8 \end{array}$$
$26 \neq 8$
No, –9 is not a solution of the equation $17 - x = 8$.

16.

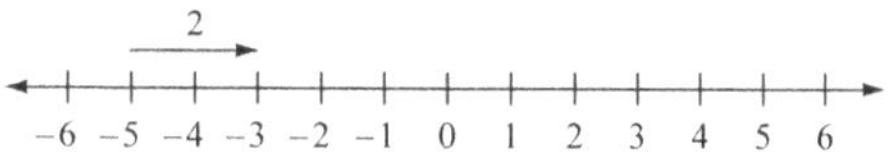

–3 is 2 units to the right of –5.

17. Strategy:
To find the difference in scores, subtract Tseng's score (–19) from Leon's score (2).

Solution:
$2 - (-19) = 2 + 19$
$= 21$
The difference in scores was 21 strokes.

18. $\frac{0}{-16} = 0$

19. $2bc - (c - a)^3$
$(2 \cdot 4 \cdot (-1)) - (-1 + (-2))^3 = -8 - (-3)^3$
$= -8 - (-27)$
$= -8 + 27$
$= 19$

20. –25

21. $-6.82 - 4.793 = -11.613$

22. $0 - 11 = -11$

23. $-96 \div (-4) = 24$

24. $16 \div 4 - 12 \div (-2) = 4 - (-6)$
$= 4 + 6$
$= 10$

25. $\frac{-x}{y}$
$\frac{-(-56)}{-8} = \frac{56}{-8}$
$= -7$

26. $3xy$
$3 \cdot (-2) \cdot (-10) = 3 \cdot 20$
$= 60$

27. $-18 \div \frac{2}{3} = -\left(\frac{18}{1} \div \frac{2}{3}\right)$
$= -\left(\frac{18}{1} \cdot \frac{3}{2}\right) = -\frac{27}{1} = -27$

28. $4 - 14 = -10$

29. $7 \div \left(\frac{1}{7} - \frac{3}{14}\right) - 9 = 7 \div \left(\frac{2}{14} - \frac{3}{14}\right) - 9$
$= 7 \div \left(-\frac{1}{14}\right) - 9 = \frac{7}{1} \div \left(-\frac{1}{14}\right) - 9$
$= \frac{7}{1} \cdot \left(-\frac{14}{1}\right) - 9 = -\frac{98}{1} - 9$
$= -98 - 9 = -107$

30. xy
$-0.3(5.1) = -1.53$

31. $-9.61 + (-5.7) = -15.31$

32. Strategy:
To find the temperature, add the increase (11) to the previous temperature (–6).

Solution:
$-6 + 11 = 5$
The temperature is 5°C.

33. The unknown number: n

the wind chill at –25°F with a 40 mph wind	is	Four times the wind chill factor at –16°F with a 40 mph wind

$n = 4 \cdot (-16)$
$n = -64$
The wind chill factor is –64°F.

34. Strategy:
To find drop in temperature, subtract the evening temperature from the morning temperature.

Solution:
$3.5 - (-1.75) = 3.5 + 1.75$
$= 5.25$
The temperature is 5.25°C

35. $d = |a - b|$
$d - |4 - (-12)| - |4 + 12|$
$= |16|$
$= 16$
The solution is 16 units.

Cumulative Review Exercises

1. $-27 - (-32) = -27 + 32$
$= 5$

2. $439 \rightarrow 400$
$28 \rightarrow 30$
$400 \cdot 30 = 12{,}000$

3.
$$\begin{array}{r} 32.3 \\ 0.5\overline{)\,16.15} \\ \underline{-15} \\ 11 \\ \underline{-10} \\ 15 \\ \underline{-15} \\ 0 \end{array}$$

4. $16 \div (3+5) \cdot 9 - 2^4 = 16 \div 8 \cdot 9 - 2^4$
$= 16 \div 8 \cdot 9 - 16$
$= 2 \cdot 9 - 16$
$= 18 - 16$
$= 18 + (-16)$
$= 2$

5. $-|-82| = -82$

6. 309,480

7. $5xy$
$5 \cdot 80 \cdot 6 = 400 \cdot 6$
$= 2{,}400$

8. $-294 \div (-14) = 21$

9. $-28 - (-17) = -28 + 17$
$= -11$

10. $-24 + 16 + (-32) = -8 + (-32)$
$= -40$

11. $44 \div 1 = 44$
$44 \div 2 = 22$
$44 \div 4 = 11$
$44 \div 11 = 4$
The factors of 44 are 1, 2, 4, 11, 22, and 44.

12. $x^4 y^2$
$\left(\frac{1}{2}\right)^4 \cdot 4^2 = \left(\frac{1}{2} \cdot \frac{1}{2} \cdot \frac{1}{2} \cdot \frac{1}{2}\right) \cdot (4 \cdot 4)$
$= \frac{1}{16} \cdot 16 = \frac{1}{16} \cdot \frac{16}{1} = \frac{16}{16} = 1$

13. 629,874 *Given place value* $8 > 5$

629,874 rounded to the nearest thousand is 630,000.

14.
$$\begin{array}{rcr} 356 & \rightarrow & 400 \\ 481 & \rightarrow & 500 \\ 294 & \rightarrow & 300 \\ 117 & \rightarrow & +100 \\ \hline & & 1{,}300 \end{array}$$

15. $-a - b$

$-(-4) - (-5) = 4 + 5$

$= 9$

16. $-100 \cdot 25 = -2{,}500$

17. $3.97 + 4.8 = 8.77$

18. $2\frac{1}{6} + 3\frac{1}{2} = 2\frac{1}{6} + 3\frac{3}{6} = 5\frac{4}{6} = 5\frac{2}{3}$

19. $(1-5)^2 \div (-6+4) + 8(-3)$

$= (-4)^2 \div (-2) + 8(-3) = 16 \div (-2) + 8(-3)$

$= -8 + 8(-3)$

$= -8 + (-24)$

$= -32$

20. $-c \div d$

$-(-32) \div (-8) = 32 \div (-8)$

$= -4$

21. $\frac{9}{10} \div \frac{3}{4} = \frac{9}{10} \cdot \frac{4}{3} = \frac{3 \cdot 3 \cdot 2 \cdot 2}{2 \cdot 5 \cdot 3} = \frac{6}{5} = 1\frac{1}{5}$

22. $-62 < 26$

23. $-18(-7) = 126$

24. $-3.312 \div (-0.8) = 4.14$

25. $2 \cdot 2 \cdot 2 \cdot 2 \cdot 2 \cdot 7 \cdot 7 = 2^5 \cdot 7^2$

26. $4a + (a-b)^3$

$4(5) + (5-2)^3 = 4(5) + 3^3$

$= 4(5) + 27$

$= 20 + 27$

$= 47$

27.
$$\begin{array}{r} 5{,}971 \\ 482 \\ +\,3{,}609 \\ \hline 10{,}062 \end{array}$$

28. $-21 - 5 = -21 + (-5)$

$= -26$

29.
$$\begin{array}{rcr} 7{,}352 & \rightarrow & 7{,}000 \\ 1{,}986 & \rightarrow & -\,2{,}000 \\ \hline & & 5{,}000 \end{array}$$

30. $3^4 \cdot 5^2 = (3 \cdot 3 \cdot 3 \cdot 3) \cdot (5 \cdot 5)$

$= 81 \cdot 25$

$= 2{,}025$

31. Strategy:
To find the land area, add the land area prior to the purchase (891,364) to the amount of land purchased (831,321).

Solution:

$$\begin{array}{r} 891{,}364 \\ +\,831{,}321 \\ \hline 1{,}722{,}685 \end{array}$$

The land area of the United States after the Louisiana purchase was 1,722,685 mi^2.

32. Strategy:
To find the age, subtract the year of the birth (1879) from the year of his death (1955).

Solution

$$\begin{array}{r} 1955 \\ -\,1879 \\ \hline 76 \end{array}$$

Albert Einstein was 76 years old when he died.

33. Strategy:
To find the amount, subtract the down payment (7,850) from the cost (35,500).

Solution:

$$\begin{array}{r} 35{,}500 \\ -\ 7{,}850 \\ \hline 27{,}650 \end{array}$$

The amount to be paid is $27,650.

34. Strategy:
To find the cost of the land, multiply the number of acres (25) times the cost per acre (11,270).

Solution:

$$\begin{array}{r} 11{,}270 \\ \times 25 \\ \hline 56\ 350 \\ 225\ 40 \\ \hline 281{,}750 \end{array}$$

The cost of the land is $281,750.

35. Strategy:
To find the temperature, add the increase (7) to the original temperature (–12).

Solution:
$-12 + 7 = -5$
The temperature is –5°C.

36. Strategy:

a. To find the difference in Arizona, subtract the record low (–40) from the record high (128).

b. To find the greatest difference, subtract the record lows from the record highs for each state and compare.

Solution:

a. $128 - (-40) = 128 + 40$
$= 168$
The difference in temperatures is 168°F.

b. $112 - (-27) = 112 + 27 = 139$
100 (80) − 100 + 80 = 180
$128 - (-40) = 128 + 40 = 168$
$120 - (-29) = 120 + 29 = 149$
$180 > 168 > 149 > 139$
The greatest difference in temperatures is in Alaska.

37. Strategy:
To find the amount:
→Add the sales figures for the first three quarters (28,550 + 34,850 + 31,700).
→Subtract the sum from the goal for the year (120,000).

Solution:

$$\begin{array}{r} 28{,}550 \\ 34{,}850 \\ +\ 31{,}700 \\ \hline 95{,}100 \end{array}$$

$$\begin{array}{r} 120{,}000 \\ -\ 95{,}100 \\ \hline 24{,}900 \end{array}$$

You must sell $24,900 in the last quarter to meet the goal.

38. Strategy:
To find the score, substitute 198 for N and 206 for P in the given formula and solve for S.

Solution:
$S = N - P$
$S = 198 - 206$
$S = -8$
The golfer's score is –8.

Chapter 4: Variable Expressions

Prep Test

1. $-12-(-15)=-12+15=3$

2. $-36\div(-9)=4$

3. $-\frac{3}{4}+\frac{5}{6}=-\frac{3}{4}\cdot\frac{3}{3}+\frac{5}{6}\cdot\frac{2}{2}$
$=-\frac{9}{12}+\frac{10}{12}=\frac{-9+10}{12}=\frac{1}{12}$

4. $-\frac{4}{9}$

5. $-\frac{3}{4}\div\left(-\frac{5}{2}\right)=-\frac{3}{4}\cdot\left(-\frac{2}{5}\right)=\frac{3\cdot \overset{1}{\cancel{2}}}{\underset{1}{\cancel{2}}\cdot 2\cdot 5}=\frac{3}{10}$

6. $-2^4=-(2\cdot 2\cdot 2\cdot 2)=-16$

7. $\left(\frac{2}{3}\right)^3=\frac{2}{3}\cdot\frac{2}{3}\cdot\frac{2}{3}=\frac{8}{27}$

8. $3\cdot 4^2=3\cdot 16=48$

9. $7-2\cdot 3=7-6=1$

10. $5-7(3-2^2)=5-7(3-4)$
$=5-7(-1)$
$=5+7$
$=12$

11. The phrase "evaluate the variable expression" means to replace each variable in the expression with its value and then to simplify the resulting numerical expression.

Section 4.1

Concept Check

1. $-3n^2$, $5x$, $\underline{-8}$

3. $-a^4$, $\underline{6}$

5. $7\underline{x^2y}$, $6\underline{xy^2}$

7. coefficient of x^2: 1
coefficient of $-9x$: -9

9. coefficient of n^3: 1

coefficient of $-4n^2$: -4
coefficient of $-n$: -1

Objective A Exercises

13. $a-2c$
$2-2(-4)=2+8=10$

15. $2c^2$
$2(-4)^2=2(16)=32$

17. $3b-3c$
$3(3)-3(-4)=9+12=21$

19. $-3c+4$
$-3(-4)+4=12+4=16$

21. $6b\div(-a)$
$6(3)\div(-2)=18\div(-2)=-9$

23. b^2-4ac
$(3)^2-4(2)(-4)=9-4(2)(-4)$
$=9-(-32)=9+32$
$=41$

25. b^2-c^2
$3^2-(-4)^2=9-16=-7$

27. a^2+b^2
$2^2+3^2=4+9=13$

29. $(b-a)^2+4c$
$(3-2)^2+4(-4)=1^2+4(-4)$
$=1+(-16)=-15$

31. $\frac{5ab}{6}-3cb$

$\frac{5(2)(3)}{6}-3(-4)(3)=\frac{30}{6}-(-36)$
$=5-(-36)=41$

33. $\frac{b+c}{d}$

$\frac{4+(-1)}{3}=\frac{3}{3}=1$

35. $\frac{2d+b}{-a}$

$\frac{2(3)+4}{-(-2)}=\frac{6+4}{2}=\frac{10}{2}=5$

37. $\frac{b-d}{c-a}$

$\frac{4-3}{-1-(-2)}=\frac{1}{1}=1$

39. $(b+d)^2-4a$

$(4+3)^2-4(-2)=7^2-4(-2)$

$=49-(-8)=57$

41. $(d-a)^2\div 5$

$\left[3-(-2)\right]^2\div 5=5^2\div 5=25\div 5=5$

43. $\frac{b-2a}{bc^2-d}$

$\frac{4-2(-2)}{4(-1)^2-3}=\frac{4-(-4)}{4(1)-3}=\frac{8}{4-3}=\frac{8}{1}=8$

45. $\frac{1}{3}d^2-\frac{3}{8}b^2$

$\frac{1}{3}(3)^2-\frac{3}{8}(4)^2=\frac{1}{3}(9)-\frac{3}{8}(16)=3-6=-3$

47. $\frac{-4bc}{2a-b}$

$\frac{-4(4)(-1)}{2(-2)-4}=\frac{16}{-4-4}=\frac{16}{-8}=-2$

49. $-\frac{2}{3}d-\frac{1}{5}(bd-ac)$

$-\frac{2}{3}(3)-\frac{1}{5}\left[4(3)-(-2)(-1)\right]=-\frac{2}{3}(3)-\frac{1}{5}\left[12-2\right]$

$=-\frac{2}{3}(3)-\frac{1}{5}(10)$

$=-2-2=-4$

51. Positive

53. Negative

Critical Thinking 4.1

55. $a=3x^2-4x-5$ when $x=-2$

$a=3(-2)^2-4(-2)-5$

$a=3(4)+8-5$

$a=12+8-5$

$a=15$

$3a-4$

$3(15)-4$

$=45-4$

$=41$

57. 3^x-x^3

$3^2-2^3=9-8=1$

59. x^x-y^y

$2^2-3^3=4-27=-23$

Projects or Group Activities 4.1

61. a. $x=3$: $x^3=3^3=27$; $3^x=3^3=27$;

$x=4$: $x^3=4^3=64$; $3^x=3^4=81$;

$81>27$

b. $x=3$: $x^4=3^4=81$; $4^x=4^3=64$

$x=4$: $x^4=4^4=256$; $4^x=4^4=256$

$x=5$: $x^4=5^4=625$; $4^x=4^5=1024$;

$1024>625$

c. $x=3$: $x^5=3^5=243$; $5^x=5^3=125$

$x=4$: $x^5=4^5=1024$; $5^x=5^4=625$

$x=5$: $x^5=5^5=3125$; $5^x=5^5=3125$

$x=6$: $x^5=6^5=7776$; $5^x=5^6=15{,}625$

$15{,}625>7776$

d. $x=3$; $x^6=3^6=729$; $6^x=6^3=216$

$x=4$; $x^6=4^6=4096$; $6^x=6^4=1296$

$x=5$; $x^6=5^6=15625$; $6^x=6^5=7776$

$x=6$; $x^6=6^6=46656$; $6^x=6^6=46656$

$x=7$; $x^6=7^6=117649$; $6^x=6^7=279936$

$279936>117649$

Section 4.2

Concept Check

1. Commutative

3. Reciprocal (or multiplicative inverse)

5. Like terms are variable terms with the same variable part. Constant terms are also like terms. Examples of like terms are $4x$ and $-9x$. Examples of terms that are not alike are $4x^2$ and $-9x$. The terms 4 and 9 are also like terms; 4 and $4x$ are not.

Objective A Exercises

7. $6x + 8x = 14x$

9. $9a - 4a = 5a$

11. $4y + (-10y) = -6y$

13. $7 - 3b = 7 - 3b$

15. $-12a + 17a = 5a$

17. $5ab - 7ab = -2ab$

19. $-12xy + 17xy = 5xy$

21. $-3ab + 3ab = 0$

23. $-\frac{1}{2}x - \frac{1}{3}x = -\frac{3}{6}x - \frac{2}{6}x = -\frac{5}{6}x$

25. $2.3x + 4.2x = 6.5x$

27. $x - 0.55x = 0.45x$

29. $5a - 3a + 5a = 7a$

31. $-5x^2 - 12x^2 + 3x^2 = -14x^2$

33. $\frac{3}{4}x - \frac{1}{3}x - \frac{7}{8}x = \frac{18}{24}x - \frac{8}{24}x - \frac{21}{21}x = -\frac{11}{24}x$

35. $7x - 3y + 10x = 17x - 3y$

37. $3a + (-7b) - 5a + b = -2a - 6b$

39. $3x + (-8y) - 10x + 4x = -3x - 8y$

41. $x^2 - 7x + (-5x^2) + 5x = -4x^2 - 2x$

43. $-10x - 10y - 10y - 10x = -20x - 20y$
   - **i.** 0 No
   - **ii.** $-20y$ No
   - **iii.** $-20y$ No
   - **iv.** $-20x - 20y$ Yes
   - **v.** $-20y - 20x$ Yes

   (iv) and (v)

Objective B Exercises

45. $12(5x) = 60x$

47. $-2(5a) = -10a$

49. $-5(-6y) = 30y$

51. $(6x)12 = 72x$

53. $(7a)(-4) = -28a$

55. $(-12b)(-9) = 108b$

57. $-8(7x^2) = -56x^2$

59. $\frac{1}{6}(6x^2) = x^2$

61. $\frac{1}{8}(8x) = x$

63. $-\frac{1}{4}(-4a) = a$

65. $-\frac{1}{9}(-9b) = b$

67. $(12x)\left(\frac{1}{12}\right) = x$

69. $(-10n)\left(-\frac{1}{10}\right) = n$

71. $\frac{1}{7}(14x) = 2x$

73. $-0.25(8x) = -2x$

75. $-\frac{5}{8}(24a^2) = -15a^2$

77. $-0.75(-8y) = 6y$

79. $(33y)\left(\frac{1}{11}\right) = 3y$

81. $(-10x)\left(\frac{1}{5}\right) = -2x$

83. $(21y)\left(-\frac{3}{7}\right)=-9y$

Objective C Exercises

85. $2(4x-3)=8x-6$

87. $-2(a+7)=-2a-14$

89. $-3(2y-8)=-6y+24$

91. $-(x+2)=-x-2$

93. $(5-3b)7=35-21b$

95. $\frac{1}{3}(6-15y)=2-5y$

97. $3(5x^2+2x)=15x^2+6x$

99. $-2(-y+9)=2y-18$

101. $(-3x-6)5=-15x-30$

103. $2(-3x^2-14)=-6x^2-28$

105. $-3(2y^2-7)=-6y^2+21$

107. $3(x^2-y^2)=3x^2-3y^2$

109. $-\frac{2}{3}(6x-18y)=-4x+12y$

111. $-(6a^2-7b^2)=-6a^2+7b^2$

113. $4(x^2-3x+5)=4x^2-12x+20$

115. $\frac{3}{4}(2x-6y+8)=\frac{3}{2}x-\frac{9}{2}y+6$

117. $4(-3a^2-5a+7)=-12a^2-20a+28$

119. $-3(-4x^2+3x-4)=12x^2-9x+12$

121. $5(2x^2-4xy-y^2)=10x^2-20xy-5y^2$

123. $-(8b^2-6b+9)=-8b^2+6b-9$

Objective D Exercises

125. $12-7(y-9)=12-7y+63=-7y+75$

i. $5(y-9)=5y-45$ No

ii. $12-7y-63=-7y-51$ No

iii. $12-7y+63=-7y+75$ Yes

iv. $12-7y-9=-7y+3$ No

127. $6a-(5a+7)=6a-5a-7=a-7$

129. $10-(11x-3)=10-11x+3$
$=-11x+13$

131. $8-(12+4y)=8-12-4y=-4y-4$

133. $2(x-4)-4(x+2)=2x-8-4x-8$
$=-2x-16$

135. $6(2y-7)-(3-2y)=12y-42-3+2y$
$=14y-45$

137. $2(a+2b)-(a-3b)=2a+4b-a+3b=a+7b$

139. $2[x+2(x+7)]=2[x+2x+14]=2[3x+14]$
$=6x+28$

141. $-5[2x+3(5-x)]=-5[2x+15-3x]$
$=-5[-x+15]=5x-75$

143. $-2[3x-(5x-2)]=-2[3x-5x+2]$
$=-2[-2x+2]=4x-4$

145. $-7x+3[x-(3-2x)]=-7x+3[x-3+2x]$
$=-7x+3[3x-3]$
$=-7x+9x-9=2x-9$

147. $0.12(2x+3)+x=0.24x+0.36+x$
$=1.24x+0.36$

149. $0.03x+0.04(1000-x)=0.03x+40-0.04x$
$=-0.01x+40$

Critical Thinking 4.2

151. Answers will vary. One example is unbuckling the seat belt and then getting out of the car.

Projects and Group Activities 4.2

153. a. Yes; $7 \otimes 5 = 5 \otimes 7$

$7 \otimes 5 = 23 \qquad 5 \otimes 7$

$= (5 \cdot 7) - (5 + 7) = 35 - 12 = 23$

b. No; $(7 \otimes 5) \otimes 2 \neq 7 \otimes (5 \otimes 2)$

$(7 \otimes 5) \otimes 2$
$= 23 \otimes 2$
$= (23 \cdot 2) - (23 + 2)$
$= 46 - 25$
$= 21$

$7 \otimes (5 \otimes 2)$
$= 7 \otimes [(5 \cdot 2) - (5 + 2)]$
$= 7 \otimes [10 - 7]$
$= 7 \otimes 3$
$= (7 \cdot 3) - (7 + 3)$
$= 21 - 10$
$= 11$

Check Your Progress: Chapter 4

1. $-2a + 3b$

$-2(3) + 3(-2) = -6 - 6 = -12$

2. $c^2 - 5ab$

$(4)^2 - 5(3)(-2) = 16 - 15(-2)$
$= 16 + 30 = 46$

3. $2c - (a + b)^2$

$2(4) - (3 + (-2))^2$
$= 8 - (1)^2 = 8 - 1 = 7$

4. $\dfrac{y + z}{x}$

$\dfrac{-3 + (-1)}{2} = \dfrac{-4}{2} = -2$

5. $(y + x)^2 - 5z$

$(-3 + 2)^2 - 5(-1) = (-1)^2 + 5 = 1 + 5 = 6$

6. $\dfrac{-yz}{2x + z}$

$\dfrac{-(-3)(-1)}{2(2) + (-1)} = \dfrac{-3}{4 - 1} = \dfrac{-3}{3} = -1$

7. $10y - 16y + 3y = -6y + 3y = -3y$

8. $-5a + 6b + 8a - 2b = 3a + 4b$

9. $(8a)(-5) = -40a$

10. $(-9z)\left(-\dfrac{1}{9}\right) = z$

11. $(12 - 8b)3 = 36 - 24b$

12. $-2(-3x^2 + 4x - 5) = 6x^2 - 8x + 10$

13. $3x - 4(2x - 5) = 3x - 8x + 20 = -5x + 20$

14. $6(3a - 7) - (4 + 9a) = 18a - 42 - 4 - 9a = 9a - 46$

15. $5 - 3[2x - (6 - 4x)] = 5 - 3(2x - 6 + 4x)$
$= 5 - 3[6x - 6] = 5 - 18x + 18$
$= -18x + 23$

Section 4.3

Concept Check

1. sum, times

3. total, divided by

5. $25 - x$

Objective A Exercises

7. the sum of 8 and y
$8 + y$

9. t increased by 10
$t + 10$

11. 14 added to z
$z + 14$

13. 20 less than the square of x

$x^2 - 20$

15. The sum of three-fourths of n and 12

$\frac{3}{4}n + 12$

17. 8 increased by the quotient of n and 4

$8 + \frac{n}{4}$

19. the product of 3 and the total of y and 7

$3(y + 7)$

21. the product of t and the sum of t and 16

$t(t + 16)$

23. 15 more than one-half of the square of x

$\frac{1}{2}x^2 + 15$

25. the total of five times the cube of n and the square of n

$5n^3 + n^2$

27. r decreased by the quotient of r and 3

$r - \frac{r}{3}$

29. the difference between the square of x and the total of x and 17

$x^2 - (x + 17)$

31. the product of 9 and the total of z and 4

$9(z + 4)$

33. Answers may vary. For example: the product of 5 and 1 more than the square of n or 5 times the sum of 1 and the square of n.

Objective B Exercises

35. the unknown number: x

$\frac{x}{18}$

37. the unknown number: x

$x + 20$

39. the unknown number: x

the product of eleven and the number: $11x$

$11x - 8$

41. the unknown number: x

the total of five and the number: $5 + x$

$\frac{7}{5 + x}$

43. the unknown number: x

the quotient of the number and twenty: $\frac{x}{20}$

$40 - \frac{x}{20}$

45. the unknown number: x

the square of the number: x^2

twice the number: $2x$

$x^2 + 2x$

47. the unknown number: x

the difference between the number and 50: $x - 50$

$10(x - 50) = 10x - 500$

49. the unknown number: x

three more than the number: $x + 3$

$x - (x + 3) = x - x - 3 = -3$

51. the unknown number: x

twice the number: $2x$

the difference between twice the number and four: $2x - 4$

$(2x - 4) + x = 2x - 4 + x = 3x - 4$

53. the unknown number: x

three times the number: $3x$

the difference between three times the number and eight: $3x - 8$

$x - (3x - 8) = x - 3x + 8 = -2x + 8$

55. the unknown number: x

the product of three and the number: $3x$

$3x + x = 4x$

57. the unknown number: x

the sum of the number and six: $x + 6$

$(x + 6) + 5 = x + 6 + 5 = x + 11$

59. the unknown number: x

the sum of the number and ten: $x + 10$

$x - (x + 10) = x - x - 10 = -10$

61. the unknown number: x

one-sixth of the number: $\frac{1}{6}x$

four-ninths of the number: $\frac{4}{9}x$

$$\frac{1}{6}x+\frac{4}{9}x=\frac{3}{18}x+\frac{8}{18}x=\frac{11}{18}x$$

Objective C Exercises

63. The number of students enrolled in fall-term science classes

65. Number of visitors to the Metropolitan Museum of Art: M
number of visitors to the Louvre: $M+3{,}800{,}000$

67. Noise level of a car horn: d
noise level of an ambulance siren: $d+10$

69. U2's concert ticket sales: T
E Street Band's concert ticket sales: $T-28{,}500{,}000$

71. number of bones in your body: N

number of bones in your foot: $\frac{1}{4}N$

73. diameter of a baseball: d
diameter of a basketball: $4d$

75. attendance at major league basketball games: B
attendance at major league baseball games: $B+50{,}000{,}000$

77. measure of the largest angle: L

measure of the smallest angle: $\frac{1}{2}L-10$

79. hours of labor: h
amount of the repair bill: $238+89h$

81. distance traveled by the slower car: x
distance traveled by the faster car: $200-x$

Critical Thinking 4.3

83. number of oxygen atoms: x
number of hydrogen atoms: $2x$

85. Both of the expressions "the difference between x and 5 and "5 less than x" translate into $x-5$. However, in translating the two expressions from words, the word *difference* maintains the order in which the two quantities are subtracted, whereas the words *less than* reverse the order in which the two quantities are subtracted.

Projects and Group Activities 4.3

87. Answers will vary. For example: 16 less than d; the difference between d and 16; d decreased by 16; 16 subtracted from d

89. Answers will vary. For example: y divided by 5; the quotient of y and 5; the ratio of y to 5

Chapter Review Exercises

1. $3(x^2-8x-7)=3(x^2)+3(-8x)+3(-7)$
$=3x^2-24x-21$

2. $7x+4x=(7+4)x=11x$

3. $6a-4b+2a=6a+2a-4b$
$=(6+2)a-4b$
$=8a-4b$

4. $(-50n)\left(\frac{1}{10}\right)=(-50)\left(\frac{1}{10}\right)n=-5n$

5. $(5c-4a)^2-b$
$[5(1)-4(-1)]^2-2$
$=[5+4]^2-2$
$=[9]^2-2$
$=81-2$
$=79$

6. $5(2x-7)=5(2x)+5(-7)=10x-25$

7. $-6(7x^2)=-6(7)x^2=-42x^2$

8. $-9(7+4x)=-9(7)-9(4x)=-63-36x$

9. $12y-17y=(12-17)y=-5y$

10. $2bc \div (a+7)$

$2(-5)(4) \div (3+7)$

$= 2(-20) \div 10$

$= -40 \div 10$

$= -4$

11. $7 - 2(3x+4) = 7 - 2(3x) - 2(4)$

$= 7 - 6x - 8$

$= -6x - 1$

12. $6 + 2[2 - 5(4a - 3)] = 6 + 2[2 - 20a + 15]$

$= 6 + 2[-20a + 17]$

$= 6 - 40a + 34$

$= -40a + 40$

13. $6(8y - 3) - 8(3y - 6)$

$= 6(8y) + 6(-3) - 8(3y) - 8(-6)$

$= 48y - 18 - 24y + 48$

$= 48y - 24y - 18 + 48$

$= 24y + 30$

14. $5c + (-2d) - 3d - (-4c) = 5c - 2d - 3d + 4c$

$= 5c + 4c - 2d - 3d$

$= 9c - 5d$

15. $5(4x) = 5(4)x = 20x$

16. $-4(2x - 9) + 5(3x + 2)$

$= -4(2x) - 4(-9) + 5(3x) + 5(2)$

$= -8x + 36 + 15x + 10$

$= -8x + 15x + 36 + 10$

$= 7x + 46$

$4x - 3x^2 + 2x - x^2 = -3x^2 - x^2 + 4x + 2x$

$= (-3 - 1)x^2 + (4 + 2)x$

$= -4x^2 + 6x$

17. $5[2 - 3(6x - 1)] = 5[2 - 18x + 3]$

$= 5[5 - 18x]$

$= 25 - 90x$

$= -90x + 25$

18. $0.4x + 0.6(250 - x) = 0.4x + 0.6(250) + 0.6(-x)$

$= 0.4x + 150 - 0.6x$

$= -0.2x + 150$

19. $\frac{2}{3}x - \frac{3}{4}x = \left(\frac{2}{3} - \frac{3}{4}\right)x$

$= \left(\frac{8}{12} - \frac{9}{12}\right)x$

$= -\frac{1}{12}x$

20. $(7a^2 - 2a + 3)4$

$= 7a^2(4) - 2a(4) + 3(4)$

$= 28a^2 - 8a + 12$

21. $a^2 - b^2$

$= (3)^2 - (4)^2$

$= 9 - 16 = -7$

22. $-3(-12y) = 36y$

23. two-thirds <u>of</u> the <u>total</u> of x and 10

$\frac{2}{3}(x + 10)$

24. 6 <u>less than</u> x

$x - 6$

25. the unknown number: x

twice the unknown number: $2x$

$x + 2x = 3x$

26. the unknown number: x

twice the number: $2x$

one-half the number: $\frac{1}{2}x$

$2x - \frac{1}{2}x = \left(2 - \frac{1}{2}\right)x = \left(\frac{4}{2} - \frac{1}{2}\right)x = \frac{3}{2}x$

27. the unknown number: x

three times the number: $3x$

$3x + 5(x - 1) = 3x + 5x - 5 = 8x - 5$

28. number of American League cards: A

number of National League cards: $5A$

29. number of ten-dollar bills: T

number of five-dollar bills: $35 - T$

30. number of calories in an apple: a

number of calories in a candy bar: $2a + 8$

31. width of Parthenon: w

length of Parthenon: $1.6w$

32. kneeling height: h
standing height: $1.3h$

Chapter Test

1. $3x - 5x + 7x = (3 - 5 + 7)x = 5x$

2. $-3(2x^2 - 7y^2) = -3(2x^2) - 3(-7y^2)$
$= -6x^2 + 21y^2$

3. $2x - 3(x - 2) = 2x - 3(x) - 3(-2)$
$= 2x - 3x + 6$
$= -x + 6$

4. $2x + 3[4 - (3x - 7)] = 2x + 3[4 - 3x + 7]$
$= 2x + 3[11 - 3x]$
$= 2x + 33 - 9x$
$= 2x - 9x + 33$
$= -7x + 33$

5. $3x - 7y - 12x = 3x - 12x - 7y$
$= -9x - 7y$

6. $b^2 - 3(a)(b)$
$(-2)^2 - 3(3)(-2)$
$= 4 - 3(3)(-2)$
$= 4 + 18$
$= 22$

7. $\frac{1}{5}(10x) = \frac{1}{5}(10)x = 2x$

8. $5(2x + 4) - 3(x - 6) = 5(2x) + 5(4) - 3(x) - 3(-6)$
$= 10x + 20 - 3x + 18$
$= 10x - 3x + 20 + 18$
$= 7x + 38$

9. $-5(2x^2 - 3x + 6) = -5(2x^2) - 5(-3x) - 5(6)$
$= -10x^2 + 15x - 30$

10. $3x + (-12y) - 5x - (-7y) = 3x - 12y - 5x + 7y$
$= 3x - 5x - 12y + 7y$
$= -2x - 5y$

11. $\frac{-2ab}{2b - a}$
$\frac{-2(-4)(6)}{2(6) - (-4)}$
$= \frac{8(6)}{12 + 4}$
$= \frac{48}{16}$
$= 3$

12. $(12x)\frac{1}{4} = 12\left(\frac{1}{4}\right)x = 3x$

13. $-7y^2 + 6y^2 - (-2y^2) = -7y^2 + 6y^2 + 2y^2$
$= (-7 + 6 + 2)y^2$
$= y^2$

14. $-2(2x - 4) = -2(2x) - 2(-4) = -4x + 8$

15. $\frac{2}{3}(-15a) = \frac{2}{3}(-15)a = -10a$

16. $-2[x - 2(x - y)] + 5y = -2[x - 2x + 2y] + 5y$
$= -2[-x + 2y] + 5y$
$= 2x - 4y + 5y$
$= 2x + y$

17. $(-3)(-12y) = (-3)(-12)y = 36y$

18. $5(3 - 7b) = 5(3) + 5(-7b) = 15 - 35b$

19. the <u>difference between</u> the <u>squares of</u> a and b
$a^2 - b^2$

20. the unknown number: x
$10(x - 3) = 10(x) + 10(-3)$
$= 10x - 30$

21. the unknown number: x
twice the square of the number: $2x^2$
$x + 2x^2$

22. the unknown number: x
the quotient of six and the number: $\frac{6}{x}$
$\frac{6}{x} - 3$

23. b <u>decreased by</u> the <u>product of</u> b and seven
$b-7b$

24. distance from Earth to the sun: d
distance from Neptune to the sun: $30d$

25. shorter piece: x
longer piece: $4x-3$

Cumulative Review Exercises

1. $-4+7+(-10)=-4-10+7$
$=-14+7$
$=-7$

2. $-16-(-25)-4=-16+25-4$
$=-16-4+25$
$=-20+25$
$=5$

3. $(-2)(3)(-4)=(-6)(-4)=24$

4. $(-60)\div 12=-5$

5. $5\overline{)55}$ with quotient 11
$2\overline{)110}$
$110=2\cdot 5\cdot 11$

6. $\frac{7}{12}-\frac{11}{16}-\left(-\frac{1}{3}\right)=\frac{7}{12}-\frac{11}{16}+\frac{1}{3}$
$=\frac{28}{48}-\frac{33}{48}+\frac{16}{48}$
$=\frac{28-33+16}{48}$
$=\frac{28+16-33}{48}$
$=\frac{44-33}{48}$
$=\frac{11}{48}$

7. $-\frac{5}{12}\div\frac{5}{2}=-\frac{5}{12}\cdot\frac{2}{5}=-\frac{\cancel{5}\cdot\cancel{2}}{\cancel{2}\cdot 2\cdot 3\cdot\cancel{5}}=-\frac{1}{6}$

8. $\left(\frac{-9}{16}\right)\left(\frac{8}{27}\right)\left(\frac{-3}{2}\right)=\frac{\cancel{3}\cdot\cancel{3}\cdot\cancel{2}\cdot\cancel{2}\cdot\cancel{2}\cdot\cancel{3}}{\cancel{2}\cdot\cancel{2}\cdot\cancel{2}\cdot 2\cdot\cancel{3}\cdot\cancel{3}\cdot\cancel{3}\cdot 2}=\frac{1}{4}$

9.

397	→	400
516	→	500
+ 408	→	+ 400
1321		1300

10. $-2^5\div(3-5)^2-(-3)=-2^5\div(-2)^2-(-3)$
$=-32\div 4+3$
$=-8+3$
$=-5$

11. $\left(-\frac{3}{4}\right)^2\div\left(\frac{3}{8}-\frac{11}{12}\right)=\frac{9}{16}\div\left(\frac{9}{24}-\frac{22}{24}\right)$
$=\frac{9}{16}\div\left(-\frac{13}{24}\right)=-\left(\frac{9}{16}\cdot\frac{24}{13}\right)$
$=-\frac{27}{26}$

12. a^2-3b
$(2)^2-3(-4)$
$=4-3(-4)$
$=4+12$
$=16$

13. $-2x^2-(-3x^2)+4x^2=-2x^2+3x^2+4x^2$
$=-2x^2+7x^2$
$=5x^2$

14. $5a-10b-12a=5a-12a-10b=-7a-10b$

15. 8.357

16. Perimeter = 4 · side
= 4(24 ft)
= 96 ft

17. $3(8-2x)=3(8)+3(-2x)=24-6x$

18. $-2(-3y+9)=-2(-3y)-2(9)=6y-18$

19.

32.76	→	30
−19.8	→	− 20
12.96		10

20. 8.667 — *Given place value* (6); $6>5$

8.667 rounded to the nearest tenth is 8.7.

21. $-4(2x^2-3y^2)=-4(2x^2)-4(-3y^2)$
$=-8x^2+12y^2$

22. $-3(3y^2-3y-7)=-3(3y^2)-3(-3y)-3(-7)$
$=-9y^2+9y+21$

23. $-3x-2(2x-7)=-3x-2(2x)-2(-7)$
$=-3x-4x+14$
$=-7x+14$

24. $4(3x-2)-7(x+5)=4(3x)+4(-2)-7(x)-7(5)$
$=12x-8-7x-35$
$=12x-7x-8-35$
$=5x-43$

25. $2x+3[x-2(4-2x)]=2x+3[x-8+4x]$
$=2x+3[5x-8]$
$=2x+3(5x)+3(-8)$
$=2x+15x-24$
$=17x-24$

26. $3[2x-3(x-2y)]+3y=3[2x-3x+6y]+3y$
$=3[-x+6y]+3y$
$=3(-x)+3(6y)+3y$
$=-3x+18y+3y$
$=-3x+21y$

27. the sum of one-half of b and b
$\frac{1}{2}b+b$

28. 10 divided by the difference between y and 2
$\frac{10}{y-2}$

29. the unknown number: x
$8-\frac{x}{12}$

30. the unknown number: x
$x+(x+2)=x+x+2=2x+2$

31. Area = side · side
= 60 ft · 60 ft
= 3600 ft^2

32. speed of the wildebeest: w
speed of the falcon: $4w$

Chapter 5: Solving Equations

Prep Test

1. $-\frac{3}{4}\left(-\frac{4}{3}\right)=\frac{3\cdot 4}{4\cdot 3}=1$

2. $\frac{-3}{-3}=1$

3. $-16+7y+16=7y+16-16=7y$

4. $8x-9-8x=8x-8x-9=-9$

5. $\frac{1}{2}x+\frac{2}{3}x=\left(\frac{1}{2}+\frac{2}{3}\right)x$
$$=\left(\frac{3}{6}+\frac{4}{6}\right)x$$
$$=\frac{7}{6}x$$

6. $6x-3(6-x)=6x-3(6)-3(-x)$
$$=6x-18+3x$$
$$=9x-18$$

7. $2x+3$
$2(-4)+3=-8+3=-5$

8. the unknown number: n
$5-2n$

9. speed of old card: s
speed of new card: $5s$

10. length of longer piece: x
length of shorter piece: $5-x$

Section 5.1

Concept Check

1. **a.** equation
b. expression
c. expression
d. equation
e. expression

3. i, ii, and iv are equations in the form $x+a=b$.
You would subtract a from both sides.

5. Keith

Objective A Exercises

7. $2x = 8$

$2(4)$	8

$8=8$

Yes, 4 is a solution.

9. $2b-1=3$

$2(-1)-1$	3
$-2-1$	3

$-3\neq 3$

No, −1 is not a solution.

11. $4-2m = 3$

$4-2(1)$	3
$4-2$	3

$2\neq 3$

No, 1 is not a solution.

13. $2x+5 = 3x$

$2(5)+5$	$3(5)$
$10+5$	15

$15=15$

Yes, 5 is a solution.

15. $3a+2 = 2-a$

$3(-2)+2$	$2-(-2)$
$-6+2$	$2+2$

$-4\neq 4$

No, −2 is not a solution.

17. $2x^2-1 = 4x-1$

$2(2)^2-1$	$4(2)-1$
$2(4)-1$	$8-1$
$8-1$	7

$7=7$

Yes, 2 is a solution.

19. $4y+1 = 3$

$4(1/2)+1$	3
$2+1$	3

$3=3$

Yes, $\frac{1}{2}$ is a solution.

21. $8x-1 = 12x+3$

$8(3/4)-1$	$12(3/4)+3$
$6-1$	$9+3$

$5\neq 12$

No, $\frac{3}{4}$ is not a solution.

Objective B Exercises

23. x will be greater than $\frac{19}{24}$ because you will add $\frac{11}{16}$ to each side of the equation to solve the equation.

25.
$$\begin{aligned} x+5&=7 \\ x+5-5&=7-5 \\ x&=2 \end{aligned}$$
The solution is 2.

27.
$$\begin{aligned} b-4&=11 \\ b-4+4&=11+4 \\ b&=15 \end{aligned}$$
The solution is 15.

29.
$$\begin{aligned} 2+a&=8 \\ 2-2+a&=8-2 \\ a&=6 \end{aligned}$$
The solution is 6.

31.
$$\begin{aligned} n-5&=-2 \\ n-5+5&=-2+5 \\ n&=3 \end{aligned}$$
The solution is 3.

33.
$$\begin{aligned} b+7&=7 \\ b+7-7&=7-7 \\ b&=0 \end{aligned}$$
The solution is 0.

35.
$$\begin{aligned} z+9&=2 \\ z+9-9&=2-9 \\ z&=-7 \end{aligned}$$
The solution is −7.

37.
$$\begin{aligned} 10+m&=3 \\ 10-10+m&=3-10 \\ m&=-7 \end{aligned}$$
The solution is −7.

39.
$$\begin{aligned} 9+x&=-3 \\ 9-9+x&=-3-9 \\ x&=-12 \end{aligned}$$
The solution is −12.

41.
$$\begin{aligned} 2&=x+7 \\ 2-7&=x+7-7 \\ -5&=x \end{aligned}$$
The solution is −5.

43.
$$\begin{aligned} 4&=m-11 \\ 4+11&=m-11+11 \\ 15&=m \end{aligned}$$
The solution is 15.

45.
$$\begin{aligned} 12&=3+w \\ 12-3&=3-3+w \\ 9&=w \end{aligned}$$
The solution is 9.

47.
$$\begin{aligned} 4&=-10+b \\ 4+10&=-10+10+b \\ 14&=b \end{aligned}$$
The solution is 14.

49.
$$\begin{aligned} m+\frac{2}{3}&=-\frac{1}{3} \\ m+\frac{2}{3}-\frac{2}{3}&=-\frac{1}{3}-\frac{2}{3} \\ m&=-1 \end{aligned}$$
The solution is −1.

51.
$$\begin{aligned} x-\frac{1}{2}&=\frac{1}{2} \\ x-\frac{1}{2}+\frac{1}{2}&=\frac{1}{2}+\frac{1}{2} \\ x&=1 \end{aligned}$$
The solution is 1.

53.
$$\begin{aligned} \frac{5}{8}+y&=\frac{1}{8} \\ \frac{5}{8}-\frac{5}{8}+y&=\frac{1}{8}-\frac{5}{8} \\ y&=-\frac{4}{8} \\ y&=-\frac{1}{2} \end{aligned}$$
The solution is $-\frac{1}{2}$.

55.
$$\begin{aligned} -\frac{5}{6}&=x-\frac{1}{4} \\ -\frac{5}{6}+\frac{1}{4}&=x-\frac{1}{4}+\frac{1}{4} \\ -\frac{10}{12}+\frac{3}{12}&=x \\ -\frac{7}{12}&=x \end{aligned}$$
The solution is $-\frac{7}{12}$.

57.
$$\begin{aligned} d+1.3619&=2.0148 \\ d+1.3619-1.3619&=2.0148-1.3619 \\ d&=0.6529 \end{aligned}$$
The solution is 0.6529.

59.
$$6.149 = -3.108 + z$$
$$6.149 + 3.108 = -3.108 + 3.108 + z$$
$$9.257 = z$$
The solution is 9.257.

Objective C Exercises

61.
$$5x = -15$$
$$\frac{5x}{5} = \frac{-15}{5}$$
$$x = -3$$
The solution is –3.

63.
$$3b = 0$$
$$\frac{3b}{3} = \frac{0}{3}$$
$$b = 0$$
The solution is 0.

65.
$$-3x = 6$$
$$\frac{-3x}{-3} = \frac{6}{-3}$$
$$x = -2$$
The solution is −2.

67.
$$-\frac{1}{6}n = -30$$
$$-6\left(-\frac{1}{6}n\right) = -6(-30)$$
$$n = 180$$
The solution is 180.

69.
$$0 = -5x$$
$$\frac{0}{-5} = \frac{-5x}{-5}$$
$$0 = x$$
The solution is 0.

71.
$$\frac{x}{3} = 2$$
$$3\left(\frac{1}{3}x\right) = 3(2)$$
$$x = 6$$
The solution is 6.

73.
$$-\frac{y}{2} = 5$$
$$-2\left(-\frac{1}{2}y\right) = -2(5)$$
$$y = -10$$
The solution is −10.

75.
$$\frac{3}{4}y = 9$$
$$\frac{4}{3}\left(\frac{3}{4}y\right) = \frac{4}{3}(9)$$
$$y = 12$$
The solution is 12.

77.
$$-\frac{2}{3}d = 8$$
$$-\frac{3}{2}\left(-\frac{2}{3}d\right) = -\frac{3}{2}(8)$$
$$d = -12$$
The solution is −12.

79.
$$\frac{2n}{3} = 0$$
$$\frac{3}{2}\left(\frac{2}{3}n\right) = \frac{3}{2}(0)$$
$$n = 0$$
The solution is 0.

81.
$$\frac{-3z}{8} = 9$$
$$-\frac{8}{3}\left(-\frac{3}{8}z\right) = -\frac{8}{3}(9)$$
$$z = -24$$
The solution is −24.

83.
$$\frac{2}{9} = \frac{2}{3}y$$
$$\frac{3}{2}\left(\frac{2}{9}\right) = \frac{3}{2}\left(\frac{2}{3}y\right)$$
$$\frac{1}{3} = y$$
The solution is $\frac{1}{3}$.

85.
$$\frac{x}{1.46} = 3.25$$
$$1.46\left(\frac{1}{1.46}x\right) = 1.46(3.25)$$
$$x = 4.745$$
The solution is 4.745.

87.
$$3.47a = 7.1482$$
$$\frac{3.47a}{3.47} = \frac{7.1482}{3.47}$$
$$a = 2.06$$
The solution is 2.06.

89. $2m+5m=49$

$7m=49$

$\frac{7m}{7}=\frac{49}{7}$

$m=7$

The solution is 7.

91. $3n+2n=20$

$5n=20$

$\frac{5n}{5}=\frac{20}{5}$

$n=4$

The solution is 4.

93. $10y-3y=21$

$7y=21$

$\frac{7y}{7}=\frac{21}{7}$

$y=3$

The solution is 3.

95. Positive

97. Negative

99. (a) equal to
(b) less than

101. Strategy:
To find the distance, solve $d=rt$ for r using $d = 20$ mi and $t=\frac{40}{60}=\frac{2}{3}$ h.

Solution:

$d=rt$

$20=r\frac{2}{3}$

$\frac{3}{2}\cdot 20=r\frac{2}{3}\cdot\frac{3}{2}$

$30=r$

The dietician traveled 30 mph.

103. Strategy:
To find the number of miles traveled, solve $d=rt$ for d using $d = 27$ mi and $t=\frac{45}{60}=\frac{3}{4}$ h.

Solution:

$d=rt$

$27=r\left(\frac{3}{4}\right)$

$\frac{4}{3}(27)=r\left(\frac{3}{4}\right)\cdot\frac{4}{3}$

$36=r$

Marcella's average rate of speed is 36 mph.

105. Strategy:
To find the number of hours to walk the course:

→Find the rate to run the course by solving $d=rt$ for r using $d = 30$ km and $t=2$ h.

→Decrease the rate by 3 km/h to find his walking rate.

→Solve for $d=rt$ for t using $d = 30$ km and r equal to his walking rate.

Solution:

$d=rt$

$30=r(2)$

$\frac{30}{2}=r$

$15=r$ His running rate

$15-3=12$ His walking rate

$d=rt$

$30=12t$

$\frac{30}{12}=t$

$2.5=t$

It would take Palmer 2.5 h to walk the course.

107. Strategy:
The distance is 8 mi. Therefore $d = 8$.
The joggers are running toward each other, one at 5 mph and one at 7 mph. The rate is the sum of the two rates, or 12 mph. So, $r = 12$. To find the time solve $d=rt$ for t. Convert the answer to minutes.

Solution:

$d=rt$

$8=12t$

$\frac{8}{12}=t$

$\frac{2}{3}=t$

$\frac{2}{3}\text{ h}=\frac{2}{3}\cdot 60\text{ min}=40\text{ min}$

The two joggers will meet 40 min after they start.

109. Strategy:
The two cyclists are traveling in opposite directions, one at 8 mph and one at 9 mph. The rate is the sum of the two rates, or 17 mph.
So, $r = 17$. The time traveled is $30 \text{ min} = \frac{1}{2}$ h.
So, $t = \frac{1}{2}$. To find the distance, solve $d = rt$ for d.

Solution:
$$d = rt$$
$$d = 17 \cdot \frac{1}{2}$$
$$d = 8.5$$
The two cyclists are 8.5 mi apart.

111. Strategy:
To find the number of miles apart:
→Find the distance the first train travels by solving $d = rt$ for d using $r = 45$ and $t = 2$.
→Find the distance the second train travels by solving $d = rt$ for d using $r = 60$ and $t = 1$.
→Find the difference between these distances.

Solution:

First train:	Second train:
$d = rt$	$d = rt$
$d = 45(2)$	$d = 60(1)$
$d = 90$	$d = 60$

$$90 - 60 = 30$$

The two trains are 30 mi apart.

Critical Thinking 5.1

113.
$$\frac{2m + m}{5} = -9$$
$$\frac{3m}{5} = -9$$
$$\frac{5}{3} \cdot \frac{3m}{5} = -9 \cdot \frac{5}{3}$$
$$m = -15$$
The solution is –15.

115.
$$\frac{1}{\frac{1}{x}} = 5$$
$$\frac{1}{x} \cdot \frac{1}{\frac{1}{x}} = \frac{1}{x} \cdot 5$$
$$1 = \frac{5}{x}$$
$$x \cdot 1 = x \cdot \frac{5}{x}$$
$$x = 5$$
The solution is 5.

117.
$$\frac{4}{\frac{3}{b}} = 8$$
$$\frac{3}{b} \cdot \frac{4}{\frac{3}{b}} = \frac{3}{b} \cdot 8$$
$$4 = \frac{24}{b}$$
$$b \cdot 4 = b \cdot \frac{24}{b}$$
$$4b = 24$$
$$\frac{4b}{4} = \frac{24}{4}$$
$$b = 6$$
The solution is 6.

Projects and Group Activities 5.1

119. Answers will vary. One possible answer is $x + 7 = 9$.

121.
$$\frac{3}{7} + \frac{1}{b} = 2$$
$$7b \cdot \frac{3}{7} + 7b \cdot \frac{1}{b} = 7b \cdot 2$$
$$3b + 7 = 14b$$
$$3b - 3b + 7 = 14b - 3b$$
$$7 = 11b$$
$$\frac{7}{11} = \frac{11b}{11}$$
$$\frac{7}{11} = b$$

The solution is $\frac{7}{11}$.

Section 5.2

Concept Check

1. a and i, b and iii, c and ii, d and iv

3. 5; 8

Objective A Exercises

5.
$$\begin{aligned} 3x+1&=10\\ 3x+1-1&=10-1\\ 3x&=9\\ \frac{3x}{3}&=\frac{9}{3}\\ x&=3 \end{aligned}$$
The solution is 3.

7.
$$\begin{aligned} 2a-5&=7\\ 2a-5+5&=7+5\\ 2a&=12\\ \frac{2a}{2}&=\frac{12}{2}\\ a&=6 \end{aligned}$$
The solution is 6.

9.
$$\begin{aligned} 5&=4x+9\\ 5-9&=4x+9-9\\ -4&=4x\\ \frac{-4}{4}&=\frac{4x}{4}\\ -1&=x \end{aligned}$$
The solution is −1.

11.
$$\begin{aligned} 2x-5&=-11\\ 2x-5+5&=-11+5\\ 2x&=-6\\ \frac{2x}{2}&=\frac{-6}{2}\\ x&=-3 \end{aligned}$$
The solution is −3.

13.
$$\begin{aligned} 4-3w&=-2\\ 4-4-3w&=-2-4\\ -3w&=-6\\ \frac{-3w}{-3}&=\frac{-6}{-3}\\ w&=2 \end{aligned}$$
The solution is 2.

15.
$$\begin{aligned} 8-3t&=2\\ 8-8-3t&=2-8\\ -3t&=-6\\ \frac{-3t}{-3}&=\frac{-6}{-3}\\ t&=2 \end{aligned}$$
The solution is 2.

17.
$$\begin{aligned} 4a-20&=0\\ 4a-20+20&=0+20\\ 4a&=20\\ \frac{4a}{4}&=\frac{20}{4}\\ a&=5 \end{aligned}$$
The solution is 5.

19.
$$\begin{aligned} 6+2b&=0\\ 6-6+2b&=0-6\\ 2b&=-6\\ \frac{2b}{2}&=\frac{-6}{2}\\ b&=-3 \end{aligned}$$
The solution is −3.

21.
$$\begin{aligned} -2x+5&=-7\\ -2x+5-5&=-7-5\\ -2x&=-12\\ \frac{-2x}{-2}&=\frac{-12}{-2}\\ x&=6 \end{aligned}$$
The solution is 6.

23.
$$\begin{aligned} -1.2x+3&=-0.6\\ -1.2x+3-3&=-0.6-3\\ -1.2x&=-3.6\\ \frac{-1.2x}{-1.2}&=\frac{-3.6}{-1.2}\\ x&=3 \end{aligned}$$
The solution is 3.

25.
$$\begin{aligned} 2&=7-5a\\ 2-7&=7-7-5a\\ -5&=-5a\\ \frac{-5}{-5}&=\frac{-5a}{-5}\\ 1&=a \end{aligned}$$
The solution is 1.

27.
$$\begin{aligned} -35 &= -6b+1 \\ -35-1 &= -6b+1-1 \\ -36 &= -6b \\ \frac{-36}{-6} &= \frac{-6b}{-6} \\ 6 &= b \end{aligned}$$
The solution is 6.

29.
$$\begin{aligned} -3m-21 &= 0 \\ -3m-21+21 &= 0+21 \\ -3m &= 21 \\ \frac{-3m}{-3} &= \frac{21}{-3} \\ m &= -7 \end{aligned}$$
The solution is −7.

31.
$$\begin{aligned} -4y+15 &= 15 \\ -4y+15-15 &= 15-15 \\ -4y &= 0 \\ \frac{-4y}{-4} &= \frac{0}{-4} \\ y &= 0 \end{aligned}$$
The solution is 0.

33.
$$\begin{aligned} 9-4x &= 6 \\ 9-9-4x &= 6-9 \\ -4x &= -3 \\ \frac{-4x}{-4} &= \frac{-3}{-4} \\ x &= \frac{3}{4} \end{aligned}$$
The solution is $\frac{3}{4}$.

35.
$$\begin{aligned} 9x-4 &= 0 \\ 9x-4+4 &= 0+4 \\ 9x &= 4 \\ \frac{9x}{9} &= \frac{4}{9} \\ x &= \frac{4}{9} \end{aligned}$$
The solution is $\frac{4}{9}$.

37.
$$\begin{aligned} 1-3x &= 0 \\ 1-1-3x &= 0-1 \\ -3x &= -1 \\ \frac{-3x}{-3} &= \frac{-1}{-3} \\ x &= \frac{1}{3} \end{aligned}$$
The solution is $\frac{1}{3}$.

39.
$$\begin{aligned} 12w+11 &= 5 \\ 12w+11-11 &= 5-11 \\ 12w &= -6 \\ \frac{12w}{12} &= \frac{-6}{12} \\ w &= -\frac{6}{12} \\ w &= -\frac{1}{2} \end{aligned}$$
The solution is $-\frac{1}{2}$.

41.
$$\begin{aligned} 8b-3 &= -9 \\ 8b-3+3 &= -9+3 \\ 8b &= -6 \\ \frac{8b}{8} &= \frac{-6}{8} \\ b &= -\frac{6}{8} \\ b &= -\frac{3}{4} \end{aligned}$$
The solution is $-\frac{3}{4}$.

43.
$$\begin{aligned} 7-9a &= 4 \\ 7-7-9a &= 4-7 \\ -9a &= -3 \\ \frac{-9a}{-9} &= \frac{-3}{-9} \\ a &= \frac{3}{9} \\ a &= \frac{1}{3} \end{aligned}$$
The solution is $\frac{1}{3}$.

45.
$$\begin{aligned} 10 &= -18x+7 \\ 10-7 &= -18x+7-7 \\ 3 &= -18x \\ \frac{3}{-18} &= \frac{-18x}{-18} \\ -\frac{3}{18} &= x \\ -\frac{1}{6} &= x \end{aligned}$$
The solution is $-\frac{1}{6}$.

47.

$$9x+\frac{4}{5}=\frac{4}{5}$$
$$9x+\frac{4}{5}-\frac{4}{5}=\frac{4}{5}-\frac{4}{5}$$
$$9x=0$$
$$\frac{9x}{9}=\frac{0}{9}$$
$$x=0$$

The solution is 0.

49.

$$0.9=10x-0.6$$
$$0.9+0.6=10x-0.6+0.6$$
$$1.5=10x$$
$$\frac{1.5}{10}=\frac{10x}{10}$$
$$0.15=x$$

The solution is 0.15.

51.

$$-4x+3=9$$
$$-4x+3-3=9-3$$
$$-4x=6$$
$$\frac{-4x}{-4}=\frac{6}{-4}$$
$$x=-\frac{6}{4}$$
$$x=-\frac{3}{2}$$

The solution is $-\frac{3}{2}$.

53.

$$\frac{1}{3}m-1=5$$
$$\frac{1}{3}m-1+1=5+1$$
$$\frac{1}{3}m=6$$
$$3\left(\frac{1}{3}m\right)=3\cdot 6$$
$$m=18$$

The solution is 18.

55.

$$\frac{3}{4}n+7=13$$
$$\frac{3}{4}n+7-7=13-7$$
$$\frac{3}{4}n=6$$
$$\frac{4}{3}\left(\frac{3}{4}n\right)=\frac{4}{3}(6)$$
$$n=8$$

The solution is 8.

57.

$$-\frac{3}{8}b+4=10$$
$$-\frac{3}{8}b+4-4=10-4$$
$$-\frac{3}{8}b=6$$
$$-\frac{8}{3}\left(-\frac{3}{8}b\right)=-\frac{8}{3}(6)$$
$$b=-16$$

The solution is −16.

59.

$$\frac{y}{5}-2=3$$
$$\frac{y}{5}-2+2=3+2$$
$$\frac{y}{5}=5$$
$$5\left(\frac{1}{5}y\right)=5\cdot 5$$
$$y=25$$

The solution is 25.

61.

$$\frac{2}{3}x-\frac{5}{6}=-\frac{1}{3}$$
$$6\left(\frac{2}{3}x-\frac{5}{6}\right)=6\left(-\frac{1}{3}\right)$$
$$4x-5=-2$$
$$4x=3$$
$$x=\frac{3}{4}$$

The solution is $\frac{3}{4}$.

63.

$$\frac{1}{2}-\frac{2}{3}x=\frac{1}{4}$$
$$12\left(\frac{1}{2}-\frac{2}{3}x\right)=12\left(\frac{1}{4}\right)$$
$$6-8x=3$$
$$-8x=-3$$
$$x=\frac{3}{8}$$

The solution is $\frac{3}{8}$.

65.

$$\frac{3}{2}=\frac{5}{6}+\frac{3x}{8}$$
$$24\left(\frac{3}{2}\right)=24\left(\frac{5}{6}\right)+24\left(\frac{3x}{8}\right)$$
$$36=20+9x$$
$$16=9x$$
$$\frac{16}{9}=x$$

The solution is $\frac{16}{9}$.

67.
$$\frac{11}{27}=\frac{4}{9}-\frac{2x}{3}$$
$$27\left(\frac{11}{27}\right)=27\left(\frac{4}{9}\right)-27\left(\frac{2x}{3}\right)$$
$$11=12-18x$$
$$-1=-18x$$
$$\frac{1}{18}=x$$
The solution is $\frac{1}{18}$.

69.
$$7=\frac{2x}{5}+4$$
$$7-4=\frac{2x}{5}+4-4$$
$$3=\frac{2x}{5}$$
$$\frac{5}{2}(3)=\frac{5}{2}\left(\frac{2}{5}x\right)$$
$$\frac{15}{2}=x$$
The solution is $\frac{15}{2}$.

71.
$$7-\frac{5}{9}y=9$$
$$7-7-\frac{5}{9}y=9-7$$
$$-\frac{5}{9}y=2$$
$$-\frac{9}{5}\left(-\frac{5}{9}y\right)=-\frac{9}{5}(2)$$
$$y=-\frac{18}{5}$$
The solution is $-\frac{18}{5}$.

73.
$$5y+9+2y=23$$
$$7y+9=23$$
$$7y+9-9=23-9$$
$$7y=14$$
$$\frac{7y}{7}=\frac{14}{7}$$
$$y=2$$
The solution is 2.

75.
$$11z-3-7z=9$$
$$4z-3=9$$
$$4z-3+3=9+3$$
$$4z=12$$
$$\frac{4z}{4}=\frac{12}{4}$$
$$z=3$$
The solution is 3.

77. Negative

79. Negative

81. Strategy:

Given: $P=557\frac{1}{4}$

$c=425$

$h=118$

Unknown: s

Solution:
$$P=c+h+\frac{1}{4}s$$
$$557\frac{1}{4}=425+118+\frac{1}{4}s$$
$$557\frac{1}{4}=543+\frac{1}{4}s$$
$$557\frac{1}{4}-543=543-543+\frac{1}{4}s$$
$$14\frac{1}{4}=\frac{1}{4}s$$
$$14\frac{1}{4}\cdot 4=4\cdot\frac{1}{4}s$$
$$\frac{57}{4}\cdot 4=s$$
$$57=s$$

The crown spread is 57 ft.

83. Strategy:
Given: $C = 174$
$f = 2$
$c = 30$
Unknown: p

Solution:

$$\begin{aligned} C &= 9f + 4p + 4c \\ 174 &= 9(2) + 4p + 4(30) \\ 174 &= 18 + 4p + 120 \\ 174 &= 138 + 4p \\ 174 - 138 &= 138 - 138 + 4p \\ 36 &= 4p \\ \frac{36}{4} &= \frac{4p}{4} \\ 9 &= p \end{aligned}$$

The yogurt contains 9 g of protein.

85. Strategy:
Given: $t = 2$
$s = 80$
Unknown: v

Solution:

$$\begin{aligned} s &= 16t^2 + vt \\ 80 &= 16(2)^2 + v(2) \\ 80 &= 16(4) + 2v \\ 80 &= 64 + 2v \\ 80 - 64 &= 64 - 64 + 2v \\ 16 &= 2v \\ \frac{16}{2} &= \frac{2v}{2} \\ 8 &= v \end{aligned}$$

The initial velocity is 8 ft/s.

87. Strategy:
Given: $V = 38{,}000$
$C = 50{,}000$
Unknown: t

Solution:

$$\begin{aligned} V &= C - 6000t \\ 38{,}000 &= 50{,}000 - 6000t \\ 38{,}000 - 50{,}000 &= 50{,}000 - 50{,}000 - 6000t \\ -12{,}000 &= -6000t \\ \frac{-12{,}000}{-6000} &= \frac{-6000t}{-6000} \\ 2 &= t \end{aligned}$$

The depreciated value will be \$38,000 after 2 years.

89. Strategy:
Given: $H = 66$
Unknown: L

Solution:

$$\begin{aligned} H &= 1.2L + 27.8 \\ 66 &= 1.2L + 27.8 \\ 66 - 27.8 &= 1.2L + 27.8 - 27.8 \\ 38.2 &= 1.2L \\ \frac{38.2}{1.2} &= \frac{1.2L}{1.2} \\ 31.8 &\approx L \end{aligned}$$

The approximate length is 31.8 in.

91. Strategy:
Given: $C = -3$
Unknown: D

Solution:

$$\begin{aligned} C &= \frac{1}{4}D - 45 \\ -3 &= \frac{1}{4}D - 45 \\ -3 + 45 &= \frac{1}{4}D - 45 + 45 \\ 42 &= \frac{1}{4}D \\ 42 \cdot 4 &= 4 \cdot \frac{1}{4}D \\ 168 &= D \end{aligned}$$

The distance the car will skid is 168 ft.

93. ii

Critical Thinking 5.2

95. $2x - 3y = 8$
$2x - 4(0) = 8$
$2x - 0 = 8$
$2x = 8$
$\frac{2x}{2} = \frac{8}{2}$
$x = 4$

97. $5x - 2y = -3$
$5(-3) - 2y = -3$
$-15 - 2y = -3$
$-2y = 12$
$\frac{-2y}{-2} = \frac{12}{-2}$
$y = -6$

99. $3x + 5 = -4$
$3x = -9$
$\frac{3x}{3} = \frac{-9}{3}$
$x = -3$
Substituting –3 for x:
$2x - 5$
$= 2(-3) - 5$
$= -6 - 5$
$= -11$

Projects or Group Activities 5.2

101. Answers will vary. For example,
2x + 5 = –1

Section 5.3

Concept Check

1. True

3. True

5. $8 - y = y + 6$
$8 - y - y = y - y - 6$
$8 - 2y = -6$
The coefficient on the variable is –2.

Objective A Exercises

7. $6y + 2 = y + 17$
$6y - y + 2 = y - y + 17$
$5y + 2 = 17$
$5y + 2 - 2 = 17 - 2$
$5y = 15$
$\frac{5y}{5} = \frac{15}{5}$
$y = 3$
The solution is 3.

9. $13b - 1 = 4b - 19$
$13b - 4b - 1 = 4b - 4b - 19$
$9b - 1 = -19$
$9b - 1 + 1 = -19 + 1$
$9b = -18$
$\frac{9b}{9} = \frac{-18}{9}$
$b = -2$
The solution is –2.

11. $7a - 5 = 2a - 20$
$7a - 2a - 5 = 2a - 2a - 20$
$5a - 5 = -20$
$5a - 5 + 5 = -20 + 5$
$5a = -15$
$\frac{5a}{5} = \frac{-15}{5}$
$a = -3$
The solution is –3.

13. $n - 2 = 6 - 3n$
$n + 3n - 2 = 6 - 3n + 3n$
$4n - 2 = 6$
$4n - 2 + 2 = 6 + 2$
$4n = 8$
$\frac{4n}{4} = \frac{8}{4}$
$n = 2$
The solution is 2.

15. $4y - 2 = -16 - 3y$
$4y + 3y - 2 = -16 - 3y + 3y$
$7y - 2 = -16$
$7y - 2 + 2 = -16 + 2$
$7y = -14$
$\frac{7y}{7} = \frac{-14}{7}$
$y = -2$
The solution is –2.

17.
$$\begin{aligned} m+0.4&=3m+0.8\\ m-3m+0.4&=3m-3m+0.8\\ -2m+0.4&=0.8\\ -2m+0.4-0.4&=0.8-0.4\\ -2m&=0.4\\ \frac{-2m}{-2}&=\frac{0.4}{-2}\\ m&=-0.2 \end{aligned}$$
The solution is –0.2.

19.
$$\begin{aligned} 5a+7&=2a+7\\ 5a-2a+7&=2a-2a+7\\ 3a+7&=7\\ 3a+7-7&=7-7\\ 3a&=0\\ \frac{3a}{3}&=\frac{0}{3}\\ a&=0 \end{aligned}$$
The solution is 0.

21.
$$\begin{aligned} 10-4n&=16-n\\ 10-4n+n&=16-n+n\\ 10-3n&=16\\ 10-10-3n&=16-10\\ -3n&=6\\ \frac{-3n}{-3}&=\frac{6}{-3}\\ n&=-2 \end{aligned}$$
The solution is –2.

23.
$$\begin{aligned} 3-2y&=15+4y\\ 3-2y-4y&=15+4y-4y\\ 3-6y&=15\\ 3-3-6y&=15-3\\ -6y&=12\\ \frac{-6y}{-6}&=\frac{12}{-6}\\ y&=-2 \end{aligned}$$
The solution is –2.

25.
$$\begin{aligned} 2b-10&=7b\\ 2b-2b-10&=7b-2b\\ -10&=5b\\ \frac{-10}{5}&=\frac{5b}{5}\\ -2&=b \end{aligned}$$
The solution is –2.

27.
$$\begin{aligned} 9y&=5y+16\\ 9y-5y&=5y-5y+16\\ 4y&=16\\ \frac{4y}{4}&=\frac{16}{4}\\ y&=4 \end{aligned}$$
The solution is 4.

29.
$$\begin{aligned} 6y-1&=2y+2\\ 6y-2y-1&=2y-2y+2\\ 4y-1&=2\\ 4y-1+1&=2+1\\ 4y&=3\\ \frac{4y}{4}&=\frac{3}{4}\\ y&=\frac{3}{4} \end{aligned}$$
The solution is $\frac{3}{4}$.

31.
$$\begin{aligned} 2y-7&=-1-2y\\ 2y+2y-7&=-1-2y+2y\\ 4y-7&=-1\\ 4y-7+7&=-1+7\\ 4y&=6\\ \frac{4y}{4}&=\frac{6}{4}\\ y&=\frac{3}{2} \end{aligned}$$
The solution is $\frac{3}{2}$.

33.
$$\begin{aligned} 5x&=3x-8\\ 5x-3x&=3x-3x-8\\ 2x&=-8\\ \frac{2x}{2}&=\frac{-8}{2}\\ x&=-4 \end{aligned}$$
$$\begin{aligned} &4x+2\\ &=4(-4)+2\\ &=-16+2\\ &=-14 \end{aligned}$$
The answer is –14.

35.
$$\begin{aligned} 2-6a&=5-3a\\ 2-6a+3a&=5-3a+3a\\ 2-3a&=5\\ 2-2-3a&=5-2\\ -3a&=3\\ \frac{-3a}{-3}&=\frac{3}{-3}\\ a&=-1 \end{aligned}$$
$$\begin{aligned} &4a^2-2a+1\\ &=4(-1)^2-2(-1)+1\\ &=4(1)-2(-1)+1\\ &=4+2+1\\ &=6+1\\ &=7 \end{aligned}$$
The answer is 7.

Objective B Exercises

37. (ii)

39.
$$\begin{aligned} 6y+2(2y+3)&=16\\ 6y+4y+6&=16\\ 10y+6&=16\\ 10y+6-6&=16-6\\ 10y&=10\\ \frac{10y}{10}&=\frac{10}{10}\\ y&=1 \end{aligned}$$
The solution is 1.

41.
$$\begin{aligned} 12x-2(4x-6)&=28\\ 12x-8x+12&=28\\ 4x+12&=28\\ 4x+12-12&=28-12\\ 4x&=16\\ \frac{4x}{4}&=\frac{16}{4}\\ x&=4 \end{aligned}$$
The solution is 4.

43.
$$\begin{aligned} 9m-4(2m-3)&=11\\ 9m-8m+12&=11\\ m+12&=11\\ m+12-12&=11-12\\ m&=-1 \end{aligned}$$
The solution is –1.

45.
$$\begin{aligned} 4(1-3x)+7x&=9\\ 4-12x+7x&=9\\ 4-5x&=9\\ 4-4-5x&=9-4\\ -5x&=5\\ \frac{-5x}{-5}&=\frac{5}{-5}\\ x&=-1 \end{aligned}$$
The solution is –1.

47.
$$\begin{aligned} 0.22(x+6)&=0.2x+1.8\\ 0.22x+1.32&=0.2x+1.8\\ 0.22x-0.2x+1.32&=0.2x-0.2x+1.8\\ 0.02x+1.32&=1.8\\ 0.02x+1.32-1.32&=1.8-1.32\\ 0.02x&=0.48\\ \frac{0.02x}{0.02}&=\frac{0.48}{0.02}\\ x&=24 \end{aligned}$$
The solution is 24.

49.
$$\begin{aligned} 0.3x+0.3(x+10)&=300\\ 0.3x+0.3x+3&=300\\ 0.6x+3&=300\\ 0.6x+3-3&=300-3\\ 0.6x&=297\\ \frac{0.6x}{0.6}&=\frac{297}{0.6}\\ x&=495 \end{aligned}$$
The solution is 495.

51.
$$\begin{aligned} 5-(9-6x)&=2x-2\\ 5-9+6x&=2x-2\\ -4+6x&=2x-2\\ -4+6x-2x&=2x-2x-2\\ -4+4x&=-2\\ -4+4+4x&=-2+4\\ 4x&=2\\ \frac{4x}{4}&=\frac{2}{4}\\ x&=\frac{1}{2} \end{aligned}$$
The solution is $\frac{1}{2}$.

53.
$$\begin{aligned} 3[2-4(y-1)]&=3(2y+8)\\ 3[2-4y+4]&=6y+24\\ 3[6-4y]&=6y+24\\ 18-12y&=6y+24\\ 18-12y-6y&=6y-6y+24\\ 18-18y&=24\\ 18-18-18y&=24-18\\ -18y&=6\\ \frac{-18y}{-18}&=\frac{6}{-18}\\ y&=-\frac{1}{3} \end{aligned}$$
The solution is $-\frac{1}{3}$.

55.
$$3a+2[2+3(a-1)]=2(3a+4)$$
$$3a+2[2+3a-3]=6a+8$$
$$3a+2[-1+3a]=6a+8$$
$$3a-2+6a=6a+8$$
$$9a-2=6a+8$$
$$9a-6a-2=6a-6a+8$$
$$3a-2=8$$
$$3a-2+2=8+2$$
$$3a=10$$
$$\frac{3a}{3}=\frac{10}{3}$$
$$a=\frac{10}{3}$$
The solution is $\frac{10}{3}$.

57.
$$-2[4-(3b+2)]=5-2(3b+6)$$
$$-2[4-3b-2]=5-6b-12$$
$$-2[2-3b]=-7-6b$$
$$-4+6b=-7-6b$$
$$-4+6b+6b=-7-6b+6b$$
$$-4+12b=-7$$
$$-4+4+12b=-7+4$$
$$12b=-3$$
$$\frac{12b}{12}=\frac{-3}{12}$$
$$b=-\frac{1}{4}$$
The solution is $-\frac{1}{4}$.

59.
$$4-3a=7-2(2a+5)$$
$$4-3a=7-4a-10$$
$$4-3a=-3-4a$$
$$4-3a+4a=-3-4a+4a$$
$$4+a=-3$$
$$4-4+a=-3-4$$
$$a=-7$$
The answer is 0.

$$a^2+7a$$
$$=(-7)^2+(7)(-7)$$
$$=49+(7)(-7)$$
$$=49-49$$
$$=0$$

Objective C Exercises

61. Strategy:
$F=14$
Unknown: m

Solution:
$$F=2.5+2.3(m-1)$$
$$14=2.5+2.3m-2.3$$
$$14=0.2+2.3m$$
$$14-0.2=0.2-0.2+2.3m$$
$$13.98=2.3m$$
$$\frac{13.98}{2.3}=\frac{2.3m}{2.3}$$
$$6.08\approx m$$
The customer drove 6 mi.

63. **a.** $8-3=5$ ft

b. The person who is 3 ft away is heavier.

c. No

65. Solution:

To find the location of the fulcrum when the system balances, replace the variables F_1, F_2, and d in the lever system equation by the given values and solve for x.

Solution:
$$F_1x=F_2(d-x)$$
$$70x=175(14-x)$$
$$70x=2450-175x$$
$$70x+175x=2450-175x+175x$$
$$245x=2450$$
$$\frac{245x}{245}=\frac{2450}{245}$$
$$x=10$$
The fulcrum is 10 ft from the child.

67. Strategy:

To find the location of the fulcrum when the system balances, replace the variables F_1, F_2, and d in the lever system equation by the given values and solve for x.

Solution:

$$\begin{aligned} F_1x &= F_2(d-x) \\ 90x &= 60(12-x) \\ 90x &= 720-60x \\ 90x+60x &= 720 \\ 150x &= 720 \\ \frac{150x}{150} &= \frac{720}{150} \\ x &= 4.8 \end{aligned}$$

The fulcrum is 4.8 ft from the 90-pound child.

69. Strategy:

To find the force when the system balances, replace the variables F_2, x, and d in the lever system equation by the given values and solve for F_1.

Solution

$$\begin{aligned} F_1x &= F_2(d-x) \\ F_1\cdot 0.15 &= 30(9-0.15) \\ F_1\cdot 0.15 &= 30(8.85) \\ 0.15F_1 &= 265.5 \\ \frac{0.15F_1}{0.15} &= \frac{265.5}{0.15} \\ F_1 &= 1770 \end{aligned}$$

A 1770-pound force is applied to the other end.

71. Strategy:

To find the break-even point, replace the variables P, C, and F in the cost equation by the given values and solve for x.

Solution:

$$\begin{aligned} Px &= Cx+F \\ 325x &= 175x+39{,}000 \\ 325x-175x &= 39{,}000 \\ 150x &= 39{,}000 \\ \frac{150x}{150} &= \frac{39{,}000}{150} \\ x &= 260 \end{aligned}$$

The break-even point is 260 barbecues.

73. Strategy:

To find the break-even point, replace the variables P, C, and F in the cost equation by the given values and solve for x.

Solution:

$$\begin{aligned} Px &= Cx+F \\ 49x &= 12x+19{,}240 \\ 49x-12x &= 19{,}240 \\ 37x &= 19{,}240 \\ \frac{37x}{37} &= \frac{19{,}240}{37} \\ x &= 520 \end{aligned}$$

The break-even point is 520 recorders.

75. Strategy:

$m = 8.3$
Unknown: C

Solution:

$$\begin{aligned} m &= \frac{1}{6}(C-5) \\ 8.3 &= \frac{1}{6}(C-5) \\ 8.3 &= \frac{1}{6}C-\frac{5}{6} \\ 6\cdot 8.3 &= 6\cdot\frac{1}{6}C-6\cdot\frac{5}{6} \\ 49.8 &= C-5 \\ 49.8+5 &= C-5+5 \\ 54.8 &= C \end{aligned}$$

The mammal consumes 54.8 ml/min.

Critical Thinking 5.3

77. $\frac{1}{5}(25-10b)+4=\frac{1}{3}(9b-15)-6$

$$\begin{aligned} 5-2b+4 &= 3b-5-6 \\ 9-2b &= 3b-11 \\ 9-2b-3b &= 3b-3b-11 \\ 9-5b &= -11 \\ 9-9-5b &= -11-9 \\ -5b &= -20 \\ \frac{-5b}{-5} &= \frac{-20}{-5} \\ b &= 4 \end{aligned}$$

The solution is 4.

79.
$$\frac{2(5x-6)-3(x-4)}{7}=x+2$$
$$\frac{10x-12-3x+12}{7}=x+2$$
$$\frac{7x}{7}=x+2$$
$$x=x+2$$
$$x-x=x-x+2$$
$$0=2$$
No solution

81. $3x-4(x-1)$ is an expression, not an equation. There must be a equals sign to have an equation. You cannot solve an expression.

83. Strategy:
Let x be the number.
Subtract 4 from the number: $x-4$
3 times the result: $3(x-4)$

Solution:
$$3(x-4)=x$$
$$3x-12=x$$
$$3x-3x-12=x-3x$$
$$-12=-2x$$
$$\frac{-12}{-2}=\frac{-2x}{-2}$$
$$6=x$$
The number is 6.

Projects or Group Activities 5.3

85. Strategy:
Let x be the population in 1990.
Population after 10,000 increase:
$x+10,000$
Population after 10% decrease:
$x+10,000-0.1(x+10,000)$
$x+10,000-0.1x-1000$
$0.9x+9000$
6000 more than the beginning:
$0.9x+9000=x+6000$

Solution:
$$0.9x+9000=x+6000$$
$$0.9x-0.9x+9000=x-0.9x+6000$$
$$9000=0.1x+6000$$
$$9000-6000=0.1x+6000-6000$$
$$3000=0.1x$$
$$\frac{3000}{0.1}=\frac{0.1x}{0.1}$$
$$30,000=x$$
The population in 1990 was 30,000.

Check Your Progress: Chapter 5

1. $2a(a-1)=3a+3$

$2(3)(3-1)$	$3(3)+3$
$6(2)$	$9+3$
12	12

Yes

2.
$$x+7=-4$$
$$x+7-7=-4-7$$
$$x=-11$$
The solution is −11.

3.
$$-3y=-27$$
$$\frac{-3y}{-3}=\frac{-27}{-3}$$
$$y=9$$
The solution is 9.

4.
$$6-4a=-10$$
$$6-6-4a=-10-6$$
$$-4a=-16$$
$$\frac{-4a}{-4}=\frac{-16}{-4}$$
$$a=4$$
The solution is 4.

5.

$8t+1=-1$	
$8\left(-\frac{1}{4}\right)+1$	-1
$-2+1$	-1
-1	-1

Yes

6.

$$\frac{1}{6}+b=-\frac{1}{3}$$
$$\frac{6}{1}\cdot\frac{1}{6}+6\cdot b=\frac{6}{1}\left(-\frac{1}{3}\right)$$
$$1+6b=-2$$
$$6b=-3$$
$$b=-\frac{1}{2}$$

The solution is $-\frac{1}{2}$.

7.

$$5x-4(3-x)=2(x-1)-3$$
$$5x-12+4x=2x-2-3$$
$$9x-12=2x-5$$
$$9x-2x-12=2x-2x-5$$
$$7x-12=-5$$
$$7x-12+12=-5+12$$
$$7x=7$$
$$\frac{7x}{7}=\frac{7}{7}$$
$$x=1$$

The solution is 1.

8.

$$6y+5-8y=3-4y$$
$$-2y+5=3-4y$$
$$-2y+4y+5=3-4y+4y$$
$$2y+5=3$$
$$2y+5-5=3-5$$
$$2y=-2$$
$$\frac{2y}{2}=\frac{-2}{2}$$
$$y=-1$$

The solution is –1.

9.

$x(x+1)=x^2+5$	
$4(4+1)$	4^2+5
$4(5)$	$16+5$
20	21

No

10.

$$84=-16+t$$
$$84+16=-16+16+t$$
$$100=t$$

The solution is 100.

11.

$$\frac{3}{4}c=\frac{3}{5}$$
$$\frac{4}{3}\cdot\frac{3}{4}c=\frac{4}{3}\cdot\frac{3}{5}$$
$$c=\frac{4}{5}$$

The solution is $\frac{4}{5}$.

12.

$$9=\frac{1}{2}d-5$$
$$9+5=\frac{1}{2}d-5+5$$
$$14=\frac{1}{2}d$$
$$2\cdot 14=2\cdot\frac{1}{2}d$$
$$28=d$$

The solution is 28.

13.

$$-\frac{8}{9}=-\frac{2}{3}y$$
$$-\frac{3}{2}\left(-\frac{8}{9}\right)=-\frac{3}{2}\left(-\frac{2}{3}y\right)$$
$$\frac{4}{3}=y$$

The solution is $\frac{4}{3}$.

14.
$$3n+2(n-4)=7$$
$$3n+2n-8=7$$
$$5n-8=7$$
$$5n-8+8=7+8$$
$$5n=15$$
$$\frac{5n}{5}=\frac{15}{5}$$
$$n=3$$
The solution is 3.

15.
$$3x-8=5x+6$$
$$3x-3x-8=5x-3x+6$$
$$-8=2x+6$$
$$-8-6=2x+6-6$$
$$-14=2x$$
$$\frac{-14}{2}=\frac{2x}{2}$$
$$-7=x$$
The solution is –7.

16.
$$2[3-5(x-1)]=7x-1$$
$$2[3-5x+5]=7x-1$$
$$2[8-5x]=7x-1$$
$$16-10x=7x-1$$
$$16-10x-7x=7x-7x-1$$
$$16-17x=-1$$
$$16-16-17x=-1-16$$
$$-17x=-17$$
$$\frac{-17x}{-17}=\frac{-17}{-17}$$
$$x=1$$
The solution is 1.

17. Strategy:
To find the total time:
→Find the time with the current to travel 24 mi ($r = 10 + 2 = 12$)
→Find the time against the current to travel 24 mi ($r = 10 - 2 = 8$)
→Add the two times and 1 hour to get the total trip time

Solution:

with the current:	against the current:
$rt = d$	$rt = d$
$12t = 24$	$8t = 24$
$\frac{12t}{12}=\frac{24}{12}$	$\frac{8t}{8}=\frac{24}{8}$
$t = 2$	$t = 3$

$2 + 3 + 1 = 6$
The total trip was 6 h.

18. Strategy:
To find the number of miles, substitute 12.35 for F in the given equation.

Solution:
$$F = 2.75 + 2.40(m - 1)$$
$$12.35 = 2.75 + 2.40(m - 1)$$
$$12.35 = 2.75 + 2.40m - 2.40$$
$$12.35 = 2.40m + 0.35$$
$$12.00 = 2.40\ m$$
$$\frac{12.00}{2.40}=\frac{2.40m}{2.40}$$
$$5 = m$$
The passenger traveled 5 miles.

19. Strategy:
Check the equation $F_1x = F_2(d-x)$ when $F_1 = 60$, $x = 3.5$, $F_2 = 50$, and $d = 8$.

Solution:

F_1x	$= F_2(d-x)$
$60 \cdot 3.5$	$50(8-3.5)$
210	$50(4.5)$
210	225

No

Section 5.4

Concept Check

1. True
3. True
5. equals
7. 1; 2; 2

Objective A Exercises

9. The unknown number: x

The difference between a number and 15	is	seven

$$x - 15 = 7$$
$$x - 15 + 15 = 7 + 15$$
$$x = 22$$

The number is 22.

11. The unknown number: x

The difference between nine and a number	is	seven

$$9 - x = 7$$
$$9 - 9 - x = 7 - 9$$
$$-x = -2$$
$$\frac{-1x}{-1} = \frac{-2}{-1}$$
$$x = 2$$

The number is 2.

13. The unknown number: x

The difference between five and twice a number	is	one

$$5 - 2x = 1$$
$$5 - 5 - 2x = 1 - 5$$
$$-2x = -4$$
$$\frac{-2x}{-2} = \frac{-4}{-2}$$
$$x = 2$$

The number is 2.

15. The unknown number: x

The sum of twice a number and five	is	fifteen

$$2x + 5 = 15$$
$$2x + 5 - 5 = 15 - 5$$
$$2x = 10$$
$$\frac{2x}{2} = \frac{10}{2}$$
$$x = 5$$

The number is 5.

17. The unknown number: x

Six less than four times a number	is	twenty - two

$$4x - 6 = 22$$
$$4x - 6 + 6 = 22 + 6$$
$$4x = 28$$
$$\frac{4x}{4} = \frac{28}{4}$$
$$x = 7$$

The number is 7.

19. The unknown number: x

Three times the difference between four times a number and seven	is	fifteen

$$3(4x - 7) = 15$$
$$12x - 21 = 15$$
$$12x - 21 + 21 = 15 + 21$$
$$12x = 36$$
$$\frac{12x}{12} = \frac{36}{12}$$
$$x = 3$$

The number is 3.

21. The smaller number: x
the larger number: $20 - x$

Three times the smaller	is equal to	two times the larger

$$3x = 2(20 - x)$$
$$3x = 40 - 2x$$
$$3x + 2x = 40 - 2x + 2x$$
$$5x = 40$$
$$\frac{5x}{5} = \frac{40}{5}$$
$$x = 8$$
$$20 - x = 20 - 8 = 12$$

The smaller number is 8.
The larger number is 12.

23. The smaller number: x
The larger number: $14 - x$

The difference between two times the smaller and the larger	is	one

$$2x - (14 - x) = 1$$
$$2x - 14 + x = 1$$
$$3x - 14 = 1$$
$$3x = 15$$
$$\frac{3x}{3} = \frac{15}{3}$$
$$x = 5$$
$$14 - x = 14 - 5 = 9$$

The smaller number is 5.
The larger number is 9.

25. First odd integer: n
Second odd integer: $n + 2$
Third odd integer: $n + 4$
The sum of the three integers is 51.

$$n + (n + 2) + (n + 4) = 51$$
$$3n + 6 = 51$$
$$3n = 45$$
$$n = 15$$
$$n + 2 = 15 + 2 = 17$$
$$n + 4 = 15 + 4 = 19$$

The three integers are 15, 17, and 19.

27. First odd integer: n
Second odd integer: $n + 2$
Third odd integer: $n + 4$
Three times the second number is one more than the sum of the first and third numbers.

$$3(n + 2) = 1 + n + (n + 4)$$
$$3n + 6 = 5 + 2n$$
$$n + 6 = 5$$
$$n = -1$$
$$n + 2 = -1 + 2 = 1$$
$$n + 4 = -1 + 4 = 3$$

The three integers are –1, 1, and 3.

29. First even integer: n
Second even integer: $n + 2$
Three times the first integer equals twice the second integer.

$$3n = 2(n + 2)$$
$$3n = 2n + 4$$
$$n = 4$$
$$n + 2 = 4 + 2 = 6$$

The integers are 4 and 6.

31. (iii)

Objective B Exercises

33.

320	is	one-fourth of the length of the Golden Gate Bridge

$$320 = \frac{1}{4}x$$
$$4 \cdot 320 = 4 \cdot \frac{1}{4}x$$
$$1280 = x$$

The Golden Gate Bridge is 1280 ft.

35.

\$1400 million	was	one-sixty-fifth of gross national product

$$1400 = \frac{1}{65}x$$
$$65 \cdot 1400 = 65 \cdot \frac{1}{65}x$$
$$91{,}000$$

91,000 million = 91 billion
The gross national product was \$91 billion.

37. Strategy:
To find the length of the sides of the triangle, write and solve an equation using x to represent the length of an equal side.

Solution:

Perimeter of 23 ft	is	x ft + x ft + $(2x-1)$ ft

$$23 = x + x + (2x - 1)$$
$$23 = 4x - 1$$
$$24 = 4x$$
$$\frac{24}{4} = x$$
$$6 = x$$
$$2x - 1 = 12 - 1 = 11$$
The length of the sides are 6 ft, 6 ft and 11 ft.

39. Strategy:
Level of tv = 70
Blender = 70 + 20 = 90
Jet engine = 2(90) – 40

Solution:
2(90) – 40 = 180 – 40 = 140
The jet engine is 140 decibels.

41. The area of Iceland: x
The area of Greenland: $21x$

The combined area	is	880,000 mi^2

$$x + 21x = 880{,}000$$
$$22x = 880{,}000$$
$$\frac{22x}{22} = \frac{880{,}000}{22}$$
$$x = 40{,}000$$
21(40,000) = 840,000
The area of Greenland is 840,000 mi^2.

43. Strategy:
To find the number of kilowatt hours, write and solve an equation using x to represent the number of kilowatt hours over 300.

Solution:

The total cost	is	\$51.95

$$0.08(300) + 0.13x = 51.95$$
$$24 + 0.13x = 51.95$$
$$0.13x = 51.95 - 24$$
$$0.13x = 27.95$$
$$x = \frac{27.95}{0.13}$$
$$x = 215$$
The total number of kilowatt hours is 300 + 215 = 515.
The family used 515 kWh.

45. Strategy:
To find the amount of time that the phone was used, write and solve an equation using x to represent the amount of time over 900 minutes.

Solution:

\$80 plus \$0.40 per minute	is	\$100.40

$$80 + 0.40x = 100.40$$
$$0.40x = 100.40 - 80$$
$$0.40x = 20.40$$
$$x = \frac{20.40}{0.40}$$
$$x = 51$$
900 + 51 = 951
The business executive used the phone for 951 min.

47. \$.15

Critical Thinking 5.4

49. Strategy:
Length of the shorter piece: x
Perimeter of the shorter square: x
Length of the longer piece: $12\ x$
Perimeter of the longer piece: $12 - x$

Solution:

perimeter of the larger square	is	twice the perimeter of the shorter square

$$12 - x = 2x$$
$$12 - x + x = 2x + x$$
$$12 = 3x$$
$$\frac{12}{3} = \frac{3x}{3}$$
$$4 = x \qquad 12 - 4 = 8$$

The perimeter of the larger square is 8 ft.

51. Strategy:
To find the time remaining:
→Find the time to complete the whole trip
→Subtract the time completed $\frac{1}{2}$ h from the whole trip to get the time remaining.

Solution:

three-fifths of the trip	in	one-half hour

$$\frac{3}{5}t = \frac{1}{2}$$
$$\frac{5}{3} \cdot \frac{3}{5}t = \frac{5}{3} \cdot \frac{1}{2}$$
$$t = \frac{5}{6}$$
$$\frac{5}{6} - \frac{1}{2} = \frac{5}{6} - \frac{3}{6} = \frac{2}{6} = \frac{1}{3}$$

$\frac{1}{3}$ h is remaining.

53. First number: n
second consecutive even number: $n + 2$
third consecutive even number: $n + 4$
fourth consecutive even number: $n + 6$
The sum of the four numbers is −36.

$$n + n + 2 + n + 4 + n + 6 = -36$$
$$4n + 12 = -36$$
$$4n + 12 - 12 = -36 - 12$$
$$4n = -48$$
$$\frac{4n}{4} = \frac{-48}{4}$$
$$n = -12$$

$$n + 2 = -12 + 2 = -10$$
$$n + 4 = -12 + 4 = -8$$
$$n + 6 = -12 + 6 = -6$$

The integers are −12, −10, −8, and −6.

55. First number: n
second consecutive odd number: $n + 2$
third consecutive odd number: $n + 4$
The sum of the first and the third is twice the second.

$$n + n + 4 = 2(n + 2)$$
$$2n + 4 = 2n + 4$$

Since an identity is true for all values, any three consecutive odd integers will make this true.

Projects or Group Activities 5.4

57. even

59. even

61. even

63. even

65. odd

Section 5.5

Concept Check

1. \$10.50

3. \$.76

5. In the formula $d = rt$, d represents the distance, r represents the rate, and t represents the time.

Objective A Exercises

7. Strategy:
→Amount of high-protein supplement: x
Amount of vitamin supplement: $5-x$

	Amount	Cost	Value
High-protein	x	6.75	$6.75x$
Vitamin	$5-x$	3.25	$3.25(5-x)$
Mixture	5	4.65	4.65(5)

→The sum of the values before mixing equals the value after mixing.

Solution:

$$\begin{aligned}6.75x+3.25(5-x)&=4.65(5)\\6.75x+16.25-3.25x&=23.25\\3.50x+16.25&=23.25\\3.50x+16.25-16.25&=23.25-16.25\\3.50x&=7.00\\x&=\frac{7.00}{3.50}\\x&=2\\5-x&=3\end{aligned}$$

To make the mixture, 2 lb of the high protein supplement and 3 lb of the vitamin supplement were used.

9. Strategy:
→Amount of chamomile tea: x
Amount of orange tea: 12

	Amount	Cost	Value
Chamomile	x	18.20	$18.20(x)$
Orange tea	12	12.25	12(12.25)
Mixture	$x+12$	14.63	$14.63(x+12)$

→The sum of the values before mixing equals the value after mixing.

Solution:

$$\begin{aligned}18.20x+12(12.25)&=14.63(x+12)\\18.20x+147&=14.63x+175.56\\18.20x-14.63x+147&=14.63x-14.63x+175.56\\3.57x+147&=175.56\\3.57x+147-147&=175.56-147\\3.57x&=28.56\\\frac{3.57x}{3.57}&=\frac{28.56}{3.57}\\x&=8\end{aligned}$$

The amount of chamomile tea needed is 8 lb.

11. Strategy:
→Cost of mixture: x

	Amount	Cost	Value
Expensive coffee	8	9.20	8(9.20)
Cheaper coffee	12	5.50	12(5.50)
Mixture	20	x	$20(x)$

→The sum of the values before mixing equals the value after mixing.

Solution:

$$\begin{aligned}8(9.20)+12(5.50)&=20x\\73.60+66&=20x\\139.60&=20x\\\frac{139.60}{20}&=\frac{20x}{20}\\6.98&=x\end{aligned}$$

The cost of the coffee mixture is \$6.98.

13. Strategy:
→Amount of \$1 herb: x

	Amount	Cost	Value
\$2 herb	30	2	2(30)
\$1 herb	x	1	$1x$
Mixture	$30+x$	1.60	$1.6(30+x)$

→The sum of the values before mixing equals the value after mixing.

Solution:

$$\begin{aligned}2(30)+x&=1.6(30+x)\\60+x&=48+1.6x\\60+x-x&=48+1.6x-x\\60&=48+0.6x\\60-48&=48-48+0.6x\\12&=0.6x\\\frac{12}{0.6}&=\frac{0.6x}{0.6}\\20&=x\end{aligned}$$

The amount of the \$1 herb is 20 oz.

15. Strategy:

→Amount of pepper cheddar cheese: x

Amount of Pennsylvania Jack: $5 - x$

	Amount	Cost	Value
Pepper cheddar	x	16	16x
Jack	$5 - x$	12	$12(5 - x)$
Mixture	5	13.20	13.20(5)

→The sum of the values before mixing equals the value after mixing.

Solution:

$$16x + 12(5 - x) = 13.20(5)$$
$$16x + 60 - 12x = 66$$
$$4x + 60 - 60 = 66 - 60$$
$$4x = 6$$
$$\frac{4x}{4} = \frac{6}{4}$$
$$x = 1.5$$

$5 - 1.5 = 3.5$

The mixture needs 1.5 kg of pepper cheese and 3.5 kg of Pennsylvania Jack.

17. Strategy:

→Amount of grain: 500

Amount of meal: x

	Amount	Cost	Value
Grain	500	1.2	1.2(500)
Meal	x	0.8	$0.8x$
Mixture	$500 + x$	1.05	$1.05(500 + x)$

→The sum of the values before mixing equals the value after mixing.

Solution:

$$1.2(500) + 0.8x = 1.05(500 + x)$$
$$600 + 0.8x = 525 + 1.05x$$
$$600 + 0.8x - 0.8x = 525 + 1.05x - 0.8x$$
$$600 = 525 + 0.25x$$
$$600 - 525 = 525 - 525 + 0.25x$$
$$75 = 0.25x$$
$$\frac{75}{0.25} = \frac{0.25x}{0.25}$$
$$300 = x$$

The mixture needs 300 lb of meal.

19. Strategy:

→Amount of almonds: x

Amount of walnuts: $100 - x$

	Amount	Cost	Value
Almonds	x	6.50	$6.50(x)$
Walnuts	$100 - x$	5.50	$5.50(100 - x)$
Mixture	100	5.87	5.87(100)

→The sum of the values before mixing equals the value after mixing.

Solution:

$$6.50x + 5.50(100 - x) = 5.87(100)$$
$$6.50x + 550 - 5.50x = 587$$
$$x + 550 = 587$$
$$x + 550 - 550 = 587 - 550$$
$$x = 37$$
$$100 - x = 63$$

The amount of almonds is 37 lb.
The amount of walnuts is 63 lb.

21. Strategy:

→Cost of mixture: x

	Amount	Cost	Value
Sugar	40	2.00	40(2.00)
Flakes	120	1.20	120(1.20)
Mixture	160	x	$160x$

→The sum of the values before mixing equals the value after mixing.

Solution:

$$40(2.00) + 120(1.20) = 160x$$
$$80 + 144 = 160x$$
$$224 = 160x$$
$$\frac{224}{160} = \frac{160x}{160}$$
$$1.40 = x$$

The cost per pound of the sugar-coated cereal is \$1.40.

23. Strategy:
→Number of bundles of seedlings: x
Number of bundles of container-grown plants: $1720 - x$

	Amount	Cost	Value
Seedlings	x	17	$17(x)$
Contain-grown	$14 - x$	45	$45(14 - x)$
Mixture	14		406

→The sum of the values of the seedlings and container-grown plants must equal the total spent.

Solution:

$$17x + 45(14 - x) = 406$$
$$17x + 630 - 45x = 406$$
$$-28x = -224$$
$$\frac{-28x}{-28} = \frac{-224}{-28}$$
$$x = 8$$
$$14 - x = 14 - 8 = 6$$

The Park's Department bought 8 bundles of seedlings and 6 bundles of container-grown plants.

25. Strategy:
→Amount of expensive lotion: 50
Amount of supplement lotion: 100

	Amount	Cost	Value
Expensive lotion	50	4.00	4(50)
Supplement lotion	100	2.50	2.5(100)
Mixture	150	x	$150x$

→The sum of the values before mixing equals the value after mixing.

Solution:

$$200 + 250 = 150x$$
$$450 = 150x$$
$$\frac{450}{150} = \frac{150x}{150}$$
$$3 = x$$

The sunscreen mixture will cost $3.00.

Objective B Exercises

27. Strategy:
→Speed of first plane: r
→Speed of second plane: $r + 25$

	Rate	Time	Distance
First plane	r	2	$2r$
Second plane	$r + 25$	2	$2(r + 25)$

→In 2 h, the planes are 470 mi apart.

Solution:

$$2r + 2(r + 25) = 470$$
$$2r + 2r + 50 = 470$$
$$4r = 420$$
$$r = 105$$
$$r + 25 = 130$$

The first plane is flying at 105 mph and the second plane is flying at 130 mph.

29. Strategy:
→Time for first skater: t
→Time for second skater: $t - 10$

	Rate	Time	Distance
First skater	8	t	$8t$
Second skater	10	$t - 10$	$10(t - 10)$

→The skaters travel the same distance.

Solution:

$$8t = 10(t - 10)$$
$$8t = 10t - 100$$
$$8t - 10t = 10t - 10t - 100$$
$$-2t = -100$$
$$\frac{-2t}{-2} = \frac{-100}{-2}$$
$$t = 50$$

Time for second skater = 50 – 10 = 40
The second skater overtakes the first 40 s after the second skater starts.

31. Strategy:
→Time the motorboat travels: t
→Time the Tour Boat travels: $t - 2$

	Rate	Time	Distance
Motorboat	9	t	$9t$
Tour Boat	18	$t-2$	$18(t-2)$

→The motorboat and the cabin cruiser travel the same distance.

Solution:

$$\begin{aligned} 9t &= 18(t-2) \\ 9t &= 18t-36 \\ -9t &= -36 \\ t &= 4 \\ t-2 &= 2 \end{aligned}$$

The Tour Boat will overtake the motorboat in 2 h.

33. Strategy:
→Time to airport: t
→Time in flight: $3 - t$

	Rate	Time	Distance
To airport	30	t	$30t$
In flight	60	$3-t$	$60(3-t)$

→The total trip is 150 mi.

Solution:

$$\begin{aligned} 30t+60(3-t) &= 150 \\ 30t+180-60t &= 150 \\ 180-30t &= 150 \\ -30t &= -30 \\ t &= 1 \\ \text{Distance} = 60(3-t) &= 60(3-1) \\ &= 60(2) = 120 \end{aligned}$$

The corporate offices are 120 mi from the airport.

35. Strategy:
→Speed for first 3 h: r
→Speed for second 3 h: $r - 5$

	Rate	Time	Distance
First 3 h	r	3	$3r$
Second 3 h	$r-5$	3	$3(r-5)$

→The total trip is 57 mi.

Solution:

$$\begin{aligned} 3r+3(r-5) &= 57 \\ 3r+3r-15 &= 57 \\ 6r-15 &= 57 \\ 6r &= 72 \\ r &= 12 \\ \text{Distance} = 3r &= 3(12) = 36 \end{aligned}$$

The sailboat traveled 36 mi in the first 3 h.

37. Strategy:
→Rate for freight train: r
→Rate for passenger train: $r + 20$

	Rate	Time	Distance
Freight	r	5	$5r$
Passenger	$r+20$	3	$3(r+20)$

→The trains travel the same distance.

Solution:

$$\begin{aligned} 5r &= 3(r+20) \\ 5r &= 3r+60 \\ 2r &= 60 \\ r &= 30 \\ r+20 &= 30+20 = 50 \end{aligned}$$

The freight train travels at 30 mph.
The passenger train travels at 50 mph.

39. Strategy:
→The time the first ship traveled: t
→The time the second ship traveled: $t-\frac{10}{25}$

	Rate	Time	Distance
First ship	25	t	$25t$
Second ship	35	$t-\frac{10}{25}$	$35\left(t-\frac{10}{25}\right)$

→The two ships travel the same distance.

Solution:

$$25t = 35\left(t-\frac{10}{25}\right)$$
$$25t = 35t - 14$$
$$-10t = -14$$
$$t = 1.4$$
$$t-\frac{10}{25} = 1.4 - 0.4 = 1$$

The second ship catches up to the first ship in 1 h.

41. Strategy:
→Rate of the second car: r
→Rate of the first car: $r + 10$
→12 min ÷ 60 = $\frac{1}{5}$ hr

	Rate	Time	Distance
First car	$r + 10$	$\frac{1}{5}$	$\frac{1}{5}(r+10)$
Second car	r	$\frac{1}{5}$	$\frac{1}{5}r$

→The total distance traveled by the two cars is 36.

Solution:

$$\frac{1}{5}(r+10)+\frac{1}{5}r = 36$$
$$\frac{1}{5}r+2+\frac{1}{5}r = 36$$
$$\frac{5}{1}\cdot\frac{1}{5}r+5\cdot 2+\frac{5}{1}\cdot\frac{1}{5}r = 5\cdot 36$$
$$r+10+r = 180$$
$$2r+10-10 = 180-10$$
$$2r = 170$$
$$\frac{2r}{2} = \frac{170}{2}$$
$$r = 85$$

85 + 10 = 95
The faster car is traveling 95 km/h.

43. Strategy:

→Time for first car: t

→Time for second car: $t-\frac{1}{4}$

	Rate	Time	Distance
First driver	90	t	$90t$
Second driver	120	$t-\frac{1}{4}$	$120\left(t-\frac{1}{4}\right)$

→The second car will overtake the first when the distances are equal.

Solution:

$$90t = 120\left(t-\frac{1}{4}\right)$$
$$90t = 120t - 30$$
$$90t - 120t = 120t - 120t - 30$$
$$-30t = -30$$
$$\frac{-30t}{-30} = \frac{-30}{-30}$$
$$t = 1$$

Distance = $90t = 90(1) = 90$ mi

The track is only 50 mi, so the second will not overtake the first.

45. Strategy:

→Time the car traveled: t

→Time the bus traveled: $t-1$

	Rate	Time	Distance
Car	45	t	$45t$
Bus	60	$t-1$	$60(t-1)$

→The bus and the car travel the same distance.

Solution:

$$45t = 60(t-1)$$
$$45t = 60t - 60$$
$$-15t = -60$$
$$t = 4$$
$$45t = 180$$

The bus overtakes the car 180 mi from the starting point.

47. Strategy:

→Time for the first part of the trip: t

→Time for the second part of the trip: $5-t$

	Rate	Time	Distance
First part of trip	115	t	$115t$
Remainder of trip	125	$5-t$	$125(5-t)$

→The total distance traveled was 605 mi.

Solution:

$$115t + 125(5-t) = 605$$
$$115t + 625 - 125t = 605$$
$$-10t + 625 - 625 = 605 - 625$$
$$-10t = -20$$
$$\frac{-10t}{-10} = \frac{-20}{-10}$$
$$t = 2$$

$5 - 2 = 3$

The plane traveled 2 h at 115 mph and 3 h at 125 mph.

Critical Thinking 5.5

49. Strategy:

→Amount of Walnuts: x

→Amount of Cashews: $50 - 20 - x = 30 - x$

	Amount	Cost	Value
Walnuts	x	5.60	$5.60(x)$
Cashews	$30-x$	7.50	$7.50(30-x)$
Peanuts	20	4.00	4.00(20)
Mixture	50	5.72	5.72(50)

→The sum of the values before mixing equals the value after mixing.

Solution:

$$5.60(x) + 7.50(30-x) + 4.00(20) = 5.72(50)$$
$$5.6x + 225 - 7.5x + 80 = 286$$
$$-1.9x + 305 = 286$$
$$-1.9x + 305 - 305 = 286 - 305$$
$$-1.9x = -19$$
$$\frac{-1.9x}{-1.9} = \frac{-19}{-1.9}$$
$$x = 10$$

$30 - 10 = 20$

The amount of walnuts is 10 lb.

The amount of cashews is 20 lb.

51. Strategy:
→Time downstream: t
→Time upstream: $1-t$

	Rate	Time	Distance
Downstream	12	t	$12t$
Upstream	4	$1-t$	$4(1-t)$

→The distance downstream is equal to the distance upstream.

Solution:

$$\begin{aligned}12t &= 4(1-t)\\ 12t &= 4-4t\\ 12t+4t &= 4-4t+4t\\ 16t &= 4\\ \frac{16t}{16} &= \frac{4}{16}\\ t &= \frac{1}{4}\end{aligned}$$

$\frac{1}{4}\cdot 60 \text{ min} = 15 \text{ min}$

10:00 + 0:15 = 10:15

The campers turned around at 10:15 A.M.

Projects and Group Activities 5.5

53. Strategy:
→Find the total distance traveled and the total time.
→Divide the total distance by the total time to determine the average speed.

	Rate	Time	Distance
Leaving	10	2	10(2) = 20
Returning	20	$\frac{20}{20}=1$	20(1) = 20
Total		3	40

Solution:

Average speed $= \frac{40}{3} = 13\frac{1}{3}$

The bicyclist's average speed is $13\frac{1}{3}$ mph.

55. Strategy:
→Time to ascend: t
→Time to descend: $12-t$

	Rate	Time	Distance
Ascend	0.5	t	$0.5t$
Descend	1	$12-t$	$1(12-t)$

→The distances are the same.

Solution:

$$\begin{aligned}0.5t &= 1(12-t)\\ 0.5t &= 12-t\\ 0.5t+t &= 12-t+t\\ 1.5t &= 12\\ \frac{1.5t}{1.5} &= \frac{12}{1.5}\\ t &= 8\end{aligned}$$

Distance: $0.5t = 0.5(8) = 4$ mi each way

The total distance was 8 mi.

Chapter Review Exercises

1.

$$\begin{aligned}x+3 &= 24\\ x &= 24-3\\ x &= 21\end{aligned}$$

The solution is 21.

2.
$$\begin{aligned} x+5(3x-20)&=10(x-4)\\ x+15x-100&=10x-40\\ 16x-100&=10x-40\\ 6x&=60\\ \frac{6x}{6}&=\frac{60}{6}\\ x&=10 \end{aligned}$$
The solution is 10.

3.
$$\begin{aligned} 5x-6&=29\\ 5x&=29+6\\ 5x&=35\\ \frac{5x}{5}&=\frac{35}{5}\\ x&=7 \end{aligned}$$
The solution is 7.

4.

$5x-2$	=	$4x+5$
$5(3)-2$	|	$4(3)+5$
$15-2$	|	$12+5$

$13\neq 17$

No, 3 is not a solution.

5.
$$\begin{aligned} \frac{3}{5}a&=12\\ a&=12\cdot\frac{5}{3}\\ a&=20 \end{aligned}$$
The solution is 20.

6.
$$\begin{aligned} 6x+3(2x-1)&=-27\\ 6x+6x-3&=-27\\ 12x-3&=-27\\ 12x&=-27+3\\ 12x&=-24\\ \frac{12x}{12}&=\frac{-24}{12}\\ x&=-2 \end{aligned}$$
The solution is −2.

7.
$$\begin{aligned} x-3&=-7\\ x&=-7+3\\ x&=-4 \end{aligned}$$

8.
$$\begin{aligned} 5x+3&=10x-17\\ 3+17&=10x-5x\\ 20&=5x\\ 4&=x \end{aligned}$$
The solution is 4.

9.
$$\begin{aligned} 7-[4+2(x-3)]&=11(x+2)\\ 7-[4+2x-6]&=11x+22\\ 7-[-2+2x]&=11x+22\\ 7+2-2x&=11x+22\\ 9-2x&=11x+22\\ 9-22&=11x+2x\\ -13&=13x\\ -1&=x \end{aligned}$$
The solution is −1.

10.
$$\begin{aligned} -6x+16&=-2x\\ -6x+2x&=-16\\ -4x&=-16\\ x&=4 \end{aligned}$$
The solution is 4.

11.
$$\begin{aligned} 7-3x&=2-5x\\ 7-3x+5x&=2\\ 7+2x&=2\\ 2x&=2-7\\ 2x&=-5\\ x&=-\frac{5}{2} \end{aligned}$$

12.
$$\begin{aligned} -\frac{3}{8}x&=-\frac{15}{32}\\ -\frac{8}{3}\left(-\frac{3}{8}x\right)&=-\frac{8}{3}\left(-\frac{15}{32}\right)\\ x&=\frac{5}{4} \end{aligned}$$

13.
$$\begin{aligned} 35-3x&=5\\ -3x&=5-35\\ -3x&=-30\\ -\frac{1}{3}(-3x)&=-\frac{1}{3}(-30)\\ x&=10 \end{aligned}$$

14.
$$\begin{aligned} 3x&=2(3x-2)\\ 3x&=6x-4\\ 3x-6x&=-4\\ -3x&=-4\\ -\frac{1}{3}(-3x)&=-\frac{1}{3}(-4)\\ x&=\frac{4}{3} \end{aligned}$$

15. Strategy:
Given: $F_1 = 120$, $x = 2$,
$d - x = 12 - 2 = 10$
Unknown: F_2

Solution:

$$F_1 x = F_2(d - x)$$
$$120(2) = F_2(10)$$
$$240 = 10F_2$$
$$24 = F_2$$

The force is 24 lb.

16. Strategy:
→Speed on winding road: r
→Speed on level road: $r + 20$

	Rate	Time	Distance
Winding road	r	3	$3r$
Level road	$r + 20$	2	$2(r + 20)$

→The total trip was 200 mi.

Solution:

$$3r + 2(r + 20) = 200$$
$$3r + 2r + 40 = 200$$
$$5r + 40 = 200$$
$$5r = 160$$
$$r = 32$$

The average speed on the winding road was 32 mph.

17. The unknown number: x

The difference between nine and twice a number	is	five

$$9 - 2x = 5$$
$$-2x = 5 - 9$$
$$-2x = -4$$
$$\frac{-2x}{-2} = \frac{-4}{-2}$$
$$x = 2$$

The number is 2.

18. The unknown number: x

The product of five and a number	is	fifty

$$5x = 50$$
$$\frac{5x}{5} = \frac{50}{5}$$
$$x = 10$$

The number is 10.

19. Strategy:
→Amount of cranberry juice: x
→Amount of apple juice: $10 - x$

	Amount	Cost	Value
Cranberry juice	x	1.79	$1.79(x)$
Apple juice	$10 - x$	1.19	$1.19(10 - x)$
Mixture	10	1.61	$1.61(10)$

→The sum of the quantities before mixing is equal to the quantity after mixing.

Solution:

$$1.79x + 1.19(10 - x) = 1.61(10)$$
$$1.79x + 11.90 - 1.19x = 16.10$$
$$0.60x = 4.2$$
$$x = 7$$
$$10 - x = 10 - 3 = 7$$

The amount of cranberry juice was 7 qt. The amount of apple juice was 3 qt.

20. Strategy:
→First integer: n
→Second integer: $n + 1$
→Third integer: $n + 2$
→Four times the second integer equals the sum of the first and third integer.

Solution:

$$4(n + 1) = n + n + 2$$
$$4n + 4 = 2n + 2$$
$$2n = -2$$
$$n = -1$$

The integers are −1, 0, and 1.

21. The unknown number: n

four less than the product of five and a number	is	sixteen

$5n-4=16$

$5n=20$

$n=4$

The number is 4.

22. The height of the Eiffel Tower: x

1472	is	654 less than twice the height of the Eiffel Tower

$1472=2x-654$

$2126=2x$

$1063=x$

The Eiffel Tower is 1063 feet tall..

23.

$$F=\frac{9}{5}C+32$$
$$100=\frac{9}{5}C+32$$
$$100-32=\frac{9}{5}C$$
$$68=\frac{9}{5}C$$
$$\frac{5}{9}(68)=\frac{5}{9}\left(\frac{9}{5}C\right)$$
$$37.8\approx C$$

The temperature is 37.8° C

24. Strategy:

→Time for jet plane: t

→Time for propeller-driven plane: $t+2$

	Rate	Time	Distance
Jet	600	t	$600t$
Propeller	200	$t+2$	$200(t+2)$

Solution:

$600t=200(t+2)$

$600t=200t+400$

$400t=400$

$t=1$

Distance = 600t = 600(1) = 600

The jet overtakes the propeller-driven plane 600 mi from the starting point.

25. The smaller number: x

The larger number: $21-x$

three times the smaller number	is	2 less than twice the larger number

$3x=2(21-x)-2$

$3x=42-2x-2$

$3x=40-2x$

$5x=40$

$x=8$

$21-x=21-8=13$

The numbers are 8 and 13.

26. Strategy:

Subtract the increase from this year's crop to find the last year's crop.

Solution:

28,336 – 3036 = 25,300

The farmer harvested 25,300 bushels of corn last year.

Chapter Test

1.

$3x-2=5x+8$

$-2-8=5x-3x$

$-10=2x$

$-5=x$

The solution is −5.

2.

$x-3=-8$

$x=-8+3$

$x=-5$

The solution is −5.

3.

$3x-5=-14$

$3x=-14+5$

$3x=-9$

$x=-3$

The solution is −3.

4.

$4-2(3-2x)=2(5-x)$

$4-6+4x=10-2x$

$-2+4x=10-2x$

$4x+2x=10+2$

$6x=12$

$x=2$

The solution is 2.

5.

$$\begin{array}{c|c} x^2-3x & 2x-6 \\ \hline (-2)^2-3(-2) & 2(-2)-6 \\ 4+6 & -4-6 \end{array}$$

$$10 \neq -10$$

No, −2 is not a solution.

6.

$$7-4x=-13$$
$$-4x=-13-7$$
$$-4x=-20$$
$$x=5$$

The solution is 5.

7.

$$5=3-4x$$
$$5-3=-4x$$
$$2=-4x$$
$$\frac{2}{-4}=x$$
$$-\frac{1}{2}=x$$

8.

$$5x-2(4x-3)=6x+9$$
$$5x-8x+6=6x+9$$
$$-3x+6=6x+9$$
$$6-9=6x+3x$$
$$-3=9x$$
$$\frac{-3}{9}=x$$
$$-\frac{1}{3}=x$$

The solution is $-\frac{1}{3}$.

9.

$$5x+3-7x=2x-5$$
$$-2x+3=2x-5$$
$$-2x-2x=-5-3$$
$$-4x=-8$$
$$x=2$$

The solution is 2.

10.

$$\frac{3}{4}x=-9$$
$$\frac{4}{3}\cdot\frac{3}{4}x=\frac{4}{3}(-9)$$
$$x=-12$$

The solution is −12.

11.

$$\frac{x}{5}-12=7$$
$$\frac{x}{5}=7+12$$
$$\frac{x}{5}=19$$
$$5\left(\frac{x}{5}\right)=5(19)$$
$$x=95$$

12.

$$8-3x=2x-8$$
$$8-3x-2x=-8$$
$$8-5x=-8$$
$$-5x=-8-8$$
$$-5x=-16$$
$$-\frac{1}{5}(-5x)=-\frac{1}{5}(-16)$$
$$x=\frac{16}{5}$$

13.

$$y-4y+3=12$$
$$-3y+3=12$$
$$-3y=12-3$$
$$-3y=9$$
$$-\frac{1}{3}(-3y)=-\frac{1}{3}(9)$$
$$y=-3$$

14.

$$2x+4(x-3)=5x-1$$
$$2x+4x-12=5x-1$$
$$6x-12=5x-1$$
$$6x-5x-12=-1$$
$$x-12=-1$$
$$x=-1+12$$
$$x=11$$

15. Strategy:

→Amount rye flour: x

→Amount wheat flour: $15-x$

	Amount	Percent	Quantity
Rye	x	0.70	$0.70(x)$
Wheat	$15-x$	0.40	$0.40(15-x)$
Mixture	15	0.60	15(0.60)

→The sum of the quantities before mixing is equal to the quantity after mixing.

Solution:

$$0.70x+0.40(15-x)=0.60(15)$$
$$0.70x+6-0.40x=9$$
$$0.30x+6=9$$
$$0.30x=3$$
$$x=10$$
$$15-x=15-10=5$$

The amount of rye flour is 10 lb.
The amount of wheat flour is 5 lb.

16. Strategy:
Given: $T = 5000$, $U = 15$, $F = 2000$
Unknown: N

Solution:
$$T = U \cdot N + F$$
$$5000 = 15N + 2000$$
$$3000 = 15N$$
$$200 = N$$
200 calculators were produced.

17. Strategy:
→First even integer: n
→Second even integer: $n + 2$
→Third even integer: $n + 4$
→The sum of the three integers is 36.

Solution:
$$n + n + 2 + n + 4 = 36$$
$$3n + 6 = 36$$
$$3n = 30$$
$$n = 10$$
$$n + 2 = 10 + 2 = 12$$
$$n + 4 = 10 + 4 = 14$$
The numbers are 10, 12, and 14.

18.
$$T = U \cdot N + F$$
$$65{,}000 = 15N + 5000$$
$$65{,}000 - 5000 = 15N$$
$$60{,}000 = 15N$$
$$\frac{60{,}000}{15} = \frac{15N}{15}$$
$$4000 = N$$
4000 clocks were made.

19. The unknown number: x

The difference between three times a number and 15	is	twenty-seven

$$3x - 15 = 27$$
$$3x = 27 + 15$$
$$3x = 42$$
$$x = 14$$
The number is 14.

20. Strategy:
→Rate of the skier: x
→Rate of the snowmobile: $x + 4$

	Rate	Time	Distance
Skier	x	3	$3x$
Snowmobile	$x + 4$	1	$1(x + 4)$

→The two traveled the same distance.

Solution:
$$3x = x + 4$$
$$2x = 4$$
$$x = 2$$
$$x + 4 = 2 + 4 = 6$$
The rate of the snowmobile is 6 mph.

21. Strategy:
Write and solve an equation letting x represent the number of LCD flat panel TVs and $140 - x$ represent the LCD rear projection TVs.

Solution:
$$3(140 - x) = x - 20$$
$$420 - 3x = x - 20$$
$$440 = 4x$$
$$110 = x$$
The company makes 110 plasma TVs each day.

22. Strategy:
→The smaller number: x
→The larger number: $18 - x$

Solution:
$$4x - 7 = 2(18 - x) + 5$$
$$4x - 7 = 36 - 2x + 5$$
$$4x - 7 = 41 - 2x$$
$$6x = 48$$
$$x = 8$$
$$18 - x = 18 - 8 = 10$$
The smaller number is 8.
The larger number is 10.

23. Strategy:
→Time for flight out: t
→Time for flight in: $7 - t$

	Rate	Time	Distance
Flight out	90	t	$90t$
Flight in	120	$7 - t$	$120(7 - t)$

→The distance traveled is the same.

Solution:

$$90t = 120(7 - t)$$
$$90t = 840 - 120t$$
$$210t = 840$$
$$t = 4$$

Distance $= 90t = 90(4) = 360$
The distance to the airport is 360 mi.

24.

$$V = V_0 + 32t$$
$$392 = 24 + 32t$$
$$392 - 24 = 32t$$
$$368 = 32t$$
$$\frac{368}{32} = \frac{32t}{32}$$
$$11.5 = t$$

It will take 11.5 seconds.

25. Strategy:
Given: $m_1 = 100$, $T_1 = 80$, $m_2 = 50$, and $T_2 = 20$
Unknown: T

Solution:

$$m_1(T_1 - T) = m_2(T - T_2)$$
$$100(80 - T) = 50(T - 20)$$
$$8000 - 100T = 50T - 1000$$
$$-150T = -9000$$
$$T = 60$$

The final temperature is 60° C.

Cumulative Review Exercises

1. $-6 - (-20) - 8 = -6 + 20 - 8 = 14 - 8 = 6$

2. $(-2)(-6)(-4) = 12(-4) = -48$

3.

$$-\frac{5}{6} - \left(-\frac{7}{16}\right) = -\frac{40}{48} - \left(-\frac{21}{48}\right)$$
$$= \frac{-40 - (-21)}{48}$$
$$= \frac{-40 + 21}{48}$$
$$= -\frac{19}{48}$$

4. $-\frac{7}{3} \div \frac{7}{6} = -\frac{7}{3} \cdot \frac{6}{7} = -\frac{7 \cdot 6}{3 \cdot 7} = -2$

5.

$$-4^2 \cdot \left(-\frac{3}{2}\right)^3 = -(4)(4) \cdot \left(-\frac{3}{2}\right)\left(-\frac{3}{2}\right)\left(-\frac{3}{2}\right)$$
$$= -16 \cdot \left(-\frac{27}{8}\right) = 16 \cdot \frac{27}{8}$$
$$= \frac{16 \cdot 27}{8} = \frac{2 \cdot 2 \cdot 2 \cdot 2 \cdot 3 \cdot 3 \cdot 3}{2 \cdot 2 \cdot 2}$$
$$= \frac{2 \cdot 3 \cdot 3 \cdot 3}{1} = \frac{54}{1} = 54$$

6.

$$25 - 3\frac{(5-2)^2}{2^3 + 1} - (-2) = 25 - \frac{3(3)^2}{9} + 2$$
$$= 25 - \frac{3(9)}{9} + 2$$
$$= 25 - 3 + 2$$
$$= 22 + 2$$
$$= 24$$

7.

$$3(a - c) - 2ab$$
$$= 3[2 - (-4)] - 2(2)(3) = 3[2 + 4] - 2(2)(3)$$
$$= 3[6] - 2(2)(3) = 18 - 2(2)(3) = 18 - 4(3)$$
$$= 18 - 12 = 6$$

8.

$$3x - 8x + (-12x) = -5x + (-12x)$$
$$= -5x - 12x = -17x$$

9.

$$2a - (-3b) - 7a - 5b = 2a + 3b - 7a - 5b$$
$$= (2a - 7a) + (3b - 5b)$$
$$= -5a + (-2b) = -5a - 2b$$

10. $(16x)\left(\frac{1}{8}\right) = \frac{1}{8}(16x) = \left(\frac{1}{8} \cdot 16\right)x = 2x$

11. $-4(-9y) = 4(9y) = (4 \cdot 9)y = 36y$

12.

$$-2(-x^2 - 3x + 2) = -2(-x^2) + (-2)(-3x) + (-2)(2)$$
$$= 2x^2 + 6x - 4$$

13.

$$-2(x - 3) + 2(4 - x) = -2x + 6 + 8 - 2x$$
$$= (-2x - 2x) + (6 + 8) = -4x + 14$$

14.
$$\begin{aligned}-3[2x-4(x-3)]+2&=-3[2x-4x+12]+2\\&=-3[-2x+12]+2\\&=6x-36+2\\&=6x-34\end{aligned}$$

15.
$$\begin{array}{c|c} x^2+6x+9 & x+3 \\ \hline (-3)^2+6(-3)+9 & -3+3 \\ 9-18+9 & 0 \\ -9+9 & 0 \end{array}$$
$$0=0$$
Yes, −3 is a solution.

16.
$$\begin{array}{c|c} 3-8x & 12x-2 \\ \hline 3-8\left(\frac{1}{2}\right) & 12\left(\frac{1}{2}\right)-2 \\ 3-4 & 6-2 \end{array}$$
$$-1\neq 4$$

No, $\frac{1}{2}$ is not a solution.

17.
$$\begin{aligned}\left(\frac{3}{8}-\frac{1}{4}\right)\div\frac{3}{4}+\frac{4}{9}&=\left(\frac{3}{8}-\frac{2}{8}\right)\div\frac{3}{4}+\frac{4}{9}\\&=\frac{1}{8}\cdot\frac{4}{3}+\frac{4}{9}\\&=\frac{1}{6}+\frac{4}{9}=\frac{3}{18}+\frac{8}{18}\\&=\frac{11}{18}\end{aligned}$$

18.
$$\begin{aligned}\frac{3}{5}x&=-15\\\frac{5}{3}\cdot\frac{3}{5}x&=\frac{5}{3}\cdot(-15)\\1\cdot x&=-25\\x&=-25\end{aligned}$$
The solution is −25.

19.
$$\begin{aligned}7x-8&=-29\\7x-8+8&=-29+8\\7x&=-21\\\frac{7x}{7}&=\frac{-21}{7}\\x&=-3\end{aligned}$$
The solution is −3.

20.
$$\begin{aligned}13-9x&=-14\\13-13-9x&=-14-13\\-9x&=-27\\\frac{-9x}{-9}&=\frac{-27}{-9}\\x&=3\end{aligned}$$
The solution is 3.

21. $9.67\times 0.0049=0.047383$

22. $13-6=7$

23.
$$\begin{aligned}8x-3(4x-5)&=-2x-11\\8x-12x+15&=-2x-11\\-4x+15&=-2x-11\\-2x&=-26\\x&=13\end{aligned}$$
The solution is 13.

24.
$$\begin{aligned}6-2(5x-8)&=3x-4\\6-10x+16&=3x-4\\-10x+22&=3x-4\\-13x&=-26\\x&=2\end{aligned}$$
The solution is 2.

25.
$$\begin{aligned}5x-8&=12x+13\\-7x&=21\\x&=-3\end{aligned}$$
The solution is −3.

26.
$$\begin{aligned}11-4x&=2x+8\\3&=6x\\\frac{3}{6}&=\frac{6x}{6}\\\frac{1}{2}&=x\end{aligned}$$
The solution is $\frac{1}{2}$.

27. Strategy:
Given: $m_1=300$, $T_1=75$, $m_2=100$, and $T_2=15$
Unknown: T

Solution:
$$\begin{aligned}m_1(T_1-T)&=m_2(T-T_2)\\300(75-T)&=100(T-15)\\22{,}500-300T&=100T-1500\\24{,}000&=400T\\60&=T\end{aligned}$$
The final temperature is 60° C.

28. The unknown number: x

The difference between twelve and the product of 5 and a number	is	negative eighteen

$12 - 5x = -18$

$-5x = -30$

$x = 6$

The number is 6.

29. Strategy:

To find the area of the garage, let x = the area.

200 ft^2 more than three times the area of the garage	is	2000 ft^2

Solution:

$3x + 200 = 2000$

$3x = 1800$

$x = 600$

The area of the garage is 600 ft^2.

30. Strategy:

→Amount of oat flour: x

	Amount	Percent	Quantity
Oat	x	0.80	$0.80x$
Wheat	40	0.50	0.50(40)
Mixture	$x + 40$	0.60	$0.60(x + 40)$

→The sum of the quantities before mixing is equal to the quantity after mixing.

Solution:

$0.80x + 0.50(40) = 0.60(x + 40)$

$0.80x + 20 = 0.60x + 24$

$0.20x = 4$

$x = 20$

20 lb of oat flour are needed for the mixture.

31. Let n be the unknown number.

$3n + 4$

32. The unknown number: x

Three less than eight times a number	is	Three more than five times a number

$8x - 3 = 5x + 3$

$8x - 5x - 3 = 3$

$3x - 3 = 3$

$3x = 3 + 3$

$3x = 6$

$\frac{3x}{3} = \frac{6}{3}$

$x = 2$

The number is 2.

33. Strategy:

→Time running: t

→Time jogging: $55 - t$

	Rate	Time	Distance
Running	8	t	$8t$
Jogging	3	$55 - t$	$3(55 - t)$

→The distance traveled is the same.

Solution:

$8t = 3(55 - t)$

$8t = 165 - 3t$

$11t = 165$

$t = 15$

Distance = 8t = 8(15) = 120

The length of the track is 120 m.

Chapter 6: Proportion and Percent

Prep Test

1. $\frac{19}{100}$

2. 0.23

3. 47

4. 2,850

5. $0.015\overline{)60.000}$ = 4000.
   -60, 00, -0, 00, -0, 0

6. $\frac{8}{10}=\frac{2\cdot 4}{2\cdot 5}=\frac{4}{5}$

7. $\frac{5}{8}\times\frac{100}{1}=\frac{5\cdot \not{2}\cdot \not{2}\cdot 5\cdot 5}{\not{2}\cdot \not{2}\cdot 2}=\frac{125}{2}=62\frac{1}{2}=62.5$

8. $66\frac{2}{3}$

9. $16\overline{)28.00}$ = 1.75

Section 6.1

Concept Check

1. are not; are

Objective A Exercises

3. $\frac{3 \text{ pints}}{15 \text{ pints}}=\frac{3}{15}=\frac{1}{5}$
   3 pints: 15 pints = 1:5
   3 pints to 15 pints = 1 to 5

5. $\frac{\$40}{\$20}=\frac{40}{20}=\frac{2}{1}$
   \$40: \$20 = 2:1
   \$40 to \$20 = 2 to 1

7. $\frac{3 \text{ miles}}{8 \text{ miles}}=\frac{3}{8}$
   3 miles: 8 miles = 3:8
   3 miles to 8 miles = 3 to 8

9. $\frac{6 \text{ min}}{6 \text{ min}}=\frac{6}{6}=\frac{1}{1}$
   6 min: 6 min = 1:1
   6 min to 6 min = 1 to 1

11. $\frac{35 \text{ cents}}{50 \text{ cents}}=\frac{35}{50}=\frac{7}{10}$
    35 cents: 50 cents = 7:10
    35 cents to 50 cents = 7 to 10

13. $\frac{160 \text{ miles}}{6 \text{ hours}}=\frac{80 \text{ miles}}{3 \text{ hours}}$

15. 1

17. 1

19. $\frac{\$51{,}000}{12 \text{ months}}=\$4250/\text{month}$

21. $\frac{3750 \text{ words}}{15 \text{ pages}}=250 \text{ words}/\text{page}$

23. $\frac{628.8 \text{ miles}}{12 \text{ hours}}=52.4 \text{ miles}/\text{hour}$

25. $\frac{\$11.05}{3.4 \text{ pounds}}=\$3.25/\text{pound}$

27. Strategy:
    To find the ratio, divide the number of rookies (50) by the number of seniors (2,800).

    Solution:
    $\frac{50}{2{,}800}=\frac{1}{56}$
    The ratio of rookies to seniors is $\frac{1}{56}$.

29. Strategy:
To find the number of messages per day, divide each year's messages per day (billions) by the number of users (millions).

Solution:

$$\frac{156{,}000}{1050} \approx 148.571$$

$$\frac{220{,}000}{1200} \approx 183.333$$

$$\frac{320{,}000}{1330} \approx 240.601$$

2009: 148.6 messages
2010: 183.3 messages
2011: 240.6 messages

Critical Thinking 6.1

31. Television advertising rates are given as dollars per second, so the numerator is dollars and the denominator is seconds.

Section 6.2

Concept Check

1. proportion; *a*; *d*; *b*; *c*; equal

3. $\frac{n \text{ mi}}{30 \text{ min}}$

Objective A Exercises

5. $\frac{27}{8} \times \frac{9}{4} \quad \begin{matrix} \to 72 \\ \to 108 \end{matrix}$

The product of the extremes does not equal the product of the means.
The proportion is not true.

7. $\frac{45}{135} \times \frac{3}{9} \quad \begin{matrix} \to 405 \\ \to 405 \end{matrix}$

The product of the extremes equals the product of the means.
The proportion is true.

9. $\frac{6 \text{ min}}{5 \text{ cents}} \times \frac{30 \text{ min}}{25 \text{ cents}} \quad \begin{matrix} \to 150 \\ \to 150 \end{matrix}$

The product of the extremes equals the product of the means.
The proportion is true.

11. $\frac{300 \text{ ft}}{4 \text{ rolls}} \times \frac{450 \text{ ft}}{7 \text{ rolls}} \quad \begin{matrix} \to 1{,}800 \\ \to 2{,}100 \end{matrix}$

The product of the extremes does not equal the product of the means.
The proportion is not true.

13. $\frac{2}{3} = \frac{n}{15}$

$2 \cdot 15 = 3 \cdot n$
$30 = 3n$
$\frac{30}{3} = \frac{3n}{3}$
$10 = n$

15. $\frac{n}{5} = \frac{12}{25}$

$n \cdot 25 = 5 \cdot 12$
$25n = 60$
$\frac{25n}{25} = \frac{60}{25}$
$n = 2.4$

17. $\frac{3}{8} = \frac{n}{12}$

$3 \cdot 12 = 8 \cdot n$
$36 = 8n$
$\frac{36}{8} = \frac{8n}{8}$
$4.5 = n$

19. $\frac{3}{n} = \frac{7}{40}$

$3 \cdot 40 = n \cdot 7$
$120 = 7n$
$\frac{120}{7} = \frac{7n}{7}$
$17.14 \approx n$

21. $\frac{16}{n} = \frac{25}{40}$

$16 \cdot 40 = n \cdot 25$
$640 = 25n$
$\frac{640}{25} = \frac{25n}{25}$
$25.6 = n$

23. $\frac{120}{n} = \frac{144}{25}$

$120 \cdot 25 = n \cdot 144$
$3{,}000 = 144n$
$\frac{3{,}000}{144} = \frac{144n}{144}$
$20.83 \approx n$

25. $\frac{0.5}{2.3}=\frac{n}{20}$
$0.5\cdot 20=2.3\cdot n$
$10=23n$
$\frac{10}{2.3}=\frac{2.3n}{2.3}$
$n\approx 4.35$

27. $\frac{0.7}{1.2}=\frac{6.4}{n}$
$0.7\cdot n=1.2\cdot 6.4$
$0.7n=7.68$
$\frac{0.7n}{0.7}=\frac{7.68}{0.7}$
$n\approx 10.97$

29. $\frac{x}{6.25}=\frac{16}{87}$
$x\cdot 87=6.25\cdot 16$
$87x=100$
$\frac{87x}{87}=\frac{100}{87}$
$x\approx 1.15$

31. $\frac{1.2}{0.44}=\frac{y}{14.2}$
$1.2\cdot 14.2=0.44\cdot y$
$17.04=0.44y$
$\frac{17.04}{0.44}=\frac{0.44y}{0.44}$
$38.73\approx y$

33. $\frac{n+2}{5}=\frac{1}{2}$
$(n+2)2=5\cdot 1$
$2n+4=5$
$2n=1$
$\frac{2n}{2}=\frac{1}{2}$
$n=\frac{1}{2}=0.5$

35. $\frac{4}{3}=\frac{n-2}{6}$
$4\cdot 6=3(n-2)$
$24=3n-6$
$30=3n$
$\frac{30}{3}=\frac{3n}{3}$
$10=n$

37. $\frac{2}{n+3}=\frac{7}{12}$
$2\cdot 12=(n+3)(7)$
$24=7n+21$
$3=7n$
$\frac{3}{7}=\frac{7n}{7}$
$0.43\approx n$

39. $\frac{7}{10}=\frac{3+n}{2}$
$7\cdot 2=10(3+n)$
$14=30+10n$
$-16=10n$
$\frac{-16}{10}=\frac{10n}{10}$
$-1.6=n$

41. $\frac{x-4}{3}=\frac{3}{4}$
$(x-4)4=3\cdot 3$
$4x-16=9$
$4x=25$
$\frac{4x}{4}=\frac{25}{4}$
$x=6.25$

43. $\frac{6}{1}=\frac{x-2}{5}$
$6\cdot 5=1(x-2)$
$30=x-2$
$32=x$

45. $\frac{5}{8}=\frac{2}{x-3}$
$5(x-3)=8\cdot 2$
$5x-15=16$
$5x=31$
$\frac{5x}{5}=\frac{31}{5}$
$x=6.2$

47. $\frac{3}{x-4}=\frac{5}{3}$
$3\cdot 3=(x-4)5$
$9=5x-20$
$29=5x$
$\frac{29}{5}=\frac{5x}{5}$
$5.8=x$

49. a. For example, $\frac{2}{9}=\frac{4}{18}$.

b. For example, $\frac{2}{4}=\frac{4}{8}$.

Objective B Exercises

51. Strategy:
To find the weight on the moon, write and solve a proportion using n to represent the weight.

Solution:
$$\frac{1 \text{ lb}}{6 \text{ lb}} = \frac{n}{174 \text{ lb}}$$
$1 \cdot 174 = 6 \cdot n$
$174 = 6n$
$$\frac{174}{6} = \frac{6n}{6}$$
$29 = n$
The weight on the moon would be 29 lb.

53. Strategy:
To find the number of robes, write and solve a proportion using n to represent the number of robes.

Solution:
$$\frac{6 \text{ robes}}{6.5 \text{ yd}} = \frac{n}{26 \text{ yd}}$$
$6 \cdot 26 = 6.5n$
$156 = 6.5n$
$$\frac{156}{6.5} = \frac{6.5n}{6.5}$$
$24 = n$
24 robes can be made.

55. Strategy:
To find the property tax, write and solve a proportion using x to represent the amount of tax.

Solution:
$$\frac{\$4,320}{\$180,000} = \frac{x}{\$280,000}$$
$4,320 \cdot 280,000 = 180,000 \cdot x$
$1,209,600,000 = 180,000x$
$$\frac{1,209,600,000}{180,000} = \frac{180,000x}{180,000}$$
$6,720 = x$
The property tax is \$6,720.

57. Strategy:
To find the distance, write and solve a proportion using n to represent the miles driven.

Solution:
$$\frac{84 \text{ mi}}{3 \text{ gal}} = \frac{n}{14.5 \text{ gal}}$$
$84 \cdot 14.5 = 3 \cdot n$
$1218 = 3n$
$$\frac{1218}{3} = \frac{3n}{3}$$
$406 = n$
The car would travel 406 mi on 14.5 gal of gasoline.

59. Strategy:
To find the number of light fixtures, write and solve a proportion using n to represent the required number of light fixtures.

Solution:
$$\frac{5 \text{ lights}}{400 \text{ ft}^2} = \frac{n}{35,000 \text{ ft}^2}$$
$5 \cdot 35,000 = 400 \cdot n$
$175,000 = 400n$
$$\frac{175,000}{400} = \frac{400n}{400}$$
$437.5 = n$
The office will require 438 light fixtures.

61. Strategy:
To find the time necessary to lose 36 lb, write and solve a proportion using n to represent the time.

Solution:
$$\frac{3 \text{ lb}}{5 \text{ weeks}} = \frac{36 \text{ lb}}{n}$$
$3 \cdot n = 5 \cdot 36$
$3n = 180$
$$\frac{3n}{3} = \frac{180}{3}$$
$n = 60$
The dieter will lose 36 lb in 60 weeks.

63. Strategy:
To find the number of miles, write and solve a proportion using n to represent the number of miles walked.

Solution:
$\frac{5 \text{ mi}}{650 \text{ calories}} = \frac{n}{3{,}500 \text{ calories}}$
$5 \cdot 3{,}500 = 650 \cdot n$
$17{,}500 = 650n$
$\frac{17{,}500}{650} = \frac{650n}{650}$
$26.92 \approx n$
A person would have to walk approximately 26.92 mi to lose one pound.

65. Strategy:
To find the number of voters, write and solve a proportion using n to represent the number of voters.

Solution:
$\frac{2 \text{ voters}}{3 \text{ voters}} = \frac{n}{240{,}000 \text{ voters}}$
$2 \cdot 240{,}000 = 3 \cdot n$
$480{,}000 = 3n$
$\frac{480{,}000}{3} = \frac{3n}{3}$
$160{,}000 = n$
160,000 people are expected to vote.

67. Strategy:
To find the distance between the two points, write and solve a proportion using n to represent the distance.

Solution:
$\frac{\frac{1}{2} \text{ in.}}{8 \text{ mi}} = \frac{1\frac{1}{4} \text{ in.}}{n}$
$\frac{1}{2} \cdot n = 8 \cdot 1\frac{1}{4}$
$\frac{1}{2}n = 8 \cdot \frac{5}{4}$
$\frac{1}{2}n = 10$
$\frac{2}{1}\left(\frac{1}{2}n\right) = \frac{2}{1}(10)$
$n = 20$
The distance between the two points is 20 mi.

Critical Thinking 6.2

69. $\frac{2}{5}$ cast a vote in favor of the amendment.
$\frac{3}{4}$ cast a vote against the amendment.
$\frac{2}{5} + \frac{3}{4} = \frac{8}{20} + \frac{15}{20} = \frac{23}{20}$
The number 1 represents the total population. The fraction $\frac{23}{20}$ indicates that there were more votes than voters. This is not possible.

71. Strategy:
To find the first person's share, first add up the amount of money the three people put in ($90) and then write and solve a proportion using n to represent the share.

Solution:
$\frac{\$25}{\$90} = \frac{n}{\$4{,}500{,}000}$
$25 \cdot 4{,}500{,}000 = 90 \cdot n$
$112{,}500{,}000 = 90n$
$\frac{112{,}500{,}000}{90} = \frac{90n}{90}$
$1{,}250{,}000 = n$
The first person's share would be $1.25 million.

Projects or Groups Activities 6.2

73. $\frac{4 \text{ earned runs}}{21 \text{ innings}} = \frac{x}{9 \text{ innings}}$
$4 \cdot 9 = 21 \cdot x$
$36 = 21x$
$\frac{36}{21} = \frac{21x}{21}$
$1.714 \approx x$
Reardon's ERA for 1979 was 1.71.

75. For 2008,

$\frac{76 \text{ earned runs}}{246 \text{ innings}} = \frac{x}{9 \text{ innings}}$

$76 \cdot 9 = 246 \cdot x$

$684 = 246x$

$\frac{684}{246} = \frac{246x}{246}$

$2.78 \approx x$

For 2009,

$\frac{74 \text{ earned runs}}{239 \text{ innings}} = \frac{x}{9 \text{ innings}}$

$74 \cdot 9 = 239 \cdot x$

$666 = 239x$

$\frac{666}{239} = \frac{239x}{239}$

$2.79 \approx x$

$2.79 - 2.78 = 0.01$

Halladay's ERA was lower in 2008 by 0.01.

77. Answers will vary

Check Your Progress: Chapter 6

1. $\frac{30 \text{ min}}{60 \text{ min}} = \frac{30}{60} = \frac{1}{2}$

30 min: 60 min = 1:2

30 min to 60 min = 1 to 2

2. $\frac{12 \text{ qt}}{18 \text{ qt}} = \frac{12}{18} = \frac{12}{3}$

12 qt: 18 qt = 2:3

12 qt to 18 qt = 2 to 3

3. $\frac{\$39}{6 \text{ T-shirts}} = \frac{\$13}{2 \text{ T-shirts}}$

4. $\frac{84 \text{ feet}}{9 \text{ seconds}} = \frac{28 \text{ feet}}{3 \text{ seconds}}$

5. $\frac{\$38{,}700}{12 \text{ months}} = \$3225 / \text{month}$

6. $\frac{\$19.08}{4.5 \text{ lb}} = \$4.24 / \text{lb}$

7. $\frac{3}{7} \times \frac{6}{14}$ → 42, 42

The product of the extremes equals the product of the means.

The proportion is true.

8. $\frac{11 \text{ miles}}{3 \text{ hours}} \times \frac{812 \text{ miles}}{22 \text{ hours}}$ → 2436, 2442

The product of the extremes does not equal the product of the means.

The proportion is not true.

9. $\frac{n}{8} = \frac{21}{56}$

$56 \cdot n = 8 \cdot 21$

$56n = 168$

$\frac{56n}{56} = \frac{168}{56}$

$n = 3$

10. $\frac{60}{n} = \frac{24}{7}$

$24 \cdot n = 60 \cdot 7$

$24n = 420$

$\frac{24n}{24} = \frac{420}{24}$

$n = 17.5$

11. $\frac{18}{20} = \frac{15}{n}$

$18 \cdot n = 20 \cdot 15$

$18n = 300$

$\frac{18n}{18} = \frac{300}{18}$

$n \approx 16.67$

12. $\frac{x+1}{2} = \frac{4}{5}$

$(x + 1)5 = 2 \cdot 4$

$5x + 5 = 8$

$5x = 3$

$\frac{5x}{5} = \frac{3}{5}$

$x = 0.6$

13. Strategy:
To find the ratio, divide the number of wins (18) by the total number of games (18 + 8 = 26).

Solution:
$\frac{18}{26} = \frac{9}{13}$
The ratio of wins to the total number of games is $\frac{9}{13}$.

14. $\frac{24 \text{ teeth}}{36 \text{ teeth}} = \frac{24}{36} = \frac{2}{3}$

15. Strategy:
To find the rate of travel, substitute 1,155 for *d* and 2.5 for *t* in the given equation and solve for *r*.

Solution:
$r = \frac{d}{t}$
$r = \frac{1{,}155}{2.5}$
$r = 462$
The rate of travel for the plane was 462 mph.

16. Strategy:
To find the annual dividend, write and solve a proportion using *n* to represent the dividend.

Solution:
$\frac{\$555}{150 \text{ stocks}} = \frac{n}{180 \text{ stocks}}$
$555 \cdot 180 = 150 \cdot n$
$99{,}900 = 150n$
$\frac{99{,}900}{150} = \frac{150n}{150}$
$666 = n$
The annual dividend would be $666.

17. Strategy:
To find the cost, write and solve a proportion using *c* to represent the cost of the life insurance.

Solution:
$\frac{\$8.52}{\$1{,}000} = \frac{c}{\$20{,}000}$
$8.52 \cdot 20{,}000 = 1{,}000 \cdot c$
$170{,}400 = 1{,}000c$
$170.4 = c$
The cost of the life insurance is $170.40.

Section 6.3

Concept Check

1. 29 figures should be circled

3. a. Students should explain that to convert a percent to a fraction, drop the percent sign and multiply by $\frac{1}{100}$.

b. Students should explain that to convert a percent to a decimal, drop the percent sign and multiply by 0.01.

5. 0.01; 0.01; 0.53

7. $\frac{1}{100}$; $\frac{1}{100}$; $\frac{4}{5}$

9. 100%; 100%; 46%

11. 100%; 100%; 30%

Objective A Exercises

13. $5\% = 5\left(\frac{1}{100}\right) = \frac{5}{100} = \frac{1}{20}$
$5\% = 5(0.01) = 0.05$

15. $30\% = 30\left(\frac{1}{100}\right) = \frac{30}{100} = \frac{3}{10}$
$30\% = 30(0.01) = 0.30$

17. $250\% = 250\left(\frac{1}{100}\right) = \frac{250}{100} = \frac{5}{2}$
$250\% = 250(0.01) = 2.50$

19. $28\% = 28\left(\frac{1}{100}\right) = \frac{28}{100} = \frac{7}{25}$
$28\% = 28(0.01) = 0.28$

21. $35\% = 35\left(\frac{1}{100}\right) = \frac{35}{100} = \frac{7}{20}$
$35\% = 35(0.01) = 0.35$

23. $29\% = 29\left(\frac{1}{100}\right) = \frac{29}{100}$
$29\% = 29(0.01) = 0.29$

25. $11\frac{1}{9}\% = 11\frac{1}{9}\left(\frac{1}{100}\right) = \frac{100}{9}\left(\frac{1}{100}\right) = \frac{1}{9}$

27. $37\frac{1}{2}\% = 37\frac{1}{2}\left(\frac{1}{100}\right) = \frac{75}{2}\left(\frac{1}{100}\right) = \frac{3}{8}$

29. $66\frac{2}{3}\% = 66\frac{2}{3}\left(\frac{1}{100}\right) = \frac{200}{3}\left(\frac{1}{100}\right) = \frac{2}{3}$

31. $6\frac{2}{3}\% = 6\frac{2}{3}\left(\frac{1}{100}\right) = \frac{20}{3}\left(\frac{1}{100}\right) = \frac{1}{15}$

33. $\frac{1}{2}\% = \frac{1}{2}\left(\frac{1}{100}\right) = \frac{1}{200}$

35. $6\frac{1}{4}\% = 6\frac{1}{4}\left(\frac{1}{100}\right) = \frac{25}{4}\left(\frac{1}{100}\right) = \frac{1}{16}$

37. $7.3\% = 7.3(0.01) = 0.073$

39. $15.8\% = 15.8(0.01) = 0.158$

41. $0.3\% = 0.3(0.01) = 0.003$

43. $121.2\% = 121.2(0.01) = 1.212$

45. $62.14\% = 62.14(0.01) = 0.6214$

47. $8.25\% = 8.25(0.01) = 0.0825$

49. $65\% = 65\left(\frac{1}{100}\right) = \frac{65}{100} = \frac{13}{20}$

51. greater than

53. $50\% = 50\left(\frac{1}{100}\right) = \frac{50}{100} = \frac{1}{2}$

Objective B Exercises

55. $0.37 = 0.37(100\%) = 37\%$

57. $0.02 = 0.02(100\%) = 2\%$

59. $0.125 = 0.125(100\%) = 12.5\%$

61. $1.36 = 1.36(100\%) = 136\%$

63. $0.96 = 0.96(100\%) = 96\%$

65. $0.07 = 0.07(100\%) = 7\%$

67. $\frac{83}{100} = \frac{83}{100}(100\%) = 83\%$

69. $\frac{1}{3} = \frac{1}{3}(100\%) \approx 33.3\%$

71. $\frac{4}{9} = \frac{4}{9}(100\%) \approx 44.4\%$

73. $\frac{9}{20} = \frac{9}{20}(100\%) = 45\%$

75. $2\frac{1}{2} = \frac{5}{2} = \frac{5}{2}(100\%) = 250\%$

77. $\frac{1}{6} = \frac{1}{6}(100\%) \approx 16.7\%$

79. $\frac{17}{25} = \frac{17}{25}(100\%) = 68\%$

81. $\frac{9}{16} = \frac{9}{16}(100\%) = 56\frac{1}{4}\%$

83. $2\frac{5}{8} = \frac{21}{8} = \frac{21}{8}(100\%) = 262\frac{1}{2}\%$

85. $2\frac{5}{6} = \frac{17}{6} = \frac{17}{6}(100\%) = 283\frac{1}{3}\%$

87. $\frac{7}{30} = \frac{7}{30}(100\%) = 23\frac{1}{3}\%$

89. $\frac{2}{9} = \frac{2}{9}(100\%) = 22\frac{2}{9}\%$

91. $\frac{3}{4}$; 75%

93. 0.375; 37.5%

95. $\frac{9}{16}$; 0.5625

97. $\frac{13}{25}$; 52%

99. 0.18; 18%

101. $\frac{3}{4} = \frac{3}{4}(100\%) = \frac{300}{4}\% = 75\%$

103. Since $\frac{4}{3}$ has a numerator greater than the denominator, the value of the fraction is *greater than* 100%.

Critical Thinking 6.3

105. Shaded: $\frac{1}{4}$; 0.25; 25%

Not shaded: $\frac{3}{4}$; 0.75; 75%

107. **a.** False

b. Answers will vary. For example, $200\% \times 4 = 2 \times 4 = 8$

Projects and Group Activities 6.3

109. Employee B's salary is the highest after the raise. Explanations will vary. One possible explanation is: Because each employee start with the same original salary, the largest resulting salary will go to the employee with the largest percent raise. Since Employee B is given the largest percent raise, that employee will have the highest salary after the raises.

Section 6.4

Concept Check

1. 0.12; n; 68

3. 0.08; 450; n

5. 36; n; 25

Objective A Exercises

7. $70\% = 0.70$
$0.70 \cdot 30 = 21$
21 of the 30 squares should be shaded.

9. $62.5\% = 0.625$
$0.625 \cdot 40 = 25$
25 of the 40 squares should be shaded.

11. $\frac{1}{6} \cdot 30 = 5$
5 of the 30 squares should be shaded.

13. $56.5\% = 0.565$
$0.565 \cdot 48 = 27$
27 of the 48 people should be circled.

15. Strategy:
To find the amount, solve the basic percent equation. Percent = 8% = 0.08, base = 100, amount = n

Solution:
Percent · base = amount
$0.08 \cdot 100 = n$
$8 = n$
8% of 100 is 8.

17. Strategy:
To find the amount, solve the basic percent equation. Percent = 0.05% = 0.0005, base = 150, amount = n.

Solution:
Percent · base = amount
$0.0005 \cdot 150 = n$
$0.075 = n$
0.05% of 150 is 0.075.

19. Strategy:
To find the amount, solve the basic percent equation. Percent = n, base = 90, amount = 15

Solution:
Percent · base = amount
$n \cdot 90 = 15$
$n = \frac{15}{90} = 0.16\frac{2}{3}$
$n = 16\frac{2}{3}\%$
15 is $16\frac{2}{3}\%$ of 90.

21. Strategy:
To find the percent, solve the basic percent equation. Percent = n, base = 16, amount = 6

Solution:
Percent · base = amount
$n \cdot 16 = 6$
$n = \frac{6}{16} = 0.375$
$n = 37.5\%$

23. Strategy:
To find the base, solve the basic percent equation. Percent = 10% = 0.10, base = n, amount = 10
Solution:
Percent · base = amount
$0.10 \cdot n = 10$
$n = \frac{10}{0.10}$
$n = 100$
10 is 10% of 100.

25. Strategy:
To find the base, solve the basic percent equation. Percent = 2.5% = 0.025, base = n, amount = 30

Solution:
Percent · base = amount
$0.025 \cdot n = 30$
$n = \frac{30}{0.025}$
$n = 1{,}200$
2.5% of 1,200 is 30.

27. Strategy:
To find the amount, solve the basic percent equation. Percent = 10.7% = 0.107, base = 485, amount = n

Solution:
Percent · base = amount
$0.107 \cdot 485 = n$
$51.895 = n$
10.7% of 485 is 51.895.

29. Strategy:
To find the amount, solve the basic percent equation. Percent = 80% = 0.80, base = 16.25, amount = n

Solution:
Percent · base = amount
$0.80 \cdot 16.25 = n$
$13 = n$
80% of 16.25 is 13.

31. Strategy:
To find the percent, solve the basic percent equation.
Percent = n, base = 2,000, amount = 54

Solution:
Percent · base = amount
$n \cdot 2{,}000 = 54$
$n = \frac{54}{2{,}000} = 0.027$
$n = 2.7\%$
54 is 2.7% of 2,000.

33. Strategy:
To find the percent, solve the basic percent equation. Percent = n, base = 4.1, amount = 16.4

Solution:
Percent · base = amount
$n \cdot 4.1 = 16.4$
$n = \frac{16.4}{4.1} = 4$
$n = 400\%$
16.4 is 400% of 4.1.

35. Strategy:
To find the percent, solve the basic percent equation. Percent = 240% = 2.40, base = n, amount = 18

Solution:
Percent · base = amount
$2.40 \cdot n = 18$
$n = \frac{18}{2.40}$
$n = 7.5$
18 is 240% of 7.5.

37. Strategy:
29.8% responded yes, so
100% − 29.8% = 70.2% did not respond yes. To find the percent, solve the basic percent equation. Percent = 70.2% = 0.702, base = 8878, amount = n

Solution:
Percent · base = amount
$0.702 \cdot 8878 = n$
$6232.356 = n$
$6232 \approx n$
6232 respondents did not answer yes.

39. Strategy:
To find the percent, solve the basic percent equation. Percent = 11% = 0.11, base = n, amount = 1.738

Solution:
Percent · base = amount
$0.11 \cdot n = 1.738$
$n = \frac{1.738}{0.11} = 15.8$
About 15.8 million travelers allowed their children to miss school to go on a trip.

41. $x < y$

Objective B Exercises

43. Percent = 18, base = 150, amount = n

$\frac{\text{percent}}{100} = \frac{\text{amount}}{\text{base}}$

$\frac{18}{100} = \frac{n}{150}$

$18 \cdot 150 = 100 \cdot n$

$2{,}700 = 100n$

$27 = n$

18% of 150 is 27.

45. Percent = n, base = 150, amount = 33

$\frac{\text{percent}}{100} = \frac{\text{amount}}{\text{base}}$

$\frac{n}{100} = \frac{33}{150}$

$n \cdot 150 = 3{,}300$

$150n = 3{,}300$

$n = \frac{3{,}300}{150}$

$n = 22$

33 is 22% of 150.

47. Percent = 84, base = n, amount = 126

$\frac{\text{percent}}{100} = \frac{\text{amount}}{\text{base}}$

$\frac{84}{100} = \frac{126}{n}$

$84 \cdot n = 100 \cdot 126$

$84n = 12{,}600$

$n = \frac{12{,}600}{84}$

$n = 150$

126 is 84% of 150.

49. Percent = n, base = 50, amount = 750

$\frac{\text{percent}}{100} = \frac{\text{amount}}{\text{base}}$

$\frac{n}{100} = \frac{750}{50}$

$n \cdot 50 = 100 \cdot 750$

$50n = 75{,}000$

$n = \frac{75{,}000}{50}$

$n = 1{,}500$

750 is 1,500% of 50.

51. Percent = 2.4, base = n, amount = 21

$\frac{\text{percent}}{100} = \frac{\text{amount}}{\text{base}}$

$\frac{2.4}{100} = \frac{21}{n}$

$2.4 \cdot n = 100 \cdot 21$

$2.4n = 2{,}100$

$n = \frac{2{,}100}{2.4}$

$n = 875$

21 is 2.4% of 875.

53. Percent = 96, base = 75, amount = n

$\frac{\text{percent}}{100} = \frac{\text{amount}}{\text{base}}$

$\frac{96}{100} = \frac{n}{75}$

$96 \cdot 75 = 100 \cdot n$

$7{,}200 = 100n$

$72 = n$

72 is 96% of 75.

55. Percent = n, base = 1,500, amount = 693

$\frac{\text{percent}}{100} = \frac{\text{amount}}{\text{base}}$

$\frac{n}{100} = \frac{693}{1{,}500}$

$n \cdot 1{,}500 = 100 \cdot 693$

$1{,}500n = 69{,}300$

$n = \frac{69{,}300}{1{,}500} = 46.2$

693 is 46.2% of 1,500.

57. Percent = 87.4, base = 225, amount = n

$\frac{\text{percent}}{100} = \frac{\text{amount}}{\text{base}}$

$\frac{87.4}{100} = \frac{n}{225}$

$87.4 \cdot 225 = 100 \cdot n$

$19{,}665 = 100n$

$196.65 = n$

196.65 is 87.4% of 225.

59. Percent = 175, base = n, amount = 14

$\frac{\text{percent}}{100} = \frac{\text{amount}}{\text{base}}$

$\frac{175}{100} = \frac{14}{n}$

$175 \cdot n = 100 \cdot 14$

$175n = 1{,}400$

$n = \frac{1{,}400}{175}$

$n = 8$

175% of 8 is 14.

61. Percent = 325, base = 4.4, amount = n

$$\frac{\text{percent}}{100} = \frac{\text{amount}}{\text{base}}$$

$$\frac{325}{100} = \frac{n}{4.4}$$

$325 \cdot 4.4 = 100 \cdot n$

$1{,}430 = 100n$

$14.3 = n$

325% of 4.4 is 14.3.

63. Percent = n, base = 38, amount = 95

$$\frac{\text{percent}}{100} = \frac{\text{amount}}{\text{base}}$$

$$\frac{n}{100} = \frac{95}{38}$$

$n \cdot 38 = 100 \cdot 95$

$38n = 9{,}500$

$$n = \frac{9{,}500}{38}$$

$n = 250$

95 is 250% of 38.

65. Strategy:
To find the percent, solve the basic percent equation. Percent = n, base = 4450, amount = 17.8

Solution:
Percent · base = amount

$n \cdot 4450 = 17.8$

$$n = \frac{17.8}{4450} = 0.004$$

$n = 0.4\%$

0.4% of the total energy production was generated by wind machines.

67. $$\frac{\frac{1}{4}}{100} = \frac{1}{4} \cdot \frac{1}{100} = \frac{1}{400}$$

The answer is (ii).

69. True. Multiplication is associative and commutative.

Objective C Exercises

71. Strategy:
To find the percent of the value of the fireworks, use the basic percent equation.
Percent = n, base = 163.1, amount = 157.2

Solution:
Percent · base = amount

$n \cdot 163.1 = 157.2$

$$n = \frac{157.2}{163.1}$$

$n \approx 0.964 = 96.4\%$

The percent of the value was 96.4%.

73. Strategy:
To find the percent, use the basic percent equation. Add all the deaths to find the base.
Percent = n, base = (total), amount = 14

Solution:
$30 + 27 + 23 + 21 + 14 = 115$

Percent · base = amount

$n \cdot 115 = 14$

$$n = \frac{14}{115}$$

$n \approx 0.122 = 12.2\%$

Of the total deaths, 12.2% occurred during training.

75. Strategy:
To find the number of students, use the basic percent equation.
Percent = 2.2% = 0.022,
base = n, amount = 1.1

Solution:
Percent · base = amount

$0.022 \cdot n = 1.1$

$$\frac{0.022n}{0.022} = \frac{1.1}{0.022}$$

$n = 50$

There were 50 million students in the United States.

77. Strategy:
To find number of couples, use the basic percent equation. Percent = 4.4% = 0.044, base = 114, amount = n

Solution:
Percent · base = amount

$0.044 \cdot 114 = n$

$5.016 = n$

There are 5 million couples.

79. Strategy:
To find the cash generated, use the basic percent equation.
a.
Percent = 25% = 0.25,
base = 700, amount = n
b.
Percent = 9% = 0.9,
base = 700, amount = n

Solution:
a.
Percent · base = amount
$0.25 \cdot 700 = n$
$175 = n$
The cash generated by the sale of Thin Mints is $175 million.
b.
Percent · base = amount
$0.09 \cdot 700 = n$
$63 = n$
The cash generated by the sale of Trefoil shortbread is $63 million.

81. Strategy:
To find the percent, use the basic percent equation. Percent = n, base = 572, amount = 291.72

Solution:
Percent · base = amount
$n \cdot 572 = 291.72$
$$n = \frac{291.72}{572}$$
$n = 0.51 = 51\%$
Wisconsin produced 51% of the cranberry crop.

83. Strategy:
To find the total revenue, use the basic percent equation. Percent = 16% = 0.16, base = n, amount = 2,240,000

Solution:
Percent · base = amount
$0.16 \cdot n = 2{,}240{,}000$
$$n = \frac{2{,}240{,}000}{0.16}$$
$n = 14{,}000{,}000$
14,000,000 oz of gold were minded.

85. Strategy:
To find the percent, use the basic percent equation. Add all the costs to find the base.
Percent = n, base = (total), amount = 4,020

Solution:
1,400 +1,070 + 1,220 + 2,960 + 3,930 + 4,020 = 14,600
Percent · base = amount
$n \cdot 14{,}600 = 4{,}020$
$$n = \frac{4{,}020}{14{,}600}$$
$n \approx 0.275 = 27.5\%$
Of the total cost, 27.5% is spent on food.

87. Strategy:
To find number of people, use the basic percent equation.
Percent = 30% = 0.30,
base = 44, amount = n

Solution:
Percent · base = amount
$0.30 \cdot 44 = n$
$13.2 = n$
About 13.2 million people aged 18 to 24 do not have health insurance.

89. Strategy:
To find the percent,
→Add the number of people who have been diagnosed and who have not been diagnosed to find the base (14.6 + 6.2 = 20.8).
→Use the basic percent equation.
Percent = n, base = 20.8, amount = 6.2

Solution:
Percent · base = amount
$n \cdot 20.8 = 6.2$
$$n = \frac{6.2}{20.8}$$
$n \approx 0.29807 \approx 29.8\%$
$n = 12{,}150$
Around 29.8% of Americans with diabetes have not been diagnosed.

91. Strategy:
To find how many fewer faculty members described their views as conservative than middle of the road,
→Subtract the percent of Conservative from the percent of Middle of the road
→Use the basic percent equation.
Percent = difference, base = 32,840, amount = n

Solution:
Percent · base = amount
$34.3 - 17.7 = 16.6\% = 0.166$
$0.166 \cdot 32{,}840 = n$
$n = 5{,}451$
There were 5,451 fewer faculty members who described their views as conservative than middle of the road.

Critical Thinking 6.4

93. Increase 100 by 10% $100 + 0.10(100) = 100 + 10 = 110$
Now decrease 110 by 10% $110 - 0.10(110) = 110 - 11 = 99$
No, the new number is not the original number. The 10% increase applied to the number 100, while the decrease in 10% applied to 110, thus the decrease was greater than the increase.

Projects or Group Activities 6.4

95. **a.** $\frac{100 \text{ lb}}{100 \text{ squares}} = 1 \text{ lb per square}$

b. Percent · base = amount
$0.01 \cdot 100 = n$
$1 = n$
1 lb is 1% of 100 lb.

c. 50 · 1 lb per square = 50 lb

d. Percent · base = amount
$0.50 \cdot 100 = n$
$50 = n$
50 lb is 50% of 100 lb.

e. 25 · 1 lb per square = 25 lb

f. Percent · base = amount
$0.25 \cdot 100 = n$
$25 = n$
25 lb is 25% of 100 lb.

g. 70 squares should be shaded as each square represents 1 lb

h. Percent · base = amount
$n \cdot 100 = 70$
$n = \frac{70}{100} = 0.70 = 70\%$
70 lb is 70% of 100 lb.

97. **a.** $\frac{20 \text{ lb}}{100 \text{ squares}} = 0.2 \text{ lb per square}$

b. Percent · base = amount
$0.01 \cdot 20 = n$
$0.22 = n$
0.2 lb is 1% of 20 lb.

c. 50 · 0.2 lb per square = 10 lb

d. Percent · base = amount
$0.50 \cdot 20 = n$
$10 = n$
10 lb is 50% of 20 lb.

e. 25 · 0.2 lb per square = 5 lb

f. Percent · base = amount
$0.25 \cdot 20 = n$
$5 = n$
5 lb is 25% of 20 lb.

g. 40 squares should be shaded as each square represents 0.2 lb

h. Percent · base = amount
$n \cdot 20 = 8$
$n = \frac{8}{20} = 0.40 = 40\%$
8 lb is 40% of 20 lb.

Section 6.5

Concept Check

1. Answers will vary

Objective A Exercises

3. $I = Prt$

1 month: $I = 5{,}000 \cdot 0.06 \cdot \frac{1}{12} = \$25$

2 month: $I = 5{,}000 \cdot 0.06 \cdot \frac{2}{12} = \$50$

3 month: $I = 5{,}000 \cdot 0.06 \cdot \frac{3}{12} = \$75$

4 month: $I = 5{,}000 \cdot 0.06 \cdot \frac{4}{12} = \$100$

5 month: $I = 5{,}000 \cdot 0.06 \cdot \frac{5}{12} = \$125$

5. To calculate the interest for a 7-month loan, multiply the interest due on a one month loan by 7.

7. Strategy:
To find the simple interest, solve the simple interest formula $I = Prt$ for I.

$P = 7{,}500$, $t = \frac{75}{365}$,
$r = 0.045$

Solution:
$I = Prt$

$I = (7{,}500)(0.045)\left(\frac{75}{365}\right)$

$I \approx 69.349$
The interest on the loan is \$69.35

9. Strategy:
To find the simple interest, solve the simple interest formula $I = Prt$ for I.

$P = 50{,}000$, $t = \frac{8}{12}$,
$r = 0.035$

Solution:
$I = Prt$

$I = (50{,}000)(0.035)\left(\frac{8}{12}\right)$

$I \approx 1{,}166.666$
The interest on the loan is \$1,166.67.

11. Strategy:
To find the simple interest, solve the simple interest formula $I = Prt$ for I.
$P = 800$,
$rt = 0.02$ (rate per month)

Solution:
$I = Prt$
$I = (800)(0.02)$
$I = 16$
The interest owed to the credit union is \$16.

13. Strategy:
To find the simple interest, solve the interest formula
$I = Prt$ for I.

$P = 1{,}500$, $t = 1\frac{1}{2}$,
$r = 0.075$

Solution:
$I = Prt$

$I = (1{,}500)(0.075)\left(\frac{3}{2}\right)$

$I = 168.75$
The interest on the loan is \$168.75.

15. Strategy:
To find the maturity value:
→Solve the formula $I = Prt$ for I.
$P = 25{,}000$, $r = 0.082$, $t = 1$
→Use the formula for the maturity value of a simple interest loan, $M = P + I$.

Solution:
$I = Prt$
$I = 25{,}000(0.082)(1)$
$I = 2{,}050$
$M = P + I$
$M = 25{,}000 + 2{,}050$
$M = 27{,}050$
The maturity value is \$27,050.

17. Strategy:
To find the maturity value:
→Solve the formula $I = Prt$ for I.

$P = 5{,}000$, $r = 0.069$, $t = \frac{18}{12}$.

→Use the formula for the maturity value of a simple interest loan, $M = P + I$.

Solution:
$I = Prt$

$I = (5{,}000)(0.069)\left(\frac{18}{12}\right)$

$I = 517.50$
$M = P + I$
$M = 5{,}000 + 517.50$
$M = 5{,}517.50$
The maturity value is \$5,517.50

19. Strategy:
To find the interest rate, solve the formula $I = Prt$ for r.
$I = 168.75, P = 3{,}000, t = \frac{9}{12}$

Solution:
$I = Prt$
$168.75 = (3{,}000)(r)\left(\frac{9}{12}\right)$
$168.75 = 2{,}250r$
$\frac{168.75}{2{,}250} = r$
$0.075 = r$
The simple interest rate is 7.5%.

21. Strategy:
To find the rate, solve the formula $I = Prt$ for r.
$I = 604.80, P = 18{,}000, t = \frac{210}{365}$

Solution:
$I = Prt$
$604.80 = 18{,}000(r)\left(\frac{210}{365}\right)$
$604.80 = \frac{756{,}000}{73}r$
$\frac{73 \cdot 604.80}{756{,}000} = r$
$0.0584 = r$
The interest rate is 5.84%.

Critical Thinking 6.5

23. a. At 8.75%,
$I = Prt$
$I = (10{,}000)(0.0875)\left(\frac{6}{12}\right)$
$I = 437.5$
$M = P + I$
$M = 10{,}000 + 437.5$
$M = 10{,}437.5$
The maturity value is $10,437.50

At 6.25%,
$I = Prt$
$I = (10{,}000)(0.0625)\left(\frac{6}{12}\right)$
$I = 312.5$
$M = P + I$
$M = 10{,}000 + 312.5$
$M = 10{,}312.50$
The maturity value is $10,312.50

The difference between the maturity values is $10,437.50 - $10,312.50 = $125.

b. The difference in the monthly payments would be the difference in the maturity values divided by the number of months:
$\frac{125}{6} \approx 20.8333$
The difference between the monthly payments would be $20.83.

Chapter Review Exercises

1. $\frac{100 \text{ lb}}{100 \text{ lb}} = \frac{1}{1}$, 1 : 1, 1 to 1

2. $\frac{18 \text{ roof supports}}{9 \text{ ft}} = \frac{2 \text{ roof supports}}{1 \text{ ft}}$

3. $\frac{\$628}{40 \text{ h}} = \$15.70/\text{h}$

4. $\frac{8 \text{ h}}{15 \text{ h}} = \frac{8}{15}$

5. $\frac{n}{3} = \frac{8}{15}$
$n \cdot 15 = 3 \cdot 8$
$15n = 24$
$n = \frac{24}{15}$
$n = 1.6$

6. $\frac{15 \text{ lb}}{12 \text{ trees}} = \frac{5 \text{ lb}}{4 \text{ trees}}$

7. $\frac{171 \text{ mi}}{3 \text{ h}} = 57 \text{ mph}$

8. $\frac{2}{3.5} = \frac{n}{12}$
$2 \cdot 12 = 3.5 \cdot n$
$24 = 3.5n$
$\frac{24}{3.5} = n$
$6.86 \approx n$

9. $32\% = 32\left(\frac{1}{100}\right) = \frac{32}{100} = \frac{8}{25}$

10. $22\% = 22(0.01) = 0.22$

11. $25\% = 25\left(\frac{1}{100}\right) = \frac{25}{100} = \frac{1}{4}$
$25\% = 25(0.01) = 0.25$

12. $3\frac{2}{5}\% = 3\frac{2}{5}\left(\frac{1}{100}\right) = \frac{17}{5}\left(\frac{1}{100}\right) = \frac{17}{500}$

13. $\frac{7}{40} = \frac{7}{40}(100\%) = 17.5\%$

14. $1\frac{2}{7} = 1\frac{2}{7}(100\%) = \frac{9}{7}(100\%) = \frac{900}{7}\%$
$\approx 128.6\%$

15. $2.8 = 2.8(100\%) = 280\%$

16. Strategy:
To find the amount, solve the basic percent equation. Percent = 42% = 0.42, base = 50, amount = n

Solution:
Percent · base = amount
$0.42 \cdot 50 = n$
$21 = n$
42% of 50 is 21.

17. Strategy:
To find the percent, solve the basic percent equation. Percent = n, base = 3, amount = 15

Solution:
Percent · base = amount
$n \cdot 3 = 15$
$n = \frac{15}{3}$
$n = 5 = 500\%$
500% of 3 is 15.

18. Strategy:
To find the percent, solve the basic percent equation. Percent = n, base = 18, amount = 12

Solution:
Percent · base = amount
$n \cdot 18 = 12$
$n = \frac{12}{18}$
$n \approx 0.667 = 66.7\%$
12 is approximately 66.7% of 18.

19. Strategy:
To find the amount, solve the basic percent equation. Percent = 150% = 1.50, base = 20, amount = n

Solution:
Percent · base = amount
$1.50 \cdot 20 = n$
$30 = n$
150% of 20 is 30.

20. Strategy:
To find the amount, solve the basic percent equation. Percent = 18% = 0.18, base = 85, amount = n

Solution:
Percent · base = amount
$0.18 \cdot 85 = n$
$15.3 = n$
18% of 85 is 15.3.

21. Strategy:
To find the base, solve the basic percent equation. Percent = 32% = 0.32, base = n, amount = 180

Solution:
Percent · base = amount
$0.32 \cdot n = 180$
$n = \frac{180}{0.32}$
$n = 562.5$
32% of 562.5 is 180.

22. Strategy:
To find the percent, solve the basic percent equation. Percent = n, base = 80, amount = 4.5

Solution:
Percent · base = amount
$n \cdot 80 = 4.5$
$n = \frac{4.5}{80}$
$n = 0.05625 = 5.625\%$
4.5 is 5.625% of 80.

23. Strategy:
To find the ratio, form the ratio of the smaller gear to the larger gear

Solution:
$125 - 75 = 50$
$\frac{10}{50} = \frac{1}{5}$
The ratio of the two meshed gears is $\frac{1}{5}$.

24. Strategy:
To find the amount invested, write and solve for proportion using n to represent the amount of money invested.

Solution:
$\frac{8{,}000}{520} = \frac{n}{780}$
$8{,}000 \cdot 780 = 520 \cdot n$
$6{,}240{,}000 = 520n$
$\frac{6{,}240{,}000}{520} = n$
$12{,}000 = n$
The amount of money invested is $12,000.

25. Strategy:
To find the amount of plant food, write and solve a proportion using n to represent the amount of plant food.

Solution:
$\frac{0.5}{50} = \frac{n}{275}$
$0.5 \cdot 275 = 50 \cdot n$
$137.5 = 50n$
$\frac{137.5}{50} = n$
$2.75 = n$
2.75 lb of plant food should be used.

26. Strategy:
To find the amount the other attorney receives, write and solve a proportion using n to represent the amount.

Solution:
$\frac{3}{2} = \frac{96{,}000}{n}$
$3 \cdot n = 2 \cdot 96{,}000$
$3n = 192{,}000$
$\frac{3n}{3} = \frac{192{,}000}{3}$
$n = 64{,}000$
The attorney receives $64,000.

27. Strategy:
To find the percent visiting in China:
→Add to find the total amount of tourists visiting the four countries.
→Use the basic percent equation to find the percent.
Percent = n, base = the total amount of tourists visiting the four countries, amount = 137 million

Solution:
$137 + 93 + 71 + 102 = 403$
Percent · base = amount
$n \cdot 403 = 137$
$n = \frac{137}{403}$
$n \approx 0.340 = 34.0\%$
About 34.0% of the tourists will be visiting China.

28. Strategy:
To find the amount of the budget spent for advertising, use the basic percent equation.
Percent = 7% = 0.07,
base = 120,000, amount = n

Solution:
Percent · base = amount
$0.07 \cdot 120{,}000 = n$
$8{,}400 = n$
$8,400 of the budget was spent for advertising.

29. Strategy:
To find the number of phones that were not defective:
→Find the number of defective phones by using the basic percent equation.
Percent = 1.2% = 0.012,
base = 4,000, amount = n
→Subtract the number of defective phones from 4,000.

Solution:
Percent · base = amount
$0.012 \cdot 4{,}000 = n$
$48 = n$
$4{,}000 - 48 = 3{,}952$
3,952 of the telephones were not defective.

30. Strategy:
To find the percent of the week, use the basic percent equation.
Percent = n, base = 168, amount = 61.35.

Solution:
Percent · base = amount
$n \cdot 168 = 61.35$
$n = \dfrac{61.35}{168}$
$n \approx 0.365 = 36.5\%$
The percent is approximately 36.5% of the week.

31. Strategy:
To find the simple interest, solve the formula
$I = Prt$ for I.
$P = 3{,}000,\ r = 0.086,\ t = \dfrac{45}{365}$

Solution:
$I = Prt$
$I = 3{,}000(0.086)\left(\dfrac{45}{365}\right)$
$I = 31.81$
The interest on the loan is $31.81.

32. Strategy:
To find the rate, solve the formula $I = Prt$ for r.
$I = 7{,}397.26,\ P = 500{,}000,\ t = \dfrac{60}{365}$

Solution:
$I = Prt$
$7{,}397.26 = 500{,}000(r)\left(\dfrac{60}{365}\right)$
$7{,}397.26 = \dfrac{6{,}000{,}000}{73}r$
$\dfrac{73 \cdot 7{,}397.26}{6{,}000{,}000} = r$
$0.09 = r$
The interest rate is 9.00%.

33. Strategy:
To find the maturity value:
→Solve the formula $I = Prt$ for I.
$P = 10{,}000,\ r = 0.054,$
$t = \dfrac{9}{12}.$
→Use the formula for the maturity value of a simple interest loan, $M = P + I$.

Solution:
$I = Prt$
$I = (10{,}000)(0.054)\left(\dfrac{9}{12}\right)$
$I = 405$
$M = P + I$
$M = 10{,}000 + 405$
$M = 10{,}405$
The maturity value is $10,405.

Chapter Test

1. $\dfrac{3\text{ yd}}{24\text{ yd}} = \dfrac{1}{8}$, 1 : 8, 1 to 8

2. $\dfrac{16\text{ oz}}{64\text{ cookies}} = \dfrac{1\text{ oz}}{4\text{ cookies}}$

3. $\dfrac{120\text{ mi}}{200\text{ min}} = 0.6\text{ mi/min}$

4. $\dfrac{200\text{ ft}}{100\text{ ft}} = \dfrac{2}{1}$

5. $\frac{n}{5}=\frac{3}{20}$
$n \cdot 20 = 5 \cdot 3$
$20n = 15$
$n = \frac{15}{20}$
$n = 0.75$

6. $\frac{8\text{ ft}}{4\text{ s}} = 2\text{ ft/s}$

7. $\frac{2{,}860\text{ ft}^2}{6\text{h}} = 476.67\text{ ft}^2/\text{h}$

8. $\frac{n}{4}=\frac{8}{9}$
$n \cdot 9 = 4 \cdot 8$
$9n = 32$
$\frac{9n}{9}=\frac{32}{9}$
$n \approx 3.56$

9. $86.4\% = 86.4(0.01) = 0.864$

10. $0.4 = 0.4(100\%) = 40\%$

11. $\frac{5}{4}=\frac{5}{4}(100\%) = 125\%$

12. $83\frac{1}{3}\% = 83\frac{1}{3}\left(\frac{1}{100}\right)=\frac{250}{3}\left(\frac{1}{100}\right)=\frac{5}{6}$

13. $32\% = 32\left(\frac{1}{100}\right)=\frac{32}{100}=\frac{8}{25}$

14. $1.18 = 1.18(100\%) = 118\%$

15. Strategy:
To find the base, solve the basic percent equation. Percent = 20% = 0.20, base = n, amount = 18

Solution:
Percent · base = amount
$0.20 \cdot n = 18$
$n = \frac{18}{0.20} = 90$
20% of 90 is 18.

16. Strategy:
To find the amount, solve the basic percent equation. Percent = 68% = 0.68, base = 73, amount = n

Solution:
Percent · base = amount
$0.68 \cdot 73 = n$
$49.64 = n$
68% of 73 is 49.64.

17. Strategy:
To find the percent, solve the basic percent equation. Percent = n, base = 320, amount = 180

Solution:
Percent · base = amount
$n \cdot 320 = 180$
$n = \frac{180}{320}$
$n = 0.5625 = 56.25\%$
56.25% of 320 is 180.

18. Strategy:
To find the base, solve the basic percent equation. Percent = 14% = 0.14, base = n, amount = 28

Solution:
Percent · base = amount
$0.14 \cdot n = 28$
$n = \frac{28}{0.14} = 200$
14% of 200 is 28.

19. Strategy:
To find the ratio, form the ratio of the lung capacity of an inactive male to an active male.

Solution:
$\frac{30}{60}=\frac{1}{2}$

The ratio of the lung capacity is $\frac{1}{2}$, 1:2, or 1 to 2.

20. Strategy:
To find the sales tax, write and solve a proportion using x to represent the amount of tax.

Solution:
$\frac{\$7.60}{\$95} = \frac{x}{\$39,200}$
$7.60 \cdot 39{,}200 = 95 \cdot x$
$297{,}920 = 95x$
$\frac{297{,}920}{95} = \frac{95x}{95}$
$3{,}136 = x$
The sales tax is \$3,136.

21. Strategy:
To find the number of registered voters that would vote, write and solve a proportion using n to represent the number of people that would vote.

Solution:
$\frac{3}{4} = \frac{n}{325{,}000}$
$3 \cdot 325{,}000 = 4n$
$975{,}000 = 4n$
$\frac{975{,}000}{4} = n$
$243{,}750 = n$
243,750 of the registered voters would vote.

22. Strategy:
To find the length of the room, write and solve a proportion using n to represent the length.

Solution:
$\frac{4}{1} = \frac{n}{12\frac{1}{2}}$
$4 \cdot \frac{25}{2} = 1 \cdot n$
$50 = n$
The length of the room is 50 ft.

23. Strategy:
To find the amount of expected accidents, use the basic percent equation.
Percent = 2.2% = 0.022,
base = 1,500, amount = n

Solution:
Percent · base = amount
$0.022 \cdot 1{,}500 = n$
$33 = n$
33 accidents are expected.

24. Strategy:
To find the percent answered correctly, use the basic percent equation.
Percent = n, base = 90,
amount = 90 – 16 = 74.

Solution:
Percent · base = amount
$n \cdot 90 = 74$
$n = \frac{74}{90}$
$n \approx 0.822 = 82.2\%$
The percent is approximately 82.2% correct.

25. Strategy:
To find the dollar increase:
→Find the increase by using the basic percent equation.
Percent = 120% = 1.20, base = n, amount = 480
→Subtract the increase from 480.

Solution:
Percent · base = amount
$1.20 \cdot n = 480$
$n = \frac{480}{1.20}$
$n = 400$
$480 - 400 = 80$
The dollar increase is \$80.

26. Strategy:
To find the percent, use the basic percent equation. Percent = n, base = 4, amount = 24.

Solution:
Percent · base = amount
$n \cdot 4 = 24$
$n = \frac{24}{4}$
$n = 6 = 600\%$
The number of grams of fat in the beef burger is 600% the number of grams in the soy burger.

27. Strategy:
To find the simple interest, solve the interest formula
$I = Prt$ for I.
$P = 5{,}000,\ t = \frac{9}{12},$
$r = 0.054$

Solution:
$I = Prt$
$I = (5{,}000)(0.054)\left(\frac{9}{12}\right)$
$I = 202.5$
The interest on the loan is \$202.50.

28. Strategy:
To find the maturity value:
→Solve the formula $I = Prt$ for I.
$P = 40{,}000,\ r = 0.0625,$
$t = \frac{150}{365}.$
→Use the formula for the maturity value of a simple interest loan, $M = P + I$.

Solution:
$I = Prt$
$I = (40{,}000)(0.0625)\left(\frac{150}{365}\right)$
$I = 1{,}027.40$
$M = P + I$
$M = 40{,}000 + 1{,}027.40$
$M = 41{,}027.40$
The maturity value is \$41,027.40.

29. Strategy:
To find the rate, solve the formula $I = Prt$ for r.
$I = 672,\ P = 12{,}000,\ t = \frac{8}{12}$

Solution:
$I = Prt$
$672 = 12{,}000(r)\left(\frac{8}{12}\right)$
$672 = 8{,}000r$
$\frac{672}{8{,}000} = r$
$0.084 = r$
The interest rate is 8.4%.

Cumulative Review Exercises

1. $18 \div \frac{6-3}{9} - (-3)$
$= 18 \div \frac{3}{9} - (-3)$
$= 18 \div \frac{1}{3} - (-3)$
$= 18 \cdot \frac{3}{1} - (-3)$
$= 54 - (-3)$
$= 54 + 3$
$= 57$

2. $5^4 = 5 \cdot 5 \cdot 5 \cdot 5$
$= 625$

3. $7\frac{5}{12} - 3\frac{5}{9} = 7\frac{15}{36} - 3\frac{20}{36}$
$= 6\frac{51}{36} - 3\frac{20}{36}$
$= 3\frac{31}{36}$

4. $\frac{4}{5} \div \frac{4}{5} + \frac{2}{3} = \frac{4}{5} \cdot \frac{5}{4} + \frac{2}{3}$
$= 1 + \frac{2}{3}$
$= 1\frac{2}{3}$

5. $342 \div (-3) = -114$

6. $2a - 3ab$
$2(2) - 3(2)(-3)$
$= 4 - 3(2)(-3)$
$= 4 - 6(-3)$
$= 4 - (-18)$
$= 4 + 18$
$= 22$

7. $5x - 20 = 0$
$5x - 20 + 20 = 0 + 20$
$5x = 20$
$\frac{5x}{5} = \frac{20}{5}$
$x = 4$
The solution is 4.

8. $3(x - 4) + 2x = 3$
$3x - 12 + 2x = 3$
$5x - 12 = 3$
$5x - 12 + 12 = 3 + 12$
$5x = 15$
$\frac{5x}{5} = \frac{15}{5}$
$x = 3$
The solution is 3.

9. $1.005 \times 10^5 = 100{,}500$

10. $-\frac{5}{8} - \left(-\frac{3}{4}\right) + \frac{5}{6} = -\frac{5}{8} + \frac{3}{4} + \frac{5}{6}$
$= \frac{-15}{24} + \frac{18}{24} + \frac{20}{24}$
$= \frac{-15+18+20}{24}$
$= \frac{23}{24}$

11. $(-5)^2 - (-8) \div (7-5)^2 \cdot 2 - 8$
$= (-5)^2 - (-8) \div 2^2 \cdot 2 - 8$
$= 25 - (-8) \div 4 \cdot 2 - 8$
$= 25 - (-2) \cdot 2 - 8$
$= 25 - (-4) - 8$
$= 25 + 4 - 8$
$= 29 - 8$
$= 21$

12. $\left(-\frac{2}{3}\right)\left(-\frac{3}{4}\right)^2 = \left(-\frac{2}{3}\right)\left(\frac{9}{16}\right)$
$= -\frac{2}{3} \cdot \frac{9}{16}$
$= -\frac{2 \cdot 9}{3 \cdot 16}$
$= -\frac{2 \cdot 3 \cdot 3}{3 \cdot 2 \cdot 2 \cdot 2 \cdot 2}$
$= -\frac{3}{8}$

13. $4 - (-3) + 5 - 8 = 4 + 3 + 5 + (-8)$
$= 7 + 5 + (-8)$
$= 12 + (-8)$
$= 4$

14. $5 - 2(1 - 3a) + 2(a - 3)$
$= 5 - 2 + 6a + 2a - 6$
$= 8a - 3$

15. $\frac{7}{8} \div \frac{5}{16} = \frac{7}{8} \cdot \frac{16}{5}$
$= \frac{7 \cdot 16}{8 \cdot 5}$
$= \frac{7 \cdot 2 \cdot 2 \cdot 2 \cdot 2}{2 \cdot 2 \cdot 2 \cdot 5}$
$= \frac{14}{5} = 2\frac{4}{5}$

16. $-3y^2 + 3y - y^2 - 6y$
$= \left(-3y^2 - y^2\right) + \left(3y - 6y\right)$
$= -4^2 - 3y$

17. $\frac{3}{4}x = -9$
$\frac{4}{3}\left(\frac{3}{4}x\right) = \frac{4}{3}(-9)$
$x = -12$
The solution is –12.

18. $\frac{30 \text{ cents}}{1 \text{ dollar}} = \frac{30 \text{ cents}}{100 \text{ cents}} = \frac{3}{10}$

19. $\frac{\$19{,}425}{5 \text{ months}} = \$3{,}885/\text{month}$

20. $a - b$
$102.5 - 77.546 = 24.954$

21. $\frac{2}{3} = \frac{n}{48}$
$2 \cdot 48 = 3 \cdot n$
$96 = 3n$
$\frac{96}{3} = n$
$32 = n$

22. $\frac{\frac{1}{2} + \frac{3}{4}}{2 - \frac{5}{8}} = \frac{\frac{2}{4} + \frac{3}{4}}{\frac{16}{8} - \frac{5}{8}}$
$= \frac{\frac{5}{4}}{\frac{11}{8}}$
$= \frac{5}{4} \div \frac{11}{8}$
$= \frac{5}{4} \cdot \frac{8}{11}$
$= \frac{5 \cdot 8}{4 \cdot 11}$
$= \frac{5 \cdot 2 \cdot 2 \cdot 2}{2 \cdot 2 \cdot 11} = \frac{10}{11}$

23. Strategy:
To find the percent, use the basic percent equation. Percent = n, base = 30, amount = 2.5

Solution:
Percent · base = amount
$n \cdot 30 = 2.5$
$n = \frac{2.5}{30} \approx 0.0833$
2.5 is approximately 8.3% of 30.

24. Strategy:
To find the amount, use the basic percent equation. Percent = 42% = 0.42, base = 160, amount = n

Solution:
Percent · base = amount
$0.42 \cdot 160 = n$
$67.2 = n$
42% of 160 is 67.2.

25. Strategy:
To find the monthly difference:
→Find the difference in annual expenses between the northeast (587) and south (243).
→Convert the difference from annual to monthly by dividing by 12.

Solution:
$587 - 243 = 344$
$344 \div 12 \approx 28.67$
The monthly difference is $28.67.

26. The unknown number: x

five less than two thirds of a number	is	three

$\frac{2}{3}x - 5 = 3$
$\frac{2}{3}x - 5 + 5 = 3 + 5$
$\frac{2}{3}x = 8$
$\frac{3}{2}\left(\frac{2}{3}x\right) = \frac{3}{2} \cdot 8$
$x = 12$
The number is 12.

27. Let the number be x.
The <u>difference</u> between four <u>times</u> a number and three <u>times</u> the sum of the number and two.
$4x - 3(x + 2)$
$4x - 3x - 6$
$x - 6$

28. Strategy:
To find the number of miles left to drive:
→Find the number of miles already driven by subtracting the original odometer reading (18,325) from the present odometer reading (18,386).
→Subtract the number of miles already drive from 125.

Solution:
$$\begin{array}{r} 18{,}386 \\ -\,18{,}325 \\ \hline 61 \end{array}$$
$$\begin{array}{r} 125 \\ -\,61 \\ \hline 64 \end{array}$$
You have 64 mi left to drive.

29. Strategy:
To find the new checking balance:
→Add the deposit (122.35) to the checking account balance (422.89).
→Subtract the check (279.76) from the new checking account balance.

Solution:
$$\begin{array}{r} 422.89 \\ +\,122.35 \\ \hline 545.24 \end{array}$$
$$\begin{array}{r} 545.24 \\ -\,279.76 \\ \hline 265.48 \end{array}$$
The new checking account balance is $265.48.

30. Strategy:
To find the part that remains to be completed:
→Add the amount already done $\left(\frac{2}{5} + \frac{1}{3}\right)$.
→Subtract the amounts already done from the total job (1).

Solution:
$\frac{2}{5} + \frac{1}{3} = \frac{6}{15} + \frac{5}{15} = \frac{11}{15}$
$1 - \frac{11}{15} = \frac{15}{15} - \frac{11}{15} = \frac{4}{15}$
$\frac{4}{15}$ of the job remains to be done.

31. Strategy:
To find the fraction, convert 10% to a fraction.

Solution:

$$10\% = 10\left(\frac{1}{100}\right) = \frac{10}{100} = \frac{1}{10}$$

$\frac{1}{10}$ of the population aged 75–84 are affected by Alzheimer's Disease.

32. Strategy:
To find the number of miles driven per gallon, divide the number of miles driven (402.5) by the number of gallons of gas used (11.5).

Solution:

$$\frac{402.5}{11.5} = 35$$

The car travels 35 mi on each gallon of gas.

33. Strategy:
To find the rpm in third gear, write and solve an equation using n to represent the rpm in third gear.

Solution:

$\frac{2}{3}$ of the rpm in third gear	is	2,500

$$\frac{2}{3}n = 2{,}500$$

$$\frac{3}{2}\left(\frac{2}{3}n\right) = \frac{3}{2}(2{,}500)$$

$$n = 3{,}750$$

The rpm of the engine is 3,750 in third gear.

Chapter 7: Geometry

Prep Test

1. $2(18)+2(10)=36+20=56$

2. abc
$=(2)(3.14)(9)$
$=(6.28)(9)$
$=56.52$

3. xyz^3
$=\left(\frac{4}{3}\right)(3.14)(3)^3$
$=113.04$

4. $x+47=90$
$x+47-47=90-47$
$x=43$

5. $32+97+x=180$
$129+x=180$
$129-129+x=180-129$
$x=51$

6. $\frac{5}{12}=\frac{6}{x}$
$5x=12\cdot 6$
$\frac{5x}{5}=\frac{72}{5}$
$x=14.4$

Section 7.1

Concept Check

1. $12 = 5 + x + 4$

3. $x + 160° + 140° = 360°$

5. a; b

7. c; d; 180°

9. **a.** $\angle a$, $\angle b$, and $\angle c$

   b. $\angle y$ and $\angle z$

   c. $\angle x$

Objective A Exercises

11. The measure of the given angle is approximately 40°. The measure of the angle is between 0° and 90°. The angle is an acute angle.

13. The measure of the given angle is approximately 115°. The measure of the angle is between 90° and 180°. The angle is an obtuse angle.

15. The measure of the given angle is approximately 90°. The angle is a right angle.

17. Strategy:
Complementary angle are two angles whose sum is 90°. To find the complement, let x represent the complement of a 62° angle. Write an equation and solve for x.

Solution:
$x + 62° = 90°$
$x = 28°$
The complement of a 62° angle is a 28° angle.

19. Strategy:
Supplementary angles are two angles whose sum is 180°. To find the supplement, let x represent the supplement of a 162° angle. Write an equation and solve for x.

Solution:
$x + 162° = 180°$
$x = 18°$
The supplement of a 162° angle is an 18° angle.

21. $AB + BC + CD = AD$
$12 + BC + 9 = 35$
$21 + BC = 35$
$BC = 14$
$BC = 14$ cm

23. $QR + RS = QS$
$QR + 3(QR) = QS$
$7 + 3 \cdot 7 = QS$
$7 + 21 = QS$
$28 = QS$
$QS = 28$ ft

25. $EF + FG = EG$
$EF+\frac{1}{2}(EF)=EG$
$20+\frac{1}{2}(20)=EG$
$20 + 10 = EG$
$30 = EG$
$EG = 30$ m

27. $\angle LOM+\angle MON=\angle LON$
$53°+\angle MON=139°$
$\angle MON=139°-53°$
$\angle MON=86°$

29. Strategy:
To find the measure of $\angle x$, write an equation using the fact that the sum of the measures of $\angle x$ and $\angle 2x$ is 90°. Solve for $\angle x$.

Solution:
$x + 2x = 90°$
$3x = 90°$
$x = 30°$
The measure of x is 30°.

31. Strategy:
To find the measure of $\angle x$, write an equation using the fact that the sum of x and $x + 18°$ is 90°. Solve for x.

Solution:
$x + x + 18° = 90°$
$2x + 18° = 90°$
$2x = 72°$
$x = 36°$
The measure of $\angle x$ is 36°.

33. Strategy:
To find the measure of $\angle x$, write an equation using the fact that the sum of the measure of $\angle x$ and 74° is 145°. Solve for $\angle x$.

Solution:
$\angle x + 74° = 145°$
$\angle x = 71°$
The measure of $\angle x$ is 71°

35. Strategy:
To find the measure of $\angle a$, write an equation using the fact that the sum of the measure of $\angle a$ and 53° is 180°. Solve for $\angle a$.

Solution:
$\angle a + 53° = 180°$
$\angle a = 127°$
The measure of $\angle a$ is 127°.

37. Strategy:
The sum of the measures of the three angles shown is 360°. To find $\angle a$, write an equation and solve for $\angle a$.

Solution:
$\angle a + 76° + 168° = 360°$
$\angle a + 244° = 360°$
$\angle a = 116°$
The measure of $\angle a$ is 116°.

39. Strategy:
The sum of the measures of the three angles shown is 180°. To find x, write an equation and solve for x.

Solution:
$3x + 4x + 2x = 180°$
$9x = 180°$
$x = 20°$
The measure of x is 20°.

41. Strategy:
The sum of the measures of the three angles shown is 180°. To find x, write an equation and solve for x.

Solution:
$5x + (x + 20°) + 2x = 180°$
$8x + 20° = 180°$
$8x = 160°$
$x = 20°$
The measure of x is 20°.

43. Strategy:
The sum of the measures of the four angles shown is 360°. To find x, write an equation and solve for x.

Solution:
$3x + 4x + 6x + 5x = 360°$
$18x = 360°$
$x = 20°$
The measure of x is 20°.

45. Strategy:

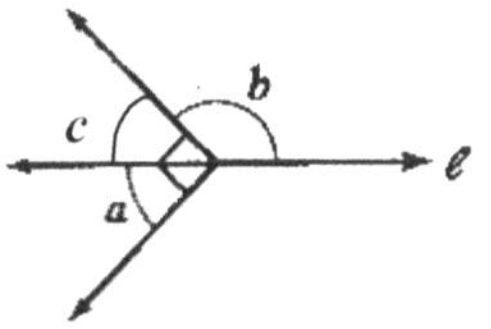

To find the measure of $\angle b$:

→Use the fact that $\angle a$ and $\angle c$ are complementary angles.

→Find $\angle b$ by using the fact that $\angle c$ and $\angle b$ are supplementary angles.

Solution:

$\angle a + \angle c = 90°$

$51° + \angle c = 90°$

$\angle c = 39°$

$\angle b + \angle c = 180°$

$\angle b + 39° = 180°$

$\angle b = 141°$

The measure of $\angle b$ is 141°.

47. $90° - x$

Objective B Exercises

49. Strategy:

The angles labeled are adjacent angles of intersecting lines and are, therefore, supplementary angles. To find x, write an equation and solve for x.

Solution:

$x + 74° = 180°$

$x = 106°$

The measure of x is 106°.

51. Strategy:

The angles labeled are vertical angles and are, therefore, equal. To find x, write an equation and solve for x.

Solution:

$5x = 3x + 22°$

$2x = 22°$

$x = 11°$

The measure of x is 11°.

53. Strategy:

→To find the measure of $\angle a$, use the fact that corresponding angles of parallel lines are equal.

→To find the measure of $\angle b$, use the fact that adjacent angles of intersecting lines are supplementary.

Solution:

$\angle a = 38°$

$\angle b + \angle a = 180°$

$\angle b + 38° = 180°$

$\angle b = 142°$

The measure of $\angle a$ is 38°.

The measure of $\angle b$ is 142°.

55. Strategy:

→To find the measure of $\angle a$, use the fact that alternate interior angles of parallel lines are equal.

→To find the measure of $\angle b$, use the fact that adjacent angles of intersecting lines are supplementary.

Solution:

$\angle a = 47°$

$\angle a + \angle b = 180°$

$47° + \angle b = 180°$

$\angle b = 133°$

The measure of $\angle a$ is 47°.

The measure of $\angle b$ is 133°.

57. Strategy:

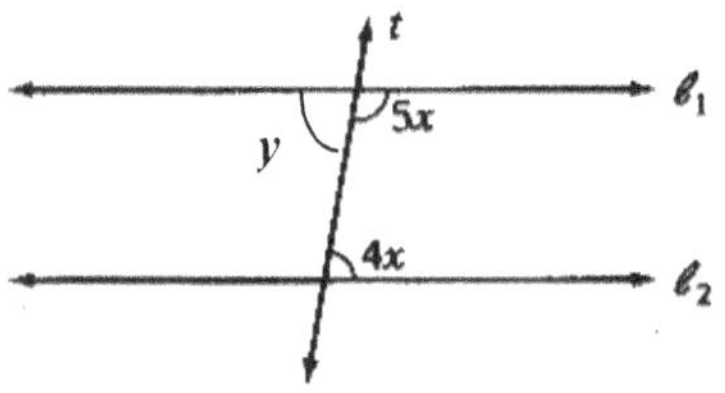

$4x = y$ because alternate interior angles have the same measure. $y + 5x = 180°$ because adjacent angles of intersecting lines are supplementary. Substitute $4x$ for y and solve for x.

Solution:

$4x + 5x = 180°$

$9x = 180°$

$x = 20°$

The measure of x is 20°.

59. False

61. True

Objective C Exercises

63. Strategy:

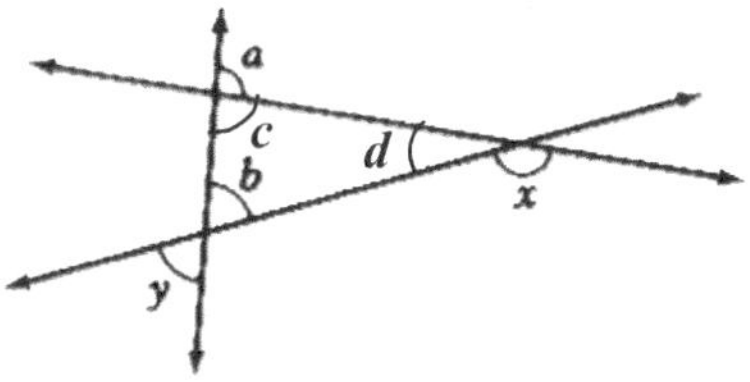

→To find the measure of ∠y, use the fact that ∠b and ∠y are vertical angles.
→To find the measure of ∠x:
Find the measure of ∠c by using the fact that the sum of an interior and exterior angle is 180°.
Find the measure of ∠d by using the fact that the sum of the interior angles of a triangle is 180°.
Find the measure of ∠x, by using the fact that the sum of an interior and exterior angle is 180°.

Solution:
$\angle y = \angle b = 70°$
$\angle a + \angle c = 180°$
$95° + \angle c = 180°$
$\angle c = 85°$
$\angle b + \angle c + \angle d = 180°$
$70° + 85° + \angle d = 180°$
$155° + \angle d = 180°$
$\angle d = 25°$
$\angle d + \angle x = 180°$
$25° + \angle x = 180°$
$\angle x = 155°$
The measure of ∠x is 155°.
The measure of ∠y is 70°.

65. Strategy:

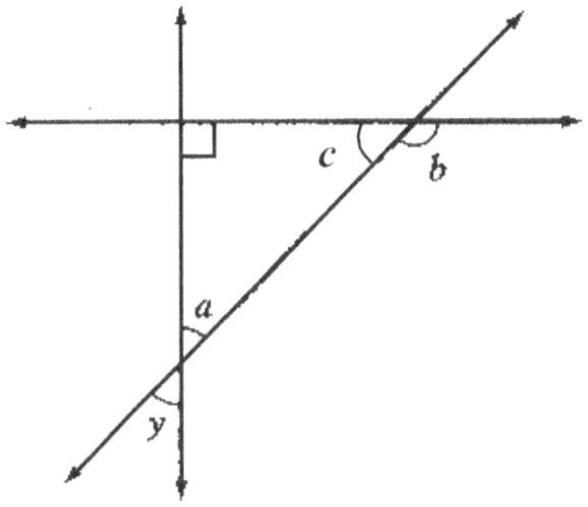

→To find the measure of ∠a, use the fact that ∠a and ∠y are vertical angles.
→To find the measure of ∠b:
Find the measure of ∠c by using the fact that the sum of the interior angles of a triangle is 180°.
Find the measure of ∠b by using the fact that the sum of an interior and exterior angle is 180°.

Solution:
$\angle a = \angle y = 45°$
$\angle a + \angle c + 90° = 180°$
$45° + \angle c + 90° = 180°$
$\angle c + 135° = 180°$
$\angle c = 45°$
$\angle c + \angle b = 180°$
$45° + \angle b = 180°$
$\angle b = 135°$
The measure of ∠a is 45°.
The measure of ∠b is 135°.

67. Strategy:
To find the measure of the third angle, use the fact that the sum of the measures of the interior angles of a triangle is 180°. Write an equation using x to represent the measure of the third angle. Solve the equation for x.

Solution:
$x + 90° + 30° = 180°$
$x + 120° = 180°$
$x = 60°$
The measure of the third angle is 60°.

69. Strategy:
To find the measure of the third angle, use the fact that the sum of the measures of the interior angles of a triangle is 180°. Write an equation using x to represent the measure of the third angle. Solve the equation for x.

Solution:
$x + 42° + 103° = 180°$
$x + 145° = 180°$
$x = 35°$
The measure of the third angle is 35°.

71. True

Critical Thinking 7.1

73. Putting the numbers in order, we have −2.5, 2, 3.5 and 5. We can see that 2 is not halfway between −2.5 and 3.5, but 3.5 is halfway between 2 and 5.

75. Let x be the measure of an angle. If the ratio of the angles of a triangle is 2 : 3 : 7, and the sum of the angles of a triangle is 180°, then we need to solve the equation
$2x + 3x + 7x = 180$
$12x = 180$
$x = \frac{180}{12} = 15$
So the measure of the largest angle would be $7x = 7(15) = 105°$.

Projects or Group Activities 7.1

77. The sum of the interior angles of a triangle is 180°; therefore, $\angle a + \angle b + \angle c = 180°$.
The sum of an interior and exterior angle is 180°; therefore, $\angle c + \angle x = 180°$.
Solving this equation for $\angle c$,
$\angle c = 180° - \angle x$.
Substitute $180° - \angle x$ for $\angle c$ in the equation
$\angle a + \angle b + \angle c = 180°$;
$\angle a + \angle b + 180° - \angle x = 180°$.
Add the measure of x to each side of the equation, and subtract 180° from each side of the equation: $\angle a + \angle b = \angle x$.
The measure of an exterior angle of a triangle is equal to the sum of the measures of the two opposite interior angles:
$\angle a + \angle c = \angle z$.

79.

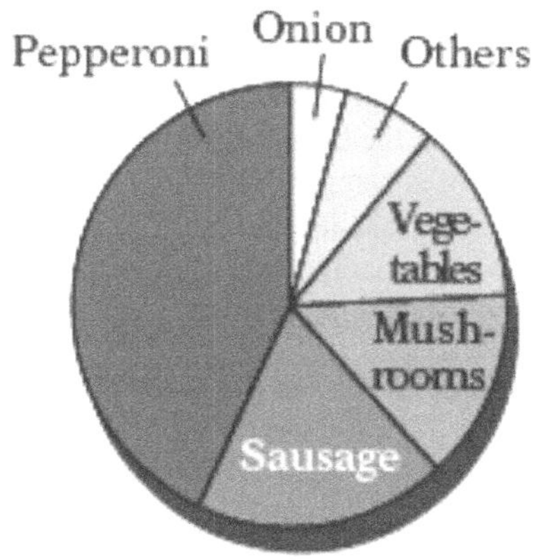

Section 7.2

Concept Check

1. The polygon has 6 sides.
The polygon is a hexagon.

3. The polygon has 5 sides.
The polygon is a pentagon.

5. The triangle has no sides equal.
The triangle is a scalene triangle.

7. The triangle has three sides equal.
The triangle is an equilateral triangle.

9. The triangle has one obtuse angle.
The triangle is an obtuse triangle.

11. The triangle has three acute angles.
The triangle is an acute triangle.

Objective A Exercises

13. Strategy:
To find the perimeter, use the formula for the perimeter of a triangle. Substitute 12 for a, 20 for b, and 24 for c. Solve for P.

Solution:
$P = a + b + c$
$P = 12 + 20 + 24$
$P = 56$
The perimeter is 56 in.

15. Strategy:
To find the perimeter, use the formula for the perimeter of a square. Substitute 3.5 for s and solve for P.

Solution:
$P = 4s$
$P = 4 \cdot 3.5$
$P = 14$
The perimeter is 14 ft.

17. Strategy:
To find the perimeter, use the formula for the perimeter of a rectangle. Substitute 13 for L and 10.5 for W. Solve for P.

Solution:
$P = 2L + 2W$
$P = 2 \cdot 13 + 2 \cdot 10.5$
$P = 26 + 21$
$P = 47$
The perimeter is 47 mi.

19. Strategy:
To find the circumference, use the circumference formula that involves the radius. For the exact answer, leave the answer in terms of π. For an approximation, use the π key on a calculator.
$r = 4$.

Solution:
$C = 2\pi r$
$C = 2\pi (4)$
$C = 8\pi$
$C \approx 25.13$
The circumference is 8π cm.
The circumference is approximately 25.13 cm.

21. Strategy:
To find the circumference, use the circumference formula that involves the radius. For the exact answer, leave the answer in terms of π. For an approximation use the π key on a calculator.
$r = 5.5$.

Solution:
$C = 2\pi r$
$C = 2\pi(5.5)$
$C = 11\pi$
$C \approx 34.56$
The circumference is 11π mi.
The circumference is approximately 34.56 mi.

23. Strategy:
To find the circumference, use the circumference formula that involves the diameter. For the exact answer leave the answer in terms of π. For an approximation use the π key on a calculator. $d = 17$.

Solution:
$C = \pi d$
$C = \pi(17)$
$C = 17\pi$
$C \approx 53.41$
The circumference is 17π ft.
The circumference is approximately 53.41 ft.

25. Strategy:
To find the perimeter, use the formula for the perimeter of a triangle. Substitute 3.8 for a, 5.2 for b, and 8.4 for c. Solve for P.

Solution:
$P = a + b + c$
$P = 3.8 + 5.2 + 8.4$
$P = 17.4$
The perimeter is 17.4 cm.

27. Strategy:
To find the perimeter, use the formula for the perimeter of a triangle. Substitute $2\frac{1}{2}$ for a and b, and 3 for c. Solve for P.

Solution:
$P = a + b + c$
$P = 2\frac{1}{2} + 2\frac{1}{2} + 3$
$P = 8$
The perimeter is 8 cm.

29. Strategy:
To find the perimeter, use the formula for the perimeter of a rectangle. Substitute 8.5 for L and 3.5 for W. Solve for P.

Solution:
$P = 2L + 2W$
$P = 2(8.5) + 2(3.5)$
$P = 17 + 7$
$P = 24$
The perimeter is 24 m.

31. Strategy:
To find the perimeter, multiply the measure of one of the equal sides (3.5) by 5.

Solution:
$P = 5(3.5)$
$P = 17.5$
The perimeter is 17.5 in.

33. Strategy:
To find the perimeter, use the formula for the perimeter of a square. Substitute 12.2 for s. Solve for P.

Solution:
$P = 4s$
$P = 4(12.2)$
$P = 48.8$
The perimeter is 48.8 cm.

35. Strategy:
To find the circumference, use the circumference formula that involves the diameter. Leave the answer in terms of π.
$d = 1.5$

Solution:
$C = \pi d$
$C = \pi(1.5)$
$C = 1.5\pi$
The circumference is 1.5π in.

37. Strategy:
To find the circumference, use the circumference formula that involves the radius. An approximation is asked for; use the π key on a calculator.
$r = 36$.

Solution:
$C = 2\pi r$
$C = 2\pi(36)$
$C = 72\pi$
$C \approx 226.19$
The circumference is approximately 226.19 cm.

39. Strategy:
To find the amount of fencing, use the formula for the perimeter of a rectangle. Substitute 18 for L and 12 for W. Solve for P.

Solution:
$P = 2L + 2W$
$P = 2(18) + 2(12)$
$P = 36 + 24$
$P = 60$
The perimeter of the garden is 60 ft.

41. Strategy:
To find the amount to be nailed down, use the formula for the perimeter of a rectangle. Substitute 12 for L and 10 for W. Solve for P.

Solution:
$P = 2L + 2W$
$P = 2(12) + 2(10)$
$P = 24 + 20$
$P = 44$
44 ft of carpet must be nailed down.

43. Strategy:
To find the length, use the formula for the perimeter of a rectangle. Substitute 440 for P and 100 for W. Solve for L.

Solution:
$P = 2L + 2W$
$440 = 2L + 2(100)$
$440 = 2L + 200$
$240 = 2L$
$120 = L$
The length is 120 ft.

45. Strategy:
To find the third side of the banner, use the formula for the perimeter of a triangle. Substitute 46 for P, 18 for a, and 18 for b. Solve for c.

Solution:
$P = a + b + c$
$46 = 18 + 18 + c$
$46 = 36 + c$
$10 = c$
The third side of the banner is 10 in.

47. Strategy:
To find the length of each side, use the formula for the perimeter of a square. Substitute 48 for P. Solve for s.

Solution:
$P = 4s$
$48 = 4s$
$12 = s$
The length of each side is 12 in.

49. Strategy:
To find the length of the diameter, use the circumference formula that involves the diameter. An approximation is asked for; use the π key on a calculator.
$C = 8$.

Solution:
$C = \pi d$
$8 = \pi d$
$\frac{8}{\pi} = d$
$2.55 \approx d$
The diameter is approximately 2.55 cm.

51. Strategy:
To find the length of molding, use the circumference formula that involves the diameter. An approximation is asked for; use the π key on a calculator.
$d = 4.2$.

Solution:
$C = \pi d$
$C = \pi(4.2)$
$C \approx 13.19$
The length of molding is approximately 13.19 ft.

53. Strategy:
To find the distance:
→Convert the diameter to feet.
→Multiply the circumference by 8. An approximation is asked for; use the π key on a calculator.

Solution:
$24 \text{ in.} = 24 \text{ in.} \cdot \frac{1 \text{ ft}}{12 \text{ in.}} = 2 \text{ ft}$
distance = $8C$
distance = $8\pi d$
distance = $8\pi(2)$
distance = 16π
distance ≈ 50.27
The bicycle travels approximately 50.27 ft.

55. Strategy:
To find the circumference of the earth, use the circumference formula that involves the radius. An approximation is asked for; use the π key on a calculator. $r = 6{,}356$.

Solution:
$C = 2\pi r$
$C = 2\pi(6{,}356)$
$C = 12{,}712\pi$
$C \approx 39{,}935.93$
The circumference of the earth is approximately 39,935.93 km.

57. A square whose side is 1 ft

Objective B Exercises

59. Strategy:
To find the area, use the formula for the area of a rectangle. Substitute 12 for L and 5 for W. Solve for A.

Solution:
$A = LW$
$A = 12(5)$
$A = 60$
The area is 60 ft^2.

61. Strategy:
To find the area, use the formula for the area of the square. Substitute 4.5 for s. Solve for A.

Solution:
$A = s^2$
$A = (4.5)^2$
$A = 20.25$
The area is 20.25 in^2.

63. Strategy:
To find the area, use the formula for the area of a triangle. Substitute 42 for b and 26 for h. Solve for A.

Solution:
$A = \frac{1}{2}bh$
$A = \frac{1}{2}(42)(26)$
$A = 546$
The area is 546 ft.^{2}.

65. Strategy:
To find the area, use the formula for the area of a circle. Substitute 4 for r. Solve for A. For the exact answer, leave the answer in terms of π. For an approximation, use the π key on a calculator.

Solution:
$A = \pi r^2$
$A = \pi(4)^2$
$A = 16\pi$
$A \approx 50.27$
The area is 16π cm^2.
The area is approximately 50.27 cm^2.

67. Strategy:
To find the area, use the formula for the area of a circle. Substitute 5.5 for r. Solve for A. For the exact answer, leave the answer in terms of π. For an approximation, use the π key on a calculator.

Solution:
$A = \pi r^2$
$A = \pi(5.5)^2$
$A = 30.25\pi$
$A \approx 95.03$
The area is 30.25π mi^2.
The area is approximately 95.03 mi^2.

69. Strategy:
To find the area:
→Find the radius of the circle.
→Use the formula for the area of a circle. For an exact answer, leave the answer in terms of π. For an approximation, use the π key on a calculator.

Solution:
$r = \frac{1}{2}d = \frac{1}{2}(17) = 8.5$
$A = \pi r^2$
$A = \pi(8.5)^2$
$A = 72.25\pi$
$A \approx 226.98$
The area is 72.25π ft^2.
The area is approximately 226.98 ft^2.

71. Strategy:
To find the area, use the formula for the area of a square. Substitute 12.5 for s. Solve for A.

Solution:
$A = s^2$
$A = (12.5)^2$
$A = 156.25$
The area is 156.25 cm^2.

73. Strategy:
To find the area, use the formula for the area of a rectangle. Substitute 38 for L and 15 for W. Solve for A.

Solution:
$A = LW$
$A = 38(15)$
$A = 570$
The area is 570 in^2.

75. Strategy:
To find the area, use the formula for the area of a parallelogram. Substitute 16 for b and 12 for h. Solve for A.

Solution:
$A = bh$
$A = 16(12)$
$A = 192$
The area is 192 in^2.

77. Strategy:
To find the area, use the formula for the area of a triangle. Substitute 6 for b and 4.5 for h. Solve for A.

Solution:
$A = \frac{1}{2}bh$
$A = \frac{1}{2}(6)(4.5)$
$A = 13.5$
The area is 13.5 ft^2.

79. Strategy:
To find the area, use the formula for the area of a trapezoid. Substitute 35 for b_1, 20 for b_2, and 12 for h. Solve for A.

Solution:
$A = \frac{1}{2}h(b_1 + b_2)$
$A = \frac{1}{2} \cdot 12(35 + 20)$
$A = 330$
The area is 330 cm^2.

81. Strategy:
To find the area, use the formula for the area of a circle. Leave the answer in terms of π. $r = 5$.

Solution:
$A = \pi r^2$
$A = \pi(5)^2$
$A = 25\pi$
The area is 25π in^2.

83. Strategy:
To approximate the area, use the formula for the area of a rectangle. Substitute 150 for L and 70 for W. Solve for A.

Solution:
$A = LW$
$A = 150(70)$
$A = 10{,}500$
The area of the reserve is approximately 10,500 mi^2.

85. Strategy:
To find the area, use the formula for the area of a circle. Leave the answer in terms of π. $r = 50$.

Solution:
$A = \pi r^2$
$A = \pi(50)^2$
$A = 2{,}500\pi$
The area is $2{,}500\pi$ ft^2.

87. Strategy:
To find the area, use the formula for the area of a square. Substitute 8.5 for s. Solve for A.

Solution:
$A = s^2$
$A = (8.5)^2$
$A = 72.25$
The area of the patio is 72.25 m^2.

89. Strategy:
To find the amount of turf, use the formula for the area of a rectangle. Substitute 100 for L and 75 for W. Solve for A.

Solution:
$A = LW$
$A = 100(75)$
$A = 7{,}500$
7,500 yd^2 of artificial turf must be purchased.

91. Strategy:
To find the width, use the formula for the area of a rectangle. Substitute 300 for A and 30 for L. Solve for W.

Solution:
$A = LW$
$300 = 30W$
$10 = W$
The width of the rectangle is 10 in.

93. Strategy:
To find the length of the base, use the formula for the area of a triangle. Substitute 50 for A and 5 for h. Solve for b.

Solution:
$A = \frac{1}{2}bh$
$50 = \frac{1}{2}b(5)$
$50 = \frac{5}{2}b$
$20 = b$
The base of the triangle is 20 m.

95. Strategy:
To find the number of quarts of stain:
→Use the formula for the area of a rectangle to find the area of the deck.
→Divide the area of the deck by the area one quart will cover (50).

Solution:
$A = LW$
$A = 10(8)$
$A = 80$
$80 \div 50 = 1.6$
Because a portion of a second quart is needed, 2 qt of stain should be purchased.

97. Strategy:
To find the cost of the wallpaper:
→Use the formula for the area of a rectangle to find the areas of the two walls.
→Add the areas of the two walls.
→Divide the total area by the area in one roll (40) to find the total number of rolls.
→Multiply the number of rolls by 37

Solution:
$A_1 = LW = 9(8) = 72$
$A_2 = LW = 11(8) = 88$
$A = A_1 + A_2 = 72 + 88 = 160$
$160 \div 40 = 4$
$4 \cdot 37 = 148$
The cost to wallpaper the two walls is $148.

99. Strategy:
Compute the area of each storage unit and compare to see which is just over the needed space (175 ft^2).

Solution:
Area of $10 \times 5 = 50$ ft^2
Area of $10 \times 10 = 100$ ft^2
Area of $10 \times 15 = 150$ ft^2
Area of $10 \times 20 = 200$ ft^2
Area of $10 \times 25 = 250$ ft^2
Area of $10 \times 30 = 300$ ft^2
You should select the 10×20 unit.

101. Strategy:
To find the increase in area:
→Use the formula for the area of a circle to find the area of a circle with $r = 6$.
→Use the formula for the area of a circle to find the area of a circle with $r = 2(6) = 12$.
→Subtract the area of the smaller circle from the area of the larger circle.
An approximation is asked for; use the π key on a calculator.

Solution:
$A_1 = \pi r^2$
$A_1 = \pi(6)^2 = 36\pi$
$A_2 = \pi(12)^2 = 144\pi$
$A_2 - A_1 = 144\pi - 36\pi$
$= 108\pi \approx 339.29$
The area is increased by 339.29 cm^2.

103. Strategy:
To find the cost of the paint:
→Use the formula for the area of a rectangle to find the area of the two walls that measure 15 by 9 and the two walls that measure 12 by 9.
→Add the area to find the total area.
→Divide the total area by the area that one gallon will cover (400).
→Multiply the number of gallons by 29.98.

Solution:
$A_1 = 2(LW) = 2[15(9)] = 270$
$A_1 = 2(LW) = 2[12(9)] = 216$
$A_1 + A_2 = 270 + 216 = 486$
$486 \div 400 = 1.22$
Because the portion of a second gallon is need, 2 gal of paint should be purchased.
$2(29.98) = 59.96$
The paint will cost $59.96.

105. Strategy:
To find the amount of material:
→Use the formula for the area of a rectangle to find the area of the drapes. Substitute $2 \cdot 2 = 4$ for W and $4 + 1 = 5$ for L.
→Multiply the area of one drape by 4.

Solution:
$A = LW = 5(4) = 20$
$4(20) = 80$
the drapes will required 80 ft^2 of material.

Critical Thinking 7.2

107. 4 times

Projects or Group Activities 7.2

109. 5 units by 5 units

111. There are a great number of quilt patterns that incorporate regular polygons, for example, Nine Patch Block, Grandmother's Flower Garden, Hour Glasses, Sunshine and Shadow, Field of Diamonds, and Trip Around the World.

Check Your Progress: Chapter 7

1. $AB + BC = AC$
$\frac{1}{3}(BC) + BC = AC$
$\frac{1}{3}(15) + 15 = AC$
$5 + 15 = AC$
$20 = AC$
$AC = 20$ ft

2. Strategy:
Supplementary angles are twoangles whose sum is 180°. To find the supplement, let x represent the supplement of a 12° angle. Write an equation and solve for x.

Solution:
$x + 12° = 180°$
$x = 168°$
The supplement of a 12° angle is an 168° angle.

3. $\angle a + \angle b = 180°$
$42° + \angle b = 180°$
$\angle b = 138°$

$\angle c = \angle a = 40°$
$\angle c = 40°$

$\angle d = \angle b = 138°$
$\angle d = 138°$

4. Strategy:
To find the measure of the third angle, use the fact that the sum of the measures of the interior angles of a triangle is 180°. Write an equation using x to represent the measure of the third angle. Solve the equation for x.

Solution:
$x + 23° + 90° = 180°$
$x + 113° = 180°$
$x = 67°$
The measure of the third angle is 67°.

5. Strategy:
To find the area, use the formula for the area of a square. Substitute 40 for s. Solve for A.

Solution:
$A = s^2$
$A = (40)^2$
$A = 1600$
The area of the patio is 1600 in^2.

6. Strategy:
The sum of the measures of the three angles shown is 180°. To find x, write an equation and solve for x.

Solution:
$(4x - 10) + 3x + (2x + 10)$
$= 180°$
$9x = 180°$
$x = 20°$
The measure of x is 20°.

7. Strategy:
To find the circumference, use the circumference formula that involves the diameter. For an approximation, use the π key on a calculator. $d = 12$.

Solution:
$C = \pi d$
$C = \pi(12)$
$C = 12\pi$
$C \approx 37.699$
The circumference is approximately 37.70 cm.

8. Strategy:
→To find the measure of $\angle a$, use the fact that alternate interior angles of parallel lines are equal.
→To find the measure of $\angle b$, use the fact that adjacent angles of intersecting lines are supplementary.

Solution:
$\angle a = 135°$
$\angle b + \angle a = 180°$
$\angle b + 135° = 180°$
$\angle b = 45°$
The measure of $\angle a$ is 135°.
The measure of $\angle b$ is 45°.

9. Strategy:
To find the length of the base, use the formula for the area of a triangle. Substitute 8 for h and 20 for A. Solve for b.

Solution:
$A = \frac{1}{2}bh$
$20 = \frac{1}{2}b(8)$
$20 = 4b$
$\frac{20}{4} = b$
$5 = b$
The length of the base is 5 m.

10. Strategy:
To find the area, use the formula for the area of parallelogram. Substitute 14 for b and 7 for h. Solve for A.

Solution:
$A = bh$
$A = 14(7)$
$A = 98$
The area is 98 m^2.

11. Strategy:
To find the length of each side, use the formula for the perimeter of a square. Substitute 38 for P. Solve for s.

Solution:
$P = 4s$
$38 = 4s$
$9.5 = s$
The length of each side is 9.5 in.

12. Strategy:

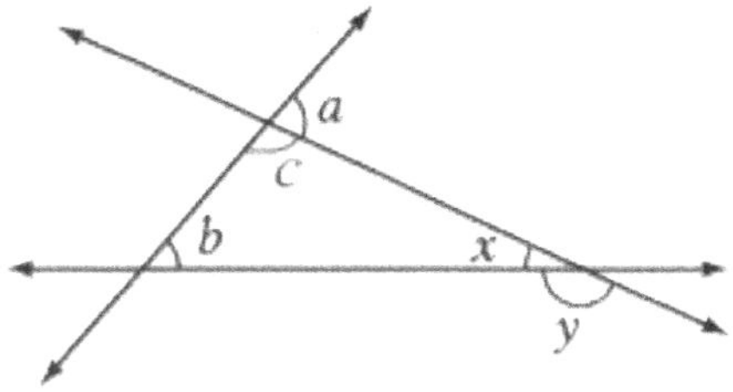

→To find the measure of $\angle x$, first find the measure of $\angle c$ by using use the fact that sum of an interior angle and exterior angle is 180° and then use the fact that the sum of the interior angles of a triangle is 180° to find $\angle x$.
→Find the measure of $\angle y$ by using the fact that the sum of an interior and exterior angle is 180°.

Solution:
$\angle a + \angle c = 180°$
$72° + \angle c = 180°$
$\angle c = 108°$
$\angle b + \angle c + \angle x = 180°$
$48° + 108° + \angle x = 180°$
$156° + \angle x = 180°$
$\angle x = 24°$
$\angle x + \angle y = 180°$
$24° + \angle y = 180°$
$\angle y = 156°$
The measure of $\angle x$ is 24°.
The measure of $\angle y$ is 156°.

13. Strategy:
To find the length, use the formula for the area of a rectangle. Substitute 8 for W and 128 for A. Solve for L.

Solution:
$A = LW$
$128 = L(8)$
$16 = L$
The length of the rectangle is 16 m.

14. Strategy:
To find the perimeter, use the formula for the perimeter of a triangle. Substitute 9 for a, 12 for b, and 15 for c. Solve for P.

Solution:
$P = a + b + c$
$P = 9 + 12 + 15$
$P = 36$
The perimeter is 36 in.

15. Strategy:
To find the area:
→Find the radius of the circle.
→Use the formula for the area of a circle. For an exact answer, leave the answer in terms of π. For an approximation, use the π key on a calculator.

Solution:
$r = \frac{1}{2}d = \frac{1}{2}(2.8) = 1.4$
$A = \pi r^2$
$A = \pi(1.4)^2$
$A = 1.96\pi$
$A \approx 6.1575$
The area is 1.96π m^2.
The area is approximately 6.16 m^2.

16. Strategy:
To find the area, use the formula for the area of a trapezoid. Substitute 14 for b_1, 10 for b_2, and 6 for h. Solve for A.

Solution:
$A = \frac{1}{2}h(b_1 + b_2)$
$A = \frac{1}{2} \cdot 6(14 + 10)$
$A = 72$
The area is 72 cm^2.

17. Strategy:
To find the length of molding, use the formula for the perimeter of a rectangle. Substitute 10 for L and $8\frac{1}{2}$ for W. Solve for P.

Solution:
$P = 2L + 2W$
$P = 2(10) + 2\left(8\frac{1}{2}\right)$
$P = 20 + 17$
$P = 37$
37 ft of decorative molding would be required.

Section 7.3

Concept Check

1. the hypotenuse; a leg

3. FD; BA

5. The ratio is $\frac{5}{10} = \frac{1}{2}$.

7. The ratio is $\frac{6}{8} = \frac{3}{4}$.

9. $\angle E$

Objective A Exercises

11. Strategy:
To find the hypotenuse, use the Pythagorean Theorem.
$a = 3, b = 4$

Solution:
$c^2 = a^2 + b^2$
$c^2 = 3^2 + 4^2$
$c^2 = 9 + 16$
$c^2 = 25$
$c = \sqrt{25}$
$c = 5$
The length of the hypotenuse is 5 in.

13. Strategy:
To find the hypotenuse, use the Pythagorean Theorem.
$a = 5, b = 7$

Solution:
$c^2 = a^2 + b^2$
$c^2 = 5^2 + 7^2$
$c^2 = 25 + 49$
$c^2 = 74$
$c = \sqrt{74}$
$c \approx 8.6$
The length of the hypotenuse if approximately 8.6 cm.

15. Strategy:
To find the measure of the other leg, use the Pythagorean Theorem.
$c = 15, a = 10$

Solution:
$a^2 + b^2 = c^2$
$10^2 + b^2 = 15^2$
$100 + b^2 = 225$
$b^2 = 125$
$b = \sqrt{125}$
$b \approx 11.2$
The measure of the other leg is approximately 11.2 ft.

17. Strategy:
To find the measure of the other leg, use the Pythagorean Theorem. $c = 6, a = 4$

Solution:
$a^2 + b^2 = c^2$
$4^2 + b^2 = 6^2$
$16 + b^2 = 36$
$b^2 = 20$
$b = \sqrt{20}$
$b \approx 4.5$
The measure of the other leg is approximately 4.5 cm.

19. Strategy:
To find the hypotenuse, use the Pythagorean Theorem.
$a = 9, b = 9$

Solution:
$c^2 = a^2 + b^2$
$c^2 = 9^2 + 9^2$
$c^2 \div 81 = 81$
$c^2 = 162$
$c = \sqrt{162}$
$c \approx 12.7$
The length of the hypotenuse is approximately 12.7 yd.

21. iii

23. Strategy:
To find the height, use the Pythagorean Theorem to find the other leg. $c = 8, a = 3$

Solution:
$a^2 + b^2 = c^2$
$3^2 + b^2 = 8^2$
$9 + b^2 = 64$
$b^2 = 55$
$b = \sqrt{55}$
$b \approx 7.4$
The ladder will reach a height of approximately 7.4 m.

25. Strategy:
To find the perimeter:
→Use the Pythagorean Theorem to find the hypotenuse of the triangle.
$a = 5, b = 9$
→Use the formula for the perimeter of a triangle to find the perimeter.

Solution:
$c^2 = a^2 + b^2$
$c^2 = 5^2 + 9^2$
$c^2 = 25 + 81$
$c^2 = 106$
$c = \sqrt{106}$
$c \approx 10.3$
$P = a + b + c$
$P - 5 + 9 + 10.3$
$P = 24.3$
The perimeter is approximately 24.3 cm.

Objective B Exercises

27. Strategy:
To find DE, write a proportion using the fact that in similar triangles, the ratios of corresponding sides are equal. Solve the proportion for DE.

Solution:
$\frac{AB}{DE} = \frac{AC}{DF}$
$\frac{4}{DE} = \frac{5}{9}$
$4(9) = (5)DE$
$36 = (5)DE$
$7.2 = DE$
The length of DE is 7.2 cm.

29. Strategy:
To find the height of triangle DEF, write a proportion using the fact that, in similar triangles, the ratio of corresponding sides equals the ratio of corresponding sides equals the ratio of corresponding heights. Solve the proportion for the height (h).

Solution:
$\frac{AC}{DF} = \frac{2}{h}$
$\frac{3}{5} = \frac{2}{h}$
$3h = 5(2)$
$3h = 10$
$h \approx 3.3$
The height of triangle DEF is approximately 3.3 m.

31. Strategy:
To find the perimeter:
→Find the side BC by writing a proportion using the fact that the ratios of corresponding sides of similar triangles are equal.
→Use the formula for the perimeter of a triangle.

Solution:
$\frac{BC}{EF} = \frac{AC}{DF}$
$\frac{BC}{6} = \frac{4}{8}$
$(8)BC = 6(4)$
$(8)BC = 24$
$BC = 3$
$P = a + b + c$
$P = 3 + 4 + 5$
$P = 12$
The perimeter of triangle ABC is 12 m.

33. Strategy:
To find the perimeter:
→Find side BC by writing a proportion using the fact that the ratios of corresponding sides of similar triangles are equal.
→Use the formula for the perimeter of a triangle.

Solution:
$\frac{BC}{EF} = \frac{AB}{DE}$
$\frac{BC}{15} = \frac{4}{12}$
$(12)BC = 15(4)$
$(12)BC = 60$
$BC = 5$
$P = a + b + c$
$P = 3 + 4 + 5$
The perimeter of the triangle is 12 in.

35. Strategy:
To find the area:
→Find the height (h) of triangle ABC by writing a proportion using the fact that, in similar triangles, the ratio of corresponding sides equals the ratio of corresponding heights.
→Use the formula for the area of a triangle. $b = 15$

Solution:
$$\frac{AB}{DE} = \frac{h}{20}$$
$$\frac{15}{40} = \frac{h}{20}$$
$$15(20) = 40h$$
$$300 = 40h$$
$$7.5 = h$$
$$A = \frac{1}{2}bh$$
$$A = \frac{1}{2}(15)(7.5)$$
$$A \approx 56.3$$
The area of triangle ABC is approximately 56.3 cm^2.

37. Strategy:
To find the height, write a proportion using the fact that, in similar triangles, the ratios of corresponding sides are equal.

Solution:
$$\frac{24}{8} = \frac{\text{height}}{6}$$
$$24(6) = 8 \cdot \text{height}$$
$$144 = 8 \cdot \text{height}$$
$$18 = \text{height}$$
The height of the flagpole is 18 ft.

39. Strategy:
To find the height, write a proportion using the fact that, in similar triangles, the ratios of corresponding sides are equal.

Solution:
$$\frac{\text{height}}{8} = \frac{8}{4}$$
$$\text{height} \cdot 4 = 8(8)$$
$$4 \cdot \text{height} = 64$$
$$\text{height} = 16$$
The height of the building is 16 m.

Objective C Exercises

41. Strategy:
To determine if the triangles are congruent, determine if one of the rules for congruence is satisfied.

Solution:
$AC = DE$, $AB = EF$, and $\angle A = \angle E$.
Two sides and the included angle of one triangle equal two sides and the included angle of the other triangle.
The triangles are congruent by the SAS rule.

43. Strategy:
To determine if the triangles are congruent, determine if one of the rules for congruence is satisfied.

Solution:
$AB = DE$, $AC = EF$, and $BC = DF$.
Three sides of one triangle equal the three sides of the other triangle. The triangles are congruent by the SSS rule.

45. Strategy:
To determine if the triangles are congruent, determine if one of the rules for congruence is satisfied.

Solution:
$AC = DF$, $\angle A = \angle D$, and $\angle C = \angle F$.
Two angles and the included side of one triangle equal two angles and the included side of the other triangle.
The triangles are congruent by the ASA rule.

47. Strategy:
Draw a sketch of the two triangles and determine if one of the rules for congruence is satisfied.

Solution:

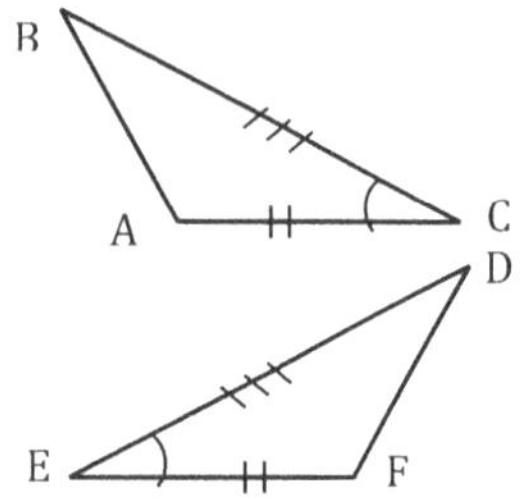

$AC = EF$, $BC = DE$, and $\angle C = \angle E$.
Because two sides and the included angle of one triangle equal two sides and the included angle of the other triangle, the triangles are congruent by the SAS rule.

49. Strategy:
Draw a sketch of the two triangles and determine if one of the rules of congruence is satisfied.

Solution:

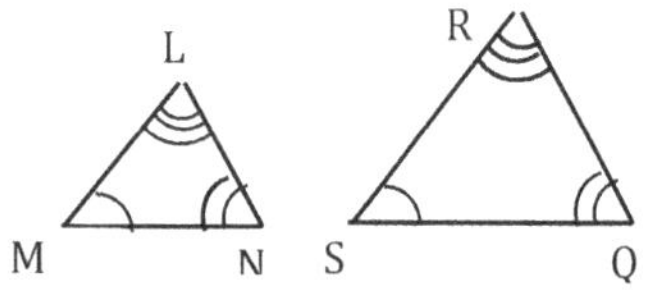

$\angle M = \angle S$, $\angle N = \angle Q$, and $\angle L = \angle R$.
The triangles do not satisfy the SSS rule, the SAS rule, or the ASA rule. The triangles are not necessarily congruent.

51. Strategy:
Draw a sketch of the two triangles and determine if one of the rules of congruence is satisfied.

Solution:

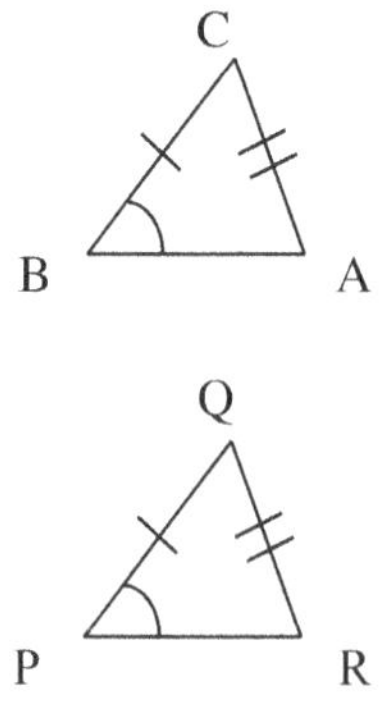

$\angle B = \angle P$, $BC = PQ$, and $AC = QR$.
The triangles do not satisfy the SSS rule, the SAS rule, or the ASA rule. The triangles are not necessarily congruent.

53. Yes

Critical Thinking 7.3

55. To determine if a 25-foot ladder is long enough to reach 24 ft up the side of the home when the bottom of the ladder is 6 ft from the base of the side of the house, use the Pythagorean Theorem to find the hypotenuse of a right triangle with legs that measure 24 ft and 6 ft.

$$c^2 = a^2 + b^2$$
$$c^2 = 24^2 + 6^2$$
$$c^2 = 576 + 36$$
$$c^2 = 612$$
$$c \approx 24.74$$

Compare the leg of the hypotenuse with 25. If the hypotenuse is shorter than 25 ft, the ladder will reach the gutter. If the hypotenuse is longer than 25 feet, the ladder will not reach the gutter.

$$24.74 < 25$$

The hypotenuse is shorter than 25 ft. The ladder will reach the gutters.

Projects and Group Activities 7.3

57.

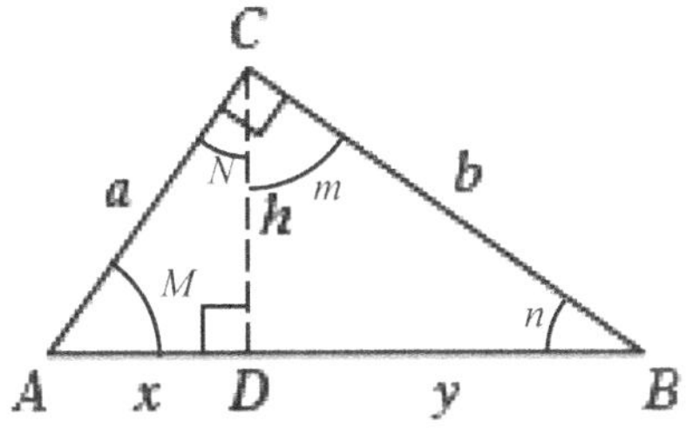

Triangles ACD and DCB are right triangles.
Angle M = angle m and angle N = angle n.
Because corresponding angles are equal,
the triangles are similar triangles.

Triangles ABC and ADC are right triangles.
Angle M = angle M and angle N = angle n.
Because corresponding angles are equal,
the triangles are similar triangles.

Triangles ABC and DCB are right triangles.
Angle n = angle n and angle M = angle m.
Because corresponding angles are equal,
the triangles are similar triangles.

Section 7.4

Concept Check

1. **a.** cone

   b. cube

   c. sphere

   d. cylinder

3. $s^2 + 2sl$; l; s

Objective A Exercises

5. Strategy:
   To find the volume, use the formula for the volume of a rectangular solid.
   $L = 14, W = 10, H = 6.$

   Solution:
   $V = LWH$
   $V = 14(10)(6)$
   $V = 840$
   The volume is 840 in^3.

7. Strategy:
   To find the volume, use the formula for the volume of a pyramid. $s = 3, h = 5.$

   Solution:
   $V = \frac{1}{3}s^2h$
   $V = \frac{1}{3}(3^2)(5)$
   $V = \frac{1}{3}(9)(5)$
   $V = 15$
   The volume is 15 ft^3.

9. Strategy:
   To find the volume:
   →Find the radius of the sphere. $d = 3.$
   →Use the formula for the volume of sphere.

   Solution:
   $r = \frac{1}{2}d = \frac{1}{2}(3) = 1.5$
   $V = \frac{4}{3}\pi r^3$
   $V = \frac{4}{3}\pi(1.5)^3$
   $V = \frac{4}{3}\pi(3.375)$
   $V = 4.5\pi$
   $V \approx 14.14$
   The volume is 4.5π cm^3.
   The volume is approximately 14.14 cm^3.

11. Strategy:
    To find the volume, use the formula for the volume of a rectangle solid.
    $L = 6.8, W = 2.5, H = 2.$

    Solution:
    $V = LWH$
    $V = 6.8(2.5)(2)$
    $V = 34$
    The volume of the rectangular solid is 34 m^3.

13. Strategy:
    To find the volume, use the formula for the volume of a cube. $s = 2.5.$

    Solution:
    $V = s^3$
    $V = (2.5)^3$
    $V = 15.625$
    The volume of the cube is 15.625 in^3.

15. Strategy:
To find the volume:
→Find the radius of the sphere. $d = 6$.
→Use the formula for the volume of a sphere.

Solution:
$r = \frac{1}{2}d = \frac{1}{2}(6) = 3$
$V = \frac{4}{3}\pi r^3$
$V = \frac{4}{3}\pi(3)^3$
$V = \frac{4}{3}\pi(27)$
$V = 36\pi$
The volume of the sphere is 36π ft^3.

17. Strategy:
To find the volume:
→Find the radius of the cylinder. $d = 24$.
→Use the formula for the volume of a cylinder. $h = 18$.

Solution:
$r = \frac{1}{2}d = \frac{1}{2}(24) = 12$
$V = \pi r^2 h$
$V = \pi(12^2)(18)$
$V = \pi(144)(18)$
$V = 2{,}592\pi$
$V \approx 8{,}143.01$
The volume of the cylinder is approximately 8,143.01 cm^3.

19. Strategy:
To find the volume:
→Find the radius of the base of the cone. $d = 10$.
→Use the formula for the volume of a cone. $h = 15$.

Solution:
$r = \frac{1}{2}d = \frac{1}{2}(10) = 5$
$V = \frac{1}{3}\pi r^2 h$
$V = \frac{1}{3}\pi(5)^2(15)$
$V = \frac{1}{3}\pi(25)(15)$
$V = 125\pi$
$V \approx 392.70$
The volume of the cone is approximately 392.70 cm^3.

21. Strategy:
To find the volume, use the formula for the volume of a pyramid. $s = 9$, $h = 8$.

Solution:
$V = \frac{1}{3}s^2 h$
$V = \frac{1}{3}(9^2)(8)$
$V = \frac{1}{3}(81)(8)$
$V = 216$
The volume of the pyramid is 216 m^3.

23. Strategy:
To find the height, use the formula for the volume of rectangular solid.
$V = 1{,}836$, $L = 18$, $W = 12$.

Solution:
$V = LWH$
$1{,}836 = 18(12)(H)$
$1{,}836 = 216H$
$8.5 = H$
The height of the aquarium is 8.5 in.

25. Strategy:
To find the height:
→Find the radius of the base of the cylinder. $d = 14$.
→Use the formula for the volume of a cylinder. $V = 2{,}310$.

Solution:
$r = \frac{1}{2}d = \frac{1}{2}(14) = 7$
$V = \pi r^2 h$
$2{,}310 = \pi(7)^2 h$
$2{,}310 = \pi(49)h$
$15.01 \approx h$
The height of the cone is approximately 15.01 cm.

27. Strategy:
To find the amount not being used for storage:
→Find the radius of the base of the cylinder. $d = 16$.
→Use the formula for the volume of a cylinder. $h = 30$.
→Multiply the volume of the cylinder by $\frac{1}{4}$, which is the portion not being used. $\left(1-\frac{3}{4}\right)$

Solution:
$r = \frac{1}{2}d = \frac{1}{2}(16) = 8$
$V = \pi r^2 h$
$V = \pi(8)^2(30)$
$V = \pi(64)(30)$
$V = 1{,}920\pi$
$\frac{1}{4}V = \frac{1}{4}(1{,}920\pi)$
$= 480\pi = 1{,}507.96$
The portion of the silo that is not used is approximately $1{,}507.96 \text{ ft}^3$.

29. Strategy:
To find the volume, use the formula for the volume of rectangular solid.
$L = 1000$, $W = 110$, $H = 60$.

Solution:
$V = LWH$
$V = 1000(110)(60)$
$V = 6{,}600{,}000$
The volume of the Panama Canal is $6{,}600{,}000 \text{ ft}^3$.

31. Yes

33. No

35. Strategy:
To find the volume, use the formula for the volume of rectangular solid.
$L = 360$, $W = 160$, $H = 3$.

Solution:
$V = LWH$
$V = 360(160)(3)$
$V = 172{,}800$
The volume of the guacamole is $172{,}800 \text{ ft}^3$.

Objective B Exercises

37. Strategy:
To find the surface area, use the formula for the surface area of a rectangular solid.
$L = 5$, $W = 4$, $H = 3$.

Solution:
$SA = 2LW + 2LH + 2WH$
$SA = 2(5)(4) + 2(5)(3)$
$+ 2(4)(3)$
$SA = 40 + 30 + 24$
$SA = 94$
The surface area of the rectangular solid is 94 m^2.

39. Strategy:
To find the surface area, use the formula for the surface area of a pyramid. $s = 4$, $l = 5$.

Solution:
$SA = s^2 + 2sl$
$SA = 4^2 + 2(4)(5)$
$SA = 16 + 40$
$SA = 56$
The surface area of the pyramid is 56 m^2.

41. Strategy:
To find the surface area, use the formula for the surface area of a cylinder. $r = 6$, $h = 2$.

Solution:
$SA = 2\pi r^2 + 2\pi rh$
$SA = 2\pi(6^2) + 2\pi(6)(2)$
$SA = 2\pi(36) + 24\pi$
$SA = 72\pi + 24\pi$
$SA = 96\pi$
$SA \approx 301.59$
The surface area of the cylinder is 96π in^2.
The surface area of the cylinder is approximately 301.59 in^2.

43. Strategy:
To find the surface area, use the formula for the surface area of a rectangular solid.
$H = 5, L = 8, W = 4$.

Solution:
$SA = 2LW + 2LH + 2WH$
$SA = 2(8)(4) + 2(8)(5)$
$+ 2(4)(5)$
$SA = 64 + 80 + 40$
$SA = 184$
The surface area of the rectangular solid is 184 ft^2.

45. Strategy:
To find the surface area, use the formula for the surface area of a cube. $s = 3.4$.

Solution:
$SA = 6s^2$
$SA = 6(3.4)^2$
$SA = 6(11.56)$
The surface area of the cube is 69.36 m^2.

47. Strategy:
To find the surface area:
→Find the radius of the sphere. $d = 15$.
→Use the formula for the surface area of a sphere.

Solution:
$r = \frac{1}{2}d = \frac{1}{2}(15) = 7.5$
$SA = 4\pi r^2$
$SA = 4\pi(7.5)^2$
$SA = 4\pi(56.25)$
$SA = 225\pi$
The surface area of the sphere is 225π cm^2.

49. Strategy:
To find the surface area, use the formula for the surface area of a cylinder. $r = 4$, $h = 12$.

Solution:
$SA = 2\pi r^2 + 2\pi rh$
$SA = 2\pi(4^2) + 2\pi(4)(12)$
$SA = 2\pi(16) + 96\pi$
$SA = 32\pi + 96\pi$
$SA = 128\pi$
$SA \approx 402.12$
The surface area of the cylinder is approximately 402.12 in^2.

51. Strategy:
To find the surface area, use the formula for the surface area of a cone.
$r = 1.5, l = 2.5$.

Solution:
$SA = \pi r^2 + \pi rl$
$SA = \pi(1.5^2) + \pi(1.5)(2.5)$
$SA = \pi(2.25) + 3.75\pi$
$SA = 6\pi$
The surface area of the cone is 6π ft^2.

53. Strategy:
To find the surface area, use the formula for the surface area of a pyramid. $s = 9$, $l = 12$.

Solution:
$SA = s^2 + 2sl$
$SA = 9^2 + 2(9)(12)$
$SA = 81 + 216$
$SA = 297$
The surface area of the pyramid is 297 in^2.

55. Strategy:
To find the width, use the formula for the surface area of a rectangular solid.
$SA = 108$, $L = 6$, and $H = 4$.

Solution:
$SA = 2LW + 2LH + 2WH$
$108 = 2(6)W + 2(6)(4) + 2W(4)$
$108 = 12W + 48 + 8W$
$108 = 20W + 48$
$60 = 20W$
$3 = W$
The width of the rectangular solid is 3 cm.

57. Strategy:
To find the amount of fabric:
→Find the radius of the sphere. $d = 32$.
→Use the formula for the surface area of a sphere.

Solution:
$r = \frac{1}{2}d = \frac{1}{2}(32) = 16$
$SA = 4\pi r^2$
$SA = 4\pi(16^2)$
$SA = 4\pi(256)$
$SA = 1024\pi$
$SA \approx 3{,}217$
Approximately 3,217 ft^2 of fabric was used to construct the balloon.

59. Strategy:
To find the amount of glass, use the formula for the surface area of a rectangular solid. Omit the top of the fish tank. The formula becomes
$SA = LW + 2LH + 2WH$.
$L = 12$, $W = 8$, $H = 9$.

Solution:
$SA = LW + 2LH + 2WH$
$SA = 12(8) + 2(12)(9)$
$+ 2(8)(9)$
$SA = 96 + 216 + 144$
$SA = 456$
The fish tank requires 456 in^2 of glass.

Critical Thinking 7.4

61. a. The distance from the edge of the base to the vertex of a regular pyramid is longer than the distance, perpendicular to the base, from the base to the vertex. The statement is always true.

b. The distance from the edge of the base of a cone to the vertex is longer than the distance, perpendicular to the base, from the base to the vertex. The statement is never true.

c. The four triangular faces of a regular pyramid could be equilateral triangles, but they could be isosceles triangles that are not equilateral. The statement is sometimes true.

63. three

Projects and Group Activities 7.4

65. Many of your students may already be familiar with M.C. Escher's work, since it now appears on everything from postcards to T-shirts. You might have each student select a different work to report on. Some examples include Relativity, House of Stairs, Whirlpools, Stars, Rind, Belvedere, Another World II, Ascending and Descending, and Waterfall.

Chapter Review Exercises

1. Strategy:
   →Label the third angle inside the triangle c.
   →To find the measure of $\angle c$, use the fact that the sum of an interior and exterior angle is 180°.
   →To find the measure of $\angle x$, use the fact that the sum of the measurements of the interior angles of a triangle is 180°.
   →To find the measure of $\angle y$, use the fact that the sum of an interior and exterior angle is 180°.

   Solution:
   $\angle a + \angle c = 180°$
   $74° + \angle c = 180°$
   $\angle c = 106°$
   $\angle b + \angle c + \angle x = 180°$
   $52° + 106° + \angle x = 180°$
   $158° + \angle x = 180°$
   $\angle x = 22°$
   $\angle x + \angle y = 180°$
   $22° + \angle y = 180°$
   $\angle y = 158°$

2. Strategy:
   To find the perimeter:
   →Find AC by writing a proportion, using the fact that in similar triangles, the ratios of corresponding sides are equal.
   →Use the formula for finding the perimeter of a triangle.

   Solution:
   $\frac{AC}{DF} = \frac{BC}{EF}$
   $\frac{AC}{12} = \frac{6}{9}$
   $9(AC) = 12(6)$
   $9(AC) = 72$
   $AC = 8$
   $P = AB + BC + AC$
   $P = 10 + 6 + 8$
   $P = 24$
   The perimeter of the triangle is 24 in.

3. Strategy:
   The volume is equal to the volume of the rectangular solid with dimensions 7 by 8 by 3.

   Solution:
   $V = LWH$
   $V = 7(8)(3)$
   $V = 168$
   The volume of the solid is 168 in^3.

4. Strategy:
   To find the measure of $\angle x$, use the fact that adjacent angles of intersecting lines are supplementary.

   Solution:
   $112° + \angle x = 180°$
   $\angle x = 68°$
   The measure of $\angle x$ is 68°.

5. Strategy:
   To determine if the triangles are congruent, determine if one of the rules for congruence is satisfied.

   Solution:
   $BC = DE, AC = DF, \angle C = \angle D$
   Two sides and the included angle of one triangle equal two sides and the included angle of the other triangle.
   The triangles are congruent by the SAS rule.

6. Strategy:
   The total surface area equals the surface area of a cylinder. The radius of the sphere is one half the diameter of the cylinder.

   Solution:
   $r = \frac{1}{2}d = \frac{1}{2}(4) = 2$
   $SA = 2\pi r^2 + 2\pi rh$
   $SA = 2\pi(2^2) + 2\pi(2)(8)$
   $SA = 2\pi(4) + 32\pi$
   $SA = 8\pi + 32\pi$
   $SA = 40\pi$
   $SA \approx 125.66$
   The surface area of the solid is approximately 125.66 m^2.

7. $AC = AB + BC$
$AC = 3(BC) + BC$
$AC = 4(BC)$
$AC = 4(11)$
$AC = 44$
The length of AC is 44 cm.

8. Strategy:
The sum of the measures of the three angles shown is 180°. To find x, write an equation and solve for x.

Solution:
$4x + 3x + (x + 28°) = 180°$
$8x + 28° = 180°$
$8x = 152°$
$x = 19°$
The measure of x is 19°.

9. Strategy:
The area is equal to the area of the rectangle.

Solution:
$A = LW$
$A = 8(4)$
$A = 32$
The area is 32 in^2.

10. Strategy:
To find the volume, use the formula for the volume of a pyramid. $s = 6, h = 8$.

Solution:
$V = \frac{1}{3}s^2h$
$V = \frac{1}{3}(6^2)(8)$
$V = \frac{1}{3}(36)(8)$
$V = 96$ cm^3.
The volume of the pyramid is 96 cm^3.

11. Strategy:
The perimeter is equal to three sides of a triangle.

Solution:
$P = a + b + c$
$P = 16 + 16 + 10$
$P = 42$
The perimeter is 42 in.

12. Strategy:
$\angle a = 138°$ because alternate interior angles of parallel lines are equal.
$\angle a + \angle b = 180°$ because adjacent angles of intersecting lines are supplementary.

Solution:
$\angle a = 138°$
$\angle a + \angle b = 180°$
$138° + \angle b = 180°$
$\angle b = 42°$
The measure of $\angle b$ is 42°.

13. Strategy:
To find the surface area, use the formula for the surface area of a rectangular solid.
$L = 10, W = 5, H = 4$.

Solution:
$SA = 2LW + 2LH + 2WH$
$SA = 2(10)(5) + 2(10)(4)$
$+ 2(5)(4)$
$SA = 100 + 80 + 40$
$SA = 220$
The surface area of the solid is 220 ft^2.

14. Strategy:
To find the measure of the other leg, use the Pythagorean Theorem.
$c = 12, a = 7$.

Solution:
$a^2 + b^2 = c^2$
$7^2 + b^2 = 12^2$
$49 + b^2 = 144$
$b^2 = 95$
$b = \sqrt{95}$
$b \approx 9.75$
The other leg is approximately 9.75 ft.

15. Strategy:
To find the volume, use the formula for the volume of a cube. $s = 3.5$.

Solution:
$V = s^3$
$V = (3.5)^3$
$V = 42.875$
The volume of the cube is 42.875 in^3.

16. Strategy:
Supplementary angles are two angles whose sum is 180°. To find the supplement, let x represent the supplement of a 32° angle. Write an equation and solve for x.

Solution:
$32° + x = 180°$
$x = 148°$
The supplement of a 32° angle is a 148° angle.

17. Strategy:
To find the volume, use the formula for the volume of a rectangular solid.
$L = 6.5$, $W = 2$, $H = 3$.

Solution:
$V = LWH$
$V = (6.5)(2)(3)$
$V = 39$
The volume of the solid is 39 ft^3.

18. Strategy:
To find the third angle, use the fact that the sum of the measures of the interior angles of a triangle is 180°. Let x = the third angle.

Solution:
$37° + 48° + x = 180°$
$85° + x = 180°$
$x = 95°$
The third angle is 95°.

19. Strategy:
To find the base, use the formula for the area of a triangle. Substitute 7 for h and 28 for A and solve for b.

Solution:
$A = \frac{1}{2}bh$
$28 = \frac{1}{2}b(7)$
$56 = 7b$
$8 = b$
The base of the triangle is 8 cm.

20. Strategy:
To find the volume, use the formula for the volume of a sphere. The radius of the sphere is one half the diameter.

Solution:
$r = \frac{1}{2}d = \frac{1}{2}(12) = 6$
$V = \frac{4}{3}\pi r^3$
$V = \frac{4}{3}\pi(6^3)$
$V = \frac{4}{3}\pi(216)$
$V = 288\pi$
The volume of the sphere is 288π mm^3.

21. Strategy:
To find the length of each side, use the formula for the perimeter of a square.
$P = 86$.

Solution:
$P = 4s$
$86 = 4s$
$21.5 = s$
A side of the square is 21.5 cm.

22. Strategy:
To find the number of cans of paint:
→Find the surface area by using the formula for the surface area of a cylinder.
→Divide the surface area by 200.

Solution:
$SA = 2\pi r^2 + 2\pi rh$
$SA = 2\pi(6^2) + 2\pi(6)(15)$
$SA = 2\pi(36) + 180\pi$
$SA = 72\pi + 180\pi$
$SA = 252\pi$
$252\pi \div 200 \approx 3.96$
Because a portion of a fourth can is needed, 4 cans of paint should purchased.

23. Strategy:
To find the amount of fencing, use the formula for the perimeter of a rectangle.

Solution:
$P = 2L + 2W$
$P = 2(56) + 2(48)$
$P = 112 + 96$
$P = 208$
208 yd of fencing are needed to fence the park.

24. Strategy:
To find the area, use the formula for the area of a square. $s = 9.5$.

Solution:
$A = s^2$
$A = (9.5)^2$
$A = 90.25$
The area of the patio is 90.25 m^2.

25. Strategy:
To find the area of the walkway:
→Find the length and width of the total area.
→Subtract the area of the plot of grass from the total area.

Solution:
$L_1 = 40 + 2 + 2 = 44$
$W_1 = 25 + 2 + 2 = 29$
$A = L_1W_1 - L_2W_2$
$A = 44(29) - 40(25)$
$A = 1{,}276 - 1{,}000$
$A = 276$
The area of the walkway is 276 m^2.

Chapter Test

1. Strategy:
To find the measure of the other leg, use the Pythagorean Theorem.
$c = 11, a = 8$.

Solution:
$a^2 + b^2 = c^2$
$8^2 + b^2 = 11^2$
$64 + b^2 = 121$
$b^2 = 57$
$b = \sqrt{57}$
$b \approx 7.55$
The other leg is approximately 7.55 cm.

2. Strategy:
To determine if the triangles are congruent, determine if one of the rules for congruence is satisfied.

Solution:
$AC = AC, AB = AB,$
$\angle A = \angle A$
Two sides and the included angle of one triangle equal two sides and the included angle of the other triangle.
The triangles are congruent by the SAS rule.

3. Strategy:
To find the area, use the formula for the area of a rectangle. Substitute 15 for L and 7.4 for W. Solve for A.

Solution:
$A = LW$
$A = 15(7.4)$
$A = 111$
The area is 111 m^2.

4. Strategy:
To find the area, use the formula for the area of a triangle. Substitute 7 for b and 12 for h. Solve for A.

Solution:
$A = \frac{1}{2}bh$
$A = \frac{1}{2}(7)(12)$
$A = 42$
The area is 42 ft^2.

5. Strategy:
To find the volume, use the formula for the volume of a cone. $r = 7, h = 16$.

Solution:
$V = \frac{1}{3}\pi r^2 h$
$V = \frac{1}{3}\pi(7)^2(16)$
$V = \frac{1}{3}\pi(49)(16)$
$V = \frac{784\pi}{3}$
The volume is $\frac{784\pi}{3}$ cm^3.

6. Strategy:
To find the surface area, use the formula for the surface area of a pyramid. $s = 3$, $l = 11$.

Solution:
$SA = s^2 + 2sl$
$SA = 3^2 + 2(3)(11)$
$SA = 9 + 66$
$SA = 75$
The surface area of the pyramid is 75 m^2.

7. Strategy:
The volume is equal to the volume of the cylinder.
$h = 30, r = 7$

Solution:
$V = \pi r^2 h$
$V = \pi(7^2)(30)$
$V = \pi(49)(30)$
$V = 1{,}470\pi$
$V \approx 4{,}618.14$
The volume of the solid is approximately 4 618.14 cm^3.

8. Strategy:
To find the area, use the formula for the area of a trapezoid. Substitute 20 for b_1, 33 for b_2, and 6 for h. Solve for A.

Solution:
$A = \frac{1}{2}h(b_1 + b_2)$
$A = \frac{1}{2} \cdot 6(20 + 33)$
$A = 159$
The area is 159 in^2.

9. Strategy:
The angles labeled are alternate interior angles of intersecting lines and are, therefore, supplementary angles. To find x, write an equation and solve for x.

Solution:
$3x + 6x = 180°$
$9x = 180°$
$x = 20°$
The measure of x is 20°.

10. Strategy:
The total surface area equals the surface area of the pyramid.

Solution:
$SA = s^2 + 2sl$
$SA = 5^2 + 2(5)(5)$
$SA = 25 + 50$
$SA = 75$
The surface area of the pyramid is 75 m^2.

11. Strategy:
The angles labeled are adjacent angles of intersecting lines and are, therefore, supplementary angles. To find x, write an equation and solve for x.

Solution:
$4x + 10° + x = 180°$
$5x = 170°$
$x = 34°$
The measure of x is 34°.

12. The polygon has 8 sides.
The polygon is an octagon.

13. Strategy:
To determine whether the triangles are congruent, determine whether one of the rules for congruence is satisfied.

Solution:
The triangles do not satisfy the SSS Rule, the SAS Rule, or the ASA Rule. The triangles are not necessarily congruent.

14. Strategy:
To find the volume, use the formula for the volume of a rectangular solid.
$L = 6, W = 7, H = 4$.

Solution:
$V = LWH$
$V = 6(7)(4) = 168$
The volume of the rectangular solid is 168 ft^3.

15. Strategy:
To find the hypotenuse, use the Pythagorean Theorem.
$a = 7, b = 4$

Solution:
$c^2 = a^2 + b^2$
$c^2 = 7^2 + 4^2$
$c^2 = 49 + 16 = 65$
$c = \sqrt{65}$
$c \approx 8.06$
The length of the hypotenuse if approximately 8.06 m.

16. Strategy:

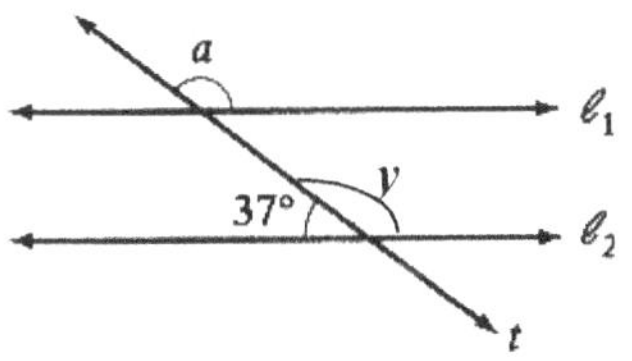

$y = a$ because corresponding angles have the same measure.
$y + 37° = 180°$ because adjacent angles of intersecting lines are supplementary angles.
Substitute a for y and solve for a.

Solution:
$a + 37° = 180°$
$a = 143°$
The measure of a is 143°.

17. Strategy:
To find the surface area, use the formula for the surface area of a cylinder. $r = 10$, $h = 15$.

Solution:
$SA = 2\pi r^2 + 2\pi rh$
$SA = 2\pi(10^2) + 2\pi(10)(15)$
$SA = 2\pi(100) + 300\pi$
$SA = 200\pi + 300\pi = 500\pi$
The surface area of the cylinder is 500π cm^2.

18. Strategy:

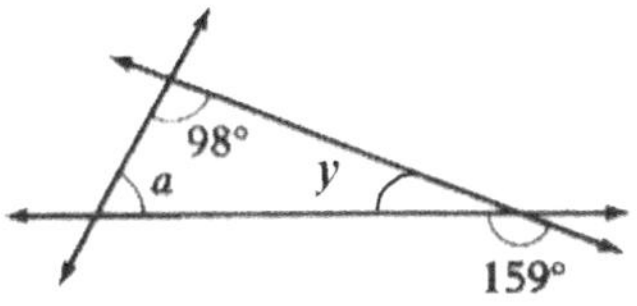

→To find the measure of $\angle y$, use the fact that $\angle y$ and 159° are supplemental angles.
→To find the measure of $\angle a$ by using the fact that the sum of the interior angles of a triangle is 180°.

Solution:
$\angle y + 159° = 180°$
$\angle y = 21°$
$21° + 98° + \angle a = 180°$
$119° + \angle a = 180°$
$\angle a = 61°$
The measure of $\angle a$ is 61°.

19. Strategy:
To find the length of side FG, write a proportion using the fact that, in similar triangles, the ratio of corresponding sides equals the ratio of corresponding heights. Solve the proportion for the height h.

Solution:
$\frac{EF}{BC} = \frac{h}{5}$
$\frac{12}{9} = \frac{h}{5}$
$5(12) = 9h$
$60 = 9h$
$6.67 \approx h$
The length of line segment FG is approximately 6.67 ft.

20. Strategy:
To find BC, write a proportion using the fact that in similar triangles, the ratios of corresponding sides are equal. Solve the proportion for BC.

Solution:
$\frac{BC}{EF} = \frac{AB}{DE}$
$\frac{BC}{8} = \frac{8}{15}$
$15BC = (8)8$
$15BC = 64$
$BC \approx 4.27$
The length of BC is approximately 4.27 ft.

21. Strategy:
To find the perimeter, use the formula for the perimeter of a square. Substitute 5 for s. Solve for P.

Solution:
$P = 4s$
$P = 4(5)$
$P = 20$
The perimeter is 20 m.

22. Strategy:
To find the perimeter, use the formula for the perimeter of a rectangle. Substitute 8 for L and 5 for W. Solve for P.

Solution:
$P = 2L + 2W$
$P = 2(8) + 2(5)$
$P = 16 + 10$
$P = 26$
The perimeter is 26 cm.

23. Strategy:
To find the perimeter:
→Use the Pythagorean Theorem to find the hypotenuse of the triangle. $a = 12$, $b = 18$
→Use the formula for the perimeter of a triangle to find the perimeter.

Solution:
$c^2 = a^2 + b^2$
$c^2 = 12^2 + 18^2$
$c^2 = 144 + 324$
$c^2 = 468$
$c = \sqrt{468}$
$c \approx 21.6$
$P = a + b + c$
$P = 12 + 18 + 21.6$
$P = 51.6$
The perimeter is approximately 51.6 ft.

24. Strategy:
To find the third angle, use the fact that the sum of the measures of the interior angles of a triangle is 180°. Let x = the third angle.

Solution:
$41° + 37° + x = 180°$
$78° + x = 180°$
$x = 102°$
The third angle is 102°.

25. Strategy:
Supplementary angles are two angles whose sum is 180°. To find the supplement, let x represent the supplement of a 41° angle. Write an equation and solve for x.

Solution:
$41° + x = 180°$
$x = 139°$
The supplement of a 41° angle is a 139° angle.

Cumulative Review Exercises

1. Strategy:
To find the amount, use the basic percent equation.
Percent = 8.5% = 0.085,
base = 2,400, amount = n.

Solution:
Percent · base = amount
$0.085(2{,}400) = n$
$204 = n$
8.5% of 2,400 is 204.

2. $78 \div 1 = 78$
$78 \div 2 = 39$
$78 \div 3 = 26$
$78 \div 6 = 13$
$78 \div 13 = 6$
The factors of 78 are 1, 2, 3, 6, 13, 26, 39, and 78.

3. $4\frac{2}{3} \div 5\frac{3}{5} = \frac{14}{3} \div \frac{28}{5}$
$= \frac{14}{3} \cdot \frac{5}{28}$
$= \frac{14 \cdot 5}{3 \cdot 28}$
$= \frac{2 \cdot 7 \cdot 5}{3 \cdot 2 \cdot 2 \cdot 7}$
$= \frac{5}{6}$

4. $|-18| = 18$

5. $82.93 \div 6.5 \approx 12.8$

6. $-6 - (-4) = -6 + 4 = -2$

7. Strategy:
To find the measure of $\angle x$, use the fact that adjacent angles of intersecting lines are supplementary.

Solution:
$\angle x + 49° = 180°$
$\angle x = 131°$
The measure of $\angle x$ is 131°.

8. Strategy:
To find the hypotenuse, use the Pythagorean Theorem. $a = 10$, $b = 24$.

Solution:
$c^2 = a^2 + b^2$
$c^2 = 10^2 + 24^2$
$c^2 = 100 + 576$
$c^2 = 676$
$c = \sqrt{676}$
$\rightarrow C$ is the square root of 676.
$c = 26$
The length of the hypotenuse is 26 cm.

9. Strategy:
To find the area, use the formula for the area of a triangle. $h = 7$, $b = 16$.

Solution:
$A = \frac{1}{2}bh$
$A = \frac{1}{2}(16)(7)$
$A = (8)(7) = 56$
The area is 56 in^2.

10. Strategy:
Corresponding angles have the same measure.
$x + 20° + 3x = 180°$ because adjacent angles of intersecting lines are supplementary angles.

Solution:
$x + 20° + 3x = 180°$
$4x + 20° = 180°$
$4x = 160°$
$x = 40°$
The measure of x is 40°.

11. $\frac{2}{3} + \frac{7}{10} = \frac{20}{30} + \frac{21}{30} = \frac{41}{30} = 1\frac{11}{30}$

12. $3(2x + 5) = 18$
$6x + 15 = 18$
$6x = 3$
$x = \frac{1}{2}$
The solution is $\frac{1}{2}$.

13. Strategy:
To find the third angle, use the fact that the sum of the measures of the interior angles of a triangle is 180°. Let x = the third angle.

Solution:
$32° + 90° + x = 180°$
$122° + x = 180°$
$x = 58°$
The third angle is 58°.

14. $-3(-25) = 75$

15. $5(2x + 4) - (3x + 2) = 10x + 20 - 3x - 2$
$= 7x + 18$

16. $2x+3y^2z$

$2(5)+3(-1)^2(-4)=2(5)+3(1)(-4)$
$=10+(-12)$
$=-2$

17. x^2y-2z

$\left(\frac{1}{2}\right)^2\left(\frac{4}{5}\right)-2\left(-\frac{3}{10}\right)=\frac{1}{4}\left(\frac{4}{5}\right)-2\left(\frac{-3}{10}\right)$
$=\frac{1}{5}-\frac{-3}{5}$
$=\frac{1}{5}+\frac{3}{5}$
$=\frac{4}{5}$

18. $3\overline{)39}$ (quotient 13)

$2\overline{)78}$

$78=2\cdot 3\cdot 13$

19. $4x+2=6x-8$
$4x-6x+2=6x-6x-8$
$-2x+2=-8$
$-2x+2-2=-8-2$
$-2x=-10$
$\frac{-2x}{-2}=\frac{-10}{-2}$
$x=5$
The solution is 5.

20. $\frac{3}{8}=\frac{3}{8}(100\%)=37.5\%$

21. the unknown number x
the product of eight and twice a number:
$8(2n)$
$8(2n)=16n$

22. Strategy:
→Rate of the first car: r
Rate of the second car: $r+5$

	Rate	Time	Distance
First car	r	2	$2r$
Second car	$r+5$	2	$2(r+5)$

→The total distance traveled by the two cars is 210 mi.

Solution:
$2r+2(r+5)=210$
$2r+2r+10=210$
$4r+10=210$
$4r=200$
$r=50$
$r+5=50+5=55$
The first car is traveling at 50 mph. The second car is traveling at 55 mph.

23. Strategy:
To find the simple interest, solve the simple interest formula $I=Prt$ for I.
$P=20{,}000,\ t=\frac{270}{365},$
$r=0.08875$

Solution:
$I=Prt$
$I=(20{,}000)(0.08875)\left(\frac{270}{365}\right)$
$I=1{,}313.01$
The interest on the loan is $1,313.01.

24. Strategy:
→Amount of \$3.50 alloy: x

	Amount	Cost	Value
\$3.50 alloy	x	3.50	$3.50x$
\$5 alloy	12	5	12(5)
Mixture	$x + 12$	4	$4(x + 12)$

→The sum of the values before mixing equals the value after mixing.

Solution:

$$3.50x + 12(5) = 4(x + 12)$$
$$3.5x + 60 = 4x + 48$$
$$60 = 4x - 3.5x + 48$$
$$60 = 0.5x + 48$$
$$60 - 48 = 0.5x$$
$$12 = 0.5x$$
$$24 = x$$

24 oz of \$3.50 silver alloy is required.

25. Strategy:
To find the number of text messages, write and solve an equation using n to represent the number of text messages.

Solution:
$\text{Cost} = 79 + 0.35n$
$86.70 = 79 + 0.35n$
$7.70 = 0.35n$
$22 = n$
22 text messages were sent or received.

26. Strategy:
To find the sales tax, write and solve a proportion using n to represent the sales tax.

Solution:

$$\frac{0.75}{12.50} = \frac{n}{75}$$
$$0.75(75) = 12.50 \cdot n$$
$$56.25 = 12.50n$$
$$\frac{56.25}{12.50} = \frac{12.50n}{12.50}$$
$$4.50 = n$$

The sales tax is \$4.50.

27. Strategy:
To find the increase:
→Subtract 1.48 from 1.56 to find the amount of increase.

Solution:
$1.56 - 1.48 = 0.08$
The increase is \$0.8 trillion.

28. Strategy:
To find the height of the box, use the formula for the volume of a rectangular solid.
$V = 144, L = 12, W = 4$.

Solution:
$V = LWH$
$144 = 12(4)H$
$144 = 48H$
$3 = H$
The height of the box is 3 ft.

29. Strategy:
To find the pressure, substitute 35 for P in the given equation and solve for D.

Solution:

$$P = 15 + \frac{1}{2}D$$
$$35 = 15 + \frac{1}{2}D$$
$$20 = \frac{1}{2}D$$
$$40 = D$$

The depth is 40 ft.

30. Strategy:
To find the amount of weight to lose:
→Add the amount already lost $\left(3\frac{1}{2} + 2\frac{1}{4}\right)$.
→Subtract the amount lost from 8.

Solution:

$$3\frac{1}{2} + 2\frac{1}{4} = 3\frac{2}{4} + 2\frac{1}{4} = 5\frac{3}{4}$$
$$8 - 5\frac{3}{4} = 7\frac{4}{4} - 5\frac{3}{4} = 2\frac{1}{4}$$

The wrestler must lose another $2\frac{1}{4}$ lb.

Chapter 8: Statistics and Probability

Prep Test

1. Strategy:
To find the percent of the emails that were spam, solve the basic percent equation for amount.

Solution:
$0.891 \times (107 \text{ trillion}) = n$

$95 \text{ trillion} \approx n$

95 trillion emails were spam.

2. Between 2005 and 2006
$\$74{,}418 - 70{,}206 = \$4212$
Between 2006 and 2007
$\$78{,}883 - 74{,}418 = \$4465$
Between 2007 and 2008
$\$83{,}616 - 78{,}883 = \$4733$
Between 2008 and 2009
$\$88{,}633 - \$83{,}616 = \$5017$
Between 2009 and 2010
$\$93{,}951 - 88{,}633 = \$5318$

a. The greatest cost increase is between 2009 and 2010.

b. The increase is \$5318.

3. $\dfrac{11 \text{ gold}}{7 \text{ silver}} = \dfrac{11}{7}$

4. $16\% = 16 \times \dfrac{1}{100} = \dfrac{16}{100} = \dfrac{4}{25}$

$\dfrac{4}{25}$ of women in the military are in the Navy.

Section 8.1

Concept Check

1. $25\% = 25 \times \dfrac{1}{100} = \dfrac{25}{100} = \dfrac{1}{4}$

3. The value of the data decreased from the first point to the second point and then increased from the second point to the third point.

Objective A Exercises

5. Strategy:
To find the ratio:
→Read the circle graph to determine the units in Finance and Accounting.
→Write in the simplest form the ratio of the numbers of units in Finance to the number of units in Accounting.

Solution:
Number of units in Finance: 15
Number of units in Accounting: 45
$\dfrac{15}{45} = \dfrac{1}{3}$
The ratio is $\dfrac{1}{3}$.

7. Strategy:
To find the amount of money spent:
→Read the circle graph to find the percent of money spent on video game hardware in 2010.
→Use the basic percent equation to find the amount.

Solution:
34% was spent on video game hardware in 2010.
Percent × base = amount

$0.34 \times 18{,}600{,}000{,}000 = n$

$6{,}324{,}000{,}000 = n$

Americans spent \$6,324,000,000 on video game hardware in 2010.

9. Strategy:
To find the fractional amount spent on accessories:
→Use the basic percent equation to find the
amount spent on accessories.
→Write the ratio of the amount spent on accessories to the total amount spent in simplest form.

Solution:

$$\text{Percent} \times \text{base} = \text{amount}$$
$$0.16 \times 18{,}600{,}000{,}000 = 2{,}976{,}000{,}000$$
$$\frac{2{,}976{,}000{,}000}{18{,}600{,}000{,}000} = \frac{4}{25}$$

The fractional amount is $\frac{4}{25}$.

11. The bar representing world population in 2030 is taller than 8 billion, so the estimated population in 2030 is more than 8 billion people.

13. Strategy:
To find the change in world population:
→Read the bar graph to determine the population
in 1990 and 2010.
→Subtract to find the difference.

Solution:
1990 population: 5.2 billion people
2010 population: 6.8 billion people
$6.8 - 5.2 = 1.6$
The change in world population between 1990 and 2010 was about 1.6 billion people.

15. Strategy:
To find what age and gender has the lowest number of recommended Calories, read the double broken-line graph and select the age and gender beneath the lowest point.

Solution:
The age and gender that has the lowest number of recommended Calories is for women age 75+.

17. True

Critical Thinking 8.1

19.

Year	Wind Power Capacity in Iowa (megawatts)	Wind Power Capacity in California (megawatts)
2000	200	1600
2001	300	1700
2002	400	1800
2003	500	2000
2004	600	2100
2005	800	2200
2006	900	2400
2007	1300	2400
2008	2800	2500
2009	3600	2800
2010	3700	3300

21. Answers will vary.

For example: The couple's largest single expense was rent.
Food represents approximately one-quarter of the month's expenditures.
Rent represents approximately one-third of the month's expenditures.
The expenditure for food is approximately the same as the expenditures for entertainment and transportation.
The couple spent more for transportation than for entertainment.

Section 8.2

Concept Check

1. The mode must be a value in the data because it is the number that occurs most often in a set of data.

3. **a.** Median

b. Mean

c. Mode

d. Median

e. Mode

f. Mean

Objective A Exercises

5. Strategy:
To find the mean value of the number of seats occupied:
→Find the sum of the number of seats occupied.
→Divide the sum by the number of flights (16).

Solution:

309
422
389
412
401
352
367
319
410
391
330
408
399
387
411
+398
6,105

$$16\overline{)6,105} = 381.5625$$

The mean of the number of seats filled is 381.5625 seats.

Strategy:
To find the median value of the number of seats occupied, arrange the numbers in order from smallest to largest. The median is the mean of the two middle numbers.

Solution:

$$\left.\begin{array}{l}309\\319\\330\\352\\367\\387\\389\end{array}\right\}\text{7 numbers}$$

$$\left.\begin{array}{l}391\\398\end{array}\right\}\text{Middle numbers}$$

$$\left.\begin{array}{l}399\\401\\408\\410\\411\\412\\422\end{array}\right\}\text{7 numbers}$$

$$\frac{391+398}{2} = 394.5$$

The median of the number of seats filled is 394.5 seats.

Strategy:
To find the mode, look at the number of seats occupied and locate the number that occurs most frequently.

Solution:
Since each number occurs only once, there is no mode.

7. Strategy:
To find the mean cost:
→Find the sum of the costs.
→Divide the total costs by the number of purchases (8).

Solution:

$$\begin{array}{r} 85.89 \\ 92.12 \\ 81.43 \\ 80.67 \\ 88.73 \\ 82.45 \\ 87.81 \\ +\ 85.82 \\ \hline 684.92 \end{array} \qquad 8\overline{)684.920} = 85.615$$

The mean cost is \$85.615.

Strategy:
To find the median cost, arrange the costs in order from smallest to largest. The median is the mean of the two middle numbers.

Solution:

$$\left.\begin{array}{l} 80.67 \\ 81.43 \\ 82.45 \end{array}\right\} \text{3 numbers}$$

$$\left.\begin{array}{l} 85.82 \\ 85.89 \end{array}\right\} \text{middle numbers}$$

$$\left.\begin{array}{l} 87.81 \\ 88.73 \\ 92.12 \end{array}\right\} \text{3 numbers}$$

$$\frac{85.82 + 85.89}{2} = 85.855$$

The median cost is \$85.855

9. Strategy:
To find the mean monthly rate:
→Find the sum of the monthly rates.
→Divide the sum by the number of plans (8).

Solution:

$$\begin{array}{r} 423 \\ 390 \\ 405 \\ 396 \\ 426 \\ 355 \\ 404 \\ +\ 430 \\ \hline 3{,}229 \end{array} \qquad 8\overline{)3{,}229.000} = 403.625$$

The mean monthly rate is \$403.625.

Strategy:
To find the median monthly rate, write the rates in order from smallest to largest. The median is the mean of the two middle terms.

Solution:

$$\left.\begin{array}{l} 355 \\ 390 \\ 396 \end{array}\right\} \text{3 numbers}$$

$$\left.\begin{array}{l} 404 \\ 405 \end{array}\right\} \text{middle numbers}$$

$$\left.\begin{array}{l} 423 \\ 426 \\ 430 \end{array}\right\} \text{3 numbers}$$

$$\frac{404 + 405}{2} = 404.50$$

The median monthly rate is \$404.50.

11. Strategy:
To find the mean life expectancy:
→Find the sum of the years.
→Divide the sum by the number of countries (10).

Solution:

$$\begin{array}{r} 73.5 \\ 79.1 \\ 79.3 \\ 75.6 \\ 71.2 \\ 76.1 \\ 74.0 \\ 70.1 \\ 77 \\ +\ 74.4 \\ \hline 750.3 \end{array} \qquad 10\overline{)750.30} = 75.03$$

The mean life expectancy is 75.03 years.

Strategy:
To find the median life expectancy, write the years in order from lowest to highest. The median is the mean of the two middle numbers.

Solution:

$$\left.\begin{array}{r} 70.1 \\ 71.2 \\ 73.5 \\ 74.0 \end{array}\right\} \text{4 numbers}$$

$$\left.\begin{array}{r} 74.4 \\ 75.6 \end{array}\right\} \text{middle numbers}$$

$$\left.\begin{array}{r} 76.1 \\ 77 \\ 79.1 \\ 79.3 \end{array}\right\} \text{4 numbers}$$

$$\frac{74.4 + 75.6}{2} = 75$$

The median life expectancy is 75 years.

13. Strategy:
To determine which average you would prefer:
→Find the mean of the test scores.
→Find the median of the test scores.
→Choose the higher score.

Solution:

Mean

$$\begin{array}{r} 78 \\ 92 \\ 95 \\ 77 \\ 94 \\ +\ 88 \\ \hline 524 \end{array} \qquad 6\overline{)524.0} = 87.3$$

Median

$$\left.\begin{array}{r} 77 \\ 78 \end{array}\right\} \text{2 numbers}$$

$$\left.\begin{array}{r} 88 \\ 92 \end{array}\right\} \text{middle numbers}$$

$$\left.\begin{array}{r} 94 \\ 95 \end{array}\right\} \text{2 numbers}$$

$$\frac{88 + 92}{2} = 90$$

You would prefer that the instructor use the higher median score 90.

15. It is 5 more than the mean of the original set.

Objective B Exercises

17. Strategy:
→Read the lowest value, the highest value, the first quartile, the third quartile, the median directly from the box-and-whiskers plot.
→Find the range by subtracting the lowest from the highest.
→Interquartile range = $Q_3 - Q_1$.

Solution:
Lowest is \$37,985.
Highest is \$66,707.
Q_1 = \$44,243
Q_3 = \$55,528
Median = \$49,177
Range: 66,707 – 37,985
= \$28,722
Interquartile range:
55,528 – 44,243 = \$11,285

19. a. Strategy:
To find the number of adults who had a cholesterol level above 217, the median, solve the basic percent equation for the amount, where the base is 80 and the percent is 50%.

Solution:
Percent ÷ base = amount

0.50 ÷ 80 = 40
There were 40 adults who had cholesterol levels above 217.

b. Strategy:
To find the number of adults who had a cholesterol level below 254, the third quartile, solve the basic percent equation for the amount, where the base is 80 and the percent is 75%.

Solution:
Percent ÷ base = amount

0.75 ÷ 80 = 60
There were 60 adults who had cholesterol levels below 254.

c. Strategy:
To find the number of cholesterol levels represented in each quartile, solve the basic percent equation for the amount, where the base is 80 and the percent is 25%.

Solution
Percent ÷ base = amount

0.25 ÷ 80 = 20
There are 20 cholesterol levels in each quartile.

d. The first quartile is at 198. So 25% of the adults had cholesterol levels not more than 198.

21. a. Strategy:

→Arrange the data from smallest to largest.
→Find the range.
→Find Q_1, the median of the lower half of the data.
→Find Q_3, the median of the upper half of the data.
→Interquartile range = $Q_3 - Q_1$.

Solution:

0.45	0.47	0.59	0.64	0.81
1.26	1.34	1.52	5.01	6.05

Range: 6.05 – 0.45 = 5.6 emissions
Q_1 = 0.59 emissions
Q_3 = 1.52 emissions

$$\begin{aligned}\text{Interquartile range} &= Q_3 - Q_1 \\ &= 1.52 - 0.59 \\ &= 0.93 \text{ emissions}\end{aligned}$$

b. $\text{Median} = \dfrac{0.81 + 1.26}{2} = 1.035$

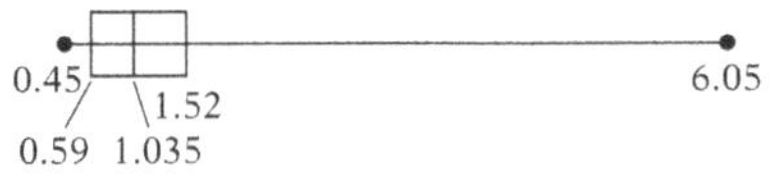

c. 6.05

23. a. Strategy:
To determine whether the difference in means is greater than 1 inch:
→Find the sum of the rainfall in Orlando.
→Divide the sum by the number of months (12) to find the mean.
→Find the sum of the rainfall in Portland.
→Divide the sum by the number of months (12) to find the mean.
→Find the difference in the means.

Solution:
Orlando

$$\begin{array}{r} 2.1 \\ 2.8 \\ 3.2 \\ 2.2 \\ 4.0 \\ 7.4 \\ 7.8 \\ 6.3 \\ 5.6 \\ 2.8 \\ 1.8 \\ +\,1.8 \\ \hline 47.8 \end{array}$$

Portland

$$\begin{array}{r} 6.2 \\ 3.9 \\ 3.6 \\ 2.3 \\ 2.1 \\ 1.5 \\ 0.5 \\ 1.1 \\ 1.6 \\ 3.1 \\ 5.2 \\ +\,6.4 \\ \hline 37.5 \end{array} \qquad 12\overline{)37.5} = 3.1$$

$4.0 - 3.1 = 0.9$
No, the difference in the means is not greater than 1 inch.

b. Strategy:
To find the difference between the medians, write the rainfall in order from lowest to highest. The median is the mean of the two middle numbers. Find the difference between the Orlando median and Portland median.

Solution:
Orlando

$$\left.\begin{array}{l} 1.8 \\ 1.8 \\ 2.1 \\ 2.2 \\ 2.8 \end{array}\right\} \text{5 numbers}$$
$$\left.\begin{array}{l} 2.8 \\ 3.2 \end{array}\right\} \text{middle}$$
$$\left.\begin{array}{l} 4.0 \\ 5.6 \\ 6.3 \\ 7.4 \\ 7.8 \end{array}\right\} \text{5 numbers}$$

Portland:

$$\left.\begin{array}{l} 0.5 \\ 1.1 \\ 1.5 \\ 1.6 \\ 2.1 \end{array}\right\} \text{5 numbers}$$
$$\left.\begin{array}{l} 2.3 \\ 3.1 \end{array}\right\} \text{middle}$$
$$\left.\begin{array}{l} 3.6 \\ 3.9 \\ 5.2 \\ 6.2 \\ 6.4 \end{array}\right\} \text{5 numbers}$$

For Orlando, $\dfrac{2.8 + 3.2}{2} = 3.0$

For Portland, $\dfrac{2.3 + 3.1}{2} = 2.7$

$3.0 - 2.7 = 0.3$
The difference in medians is 0.3 inch.

c. Strategy:
To draw box-and-whiskers:
→Find Q_1 and Q_3 in Orlando.
→Find Q_1 and Q_3 in Portland.

Solution:
For Orlando,

$$Q_1 = \frac{2.1 + 2.2}{2} = 2.15,$$

$$Q_3 = \frac{5.6 + 6.3}{2} = 5.95$$

For Portland,

$$Q_1 = \frac{1.5 + 1.6}{2} = 1.55,$$

$$Q_3 = \frac{3.9 + 5.2}{2} = 4.55$$

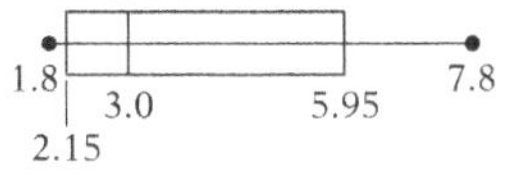

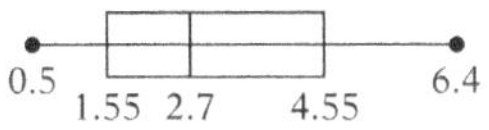

d. Answers will vary. For example, the distribution of the data is relatively similar for the two cities. However, the value of each of the 5 points on the boxplot for the Portland data is less than the corresponding value on the boxplot for the Orlando data. The average monthly rainfall in Portland is less than the average monthly rainfall in Orlando.

25. Answers will vary. For example, 55, 55, 55, 55, 55, or 50, 55, 55, 55, 60

Critical Thinking 8.2

27. Answers will vary. For example, 20, 21, 22, 24, 26, 27, 29, 31, 31, 32, 32, 33, 33, 36, 37, 37, 39, 40, 41, 43, 45, 46, 50, 54, 57

Check Your Progress: Chapter 8

1. a. Chocolate

b. Strategy:
To find the number of people who preferred mint chocolate chip:
→Read the circle graph to find the percent of people who named mint chocolate chip as their favorite flavor.
→Solve the basic percent equation for *amount*.

Solution:
$0.15 \times 1200 = n$

$180 = n$

180 people preferred mint chocolate chip.

c. Strategy:
To find how many more people chose cookie dough than chose mint chocolate chip:
→Read the circle graph to find the percent of people who named cookie dough as their favorite flavor.
→Solve the basic percent equation for *amount*.
→Use the result from part b and subtract to find the difference.

Solution
$0.20 \times 1200 = n$

$240 = n$

$240 - 180 = 60$
60 more people chose cookie dough.

3. a. \$1.60

b. 2008

c. 2006, 2008, 2009, and 2010

5. Strategy:
To find the modal response, write down the number that received the most responses.

Solution:
Because 23 bags recorded most frequently, the modal response was 23 bags of flour.

Section 8.3

Concept Check

1. {3, 6, 9}

Objective A Exercises

3. The possible outcomes of tossing a coin four times: {(HHHH), (HHHT), (HHTT), (HHTH), (HTTT), (HTHH), (HTTH), (HTHT), (TTTT), (TTTH), (TTHH), (THHH), (TTHT), (THHT), (THTT), (THTH)}

5. The possible outcomes of tossing two tetrahedral dice: {(1, 1), (1, 2), (1, 3), (1, 4), (2, 1), (2, 2), (2, 3), (2, 4), (3, 1), (3, 2), (3, 3), (3, 4), (4, 1), (4, 2), (4, 3), (4, 4)}

7. The sample space is {1, 2, 3, 4, 5, 6, 7, 8}.

9. a. Strategy:
To calculate the probability:
→Count the number of possible outcomes. See the table on p. 328.
→Count the number of favorable outcomes.
→Use the probability formula.

Solution:
There are 36 possible outcomes.
There are 4 favorable outcomes: (1, 4), (4, 1), (2, 3), (3, 2).

$$\text{Probability} = \frac{4}{36} = \frac{1}{9}$$

The probability that the sum is 5 is $\frac{1}{9}$.

b. Strategy:
To calculate the probability:
→Count the number of possible outcomes. See the table on p. 328.
→Count the number of favorable outcomes.
→Use the probability formula.

Solution:
There are 36 possible outcomes.
There are 0 favorable outcomes.

$$\text{Probability} = \frac{0}{36} = 0$$

The probability that the sum is 15 is 0.

c. Strategy:
To calculate the probability:
→Count the number of possible outcomes. See the table on p. 328.
→Count the number of favorable outcomes.
→Use the probability formula.

Solution:
There are 36 possible outcomes.
There are 36 favorable outcomes.

$$\text{Probability} = \frac{36}{36} = 1$$

The probability that the sum is less than 15 is 1.

11. a. Strategy:
To calculate the probability:
→Count the number of possible outcomes. Refer to Exercise 3.
→Count the number of favorable outcomes.
→Use the probability formula.

Solution:
There are 16 possible outcomes.
There are 3 favorable outcomes: (1, 3), (3, 1), (2, 2).

Probability $= \frac{3}{16}$

The probability is $\frac{3}{16}$ that the sum of the dots on the two dice is 4.

b. Strategy:
To calculate the probability:
→Count the number of possible outcomes. Refer to Exercise 3.
→Count the number of favorable outcomes.
→Use the probability formula.

Solution:
There are 16 possible outcomes.
There are 3 favorable outcomes: (2, 4), (4, 2), (3, 3).

Probability $= \frac{3}{16}$

The probability is $\frac{3}{16}$ that the sum of the dots on the two dice is 6.

13. Strategy:
To calculate the probability:
→Count the number of possible outcomes.
See the table on p. 328.
→Count the number of favorable outcomes.
→Use the probability formula.
→Compare the probabilities.

Solution:
There are 36 possible outcomes.
For a sum of 10, there are 3 favorable outcomes: (5, 5), (4, 6), (6, 4).

Probability $= \frac{3}{36}$

For a sum of 5, there are 4 favorable outcomes:
(1, 4), (4, 1), (2, 3), (3, 2).

Probability $= \frac{4}{36}$

$\frac{4}{36} > \frac{3}{36}$

The probability of throwing a sum of 5 is greater.

15. Strategy:
To calculate the probability:
→Count the number of possible outcomes.
→Count the number of favorable outcomes.
→Use the probability formula.
→Compare the probabilities.

Solution:
There are 12 possible outcomes.
There are 3 favorable outcomes of choosing a blue marble.

Probability $= \frac{3}{12} = \frac{1}{4}$

There are 5 favorable outcomes of choosing a red marble.

Probability $= \frac{5}{12}$

$\frac{5}{12} > \frac{3}{12}$

The probability of choosing a red marble is greater.

17. Strategy:
To calculate the empirical probability, use the probability formula and divide the number of observations (37) by the total number of observations (95).

Solution:

$$\text{Probability} = \frac{37}{95} \approx 0.39$$

The probability is 0.39 that a person prefers a cash discount.

19. Strategy:
To calculate the probability:
→Count the number of possible outcomes.
→Count the number of favorable outcomes.
→Use the probability formula.

Solution:
There are 377 (98 + 87 + 129 + 42 + 21) possible outcomes.
There are 185 (98 + 87) favorable outcomes.

$$\text{Probability} = \frac{185}{377}$$

The probability is $\frac{185}{377}$ that the customer rated the cable service satisfactory or excellent.

Projects or Group Activities 8.3

21. a. 56

b. 2 tails means 6 heads occurred; 28

c. $1 + 8 + 28 + 56 + 70 + 56 + 28 + 8 + 1$
$= 256$

d. $\frac{70}{256} = \frac{35}{128}$

e. 7 tails means 1 head occurred; $\frac{8}{256} = \frac{1}{32}$

f. $56 + 28 + 8 + 1 = 93$; $\frac{93}{256}$

g. $8 + 28 + 56 + 70 + 56 + 28 + 8 + 1 = 255$;
$\frac{255}{256}$

h. 1

Chapter Review

1. Strategy:
To find the amount of money:
→Read the circle graph to determine the amounts of money spent.
→Add the amounts.

Solution:

Defense:	148 million
Agriculture:	15 million
EPA:	24 million
Commerce:	27 million
NASA:	31 million
Other:	+ 104 million
	349 million

The agencies spent $349 million in maintaining websites.

2. Strategy:
To find the ratio:
→Read the circle graph to find the amount spent by the Department of Commerce and by the EPA.
→Write, in simplest form, the ratio of the amount spent by the Department of Commerce to the amount spent by the EPA.

Solution:
Commerce:
\$27 million
EPA:
\$24 million

$$\frac{\$27 \text{ million}}{\$24 \textit{million}} = \frac{9}{8}$$

The ratio is $\frac{9}{8}$.

3. Strategy:
To find the percent, solve the basic percent equation for percent. The base is the total amount spent (\$349 million) and the amount is the amount spent by NASA (\$31 million).

Solution:
Percent ÷ base = amount

$$n \div 349 \text{ million} = 31 \text{ million}$$
$$n = \frac{31 \text{ million}}{349 \text{ million}}$$
$$n \approx 0.0888$$

8.9% of the total mount of money was spent by NASA.

4. Texas had the larger population.

5. Strategy:
To find the difference in populations:
→Read the double broken-line graph to find the Texas population and the California population in 2000.
→Subtract the population of Texas from the population of California.

Solution:

California :	32.5
Texas :	−20.0
	12.5

The population of California is 12.5 million people more than the population of Texas.

6. Strategy:
To find which 25-year period Texas had the smallest increase in population.
→Read the double-line graph to find the population for each 25-year period.
→Subtract the two numbers.

Solution:
1900 to 1925:
6 – 2.5 = 3.5 million
1925 to 1950:
8 – 6 = 2 million
1950 to 1975:
12 – 8 = 4 million
1975 to 2000:
21 – 12 = 9 million
The Texas population increased the least from 1925 to 1950.

7. Strategy:
To find the difference between the total days of operation and days of full operation of the Midwest ski areas:
→Read the double-bar graph for the number of days that the resorts were open and the days of full operation.
→Subtract the two numbers.

Solution:

Days open:	90
Days of full operation:	– 40
	50

The difference was 50 days.

8. Strategy:
To find the percent:
→Read the double-bar graph to find the number of days that the Rocky Mountain ski areas were open and the number of days of full operation.
→Solve the basic percent equation for percent. The base is the number of days open and the amount is the number of days of full operation.

Solution:
Days open: 140
Days of full operation: 70
Percent ÷ base = amount

$$n \div 140 = 70$$
$$n = 70 \text{ ı } 140$$
$$n = 0.5$$

The percent is 50%.

9. Strategy:
To determine which region has the lowest number of days of full operation, read the bar graph and select the lowest graph that shows days of full operation.

Solution:
The Southeast had the lowest number of days of full operation.

Strategy:
Read the number of days from the lowest graph.

Solution:
This region had 30 days of full operation.

10. Strategy:
To calculate the mean:
→Calculate the sum of the weights.
→Divide by 10.
To calculate the median:
→Arrange the numbers from smallest to largest.
→Because there is an even number of values, the median is the sum of the two middle numbers divided by 2.

Solution:
The sum of the numbers is 71.7.

$\bar{x} = \frac{71.7}{10} = 7.17$

The mean weight of the babies is 7.17 lb.

5.6 5.9 6.3 6.5 6.9
7.2 7.2 8.1 8.9 9.1

median $= \frac{6.9+7.2}{2} = 7.05$

The median weight of the babies is 7.05 lb.

11. Strategy
To prepare the box-and-whiskers plot:
→Arrange the data from smallest to largest. Then find the median.
→Find Q_1, the median of the lower half of the data.
→Find Q_3, the median of the upper half of the data.
→Determine the smallest and largest data values.
→Draw the box-and-whiskers plot.

Solution:

89 99 102 105 109
110 110 111 116 120
121 124 124 131 134

median = 111

$Q_1 = 105$

$Q_3 = 124$

Smallest value: 89
Largest value: 134

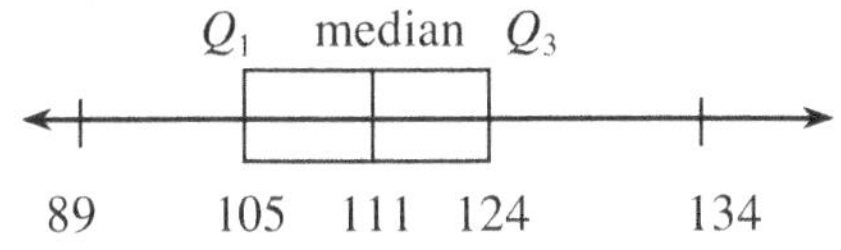

12. The possible outcomes of tossing a coin and then a regular die: (H, 1), (H, 2), (H, 3), (H, 4), (H, 5), (H, 6), (T, 1), (T, 2), (T, 3), (T, 4), (T, 5), (T, 6). There are 12 elements in the sample space.

13. Strategy:
Use the probability formula.

Solution:
There are 2500 possible outcomes.
There are 5 outcomes favorable to *E*.

$$P(E)=\frac{\text{number of favorable outcomes}}{\text{total number of outcomes}}$$

$$P(E)=\frac{5}{2500}=\frac{1}{500}$$

The probability of winning the television is $\frac{1}{500}$.

14. Strategy:
Arrange the data from smallest to largest. Find the median. Find Q_1, the median of the lower half of the data. Find Q_3, the median of the upper half of the data.

Solution:

0	0	1	1	1
2	3	3	4	5

$$\text{median}=\frac{1+2}{2}=\frac{3}{2}$$

$Q_1=1$ $\quad Q_3=3$

15. Strategy:
To calculate the probability:
→Count the number of possible outcomes.
→Count the number of favorable outcomes of the experiment *E*.
→Use the probability formula.

Solution:
There are 12 possible outcomes of the experiment.
There are 2 outcomes favorable to *E*: 6, 12

$$P(E)=\frac{2}{12}=\frac{1}{6}$$

The probability is $\frac{1}{6}$ that the number will be divisible by 6.

16. Strategy:
To calculate the probability:
→Find the number of possible outcomes.
→Find the number of favorable outcomes.
→Use the probability formula.

Solution:
There are 14(3 + 4 + 5 + 2) possible outcomes.
There are 5 favorable outcomes.

$$P(E)=\frac{5}{14}$$

The probability is $\frac{5}{14}$ that the student is a junior.

17. a. Strategy:
To find the mean heart rates:
→Find the sum of the heart rates.
→Divide the sum by the number of women (24).

Solution:

$$\begin{array}{r} 80 \\ 82 \\ 99 \\ 91 \\ 93 \\ 87 \\ 103 \\ 94 \\ 73 \\ 96 \\ 86 \\ 80 \\ 97 \\ 94 \\ 108 \\ 81 \\ 100 \\ 109 \\ 91 \\ 84 \\ 78 \\ 96 \\ 96 \\ +\ 100 \\ \hline 2198 \end{array}$$

$$91.58 \approx 91.6$$

$$\begin{array}{r} 24\overline{)2198.00} \\ -216 \\ \hline 38 \\ -24 \\ \hline 140 \\ -120 \\ \hline 200 \\ -192 \\ \hline 8 \end{array}$$

The mean heart rate is 91.6 heartbeats per minute.

Strategy:
To find the median heart rate: write the heart rates in order from smallest to largest. The median is the mean of the two middle numbers.

Solution:

$$\left.\begin{array}{r} 73 \\ 78 \\ 80 \\ 80 \\ 81 \\ 82 \\ 84 \\ 86 \\ 87 \\ 91 \\ 91 \end{array}\right\} \text{11 numbers}$$

$$\left.\begin{array}{r} 93 \\ 94 \end{array}\right\} \text{middle numbers}$$

$$\left.\begin{array}{r} 94 \\ 96 \\ 96 \\ 96 \\ 97 \\ 99 \\ 100 \\ 100 \\ 103 \\ 108 \\ 109 \end{array}\right\} \text{11 numbers}$$

$$\frac{93 + 94}{2} = 93.5$$

The median heart rate is 93.5 heartbeats per minute.

Strategy:
To find the mode, look at the heart rates and identify the number that occurs most frequently.

Solution:
The mode is 96 heartbeats per minute, the number that occurs most frequently.

b. Strategy:
→Arrange the data from smallest to largest. Then find the range.
→Find Q_1, the median of the lower half of the data.
→Find Q_3, the median of the upper half of the data.
→Interquartile range = $Q_3 - Q_1$.

Solution:
Use the list in part a.
Range = 109 – 73 = 36
The range is 36 heartbeats per minute.

$$Q_1 = \frac{82 + 84}{2} = 83$$

$$Q_3 = \frac{97 + 99}{2} = 98$$

$$Q_3 - Q_1 = 98 - 83 = 15$$

The interquartile range is 15 heartbeats per minute.

Chapter Test

1. Strategy:
Read the bar graph to find the two consecutive years that the number of fatalities were the same.

Solution:
During 1995 and 1996, the number of fatalities was the same.

2. Strategy:
To find the total fatalities on amusement rides during 1991 to 1999:
→Read the bar graph to determine the number of fatalities for each year.
→Add the nine numbers.

Solution:

3	1991
2	1992
4	1993
2	1994
3	1995
3	1996
4	1997
5	1998
+ 6	1999
32	

There were 32 fatal accidents from 1991 to 1999.

3. Strategy:
To find how many more fatalities in 1995 to 1998 than 1991 to 1994:
→Add the number of fatalities for 1995 to 1998.
→Add the number of fatalities for 1991 to 1994.
→Subtract the two numbers.

Solution:

3	1995	3	1991
3	1996	2	1992
4	1997	4	1993
+ 5	1998	+ 2	1994
15		11	

15 – 11 = 4
There were 4 more fatalities from 1995 to 1998.

4. Strategy:
To find how many more R-rated films than PG:
→Read the circle graph to find the number of films rated R and PG.
→Subtract the two numbers.

Solution:

$$\begin{array}{lr} \text{R:} & 427 \\ \text{PG:} & \underline{-\ 72} \\ & 355 \end{array}$$

There were 355 more films rated R.

5. Strategy:
To find how many times more PG-13 films were released than NC-17:
→Read the circle graph to find the number of films rated PG-13 and NC-17.
→Divide the two numbers.

Solution:
PG-13: 112
NC-17: 7

$$7\overline{)112} = 16$$

There were 16 times more films rated PG-13.

6. Strategy:
To find the percent of films rated G:
→Read the circle graph to find the number of G-rated films.
→Write and solve the basic percent equation for the percent. The base is the total number of films (655) and the amount is the number of G-rated films.

Solution:
G: 37

$$\text{Percent} \cdot \text{base} = \text{amount}$$
$$n \cdot 655 = 37$$
$$n = 37 \div 655$$
$$n \approx 0.056$$

The percent of films rated G was 5.6%.

7. Strategy:
To find which decade had the smallest increase in enrollment.
→Read the line graph to find the enrollment for each decade.
→Subtract the two numbers.

Solution:
1960 to 1970:
$8 - 4 = 4$ million
1970 to 1980:
$12 - 8 = 4$ million
1980 to 1990:
$14 - 12 = 2$ million
1990 to 2000:
$15 - 14 = 1$ million
2000 to 2010:
$21 - 15 = 6$ million

The student enrollment increased the least during the 1990s.

8. Strategy
To approximate the increase in enrollment:
→Read the enrollment for 1960 and 2010.
→Subtract the two numbers.

Solution:

$$\begin{array}{lr} 2010: & 21 \text{ million} \\ 1960: & \underline{-\ 4 \text{ million}} \\ & 17 \text{ million} \end{array}$$

The increase in enrollment was 17 million students.

9. Strategy:
To find the median response time:
→Arrange the numbers from smallest to largest.
→The median is the middle term.

Solution:
8 8 11 11 14 15 17 21 22
median = 14
The median response time is 14 min.

10. Strategy:
To find the modal response, write down the category that received the most responses.

Solution:
Because a response of very good was recorded most frequently, the modal response is very good.

11. Strategy:
To find the range, subtract the smallest value from the largest value.
To find the median, look at the median of the box-and-whisker plot.

Solution:
26 – 4 = 22

a. The range is 22 days.

b. The median is 14 vacation days.

12. Strategy:
To draw the box-and-whiskers plot:
→Arrange the data from smallest to largest, and then find the median.
→Find Q_1, the median of the lower half of the data.
→Find Q_3, the median of the upper half of the data.
→Determine the smallest and largest data values.
→Draw the box-and-whiskers plot.

Solution:

68	69	70	70	70	71	72
73	73	74	74	75	76	80

median $= \frac{72+73}{2} = 72.5$

$Q_1 = 70$
$Q_3 = 74$
Smallest value: 68
Largest value: 80

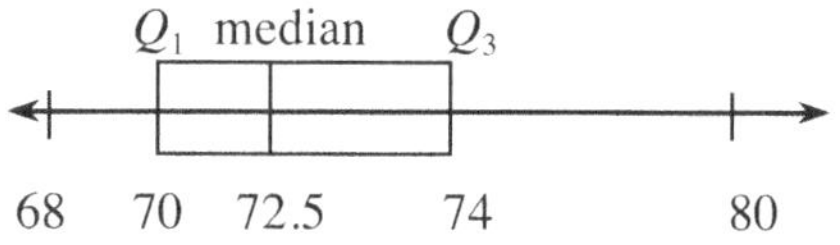

13. Strategy:
To calculate the probability:
→Count the number of possible outcomes.
→Count the number of favorable outcomes.
→Use the probability formula.

Solution:
There are 50 possible outcomes.
There are 15 favorable outcomes.

Probability $= \frac{15}{50} = \frac{3}{10}$

The probability is $\frac{3}{10}$ that the ball chosen is red.

14. Strategy:
To calculate the probability:
→Find the number of possible outcomes.
→Find the number of favorable outcomes.
→Use the probability formula.

Solution:
There are 248(14 + 32 + 202) possible outcomes.
There are 32 favorable outcomes.

$P(E) = \frac{32}{248} = \frac{4}{31}$

The probability is $\frac{4}{31}$ that the person is in business class.

15. Strategy:
To calculate the probability:
→Find the number of possible outcomes.
→Find the number of favorable outcomes.
→Use the probability formula.

Solution:
There are 6 possible outcomes: (A, K, Q), (A, Q, K),
(K, A, Q), (K, Q, A),
(Q, A, K), (Q, K, A).
There are 2 favorable outcomes.

$P(E) = \frac{2}{6} = \frac{1}{3}$

The probability is $\frac{1}{3}$ that the ace is on top of the stack.

16. Strategy:
To calculate the probability:
→Find the number of possible outcomes.
→Find the number of favorable outcomes.
→Use the probability formula.

Solution:
There are 8 possible outcomes.
There is 1 favorable outcomes.

$P(E) = \frac{1}{8}$

The probability is $\frac{1}{8}$ that the student will answer all three questions correctly.

17. a. Strategy:
To find the mean lifetime of the batteries:
→Find the sum of the times.
→Divide the sum by the number of batteries tested (20).

Solution:

2.9
2.4
3.1
2.5
2.6
2.0
3.0
2.3
2.4
2.7
2.0
2.4
2.6
2.7
2.1
2.9
2.8
2.4
2.0
+ 2.8

50.6

$$20\overline{)50.60} = 2.53$$

The mean time is 2.53 hours.

b. Strategy:
To find the median lifetime of the batteries, write times in order from lowest to highest. The median is the mean of the two middle numbers.

Solution:

2.0, 2.0, 2.0, 2.1, 2.3, 2.4, 2.4, 2.4, 2.4 } 9 numbers
2.5, 2.6 } middle numbers
2.6, 2.7, 2.7, 2.8, 2.8, 2.9, 2.9, 3.0, 3.1 } 9 numbers

$$\frac{2.5 + 2.6}{2} = 2.55$$

The median time is 2.55 hours.

c. Strategy:
The data is arranged from smallest to largest in part b.
→Find Q_1, the median of the lower half of the data.
→Find Q_3, the median of the upper half of the data.
→Draw the box-and-whiskers plot.

Solution:

$$Q_1 = \frac{2.3 + 2.4}{2} = 2.35$$

$$Q_3 = \frac{2.8 + 2.8}{2} = 2.8$$

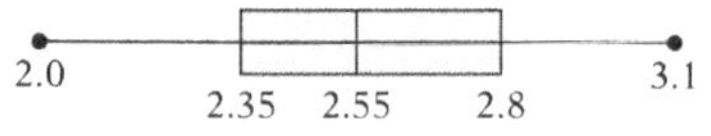

Cumulative Review Exercises

1. $2^2 \cdot 3^3 \cdot 5 = (2 \cdot 2) \cdot (3 \cdot 3 \cdot 3) \cdot (5)$
$= 4 \cdot 27 \cdot 5 = 540$

2. $3^2 \cdot (5 - 2) \div 3 + 5$
$= 9 \cdot (3) \div 3 + 5$
$= 27 \div 3 + 5$
$= 9 + 5$
$= 14$

3.

	2	3	5
24 =	(2 · 2 · 2)	(3)	
40 =	2 · 2 · 2		(5)

LCM = 2 · 2 · 2 · 3 · 5 = 120

4. $\dfrac{60}{144} = \dfrac{\cancel{2} \cdot \cancel{2} \cdot \cancel{3} \cdot 5}{\cancel{2} \cdot \cancel{2} \cdot 2 \cdot 2 \cdot \cancel{3} \cdot 3} = \dfrac{5}{12}$

5.
$4\dfrac{1}{2} = 4\dfrac{20}{40}$
$2\dfrac{3}{8} = 2\dfrac{15}{40}$
$+5\dfrac{1}{5} = 5\dfrac{8}{40}$
$11\dfrac{43}{40} = 12\dfrac{3}{40}$

6.
$12\dfrac{5}{8} = 12\dfrac{15}{24} = 11\dfrac{39}{24}$
$-7\dfrac{11}{12} = 7\dfrac{22}{24} = 7\dfrac{22}{24}$
$4\dfrac{17}{24}$

7. $\dfrac{5}{8} \times 3\dfrac{1}{5} = \dfrac{5}{8} \times \dfrac{16}{5}$
$= \dfrac{5 \cdot 16}{8 \cdot 5}$
$= \dfrac{\cancel{5} \cdot \cancel{2} \cdot \cancel{2} \cdot \cancel{2} \cdot 2}{\cancel{2} \cdot \cancel{2} \cdot \cancel{2} \cdot \cancel{5}} = 2$

8. $3\dfrac{1}{5} \div 4\dfrac{1}{4} = \dfrac{16}{5} \div \dfrac{17}{4} = \dfrac{16}{5} \times \dfrac{4}{17} = \dfrac{16 \cdot 4}{5 \cdot 17} = \dfrac{64}{85}$

9. $\dfrac{5}{8} \div \left(\dfrac{3}{4} - \dfrac{2}{3}\right) + \dfrac{3}{4} = \dfrac{5}{8} \div \left(\dfrac{9}{12} - \dfrac{8}{12}\right) + \dfrac{3}{4}$
$= \dfrac{5}{8} \div \dfrac{1}{12} + \dfrac{3}{4}$
$= \dfrac{5}{8} \times \dfrac{12}{1} + \dfrac{3}{4}$
$= \dfrac{5 \cdot \cancel{2} \cdot \cancel{2} \cdot 3}{\cancel{2} \cdot \cancel{2} \cdot 2} + \dfrac{3}{4}$
$= \dfrac{15}{2} + \dfrac{3}{4} = \dfrac{30}{4} + \dfrac{3}{4} = \dfrac{33}{4} = 8\dfrac{1}{4}$

10. 209.305

11.
```
    4.092
  × 0.69
   36828
  24552
 2.82348
```

12. $16\dfrac{2}{3} = \dfrac{50}{3} = 16.\overline{6}$

```
   16.666
3)50.000
 -3
  20
 -18
   20
  -18
    20
   -18
     20
    -18
      20
```

13. $\dfrac{330 \text{ miles}}{12.5 \text{ gal}} = 26.4$ miles/gallon

14. $\dfrac{n}{5} = \dfrac{16}{25}$
$n \times 25 = 5 \times 16$
$n \times 25 = 80$
$n = 80 \div 25 = 3.2$

15. $\dfrac{4}{5} \times 100\% = 80\%$

16. $10\% \times n = 8$

$0.10 \times n = 8$

$n = 8 \div 0.10 = 80$

17. $38\% \times 43 = n$

$0.38 \times 43 = n$

$16.34 = n$

18. $n \times 75 = 30$

$n = 30 \div 75 = 0.40 = 40\%$

19. Strategy:
To find the income for the week:
→Find the commission earned on sales by solving the basic percent equation for amount. The base is $27,500 and the percent is 2%.
→Find the total income by adding the base salary ($100) to the commission.

Solution:

$2\% \times 27{,}500 = n$

$0.02 \times 27{,}500 = n$

$550 = n$

$100 + 550 = 650$

The salesperson's income for the week was $650.

20. Strategy:
To find the cost, write and solve a proportion.

Solution:

$$\frac{8.15}{1000} = \frac{n}{50{,}000}$$

$8.15 \times 50{,}000 = n \times 1000$

$407{,}500 = n \times 1000$

$407{,}500 \div 1000 = n$

$407.50 = n$

The cost is $407.50.

21. Strategy
To find the interest due, multiply the principal by the annual interest rate and the time (in years).

Solution:

$$125{,}000 \times 0.06 \times \frac{6}{12} = 3750$$

The interest due is $3750.

22. Strategy:
To find the markup rate of the compact disc player:
→Find the markup amount by subtracting the cost ($180) from the selling price ($279).
→Solve the basic percent equation for percent. The base is $180 and the amount is the amount of the markup.

Solution:

$279 - 180 = 99$

Percent ÷ base = amount

$n \times 180 = 99$

$n = 99 \div 180$

$= 0.55 = 55\%$

The markup rate is 55%.

23. Strategy:
To find how much is budgeted for food:
→Read the circle graph to find what percent of the budget is spent on food.
→Solve the basic percent equation for amount. The base is $4500 and the rate is the percent of the budget that is spent on food.

Solution:

Amount spent on food: 19%

Percent × base = amount

$19\% \times 4500 =$ amount

$0.19 \times 4500 = 855$

The amount budgeted for food is $855.

24. Strategy:
To find the difference:
→Read the double-broken-line graph to find the number of problems student 1 answered correctly on test 1 and the number of problems student 2 answered correctly on test 1.
→Subtract the student 1 total from the student 2 total to find the difference.

Solution:

student 2: 27 answered correctly

student 1: −15 answered correctly

12 answered correctly

The difference in the number answered correctly is 12 problems.

25. Strategy:
To find the mean high temperature:
→Find the sum of the high temperatures.
→Divide the sum of the high temperatures by the number of temperatures (7).

Solution:

$$\begin{array}{r} 56^\circ \\ 72^\circ \\ 80^\circ \\ 75^\circ \\ 68^\circ \\ 62^\circ \\ +\ 74^\circ \\ \hline 487^\circ \end{array} \quad \text{sum of high temps}$$

$$7\overline{)487.00} = 69.57$$

The mean high temperature is 69.6°F.

26. Strategy:
To calculate the probability:
→Count the number of possible outcomes.
→Count the number of favorable outcomes.
→Use the probability formula.

Solution:
There are 36 possible outcomes. There are 5 favorable outcomes: (2, 6), (6, 2), (3, 5), (5, 3), (4, 4).

Probability = $\frac{5}{36}$

The probability is $\frac{5}{36}$ that the sum of the dots on the two dice is 8.

Chapter 9: Polynomials

Prep Test

1. $-2-(-3)=-2+3=1$

2. $-3(6)=-18$

3. $-\frac{24}{-36}=\frac{2}{3}$

4. $3n^4;\ n=-2$
$3(-2)^2=3\cdot 16$
$=48$

5. 0

6. No. $2x^2$ and $2x$ are not like terms since the exponent on x is different.

7. $3x^2-4x+1+2x^2-5x-7$
$=3x^2+2x^2-4x-5x+1-7$
$=5x^2-9x-6$

8. $-4y+4y=0$

9. $-3(2x-8)=-6x+24$

10. $3xy-4y-2(5xy-7y)=3xy-4y-10xy+14y$
$=3xy-10xy-4y+14y$
$=-7xy+10y$

Section 9.1

Concept Check

1. Yes
3. No
5. No
7. Binomial
9. Trinomial
11. None of these
13. Binomial

Objective A Exercises

15. $(4x^2+2x)+(x^2+6x)=(4x^2+x^2)+(2x+6x)$
$=5x^2+8x$

17. $(4x^2-5xy)+(3x^2+6xy-4y^2)$
$=(4x^2+3x^2)+(-5xy+6xy)-4y^2$
$=7x^2+xy-4y^2$

19. $(2a^2-7a+10)+(a^2+4a+7)$
$=(2a^2+a^2)+(-7a+4a)+(10+7)$
$=3a^2-3a+17$

21. $(7x+5x^3-7)+(10x^2-8x+3)$
$=5x^3+10x^2+(7x-8x)+(-7+3)$
$=5x^3+10x^2-x-4$

23. $(7-5r+2r^2)+(3r^3-6r)$
$=3r^3+2r^2+(-5r-6r)+7$
$=3r^3+2r^2-11r+7$

25.
$$\begin{array}{r} x^2+7x \\ +-3x^2-4x \\ \hline -2x^2+3x \end{array}$$

27.
$$\begin{array}{r} 2x^2+6x+12 \\ +3x^2+x+8 \\ \hline 5x^2+7x+20 \end{array}$$

29.
$$\begin{array}{r} x^3 \qquad\quad -7x+4 \\ +\quad 2x^2+x-10 \\ \hline x^3+2x^2-6x-6 \end{array}$$

31.
$$\begin{array}{r} 2a^3 \qquad\quad -7a+1 \\ +\quad -3a^2-4a+1 \\ \hline 2a^3-3a^2-11a+2 \end{array}$$

33. (i)

$P+Q$

$=(ax^3+bx^2-cx+d)+(-ax^3-bx^2+cx-d)$

$=(ax^3-ax^3)+(bx^2-bx^2)+(-cx+cx)+(d-d)$

$=0$

Monomial

(ii)

$Q+R$

$=(-ax^3-bx^2+cx-d)+(-ax^3+bx^2+cx+d)$

$=(-ax^3-ax^3)+(-bx^2+bx^2)+(cx+cx)+(-d+d)$

$=-2ax^3+2cx$

Binomial

(iii)

$P+R$

$=(ax^3+bx^2-cx+d)+(-ax^3+bx^2+cx+d)$

$=(ax^3-ax^3)+(bx^2+bx^2)+(-cx+cx)+(d+d)$

$=2bx^2+2d$

Binomial

None of the above is a trinomial. (iv)

Objective B Exercises

35. $(y^2-10xy)-(2y^2+3xy)$

$=(y^2-10xy)+(-2y^2-3xy)$

$=-y^2-13xy$

37. $(3x^2+x-3)-(x^2+4x-2)$

$=(3x^2+x-3)+(-x^2-4x+2)$

$=2x^2-3x-1$

39. $(-2x^3+x-1)-(-x^2+x-3)$

$=(-2x^3+x-1)+(x^2-x+3)$

$=-2x^3+x^2+2$

41. $(1-2a+4a^3)-(a^3-2a+3)$

$=(1-2a+4a^3)+(-a^3+2a-3)$

$=3a^3-2$

43. $(-1-y+4y^3)-(3-3y-2y^2)$

$=(-1-y+4y^3)+(-3+3y+2y^2)$

$=4y^3+2y^2+2y-4$

45. The opposite of $-y^2+2$ is y^2-2.

$$\begin{array}{r} 2y^2-4y \\ \underline{y^2-2} \\ 3y^2-4y-2 \end{array}$$

47. The opposite of x^2+5x+8 is $-x^2-5x-8$.

$$\begin{array}{r} x^2-2x+1 \\ \underline{-x^2-5x-8} \\ -7x-7 \end{array}$$

49. The opposite of $-3x^2+2x+1$ is $3x^2-2x-1$.

$$\begin{array}{r} 4x^3+5x+2 \\ \underline{3x^2-2x-1} \\ 4x^3+3x^2+3x+1 \end{array}$$

51. The opposite of y^3+y^2+4 is $-y^3-y^2-4$.

$$\begin{array}{r} 2y^3+6y^2-2y \\ \underline{-y^3-y^2-4} \\ y^3+5y^2-2y-4 \end{array}$$

53. Subtract the polynomial from the sum.

$(4x^2+3x-2)-(3x^2-6x+9)$

$=(4x^2+3x-2)+(-3x^2+6x-9)$

$=x^2+9x-11$

Critical Thinking 9.1

55. Subtract the polynomial from the sum.

$\left(x^2-x-1\right)-\left(-2x^3+4x-7\right)$
$=x^2-x-1+2x^3-4x+7$
$=2x^3+x^2-5x+6$

57. Subtract the difference from the polynomial.

$\left(2x^3-x^2+4x-2\right)-\left(x^3+2x-8\right)$
$=2x^3-x^2+4x-2-x^3-2x+8$
$=x^3-x^2+2x+6$

Projects and Group Activities 9.1

59. Answers will vary. For example: $3x^2-4x+7$ and $-3x^2+5x-2$

Section 9.2

Concept Check

1. Product

3. Power

5. Power

7. Product

Objective A Exercises

9. No

11. No

13. $\left(6x^2\right)(5x)=(6\cdot5)\left(x^2\cdot x\right)=30x^3$

15. $\left(7c^2\right)\left(-6c^4\right)=\left[7(-6)\right]\left(c^2\cdot c^4\right)=-42c^6$

17. $\left(-3a^3\right)\left(-3a^4\right)=(-3\cdot-3)\left(a^3\cdot a^4\right)=9a^7$

19. $\left(x^2\right)\left(xy^4\right)=\left(x^2\cdot x\right)y^4=x^3y^4$

21. $\left(-2x^4\right)\left(5x^5y\right)=(-2\cdot5)\left(x^4\cdot x^5\right)y=-10x^9y$

23. $\left(-4x^2y^4\right)\left(-3x^5y^4\right)=(-4\cdot-3)\left(x^2\cdot x^5\right)\left(y^4\cdot y^4\right)$
$=12x^7y^8$

25. $(2xy)\left(-3x^2y^4\right)=(2\cdot-3)\left(x\cdot x^2\right)\left(y\cdot y^4\right)=-6x^3y^5$

27. $\left(x^2yz\right)\left(x^2y^4\right)=\left(x^2\cdot x^2\right)\left(y\cdot y^4\right)z=x^4y^5z$

29. $\left(-a^2b^3\right)\left(-ab^2c^4\right)=(-1\cdot-1)\left(a^2\cdot a\right)\left(b^3\cdot b^2\right)c^4$
$=a^3b^5c^4$

31. $\left(-5a^2b^2\right)\left(6a^3b^6\right)=(-5\cdot6)\left(a^2\cdot a^3\right)\left(b^2\cdot b^6\right)$
$=-30a^5b^8$

33. $\left(-6a^3\right)\left(-a^2b\right)=(-6\cdot-1)\left(a^3\cdot a^2\right)b=6a^5b$

35. $\left(-5y^4z\right)\left(-8y^6z^5\right)=(-5\cdot-8)\left(y^4\cdot y^6\right)\left(z\cdot z^5\right)$
$=40y^{10}z^6$

37. $\left(x^2y\right)(yz)(xyz)=\left(x^2\cdot x\right)(y\cdot y\cdot y)(z\cdot z)=x^3y^3z^2$

39. $\left(3ab^2\right)(-2abc)\left(4ac^2\right)$
$=(3\cdot-2\cdot4)(a\cdot a\cdot a)\left(b^2\cdot b\right)\left(c\cdot c^2\right)$
$=-24a^3b^3c^3$

41. $\left(4x^4z\right)\left(-yz^3\right)\left(-2x^3z^2\right)$
$=(4\cdot-1\cdot-2)\left(x^4\cdot x^3\right)(y)\left(z\cdot z^3\cdot z^2\right)$
$=8x^7yz^6$

43. $\left(-2x^2y^3\right)(3xy)\left(-5x^3y^4\right)$
$=(-2\cdot3\cdot-5)\left(x^2\cdot x\cdot x^3\right)\left(y^3\cdot y\cdot y^4\right)$
$=30x^6y^8$

45. $\left(3a^2b\right)(-6bc)\left(2ac^2\right)$
$=(3\cdot-6\cdot2)\left(a^2\cdot a\right)(b\cdot b)\left(c\cdot c^2\right)$
$=-36a^3b^2c^3$

Objective B Exercises

47. No

49. Yes

51. $\left(x^3\right)^5=x^{3\cdot5}=x^{15}$

53. $\left(x^7\right)^2=x^{7\cdot2}=x^{14}$

55. $\left(-x^2\right)^4=x^{2\cdot4}=x^8$

57. $\left(-y^3\right)^4=y^{3\cdot4}=y^{12}$

59. $\left(-2x^2\right)^3=(-2)^3\left(x^{2\cdot3}\right)=-8x^6$

61. $(x^2y^3)^2 = x^{2\cdot2}y^{3\cdot2} = x^4y^6$

63. $(3x^2y)^2 = 3^2x^{2\cdot2}y^2 = 9x^4y^2$

65. $(-3x^3y^2)^5 = (-3)^5x^{3\cdot5}y^{2\cdot5} = -243x^{15}y^{10}$

67. $(-2x)(2x^3)^2 = (-2x)(2^2x^{3\cdot2}) = (-2x)(4x^6)$
$= -2\cdot4x^{1+6} = -8x^7$

69. $(3x^2y)(2x^2y^2)^3 = (3x^2y)(2^3x^{2\cdot3}y^{2\cdot3})$
$= (3x^2y)(8x^6y^6) = 3\cdot8x^{2+6}y^{1+6}$
$= 24x^8y^7$

71. $(ab^2)^2(ab)^2 = (a^{1\cdot2}b^{2\cdot2})(a^{1\cdot2}b^{1\cdot2}) = (a^2b^4)(a^2b^2)$
$= a^{2+2}b^{4+2} = a^4b^6$

73. $(-2x)^3(-2x^3y)^3 = (-2^3x^3)(-2^3x^{3\cdot3}y^3)$
$= (-8x^3)(-8x^9y^3)$
$= -8\cdot-8x^{3+9}y^3 = 64x^{12}y^3$

75. $(-2x)(-3xy^2)^2 = (-2x)(3^2x^2y^{2\cdot2}) = (-2x)(9x^2y^4)$
$= -2\cdot9x^{1+2}y^4 = -18x^3y^4$

77. $(ab^2)(-2a^2b)^3 = (ab^2)(-2^3a^{2\cdot3}b^3)$
$= (ab^2)(-8a^6b^3)$
$= -8a^{1+6}b^{2+3} = -8a^7b^5$

79. $(-2a^3)(3a^2b)^3 = (-2a^3)(3^3a^{2\cdot3}b^3)$
$= (-2a^3)(27a^6b^3)$
$= -2\cdot27a^{3+6}b^3 = -54a^9b^3$

Critical Thinking 9.2

81. $3x^2 + (3x)^2 = 3x^2 + 9x^2 = 12x^2$

83. $2x^6y^2 + (3x^2y)^2 = 2x^6y^2 + 9x^4y^2$

85. $(2a^3b^2)^3 - 8a^9b^6 = 8a^9b^6 - 8a^9b^6 = 0$

87. $(x^2y^4)^2 + (2xy^2)^4 = x^4y^8 + 16x^4y^8 = 17x^4y^8$

Projects and Group Activities 9.2

89. $x_1 = -1x^1$
$x_2 = -2x^2$
$x_3 = -3x^3$
$x_4 = -4x^4$
$x_5 = -5x^5$
$(x_1)(x_2)(x_3)(x_4)(x_5)$
$= (-1x^1)(-2x^2)(-3x^3)(-4x^4)(-5x^5)$
$= (2x^3)(-3x^3)(-4x^4)(-5x^5)$
$= (-6x^6)(-4x^4)(-5x^5)$
$= (24x^{10})(-5x^5)$
$= -120x^{15}$

Section 9.3

Concept Check

1. Always true

3. Sometimes true

5. Always true

7. Sometimes true

9. Sometime true

Objective A Exercises

11. $x(x-2) = x^2 - 2x$

13. $-x(x+7) = -x^2 - 7x$

15. $3a^2(a-2) = 3a^3 - 6a^2$

17. $-5x^2(x^2 - x) = -5x^4 + 5x^3$

19. $-x^3(3x^2 - 7) = -3x^5 + 7x^3$

21. $2x(6x^2 - 3x) = 12x^3 - 6x^2$

23. $(2x-4)(3x) = 6x^2 - 12x$

25. $(3x+4)x = 3x^2 + 4x$

27. $-xy(x^2 - y^2) = -x^3y + xy^3$

29. $x(2x^3 - 3x + 2) = 2x^4 - 3x^2 + 2x$

31. $-a\left(-2a^2-3a-2\right)=2a^3+3a^2+2a$

33. $x^2\left(3x^4-3x^2-2\right)=3x^6-3x^4-2x^2$

35. $2y^2\left(-3y^2-6y+7\right)=-6y^4-12y^3+14y^2$

37. $\left(a^2+3a-4\right)(-2a)=-2a^3-6a^2+8a$

39. $-3y^2\left(-2y^2+y-2\right)=6y^4-3y^3+6y^2$

41. $xy\left(x^2-3xy+y^2\right)=x^3y-3x^2y^2+xy^3$

43. $4x-x(3x-1)=4x-3x^2+x=-3x^2+5x$

(i) $4x-3x^2-x=-3x^2+3x$ No

(ii) $-3x^2+5x$ Yes

(iii) $4x-3x^2+x=-3x^2+5x$ Yes

(iv) $9x^2-3x$ No

(v) $3x(3x-1)=9x^2-3x$ No

Objective B Exercises

45.
$$\begin{array}{r} x^2-2x+7 \\ \times \qquad x-2 \\ \hline -2x^2+4x-14 \\ x^3-2x^2+7x \quad \\ \hline x^3-4x^2+11x-14 \end{array}$$

47.
$$\begin{array}{r} x^2-3x+5 \\ \times \qquad 2x-3 \\ \hline -3x^2+9x-15 \\ 2x^3-6x^2+10x \quad \\ \hline 2x^3-9x^2+19x-15 \end{array}$$

49.
$$\begin{array}{r} -a^2+3a-2 \\ \times \qquad 2a-1 \\ \hline a^2-3a+2 \\ -2a^3+6a^2-4a \quad \\ \hline -2a^3+7a^2-7a+2 \end{array}$$

51.
$$\begin{array}{r} -a^2-2a+3 \\ \times \qquad 2a-1 \\ \hline a^2+2a-3 \\ -2a^3-4a^2+6a \quad \\ \hline -2a^3-3a^2+8a-3 \end{array}$$

53.
$$\begin{array}{r} y^2-2y \\ \times \qquad 2y+5 \\ \hline 5y^2-10y \\ 2y^3-4y^2 \quad \\ \hline 2y^3+y^2-10y \end{array}$$

55.
$$\begin{array}{r} y^3+4y^2 \qquad -8 \\ \times \qquad 2y-1 \\ \hline -y^3-4y^2 \qquad +8 \\ 2y^4+8y^3 \qquad -16y \quad \\ \hline 2y^4+7y^3-4y^2-16y+8 \end{array}$$

57.
$$\begin{array}{r} 3y^2+3y-5 \\ \times \qquad 4y-3 \\ \hline -9y^2-9y+15 \\ 12y^3+12y^2-20y \quad \\ \hline 12y^3+3y^2-29y+15 \end{array}$$

59.
$$\begin{array}{r} 3b^3-5b^2 \qquad +7 \\ \times \qquad 6b-1 \\ \hline -3b^3+5b^2 \qquad -7 \\ 18b^4-30b^3 \qquad +42b \quad \\ \hline 18b^4-33b^3+5b^2+42b-7 \end{array}$$

61.
$$\begin{array}{r} 2a^3-3a^2+2a-1 \\ \times \qquad 2a-3 \\ \hline -6a^3+9a^2-6a+3 \\ 4a^4-6a^3+4a^2-2a \quad \\ \hline 4a^4-12a^3+13a^2-8a+3 \end{array}$$

Objective C Exercises

63. $(x+1)(x+3)=x^2+3x+x+3$
$=x^2+4x+3$

65. $(a-3)(a+4)=a^2+4a-3a-12$
$=a^2+a-12$

67. $(y+3)(y-8)=y^2-8y+3y-24$
$=y^2-5y-24$

69. $(y-7)(y-3)=y^2-3y-7y+21$
$=y^2-10y+21$

71. $(2x+1)(x+7)=2x^2+14x+x+7$
$=2x^2+15x+7$

73. $(3x-1)(x+4)=3x^2+12x-x-4$
$=3x^2+11x-4$

75. $(4x-3)(x-7)=4x^2-28x-3x+21$
$=4x^2-31x+21$

77. $(3y-8)(y+2)=3y^2+6y-8y-16$
$=3y^2-2y-16$

79. $(3x+7)(3x+11)=9x^2+33x+21x+77$
$=9x^2+54x+77$

81. $(7a-16)(3a-5)=21a^2-35a-48a+80$
$=21a^2-83a+80$

83. $(3a-2b)(2a-7b)=6a^2-21ab-4ab+14b^2$
$=6a^2-25ab+14b^2$

85. $(a-9b)(2a+7b)=2a^2+7ab-18ab-63b^2$
$=2a^2-11ab-63b^2$

87. $(10a-3b)(10a-7b)=100a^2-70ab-30ab+21b^2$
$=100a^2-100ab+21b^2$

89. $(5x+12y)(3x+4y)=15x^2+20xy+36xy+48y^2$
$=15x^2+56xy+48y^2$

91. $(2x-15y)(7x+4y)=14x^2+8xy-105xy-60y^2$
$=14x^2-97xy-60y^2$

93. $(8x-3y)(7x-5y)=56x^2-40xy-21xy+15y^2$
$=56x^2-61xy+15y^2$

95. To find the polynomial, multiply.
$(3x-4)(4x+5)=12x^2+15x-16x-20$
$=12x^2-x-20$

Objective D Exercises

97. $(y+6)(y-6)=y^2-36$

99. $(4x-7)(4x+7)=16x^2-49$

101. $(9x-2)(9x+2)=81x^2-4$

103. $(4x-9y)(4x+9y)=16x^2-81y^2$

105. $(y-3)^2=y^2-6y+9$

107. $(6x-5)^2=36x^2-60x+25$

109. $(x-2y)^2=x^2-4xy+4y^2$

111. $(2a-9b)^2=4a^2-36ab+81b^2$

113 negative

115. positive

Objective E Exercises

117. Strategy:
→Width: $3x+1$
→Length: $2(3x+1)$
→Use the formula for the area of a triangle.

Solution:
$A=LW$
$A=2(3x+1)(3x+1)$
$A=(6x+2)(3x+1)$
$A=18x^2+12x+2$
The area of the rectangle is $(18x^2+12x+2)$ in^2.

119. Strategy:
→Radius: $x+4$
→Use the formula for the area of a circle.

Solution:
$A=\pi r^2$
$A=\pi(x+4)^2$
$A=\pi(x^2+8x+16)$
$A=\pi x^2+8\pi x+16\pi$
The area of the circle is $(\pi x^2+8\pi x+16\pi)$ cm^2.

121. Strategy:
To find the total area, find the sums of the areas of the two rectangular paths and the area of a square 45 ft by 45 ft.

Solution:
$A=2LW+S^2$
$A=2\cdot 45\cdot x+45^2$
$A=90x+2025$
The area is $(90x+2025)$ft^2.

123. **(a)** Height: h
Length of the wall: $5h+22$
(b)
$A=LW$
$A=h(5h+22)$
$A=5h^2+22h$
The area of 1 wall is $(5h^2+22h)$ ft^2.

Critical Thinking 9.3

125. $(a+b)^2-(a-b)^2$

$=(a+b)(a+b)-(a-b)(a-b)$

$=(a^2+ab+ab+b^2)-(a^2-ab-ab+b^2)$

$=(a^2+2ab+b^2)-(a^2-2ab+b^2)$

$=a^2+2ab+b^2-a^2+2ab-b^2$

$=4ab$

127. $(3a^2-4a+2)^2$

$=(3a^2-4a+2)(3a^2-4a+2)$

$$\begin{array}{r} 3a^2-4a+2 \\ \times\, 3a^2-4a+2 \\ \hline 6a^2-8a+4 \\ -12a^3+16a^2-8a \quad \\ 9a^4-12a^3+6a^2 \qquad\qquad \\ \hline 9a^4-24a^3+28a^2-16a+4 \end{array}$$

129. $3x^2(2x^3+4x-1)-6x^3(x^2-2)$

$=6x^5+12x^3-3x^2-6x^5+12x^3$

$=24x^3-3x^2$

131. Strategy:
To evaluate the expression, solve the equation for n.

Solution:

$2n-3=4n-7$

$2n-2n-3=4n-2n-7$

$-3=2n-7$

$-3+7=2n-7+7$

$4=2n$

$\frac{4}{2}=\frac{2n}{2}$

$2=n$

$(4n^3)^2$

$[4(2^3)]^2$

$=[4(8)]^2$

$=[32]^2=1024$

133. Multiply the quotient by $4x+5$.

$(3x-4)(4x+5)$

$=12x^2+15x-16x-20$

$=12x^2-x-20$

Projects or Group Activities 9.3

135. $(x+1)(-x^2+x-1)$

$=-x^3+x^2-x-x^2+x-1$

$=-x^3-1$

137. $(x+1)(-x^4+x^3-x^2+x-1)$

$=-x^5+x^4-x^3+x^2-x-x^4+x^3-x^2+x-1$

$=-x^5-1$

139. $(x+1)(-x^6+x^5-x^4+x^3-x^2+x-1)=-x^7-1$

Check Your Progress: Chapter 9

1. $(4x+3x^2-6)+(x^2-5x+1)$

$=4x+3x^2-6+x^2-5x+1=4x^2-x-5$

2. $(3y-7)(y+5)=3y^2+15y-7y-35$

$=3y^2+8y-35$

3. $4y(5y-3y^3)=20y^2-12y^4$

4. $(-2a^4)(3ab^5)=-6a^5b^5$

5. $(10x-3)(10x+3)=100x^2-9$

6.

$$\begin{array}{r} 2x^2-3x+5 \\ \times \qquad x-6 \\ \hline -12x^2+18x-30 \\ 2x^3-3x^2+5x \qquad \\ \hline 2x^3-15x^2+23x-30 \end{array}$$

7. $3x^4y(-x^6y^2)(-2xy^5)$

$=-3x^{10}y^3(-2xy^5)=6x^{11}y^8$

8. $(3x^6y^5)^4=3^4x^{24}y^{20}=81x^{24}y^{20}$

9.

$$\begin{array}{r} x^3-3x^2 \quad +6 \\ \times \qquad 2x+1 \\ \hline x^3-3x^2 \quad +6 \\ 2x^4-6x^3 \quad +12x \qquad \\ \hline 2x^4-5x^3-3x^2+12x+6 \end{array}$$

10. $(b-11)^2=(b-11)(b-11)$

$=b^2-11b-11b+121=b^2-22b+121$

11. $(-2b)(-4b^3)^2=(-2b)(16b^6)=-32b^7$

12. $(2x^3+4x-5)-(1+6x-x^2)$

$=2x^3+4x-5-1-6x+x^2$

$=2x^3+x^2-2x-6$

13. $5x^2(3x^2 - x + 7) = 15x^4 - 5x^3 + 35x^2$

14. $(4a-9)(3a-2) = 12a^2 - 8a - 27a + 18$
$= 12a^2 - 35a + 18$

15. $(5-6y)(5+6y) = 25 - 36y^2$

16. $(3a-5b)^2 = (3a-5b)(3a-5b)$
$= 9a^2 - 15ab - 15ab + 25b^2$
$= 9a^2 - 30ab + 25b^2$

17. Strategy:
→Width: $3x - 2$
→Length: $4x$
→Use the formula for the area of a rectangle.

Solution:
$A = LW$
$A = 4x(3x-2)$
$A = 12x^2 - 8x$
The area of the rectangle is $(12x^2 - 8x)$ in^2.

18. Strategy:
→Length of each side: $4x + 5$
→Use the formula for the area of a square.

Solution:
$A = s^2$
$A = (4x+5)^2$
$A = (4x+5)(4x+5)$
$A = 16x^2 + 20x + 20x + 25$
$A = 16x^2 + 40x + 25$
The area of the square is $(16x^2 + 40x + 25)$ in^2.

Section 9.4

Concept Check

1. If the negative exponent is the numerator, move the variable or number from the numerator to the denominator and change the exponent to positive. If the negative exponent is in the denominator, move the number or variable from the denominator to the numerator and change the sign of the exponent to positive.

3. False

5. True

7. False

9. 1; 1; 1; –2

11. No, 39.4 is not between 1 and 10.

13. No, 2.4 is an integer.

Objective A Exercises

15. $\frac{y^7}{y^3} = y^{7-3} = y^4$

17. $\frac{a^8}{a^5} = a^{8-5} = a^3$

19. $\frac{p^5}{p} = p^{5-1} = p^4$

21. $\frac{4x^8}{2x^5} = 2x^{8-5} = 2x^3$

23. $\frac{22k^5}{11k^4} = 2k^{5-4} = 2k$

25. $\frac{m^9n^7}{m^4n^5} = m^{9-4}n^{7-5} = m^5n^2$

27. $\frac{6r^4}{4r^2} = \frac{3r^{4-2}}{2} = \frac{3r^2}{2}$

29. $\frac{-16a^7}{24a^6} = -\frac{2a^{7-6}}{3} = -\frac{2a}{3}$

31. $\frac{y^3}{y^8} = y^{3-8} = y^{-5} = \frac{1}{y^5}$

33. $\frac{a^5}{a^{11}} = a^{5-11} = a^{-6} = \frac{1}{a^6}$

35. $\frac{4x^2}{12x^5} = \frac{x^{2-5}}{3} = \frac{x^{-3}}{3} = \frac{1}{3x^3}$

37. $\frac{-12x}{-18x^6} = \frac{2x^{1-6}}{3} = \frac{2x^{-5}}{3} = \frac{2}{3x^5}$

39. $\frac{x^6y^5}{x^8y} = x^{6-8}y^{5-1} = x^{-2}y^4 = \frac{y^4}{x^2}$

41. $\frac{2m^6n^2}{5m^9n^{10}} = \frac{2m^{6-9}n^{2-10}}{5} = \frac{2m^{-3}n^{-8}}{5} = \frac{2}{5m^3n^8}$

43. $\frac{pq^3}{p^4q^4} = p^{1-4}q^{3-4} = p^{-3}q^{-1} = \frac{1}{p^3q}$

45. $\dfrac{3x^4y^5}{6x^4y^8}=\dfrac{x^{4-4}y^{5-8}}{2}=\dfrac{x^0y^{-3}}{2}=\dfrac{1}{2y^3}$

47. $\dfrac{14x^4y^6z^2}{16x^3y^9z}=\dfrac{7x^{4-3}y^{6-9}z^{2-1}}{8}=\dfrac{7xy^{-3}z}{8}=\dfrac{7xz}{8y^3}$

49. $\dfrac{15mn^9p^3}{30m^4n^9p}=\dfrac{m^{1-4}n^{9-9}p^{3-1}}{2}=\dfrac{m^{-3}n^0p^2}{2}=\dfrac{p^2}{2m^3}$

51. $5^{-2}=\dfrac{1}{5^2}=\dfrac{1}{25}$

53. $\dfrac{1}{8^{-2}}=8^2=64$

55. $\dfrac{3^{-2}}{3}=\dfrac{1}{3^3}=\dfrac{1}{27}$

57. $\dfrac{2^{-2}}{2^{-3}}=2^{-2-(-3)}=2^1=2$

59. $x^{-2}=\dfrac{1}{x^2}$

61. $\dfrac{1}{a^{-6}}=a^6$

63. $4x^{-7}=4\cdot\dfrac{1}{x^7}=\dfrac{4}{x^7}$

65. $\dfrac{2}{3}z^{-2}=\dfrac{2}{3}\cdot\dfrac{1}{z^2}=\dfrac{2}{3z^2}$

67. $\dfrac{5}{b^{-8}}=5\cdot\dfrac{1}{b^{-8}}=5b^8$

69. $\dfrac{1}{3x^{-2}}=\dfrac{1}{3}\cdot\dfrac{1}{x^{-2}}=\dfrac{1}{3}\cdot x^2=\dfrac{x^2}{3}$

71. $\left(ab^5\right)^0=1$

73. $-\left(3p^2q^5\right)^0=-(1)=-1$

75. $\left(-2xy^{-2}\right)^3=-2^3x^3y^{-6}=-\dfrac{8x^3}{y^6}$

77. $\left(3x^{-1}y^{-2}\right)^2=3^2x^{-2}y^{-4}=\dfrac{9}{x^2y^4}$

79. $\left(2x^{-1}\right)\left(x^{-3}\right)=2x^{-4}=\dfrac{2}{x^4}$

81. $\left(-5a^2\right)\left(a^{-5}\right)^2=\left(-5a^2\right)\left(a^{-10}\right)=-5a^{-8}=-\dfrac{5}{a^8}$

83. $\left(-2ab^{-2}\right)\left(4a^{-2}b\right)^{-2}=\left(-2ab^{-2}\right)\left(4^{-2}a^4b^{-2}\right)$
$=\dfrac{-2}{4^2}a^5b^{-4}=\dfrac{-2a^5}{16b^4}=-\dfrac{a^5}{8b^4}$

85. $\left(-5x^{-2}y\right)\left(-2x^{-2}y^2\right)=10x^{-4}y^3=\dfrac{10y^3}{x^4}$

87. $\dfrac{3x^{-2}y^2}{6xy^2}=\dfrac{1}{2x^3}$

89. $\dfrac{3x^{-2}y}{xy}=\dfrac{3}{x^3}$

91. $\dfrac{2x^{-1}y^{-4}}{4xy^2}=\dfrac{1}{2x^2y^6}$

93. $\dfrac{\left(x^{-2}y\right)^2}{x^2y^3}=\dfrac{x^{-4}y^2}{x^2y^3}=\dfrac{1}{x^6y}$

95. $\dfrac{\left(a^{-2}y^3\right)^{-3}}{a^2y}=\dfrac{a^6y^{-9}}{a^2y}=\dfrac{a^4}{y^{10}}$

97. $\dfrac{-16xy^4}{96x^4y^4}=-\dfrac{1}{6x^3}$

99. $\dfrac{22a^2b^4}{-132b^3c^2}=-\dfrac{a^2b}{6c^2}$

101. $\dfrac{-\left(14ab^4\right)^2}{28a^4b^2}=-\dfrac{196a^2b^8}{28a^4b^2}=-\dfrac{7b^6}{a^2}$

103. $\dfrac{\left(3^{-1}r^4s^{-3}\right)^{-2}}{\left(6r^2t^{-2}s^{-1}\right)^2}=\dfrac{3^2r^{-8}s^6}{6^2r^4t^{-4}s^{-2}}=\dfrac{9t^4s^8}{36r^{12}}=\dfrac{s^8t^4}{4r^{12}}$

105. $\left(\dfrac{15m^3n^{-2}p^{-1}}{25m^{-2}n^{-4}}\right)^{-3}=\dfrac{15^{-3}m^{-9}n^6p^3}{25^{-3}m^6n^{12}}=\dfrac{25^3p^3}{15^3m^{15}n^6}$
$=\dfrac{5\cdot5\cdot5p^3}{3\cdot3\cdot3m^{15}n^6}=\dfrac{125p^3}{27m^{15}n^6}$

107. $\dfrac{a^{4n}}{a^n}=a^4$

a^{4n-n}	a^4

$a^{3n}\neq a^4$

False

109. $$\begin{array}{c|c} a^{-n}a^{n} & 1 \\ \hline a^{-n+n} & 1 \\ a^{0} & 1 \end{array}$$

$1 = 1$

True

Objective B Exercises

111. 8; left; 8

113. $0.00000000324 = 3.24\times10^{-9}$

115. $0.000000000000000003 = 3\times10^{-18}$

117. $32{,}000{,}000{,}000{,}000{,}000 = 3.2\times10^{16}$

119. $0.000000000000000000122 = 1.22\times10^{-19}$

121. $547{,}000{,}000 = 5.47\times10^{8}$

123. $1.67\times10^{-4} = 0.000167$

125. $6.8\times10^{7} = 68{,}000{,}000$

127. $3.05\times10^{-5} = 0.0000305$

129. $1.02\times10^{-9} = 0.00000000102$

131. $-n-1$

133. $0.000000015 = 1.5\times10^{-8}$

135. $5{,}980{,}000{,}000{,}000{,}000{,}000{,}000{,}000$
$= 5.98\times10^{24}$

137. $0.000000000000665 = 6.65\times10^{-13}$

139. $2{,}450{,}000{,}000 = 2.45\times10^{9}$

141. $0.00000000000000000039 = 3.9\times10^{-19}$

143. $16{,}000{,}000{,}000 = 1.6\times10^{10}$

Critical Thinking 9.4

145. $8^{-2}+2^{-5} = \frac{1}{8^2}+\frac{1}{2^5} = \frac{1}{64}+\frac{1}{32} = \frac{1}{64}+\frac{2}{64} = \frac{3}{64}$

147.

x	2^x	2^{-x}
-2	$2^{-2} = \frac{1}{4}$	$2^{-(-2)} = 2^2 = 4$
-1	$2^{-1} = \frac{1}{2}$	$2^{-(-1)} = 2^1 = 2$
0	$2^0 = 1$	$2^{-0} = 1$
1	$2^1 = 2$	$2^{-1} = \frac{1}{2}$
2	$2^2 = 4$	$2^{-2} = \frac{1}{4}$

149. $2^{-4} = \frac{1}{2^4} = \frac{1}{16} = 0.0625$

151. $\frac{a^m}{a^n} = a^{m-n} = a^0 = 1$

153. $(-4.8)^x = 1$

$x = 0$

155. $x^{-2} = \frac{1}{x^2}$, since x^2 is always positive, $\frac{1}{x^2}$ will be always positive.

Project or Group Activities 9.4

157. a. Strategy:

To find the number of square feet per person,

→Divide the number of square mi in Texas (2.619×10^5) by the total number of people (7×10^9).

→Multiply the number of sq mi per person by the number of sq ft in a sq mi (2.8×10^7).

Solution:

$(2.619\times10^5)\div(7\times10^9)$
$= 2{,}619{,}000\div7{,}000{,}000{,}000$
$= 0.000037414$

$0.000037414\times(2.8\times10^7)$
$= 0.000037414\times28{,}000{,}000$
$= 1047.6$

There would be approximately 1047.6 ft^2 per person.

b. Strategy:
To find the number of square feet per person,
→Divide the number of square mi in Rhode Island (1.0×10^3) by the total number of people in the U.S. (3.1×10^8).
→Multiply the number of sq mi per person by the number of sq ft in a sq mi (2.8×10^7).

Solution:
$(1.0\times10^3)\div(3.1\times10^8)$
$=1{,}000\div310{,}000{,}000$
$=0.000003226$

$0.000003226\times(2.8\times10^7)$
$=0.000003226\times28{,}000{,}000$
≈90
There would be approximately 90 ft^2 per person.

c. i. Strategy:
To find the number of people per square mile, divide the number of sq ft in a square mi (2.8×10^7) by the number of ft per person $(22\text{ ft}\times22\text{ ft}=484\text{ sq ft})$.

Solution:
$2.8\times10^7\div484$
$=28{,}000{,}000\div484$
$\approx57{,}900$
There would be approximately 57,900 people per square mi.

ii. Strategy:
To find the number of square mi, divide the world's population (7×10^7) by the number of people per square mi(57,900).

Solution:
$7\times10^9\div57{,}900$
$=7{,}000{,}000{,}000\div57{,}900$
$=120{,}900$
120,900 mi^2 would be required.

d. Strategy:
To find the number of acres per person,
→Multiply the number of sq mi (5.7×10^7) by the number of ft^2 per sq mi.
→Divide the number of ft^2 by the number of sq ft per acre (43,560).
→Divide the number of acres by the number of people (7×10^9).

Solution:
$(5.7\times10^7)\times(2.8\times10^7)$
$=(5.7\times2.9)\times(10^7\times10^7)$
$=15.96\times10^{14}$
$=1.596\times10^{15}$

$(1.596\times10^{15})\div43{,}560=3.664\times10^{10}$

$(3.664\times10^{10})\div(7\times10^9)=5.2$
There would be 5.2 acres per person.

e. Strategy:
To find the capacity
→Multiply the number of sq mi (5.7×10^7) by the number of sq ft in a sq mi (2.8×10^7).
→Divide the number of sq ft by 484.

Solution:
$(5.7\times10^7)\times(2.8\times10^7)$
$=(5.7\times2.8)\times(10^7\times10^7)$
$=15.96\times10^{14}$
$=1.596\times10^{15}$

$(1.596\times10^{15})\div(484)$
$=3.3\times10^{12}$
$\approx3{,}300{,}000{,}000{,}000$
$=3{,}300$ billion
The capacity would be 3,300 billion.

Section 9.5

Concept Check

1. $15x^2+12x=3x(5x+4)$

3. True

5. $3y$; $3y$; $6y^4$; 1

Objective A Exercises

7. $\dfrac{10a-25}{5}=\dfrac{10a}{5}-\dfrac{25}{5}=2a-5$

9. $\dfrac{3a^2+2a}{a}=\dfrac{3a^2}{a}+\dfrac{2a}{a}=3a+2$

11. $\dfrac{3x^2-6x}{3x}=\dfrac{3x^2}{3x}-\dfrac{6x}{3x}=x-2$

13. $\dfrac{5x^2-10x}{-5x}=\dfrac{5x^2}{-5x}+\dfrac{-10x}{-5x}=-x+2$

15. $\dfrac{x^3+3x^2-5x}{x}=\dfrac{x^3}{x}+\dfrac{3x^2}{x}-\dfrac{5x}{x}=x^2+3x-5$

17. $\dfrac{x^6-3x^4-x^2}{x^2}=\dfrac{x^6}{x^2}-\dfrac{3x^4}{x^2}-\dfrac{x^2}{x^2}=x^4-3x^2-1$

19. $\dfrac{5x^2y^2+10xy}{5xy}=\dfrac{5x^2y^2}{5xy}+\dfrac{10xy}{5xy}=xy+2$

21. $\dfrac{9y^6-15y^3}{-3y^3}=\dfrac{9y^6}{-3y^3}+\dfrac{-15y^3}{-3y^3}=-3y^3+5$

23. $\dfrac{3x^2-2x+1}{x}=\dfrac{3x^2}{x}-\dfrac{2x}{x}+\dfrac{1}{x}=3x-2+\dfrac{1}{x}$

25. $\dfrac{-3x^2+7x-6}{x}=\dfrac{-3x^2}{x}+\dfrac{7x}{x}-\dfrac{6}{x}=-3x+7-\dfrac{6}{x}$

27. $\dfrac{16a^2b-20ab+24ab^2}{4ab}=\dfrac{16a^2b}{4ab}-\dfrac{20ab}{4ab}+\dfrac{24ab^2}{4ab}$
$=4a-5+6b$

29. $\dfrac{9x^2y+6xy-3x}{xy}=\dfrac{9x^2y}{xy}+\dfrac{6xy}{xy}-\dfrac{3x}{xy}=9x+6-\dfrac{3}{y}$

31. Multiply $4x$ by $2x^2-3x-1$.

$4x\left(2x^2-3x-1\right)=8x^3-12x^2-4x$

Objective B Exercises

33.
$$\begin{array}{r} x+2 \\ x-3\overline{\smash{)}x^2-x-6} \\ \underline{x^2-3x} \\ 2x-6 \\ \underline{2x-6} \\ 0 \end{array}$$
$\left(x^2-x-6\right)\div\left(x-3\right)=x+2$

35.
$$\begin{array}{r} 2x+1 \\ x+2\overline{\smash{)}2x^2+5x+2} \\ \underline{2x^2+4x} \\ x+2 \\ \underline{x+2} \\ 0 \end{array}$$
$\left(2x^2+5x+2\right)\div\left(x+2\right)=2x+1$

37.
$$\begin{array}{r} x+1 \\ x-1\overline{\smash{)}x^2+0+1} \\ \underline{x^2-x} \\ x+1 \\ \underline{x-1} \\ 2 \end{array}$$
$\left(x^2+1\right)\div\left(x-1\right)=x+1+\dfrac{2}{x-1}$

39.
$$\begin{array}{r} 2x-1 \\ 3x-2\overline{\smash{)}6x^2-7x+0} \\ \underline{6x^2-4x} \\ -3x+0 \\ \underline{-3x+2} \\ -2 \end{array}$$
$\left(6x^2-7x\right)\div\left(3x-2\right)=2x-1-\dfrac{2}{3x-2}$

41.
$$\begin{array}{r} b-5 \\ b-3\overline{\smash{)}b^2-8b-9} \\ \underline{b^2-3b} \\ -5b-9 \\ \underline{-5b+15} \\ -24 \end{array}$$
$(b^2-8b-9)\div(b-3)=b-5-\dfrac{24}{b-3}$

43.

$$\begin{array}{r} 3x+17 \\ x-4 \overline{\smash{\big)}\,3x^2+5x-4} \\ \underline{3x^2-12x} \\ 17x-4 \\ \underline{17x-68} \\ 64 \end{array}$$

$$\left(3x^2+5x-4\right)\div(x-4)=3x+17+\frac{64}{x-4}$$

45.

$$\begin{array}{r} 5y+3 \\ 2y+3 \overline{\smash{\big)}\,10y^2+21y+10} \\ \underline{10y^2+15y} \\ 6y+10 \\ \underline{6y+9} \\ 1 \end{array}$$

$$(10y^2+21y+10)\div(2y+3)=5y+3+\frac{1}{2y+3}$$

47.

$$\begin{array}{r} 3x-5 \\ 4x-1 \overline{\smash{\big)}\,12x^2-23x+5} \\ \underline{12x^2-3x} \\ -20x+5 \\ \underline{-20x+5} \\ 0 \end{array}$$

$$\left(12x^2-23x+5\right)\div(4x-1)=3x-5$$

49.

$$\begin{array}{r} x^2+2x+3 \\ x+1 \overline{\smash{\big)}\,x^3+3x^2+5x+8} \\ \underline{x^3+x^2} \\ 2x^2+5x \\ \underline{2x^2+2x} \\ 3x+8 \\ \underline{3x+3} \\ 5 \end{array}$$

$$(x^3+3x^2+5x+3)\div(x+1)=x^2+2x+3+\frac{5}{x+1}$$

51.

$$\begin{array}{r} x^2 \quad -3 \\ x^2+0x+2 \overline{\smash{\big)}\,x^4+0x^3-x^2+0x-6} \\ \underline{x^4+0x^3+2x^2} \\ -3x^2+0x-6 \\ \underline{-3x^2+0x-6} \\ 0 \end{array}$$

$$(x^4-x^2-6)\div(x^2+2)=x^2-3$$

53. False. Dividing a sixth-degree polynomial by a third-degree polynomial will result in a third-degree polynomial since when dividing, the exponents are subtracted. Example:

$$\frac{x^6}{x^3}=x^3$$

Critical Thinking 9.5

55. Divide the result by $6x$.

$$\frac{24xy^2}{6x}=4y^2$$

57. Multiply $\left(x^2-x+8\right)$ by $x-3$. Add the remainder (22)

$$\begin{array}{r} x^2-x+8 \\ \underline{\times \quad x-3} \\ -3x^2+3x-24 \\ \underline{x^3-x^2+8x \qquad} \\ x^3-4x^2+11x-24 \\ +22 \\ x^3-4x^2+11x-2 \end{array}$$

Projects or Group Activities 9.5

59. Divide $\left(4x^3+9x^2-10x-3\right)$ by $4x+1$.

$$\begin{array}{r} x^2+2x-3 \\ 4x+1 \overline{\smash{\big)}\,4x^3+9x^2-10x-3} \\ \underline{4x^3+x^2} \\ 8x^2-10x \\ \underline{8x^2+2x} \\ -12x-3 \\ \underline{-12x-3} \end{array}$$

Chapter Review

1. $(2b-3)(4b+5)=8b^2+10b-12b-15$

$$=8b^2-2b-15$$

2. $\left(12y^2+17y-4\right)+\left(9y^2-13y+3\right)$

$$\begin{array}{r} 12y^2+17y-4 \\ \underline{+9y^2-13y+3} \\ 21y^2+4y-1 \end{array}$$

3. $(xy^5z^3)(x^3y^3z) = x^4y^8z^4$

4. $\dfrac{8x^{12}}{12x^9} = \dfrac{2x^{12-9}}{3} = \dfrac{2x^3}{3}$

5. $-2x(4x^2+7x-9) = -2x(4x^2) - 2x(7x) - 2x(-9)$
$= -8x^3 - 14x^2 + 18x$

6. $\dfrac{3ab^4}{-6a^2b^4} = -\dfrac{1}{2a}$

7. $(-2u^3v^4)^4 = (-2)^4u^{12}v^{16} = 16u^{12}v^{16}$

8. $(2^3)^2 = 2^6 = 64$

9. $(5x^2-2x-1)-(3x^2-5x+7)$

$$\begin{array}{r} 5x^2-2x-1 \\ \underline{+-3x^2+5x-7} \\ 2x^2+3x-8 \end{array}$$

10. $\dfrac{a^{-1}b^3}{a^3b^{-3}} = \dfrac{b^6}{a^4}$

11. $(-2x^3)^2(-3x^4)^3 = (-2)^2\cdot x^6(-3)^3x^{12}$
$= (4x^6)(-27x^{12})$
$= -108x^{18}$

12. $(5y-7)^2 = (5y-7)(5y-7)$
$= 25y^2 - 35y - 35y + 49$
$= 25y^2 - 70y + 49$

13. $(5a^7b^6)^2(4ab) = (5^2a^{14}b^{12})(4ab)$
$= (25a^{14}b^{12})(4ab)$
$= 100a^{15}b^{13}$

14. $\dfrac{12b^7+36b^5-3b^3}{3b^3} = \dfrac{12b^7}{3b^3} + \dfrac{36b^5}{3b^3} - \dfrac{3b^3}{3b^3}$
$= 4b^4 + 12b^2 - 1$

15. $-4^{-2} = -\dfrac{1}{4^2} = -\dfrac{1}{16}$

16. $(13y^3-7y-2)-(12y^2-2y-1)$

$$\begin{array}{r} 13y^3 \qquad\quad -7y-2 \\ \underline{+\quad -12y^2+2y+1} \\ 13y^3-12y^2-\ 5y-1 \end{array}$$

17.

$$\begin{array}{r} -x+2 \\ x+3\overline{)-x^2\quad -x+7} \\ \underline{-x^2-3x} \\ 2x+7 \\ \underline{2x+6} \\ 1 \end{array}$$

$(-x^2-x+7)\div(x+3) = -x+2+\dfrac{1}{x+3}$

18. $(2a-b)(x-2y) = 2ax - 4ay - bx + 2by$

19. $(3y^2+4y-7)(2y+3)$
$= 3y^2(2y+3) + 4y(2y+3) - 7(2y+3)$
$= 6y^3 + 9y^2 + 8y^2 + 12y - 14y - 21$
$= 6y^3 + 17y^2 - 2y - 21$

20.

$$\begin{array}{r} b^2+5b+2 \\ b-7\overline{)b^3-2b^2-33b-7} \\ \underline{b^3-7b^2} \\ 5b^2-33b \\ \underline{5b^2-35b} \\ 2b-\ 7 \\ \underline{2b-14} \\ 7 \end{array}$$

$(b^3-2b^2-33b-7)\div(b-7)$
$= b^2+5b+2+\dfrac{7}{b-7}$

21. $2ab^3(4a^2-2ab+3b^2)$
$= 2ab^3(4a^2) + 2ab^3(-2ab) + 2ab^3(3b^2)$
$= 8a^3b^3 - 4a^2b^4 + 6ab^5$

22. $(2a-5b)(2a+5b) = 4a^2 + 10ab - 10ab - 25b^2$
$= 4a^2 - 25b^2$

23. $(6b^3-2b^2-5)(2b^2-1)$
$= 6b^3(2b^2-1) - 2b^2(2b^2-1) - 5(2b^2-1)$
$= 12b^5 - 6b^3 - 4b^4 + 2b^2 - 10b^2 + 5$
$= 12b^5 - 4b^4 - 6b^3 - 8b^2 + 5$

24. $(2x^3+7x^2+x)+(2x^2-4x-12)$

$$\begin{array}{r} 2x^3+7x^2\quad +x \\ \underline{+\qquad 2x^2-4x-12} \\ 2x^3+9x^2-3x-12 \end{array}$$

25. $\dfrac{16y^2-32y}{-4y}=\dfrac{16y^2}{-4y}+\dfrac{-32y}{-4y}=-4y+8$

26. $(a+7)(a-7)=a^2-7a+7a-49=a^2-49$

27. $37{,}560{,}000{,}000=3.756\times10^{10}$

28. $1.46\times10^7=14{,}600{,}000$

29. $(2a^{12}b^3)(-9b^2c^6)(3ac)=-54a^{13}b^5c^7$

30.

$$\begin{array}{r} 2y-9 \\ 3y-4\overline{)6y^2-35y+36} \\ \underline{6y^2-8y} \\ -27y+36 \\ \underline{-27y+36} \\ 0 \end{array}$$

$(6y^2-35y+36)\div(3y-4)=2y-9$

31. $(-3x^{-2}y^{-3})^{-2}=(-3)^{-2}x^4y^6=\dfrac{x^4y^6}{(-3)^2}=\dfrac{x^4y^6}{9}$

32. $(5a-7)(2a+9)=10a^2+45a-14a-63$
$=10a^2+31a-63$

33. $0.000000127=1.27\times10^{-7}$

34. $3.2\times10^{-12}=0.0000000000032$

35. Strategy:
→Width: w
→Length: $2w-1$
→Use the formula for the area of a rectangle.

Solution:
$A=LW$
$A=(2w-1)w$
$A=2w^2-w$
The area of the rectangle is $(2w^2-w)\text{ft}^2$.

36. Strategy:
→Side of a square: $3x-2$
→Use the formula for the area of a square.

Solution:
$A=S^2$
$A=(3x-2)^2$
$A=9x^2-6x-6x+4$
$A=9x^2-12x+4$
The area of the square is $(9x^2-12x+4)\text{in}^2$.

Chapter Test

1. $2x(2x^2-3x)=2x(2x^2)+2x(-3x)$
$=4x^3-6x^2$

2. $\dfrac{12x^3-3x^2+9}{3x^2}=\dfrac{12x^3}{3x^2}-\dfrac{3x^2}{3x^2}+\dfrac{9}{3x^2}$
$=4x-1+\dfrac{3}{x^2}$

3. $\dfrac{12x^2}{-3x^8}=-\dfrac{4}{x^6}$

4. $(-2xy^2)(3x^2y^4)=-6x^3y^6$

5.

$$\begin{array}{r} x-1 \\ x+1\overline{)x^2+0x+1} \\ \underline{x^2+x} \\ -x+1 \\ \underline{-x-1} \\ 2 \end{array}$$

$(x^2+1)\div(x+1)=x-1+\dfrac{2}{x+1}$

6. $(x-3)(x^2-4x+5)=x(x^2-4x+5)-3(x^2-4x+5)$
$=x^3-4x^2+5x-3x^2+12x-15$
$=x^3-7x^2+17x-15$

7. $(-2a^2b)^3=(-2)^3a^6b^3=-8a^6b^3$

8. $\dfrac{(3x^{-2}y^3)^3}{3x^4y^{-1}}=\dfrac{27x^{-6}y^9}{3x^4y^{-1}}=\dfrac{9y^{10}}{x^{10}}$

9. $(a-2b)(a+5b)=a^2+5ab-2ab-10b^2$
$=a^2+3ab-10b^2$

10. $\frac{16x^5-8x^3+20x}{4x}=\frac{16x^5}{4x}-\frac{8x^3}{4x}+\frac{20x}{4x}$
$=4x^4-2x^2+5$

11.
$$\begin{array}{r} x+7 \\ x-1\overline{)x^2+6x-7} \\ \underline{x^2-\ x} \\ 7x-7 \\ \underline{7x-7} \\ 0 \end{array}$$
$(x^2+6x-7)\div(x-1)=x+7$

12. $-3y^2(-2y^2+3y-6)$
$=-3y^2(-2y^2)-3y^2(3y)-3y^2(-6)$
$=6y^4-9y^3+18y^2$

13. $(-2x^3+x^2-7)(2x-3)$
$=-2x^3(2x-3)+x^2(2x-3)-7(2x-3)$
$=-4x^4+6x^3+2x^3-3x^2-14x+21$
$=-4x^4+8x^3-3x^2-14x+21$

14. $(4y-3)(4y+3)=16y^2+12y-12y-9$
$=16y^2-9$

15. $(ab^2)(a^3b^5)=a^4b^7$

16. $\frac{2a^{-1}b}{2^{-2}a^{-2}b^{-3}}=2^3ab^4=8ab^4$

17. $\frac{20a-35}{5}=\frac{20a}{5}-\frac{35}{5}=4a-7$

18. $(3a^2-2a-7)-(5a^3+2a-10)$
$$\begin{array}{r} 3a^2-2a\ -7 \\ \underline{+-5a^3\qquad\quad -2a+10} \\ -5a^3+3a^2-4a\ +3 \end{array}$$

19. $(2x-5)^2=(2x-5)(2x-5)$
$=4x^2-10x-10x+25$
$=4x^2-20x++25$

20.
$$\begin{array}{r} 2x+3 \\ 2x-3\overline{)4x^2+0x-7} \\ \underline{4x^2-6x} \\ 6x-7 \\ \underline{6x-9} \\ 2 \end{array}$$
$(4x^2-7)\div(2x-3)=2x+3+\frac{2}{2x-3}$

21. $\frac{-(2x^2y)^3}{4x^3y^3}=\frac{-8x^6y^3}{4x^3y^3}=-2x^3$

22. $(2x-7y)(5x-4y)=10x^2-8xy-35xy+28y^2$
$=10x^2-43xy+28y^2$

23. $(3x^3-2x^2-4)+(8x^2-8x+7)$
$$\begin{array}{r} 3x^3-2x^2\qquad -4 \\ \underline{+\qquad 8x^2-8x+7} \\ 3x^3+6x^2-8x+3 \end{array}$$

24. $0.00000000302=3.02\times10^{-9}$

25. Strategy:
→Radius of a circle: $x-5$
→Use the formula for the area of a circle.

Solution:
$A=\pi r^2$
$=\pi(x-5)^2$
$=\pi(x^2-10x+25)$
$=\pi x^2-10\pi x+25\pi$

The area of the circle is $(\pi x^2-10x\pi+25\pi)\text{m}^2$.

Cumulative Review Exercises

1. $\frac{3}{16}-\left(-\frac{5}{8}\right)-\frac{7}{9}=\frac{3}{16}+\frac{5}{8}-\frac{7}{9}$
$=\frac{27}{144}+\frac{90}{144}-\frac{112}{144}$
$=\frac{5}{144}$

2. $-3^2\left(\frac{2}{3}\right)^3\left(-\frac{5}{8}\right)=-9\left(\frac{8}{27}\right)\left(-\frac{5}{8}\right)=-\frac{8}{3}\left(-\frac{5}{8}\right)=\frac{5}{3}$

3. $\left(-\frac{1}{2}\right)^3 \div \left(\frac{3}{8}-\frac{5}{6}\right)+2 = \left(-\frac{1}{8}\right) \div \left(\frac{9}{24}-\frac{20}{24}\right)+2$

$= \left(-\frac{1}{8}\right) \div \left(-\frac{11}{24}\right)+2$

$= \left(-\frac{1}{8}\right) \cdot \left(-\frac{24}{11}\right)+2$

$= \frac{3}{11}+\frac{22}{11}$

$= \frac{25}{11}$

4. $\frac{b-(a-b)^2}{b^2}$

$\frac{3-(-2-3)^2}{3^2} = \frac{3-(-5)^2}{9}$

$= \frac{3-25}{9} = -\frac{22}{9}$

5. $-2x-(-xy)+7x-4xy = -2x+7x+xy-4xy$

$= 5x-3xy$

6. $(12x)\left(-\frac{3}{4}\right) = -\frac{12x}{1}\cdot\frac{3}{4} = -\frac{36x}{4} = -9x$

7. $-2[3x-2(4-3x)+2] = -2[3x-8+6x+2]$

$= -2[9x-6]$

$= -18x+12$

8. $12 = -\frac{3}{4}x$

$-\frac{4}{\cancel{3}_1}\left(\cancel{12}^4\right) = \frac{-3}{4}x\left(\frac{-4}{3}\right)$

$-16 = x$

The solution is −16.

9. $2x-9 = 3x+7$

$-x = 16$

$x = -16$

The solution is −16.

10. $2-3(4-x) = 2x+5$

$2-12+3x = 2x+5$

$-10+3x = 2x+5$

$x = 15$

The solution is 15.

11. $PB = A$

$P160 = 35.2$

$P = \frac{35.2}{160}$

$P = 0.22 = 22\%$

12. $(4b^3-7b^2-7)+(3b^2-8b+3)$

$$\begin{array}{r} 4b^3-7b^2-7 \\ +\ \ 3b^2-8b+3 \\ \hline 4b^3-4b^2-8b-4 \end{array}$$

13. $(3y^3-5y+8)-(-2y^2+5y+8)$

$$\begin{array}{r} 3y^3-5y+8 \\ +\ \ 2y^2-5y-8 \\ \hline 3y^3+2y^2-10y \end{array}$$

14. $(a^3b^5)^3 = a^9b^{15}$

15. $(4xy^3)(-2x^2y^3) = -8x^3y^6$

16. $-2y^2(-3y^2-4y+8)$

$= -2y^2(-3y^2)-2y^2(-4y)-2y^2(8)$

$= 6y^4+8y^3-16y^2$

17. $(2a-7)(5a^2-2a+3)$

$= 2a(5a^2-2a+3)-7(5a^2-2a+3)$

$= 10a^3-4a^2+6a-35a^2+14a-21$

$= 10a^3-39a^2+20a-21$

18. $(3b-2)(5b-7) = 15b^2-21b-10b+14$

$= 15b^2-31b+14$

19. $\frac{(-2a^2b^3)^2}{8a^4b^8} = \frac{(-2)^2a^4b^6}{8a^4b^8}$

$= \frac{4a^4b^6}{8a^4b^8} = \frac{1}{2b^2}$

20.

$$\begin{array}{r} a-7 \\ a+3\overline{)a^2-4a-21} \\ \underline{a^2+3a} \\ -7a-21 \\ \underline{-7a-21} \\ 0 \end{array}$$

$(a^2-4a-21)\div(a+3) = a-7$

21. $6.09 \times 10^{-5} = 0.0000609$

22. The unknown number: x

The difference between eight times a number and twice the number	is	eighteen.

$8x - 2x = 18$

$6x = 18$

$x = 3$

The unknown number is 3.

23. Strategy:

→Determine the total cost of the fruit juice and the soda

→Divide this cost by the total number of ounces (500)

Solution:

Cost of juice: 200($.25) = $50

Cost of soda: 300($.05) = $15

Total cost: $50 + $15 = $65

$\dfrac{\$65}{500 \text{ ounces}} = 0.13$

The cost is $.13 per ounce.

24. Strategy:

→Time for car: x

→Time for cyclist: $x + 2$

	Rate	Time	Distance
Car	50	x	$50x$
Cyclist	10	$x + 2$	$10(x + 2)$

→The car and the cyclist travel the same distance.

Solution:

$50x = 10(x+2)$

$50x = 10x + 20$

$40x = 20$

$x = \dfrac{1}{2}$

Distance = $50x = 50\left(\dfrac{1}{2}\right) = 25$

The car overtakes the cyclist 25 mi from the starting point.

25. Strategy:

→Length: x

→Width: $40\% \cdot x = 0.40x$

→Use the equation for the perimeter of a rectangle.

Solution:

$P = 2L + 2W$

$42 = 2(x) + 2(0.40x)$

$42 = 2x + 0.8x$

$42 = 2.8x$

$15 = x$

$40\% \cdot x = 0.40(15) = 6$

The length is 15 m and the width is 6 m.

Chapter 10: Factoring

Prep Test

1. $30 = 2 \cdot 3 \cdot 5$

2. $-3(4y-5) = -12y+15$

3. $-(a-b) = -a+b$

4. $$\begin{aligned} 2(a-b)-5(a-b) &= 2a-2b-5a+5b \\ &= -3a+3b \end{aligned}$$

5. $$\begin{aligned} 4x &= 0 \\ \frac{4x}{4} &= \frac{0}{4} \\ x &= 0 \end{aligned}$$
   The solution is 0.

6. $$\begin{aligned} 2x+1 &= 0 \\ 2x+1-1 &= 0-1 \\ 2x &= -1 \\ \frac{2x}{2} &= \frac{-1}{2} \\ x &= -\frac{1}{2} \end{aligned}$$
   The solution is $-\frac{1}{2}$.

7. $$\begin{aligned} (x+4)(x-6) &= x^2-6x+4x-24 \\ &= x^2-2x-24 \end{aligned}$$

8. $$\begin{aligned} (2x-5)(3x+2) &= 6x^2+4x-15x-10 \\ &= 6x^2-11x-10 \end{aligned}$$

9. $\frac{x^5}{x^2} = x^{5-2} = x^3$

10. $\frac{6x^4y^3}{2xy^2} = 3x^{4-1}y^{3-2} = 3x^3y$

Section 10.1

Concept Check

1. 4

3. **a.** x **b.** $2x-1$

5. $(2x^3-x^2)+(6x-3)$

Objective A Exercises

7. $5a+5 = 5(a)+5(1) = 5(a+1)$

9. $16-8a^2 = 8(2)+8(-a^2) = 8(2-a^2)$

11. $8x+12 = 4(2x)+4(3) = 4(2x+3)$

13. $7x^2-3x = x(7x)+x(-3) = x(7x-3)$

15. $3a^2+5a^5 = a^2(3)+a^2(5a^3) = a^2(3+5a^3)$

17. $2x^4-4x = 2x(x^3)+2x(-2) = 2x(x^3-2)$

19. $10x^4-12x^2 = 2x^2(5x^2)+2x^2(-6) = 2x^2(5x^2-6)$

21. $8a^8-4a^5 = 4a^5(2a^3)+4a^5(-1) = 4a^5(2a^3-1)$

23. $x^2y^2-xy = xy(xy)+xy(-1) = xy(xy-1)$

25. $3x^2y^4-6xy = 3xy(xy^3)+3xy(-2) = 3xy(xy^3-2)$

27. $$\begin{aligned} 3x^3+6x^2+9x &= 3x(x^2)+3x(2x)+3x(3) \\ &= 3x(x^2+2x+3) \end{aligned}$$

29. $$\begin{aligned} 2x^4-4x^3+6x^2 &= 2x^2(x^2)+2x^2(-2x)+2x^2(3) \\ &= 2x^2(x^2-2x+3) \end{aligned}$$

31. $$\begin{aligned} 2x^3+6x^2-14x &= 2x(x^2)+2x(3x)+2x(-7) \\ &= 2x(x^2+3x-7) \end{aligned}$$

33. $$\begin{aligned} 2y^5-3y^4+7y^3 &= y^3(2y^2)+y^3(-3y)+y^3(7) \\ &= y^3(2y^2-3y+7) \end{aligned}$$

35. $$\begin{aligned} x^3y-3x^2y^2+7xy^3 &= xy(x^2)+xy(-3xy)+xy(7y^2) \\ &= xy(x^2-3xy+7y^2) \end{aligned}$$

37. $$\begin{aligned} 5y^3+10y^2-25y &= 5y(y^2)+5y(2y)+5y(-5) \\ &= 5y(y^2+2y-5) \end{aligned}$$

39. $3a^2b^2 - 9ab^2 + 15b^2$
$= 3b^2(a^2) + 3b^2(-3a) + 3b^2(5)$
$= 3b^2(a^2 - 3a + 5)$

41. The GCF of $x^a + x^b + x^c$ is x^c, where c is the smallest exponent.

Objective B Exercises

43. $x(b+4) + 3(b+4) = (b+4)(x+3)$

45. $a(y-x) - b(y-x) = (y-x)(a-b)$

47. $x(x-2) + y(2-x) = x(x-2) - y(x-2)$
$= (x-2)(x-y)$

49. $8c(2m-3n) + (3n-2m)$
$= 8c(2m-3n) - 1(2m-3n)$
$= (2m-3n)(8c-1)$

51. $x^2 + 2x + 2xy + 4y = (x^2 + 2x) + (2xy + 4y)$
$= x(x+2) + 2y(x+2)$
$= (x+2)(x+2y)$

53. $p^2 - 2p - 3rp + 6r = (p^2 - 2p) + (-3rp + 6r)$
$= p(p-2) - 3r(p-2)$
$= (p-2)(p-3r)$

55. $ab + 6b - 4a - 24 = (ab + 6b) + (-4a - 24)$
$= b(a+6) - 4(a+6)$
$= (a+6)(b-4)$

57. $2z^2 - z + 2yz - y = (2z^2 - z) + (2yz - y)$
$= z(2z-1) + y(2z-1)$
$= (2z-1)(z+y)$

59. $2x^2 - 5x - 6xy + 15y = (2x^2 - 5x) + (-6xy + 15y)$
$= x(2x-5) - 3y(2x-5)$
$= (2x-5)(x-3y)$

61. $3y^2 - 6y - ay + 2a = (3y^2 - 6y) + (-ay + 2a)$
$= 3y(y-2) - a(y-2)$
$= (y-2)(3y-a)$

63. $3xy - y^2 - y + 3x = (3xy - y^2) + (-y + 3x)$
$= y(3x-y) + 1(-y+3x)$
$= y(3x-y) + 1(3x-y)$
$= (3x-y)(y+1)$

65. $3st + t^2 - 2t - 6s = (3st + t^2) + (-2t - 6s)$
$= t(3s+t) - 2(t+3s)$
$= t(3s+t) - 2(3s+t)$
$= (3s+t)(t-2)$

Critical Thinking 10.1

67. $a - 3 = \underline{-1}(3-a)$

69. $4x + (3a-b) = 4x - (-3a+b) = 4x - (\underline{b-3a})$

Section 10.2

Concept Check

1. –8

3. –2 and 6

5. Different

Objective A Exercises

7.

Factors	*Sum*
+1, +2	+3

$x^2 + 3x + 2 = (x+1)(x+2)$

9.

Factors	*Sum*
−1, +2	+1
+1, −2	−1

$x^2 - x - 2 = (x+1)(x-2)$

11.

Factors	*Sum*
−1, +12	+11
+1, −12	−11
−2, +6	+4
+2, −6	−4
−3, +4	+1
+3, −4	−1

$a^2 + a - 12 = (a+4)(a-3)$

13.

Factors	*Sum*
−1, −2	−3

$a^2 - 3a + 2 = (a-1)(a-2)$

15.

Factors	*Sum*
−1, +2	+1
+1, −2	−1

$a^2+a-2=(a+2)(a-1)$

17.

Factors	*Sum*
−1, −9	−10
−3, −3	−6

$b^2-6b+9=(b-3)(b-3)$

19.

Factors	*Sum*
−1, +8	+7
+1, −8	−7
−2, +4	+2
+2, −4	−2

$b^2+7b-8=(b+8)(b-1)$

21.

Factors	*Sum*
−1, +55	+54
+1, −55	−54
−5, +11	+6
+5, −11	−6

$y^2+6y-55=(y+11)(y-5)$

23.

Factors	*Sum*
−1, −6	−7
−2, −3	−5

$y^2-5y+6=(y-2)(y-3)$

25.

Factors	*Sum*
−1, −45	−46
−3, −15	−18
−5, −9	−14

$z^2-14z+45=(z-5)(z-9)$

27.

Factors	*Sum*
−1, +160	+159
+1, −160	−159
−2, +80	+78
+2, −80	−78
−4, +40	+36
+4, −40	−36
−5, +32	+27
+5, −32	−27
−8, +20	+12
+8, −20	−12
−10, +16	+6
+10, −16	−6

$z^2-12z-160=(z+8)(z-20)$

29.

Factors	*Sum*
+1, +27	+28
+3, +9	+12

$p^2+12p+27=(p+3)(p+9)$

31.

Factors	*Sum*
+1, +100	+101
+2, +50	+52
+4, +25	+29
+5, +20	+25
+10, +10	+20

$x^2+20x+100=(x+10)(x+10)$

33.

Factors	*Sum*
−1, +20	+19
+1, −20	−19
−2, +10	+8
+2, −10	−8
−4, +5	+1
+4, −5	−1

$b^2-b-20=(b+4)(b-5)$

35.

Factors	*Sum*
−1, +51	+50
+1, −51	−50
−3, +17	+14
+3, −17	−14

$y^2-14y-51=(y+3)(y-17)$

37.

Factors	*Sum*
−1, +21	+20
+1, −21	−20
−3, +7	+4
+3, −7	−4

$p^2-4p-21=(p+3)(p-7)$

39.

Factors	*Sum*
−1, −32	−33
−2, −16	−18
−4, −8	−12

$y^2-8y+32$ is nonfactorable over the integers.

41.

Factors	*Sum*
−1, −75	−76
−3, −25	−28
−5, −15	−20

$x^2-20x+75=(x-5)(x-15)$

43.

Factors	*Sum*
+1, +63	+64
+3, +21	+24

$p^2+24p+63=(p+3)(p+21)$

45.

Factors	*Sum*
+1, +38	+39
+2, +19	+21

$x^2 + 21x + 38 = (x+2)(x+19)$

47.

Factors	*Sum*
+1, −3	−2
−1, +3	+2

$x^2 + 5x - 3$ is nonfactorable over the integers.

49.

Factors	*Sum*
−1, +44	+43
+1, −44	−43
−2, +22	+20
+2, −22	−20
−4, +11	+7
+4, −11	−7

$a^2 - 7a - 44 = (a+4)(a-11)$

51.

Factors	*Sum*
−1, −54	−55
−2, −27	−29
−3, −18	−21
−6, −9	−15

$a^2 - 21a + 54 = (a-3)(a-18)$

53.

Factors	*Sum*
−1, +147	+146
+1, −147	−146
−3, +49	+46
+3, −49	−46
−7, +21	+14
+7, −21	−14

$z^2 + 14z - 147 = (z+21)(z-7)$

55.

Factors	*Sum*
−1, +180	+179
+1, −180	−179
−2, +90	+88
+2, −90	−88
−3, +60	+57
+3, −60	−57
−4, +45	+41
+4, −45	−41
−5, +36	+31
+5, −36	−31
−6, +30	+24
+6, −30	−24
−9, +20	+11
+9, −20	−11
−10, +18	+8
+10, −18	−8
−12, +15	+3
+12, −15	−3

$c^2 - 3c - 180 = (c+12)(c-15)$

57.

Factors	*Sum*
+1, +135	+136
+3, +45	+48
+5, +27	+32
+9, +15	+24

$p^2 + 24p + 135 = (p+9)(p+15)$

59.

Factors	*Sum*
+1, +18	+19
+2, +9	+11
+3, +6	+9

$c^2 + 11c + 18 = (c+2)(c+9)$

61.

Factors	*Sum*
−1, +75	+74
+1, −75	−74
−3, +25	+22
+3, −25	−22
−5, +15	+10
+5, −15	−10

$x^2 + 10x - 75 = (x+15)(x-5)$

63.

Factors	*Sum*
−1, +100	+99
+1, −100	−99
−2, +50	+48
+2, −50	−48
−4, +25	+21
+4, −25	−21
−5, +20	+15
+5, −20	−15
−10, +10	0

$x^2 + 21x - 100 = (x + 25)(x - 4)$

65.

Factors	*Sum*
−1, −72	−73
−2, −36	−38
−3, −24	−27
−4, −18	−22
−6, −12	−18
−8, −9	−17

$b^2 - 22b + 72 = (b - 4)(b - 18)$

67.

Factors	*Sum*
−1, +135	+134
+1, −135	−134
−3, +45	+42
+3, −45	−42
−5, +27	+22
+5, −27	−22
−9, +15	+6
+9, −15	−6

$a^2 + 42a - 135 = (a + 45)(a - 3)$

69.

Factors	*Sum*
−1, −126	−127
−2, −63	−65
−3, −43	−46
−6, −21	−27
−7, −18	−25
−9, −14	−23

$b^2 - 25b + 126 = (b - 7)(b - 18)$

71.

Factors	*Sum*
+1, +144	+145
+2, +72	+74
+3, +48	+51
+4, +36	+40
+6, +24	+30
+8, +18	+26
+9, +16	+25
+12, +12	+24

$z^2 + 24z + 144 = (z + 12)(z + 12)$

73.

Factors	*Sum*
−1, −100	−101
−2, −50	−52
−4, −25	−29
−5, −20	−25
−10, −10	−20

$x^2 - 29x + 100 = (x - 4)(x - 25)$

75.

Factors	*Sum*
−1, +112	+111
+1, −112	−111
−2, +56	+54
+2, −56	−54
−4, +28	+24
+4, −28	−24
−7, +16	+9
+7, −16	−9
−8, +14	+6
+8, −14	−6

$x^2 + 9x - 112 = (x + 16)(x - 7)$

77. b is positive. The sum of two positive numbers is positive.

Objective B Exercises

79. The GCF is 3.

$3x^2 + 15x + 18 = 3(x^2 + 5x + 6)$

Factor the trinomial $x^2 + 5x + 6$.

Factors	*Sum*
+1, +6	+7
+2, +3	+5

$3x^2 + 15x + 18 = 3(x + 2)(x + 3)$

81. The GCF is −1.

$-x^2-4x+12=-\left(x^2+4x-12\right)$

Factor the trinomial $x^2+4x-12$.

Factors	Sum
−1, +12	+11
+1, −12	−11
−2, +6	+4
+2, −6	−4
−3, +4	+1
+3, −4	−1

$12-4x-x^2=-(x-2)(x+6)$

83. The GCF is a.

$ab^2+7ab-8a=a\left(b^2+7b-8\right)$

Factor the trinomial b^2+7b-8.

Factors	Sum
−1, +8	+7
+1, −8	−7
−2, +4	+2
+2, −4	−2

$ab^2+7ab-8a=a(b+8)(b-1)$

85. The GCF is x.

$xy^2+8xy+15x=x\left(y^2+8y+15\right)$

Factor the trinomial

$y^2+8y+15$.

Factors	Sum
+1, +15	+16
+3, +5	+8

$xy^2+8xy+15x=x(y+3)(y+5)$

87. The GCF is $-2a$.

$-2a^3-6a^2-4a=-2a\left(a^2+3a+2\right)$

Factor the trinomial a^2+3a+2.

Factors	Sum
+1, +2	+3

$-2a^3-6a^2-4a=-2a(a+1)(a+2)$

89. The GCF is $4y$.

$4y^3+12y^2-72y=4y\left(y^2+3y-18\right)$

Factor the trinomial $y^2+3y-18$.

Factors	Sum
−1, +18	+17
+1, −18	−17
−2, +9	+7
+2, −9	−7
−3, +6	+3
+3, −6	−3

$4y^3+12y^2-72y=4y(y+6)(y-3)$

91. The GCF is $2x$.

$2x^3-2x^2+4x=2x\left(x^2-x+2\right)$

Factor the trinomial

x^2-x+2.

Factors	Sum
−1, −2	−3

x^2-x+2 is nonfactorable over the integers.

$2x^3-2x^2+4x=2x\left(x^2-x+2\right)$

93. The GCF is 6.

$6z^2+12z-90=6\left(z^2+2z-15\right)$

Factor the trinomial $z^2+2z-15$.

Factors	Sum
−1, +15	+14
+1, −15	−14
−3, +5	+2
+3, −5	−2

$6z^2+12z-90=6(z+5)(z-3)$

95. The GCF is $3a$.

$3a^3-9a^2-54a=3a\left(a^2-3a-18\right)$

Factor the trinomial

$a^2-3a-18$.

Factors	Sum
−1, +18	+17
+1, −18	−17
−2, +9	+7
+2, −9	−7
−3, +6	+3
+3, −6	−3

$3a^3-9a^2-54a=3a(a+3)(a-6)$

97. There is no common factor.

Factor the trinomial $x^2 + 4xy - 21y^2$.

Factors	*Sum*
$-1, +21$	$+20$
$+1, -21$	-20
$-3, +7$	$+4$
$+3, -7$	-4

$x^2 + 4xy - 21y^2 = (x + 7y)(x - 3y)$

99. There is no common factor.

Factor the trinomial $a^2 - 15ab + 50b^2$.

Factors	*Sum*
$-1, -50$	-51
$-2, -25$	-27
$-5, -10$	-15

$a^2 - 15ab + 50b^2 = (a - 5b)(a - 10b)$

101. There is no common factor.

Factor the trinomial $s^2 + 2st - 48t^2$.

Factors	*Sum*
$-1, +48$	$+47$
$+1, -48$	-47
$-2, +24$	$+22$
$+2, -24$	-22
$-3, +16$	$+13$
$+3, -16$	-13
$-4, +12$	$+8$
$+4, -12$	-8
$-6, +8$	$+2$
$+6, -8$	-2

$s^2 + 2st - 48t^2 = (s + 8t)(s - 6t)$

103. There is no common factor.

Factor the trinomial $y^2 + 85yz + 36z^2$.

Factors	*Sum*
$+1, +36$	$+37$
$+2, +18$	$+20$
$+3, +12$	$+15$
$+4, +9$	$+13$
$+6, +6$	$+12$

$y^2 + 85yz + 36z^2$ is nonfactorable over the integers.

105. The GCF is z^2.

$z^4 + 2z^3 - 80z^2 = z^2(z^2 + 2z - 80)$

Factor the trinomial $z^2 + 2z - 80$.

Factors	*Sum*
$-1, +80$	$+79$
$+1, -80$	-79
$-2, +40$	$+38$
$+2, -40$	-38
$-4, +20$	$+16$
$+4, -20$	-16
$-5, +16$	$+11$
$+5, -16$	-11
$-8, +10$	$+2$
$+8, -10$	-2

$z^4 + 2z^3 - 80z^2 = z^2(z + 10)(z - 8)$

107. The GCF is b^2.

$b^4 - 3b^3 - 10b^2 = b^2(b^2 - 3b - 10)$

Factor the trinomial $b^2 - 3b - 10$.

Factors	*Sum*
$-1, +10$	$+9$
$+1, -10$	-9
$-2, +5$	$+3$
$+2, -5$	-3

$b^4 - 3b^3 - 10b^2 = b^2(b + 2)(b - 5)$

109. The GCF is $3y^2$.

$3y^4 + 54y^3 + 135y^2 = 3y^2(y^2 + 18y + 45)$

Factor the trinomial $y^2 + 18y + 45$.

Factors	*Sum*
$+1, +45$	$+46$
$+3, +15$	$+18$
$+5, +9$	$+14$

$3y^4 + 54y^3 + 135y^2 = 3y^2(y + 3)(y + 15)$

111. The GCF is $-x^2$.

$-x^4 + 11x^3 + 12x^2 = -x^2(x^2 - 11x - 12)$

Factor the trinomial $x^2 - 11x - 12$.

Factors	*Sum*
−1, +12	+11
+1, −12	−11
−2, +6	+4
+2, −6	−4
−3, +4	+1
+3, −4	−1

$-x^4 + 11x^3 + 12x^2 = -x^2(x+1)(x-12)$

113. The GCF is $3y$.

$3x^2y - 6xy - 45y = 3y(x^2 - 2x - 15)$

Factor the trinomial $x^2 - 2x - 15$.

Factors	*Sum*
−1, +15	+14
+1, −15	−14
−3, +5	+2
+3, −5	−2

$3x^2y - 6xy - 45y = 3y(x+3)(x-5)$

115. The GCF is $-3x$.

$-3x^2 + 36x^2 - 81x = -3x(x^2 - 12x + 27)$

Factor the trinomial $x^2 - 12x + 27$.

Factors	*Sum*
−1, −27	−28
−3, −9	−12

$-3x^3 + 36x^2 - 81x = -3x(x-3)(x-9)$

117. There is no common factor.

Factor the trinomial $x^2 - 8xy + 15y^2$.

Factors	*Sum*
−1, −15	−16
−3, −5	−8

$x^2 - 8xy + 15y^2 = (x-3y)(x-5y)$

119. There is no common factor.

Factor the trinomial $a^2 - 13ab + 42b^2$.

Factors	*Sum*
−1, −42	−43
−2, −21	−23
3, 14	−17
−6, −7	−13

$a^2 - 13ab + 42b^2 = (a-6b)(a-7b)$

121. There is no common factor.

Factor the trinomial $y^2 + 8yz + 7z^2$.

Factors	*Sum*
+1, +7	+8

$y^2 + 8yz + 7z^2 = (y+z)(y+7z)$

123. The GCF is $3y$.

$3x^2y + 60xy - 63y = 3y(x^2 + 20x - 21)$

Factor the trinomial $x^2 + 20x - 21$.

Factors	*Sum*
−1, +21	+20
+1, −21	−20
−3, +7	+4
+3, −7	−4

$3x^2y + 60xy - 63y = 3y(x+21)(x-1)$

125. The GCF is $3x$.

$3x^3 + 3x^2 - 36x = 3x(x^2 + x - 12)$

Factor the trinomial $x^2 + x - 12$.

Factors	*Sum*
−1, +12	+11
+1, −12	−11
−2, +6	+4
+2, −6	−4
−3, +4	+1
+3, −4	−1

$3x^3 + 3x^2 - 36x = 3x(x+4)(x-3)$

127. The GCF is 2.

$2t^2 - 24ts + 70s^2 = 2(t^2 - 12ts + 35t)$

Factor the trinomial $t^2 - 12ts + 35s^2$.

Factors	*Sum*
−1, −35	−36
−5, −7	−12

$2t^2 - 24ts + 70s^2 = 2(t-5s)(t-7s)$

129. The GCF is 3.

$3a^2 - 24ab - 99b^2 = 3(a^2 - 8ab - 33b^2)$

Factor the trinomial $a^2 - 8ab - 33b^2$.

Factors	*Sum*
−1, +33	32
+1, −33	−32
+3, −11	−8
−3, +11	8

$3a^2 - 24ab - 99b^2 = 3(a+3b)(a-11b)$

131. The GCF is $5x$.

$5x^3+30x^2y+40xy^2=5x\left(x^2+6xy+8y^2\right)$

Factor the trinomial $x^2+6xy+8y^2$.

Factors	*Sum*
+1, +8	+9
+2, +4	+6

$5x^3+30x^2y+40xy^2=5x(x+2y)(x+4y)$

133. **a.** The GCF is 2.

$2x^2-2xy-4y^2=2\left(x^2-xy-2y^2\right)$

Factor the trinomial $x^2-xy-2y^2$.

Factors	*Sum*
+1, −2	−1
+2, −1	1

$2x^2-2xy-4y^2=2(x+y)(x-2y)$

Yes, $x+y$ is a factor.

b. The GCF is $2y$.

$2x^2y-4xy-4y=2y\left(x^2-2x-2\right)$

Factor the trinomial x^2-2x-2.

Factors	*Sum*
+1, −2	−1
+2, −1	1

x^2-2x-2 is nonfactorable over the integers.

$2x^2y-4xy-4y=2y\left(x^2-2x-2\right)$

No, $x+y$ is not a factor.

Critical Thinking 10.2

135. The GCF is $-2x$.

$-2x^3-6x^2-4x=-2x\left(x^2+3x+2\right)$

Factor the trinomial $\left(x^2+3x+2\right)$.

Factors	*Sum*
+1, +2	3
−1, −2	−3

$x^2+3x+2=(x+1)(x+2)$

$-2x^3-6x^2-4x=-2x(x+1)(x+2)$

$a(x+1)(x+2)=-2x(x+1)(x+2)$

$a=-2x$

137. The GCF is y. Then write in standard form.

$x^2y-54y-3xy=y\left(x^2-54-3x\right)=y\left(x^2-3x-54\right)$

$x^2-3x-54$

Factors	*Sum*
+1, −54	−53
−1, +54	+53
+2, −27	−25
−2, +27	+25
+3, −18	−15
−3, +18	+15
+6, −9	−3
−6, +9	+3

$x^2-3x-54=(x+6)(x-9)$

$x^2y-54y-3xy=y(x+6)(x-9)$

139. The GCF is $3p$. Then write in standard form.

$12p^2-96p+3p^3=3p\left(4p-32+3p^2\right)$
$=3p\left(3p^2+4p-32\right)$

$p^2+4p-32$

Factors	*Sum*
+1, −32	−31
−1, +32	+31
+2, −16	−14
−2, +16	+14
+4, −8	−4
−4, +8	+4

$p^2+4p-32=(p-4)(p+8)$

$12p^2-96p+3p^3=3p(p-4)(p+8)$

Projects or Group Activities 10.2

141. $x^2+kx+18$. The factors of 18 must sum to k.

Factors	*Sum*
−1, −18	−19
+1, +18	+19
−2, −9	−11
+2, +9	+11
−3, −6	−9
+3, +6	+9

k can be −19, 19, −11, 11, −9, or 9.

143. $x^2 - kx + 14$. The factors of 14 must sum to k.

Factors	*Sum*
$+1, +14$	$+15$
$-1, -14$	-15
$+2, +7$	$+9$
$-2, -7$	-9

k can be 15, –15, 9, or –9.

145. $z^2 + 7z + k,\ k > 0$

Find two positive integers that sum to 7. Their product is k.

Integers	*Product*
$+1, +6$	6
$+2, +5$	10
$+3, +4$	12

k can be 6, 10, or 12.

147. $c^2 - 7c + k,\ k > 0$

Find two negative integers that sum to –7. Their product is k.

Integers	*Product*
$-1, -6$	6
$-2, -5$	10
$-3, -4$	12

k can be 6, 10, or 12.

149. $y^2 + 5y + k,\ k > 0$

Find two positive integers that sum to 5. Their product is k.

Integers	*Product*
$+2, +3$	6
$+1, +4$	4

k can be 4 or 6.

Section 10.3

Concept Check

1. $2x + 5$

3. $4x - 3$

5. 4; –5

7. $-2x - 6x$

Objective A Exercises

9.

Positive	Positive
Factors of 2: 1, 2	Factors of 1: +1, +1
Trial Factors	*Middle Term*
$(1x+1)(2x+1)$	$x + 2x = 3x$

$2x^2 + 3x + 1 = (x+1)(2x+1)$

11.

Positive	Positive
Factors of 2: 1, 2	Factors of 3: +1, +3
Trial Factors	*Middle Term*
$(1y+1)(2y+3)$	$3y + 2y = 5y$
$(1y+3)(2y+1)$	$y + 6y = 7y$

$2y^2 + 7y + 3 = (y+3)(2y+1)$

13.

Positive	Negative
Factors of 2: 1, 2	Factors of 1: –1, –1
Trial Factors	*Middle Term*
$(1a-1)(2a-1)$	$-a - 2a = -3a$

$2a^2 - 3a + 1 = (a-1)(2a-1)$

15.

Positive	Negative
Factors of 2: 1, 2	Factors of 5: –1, –5
Trial Factors	*Middle Term*
$(1b-1)(2b-5)$	$-5b - 2b = -7b$
$(1b-5)(2b-1)$	$-b - 10b = -11b$

$2b^2 - 11b + 5 = (b-5)(2b-1)$

17.

Positive	
Factors of 2: 1, 2	Factors of –1: –1, +1
Trial Factors	*Middle Term*
$(1x-1)(2x+1)$	$x - 2x = -x$
$(1x+1)(2x-1)$	$-x + 2x = x$

$2x^2 + x - 1 = (x+1)(2x-1)$

19.

Positive	
Factors of 2: 1, 2	Factors of –3: –1, +3 +1, –3
Trial Factors	*Middle Term*
$(1x-1)(2x+3)$	$3x - 2x = x$
$(1x+3)(2x-1)$	$-x + 6x = 5x$
$(1x+1)(2x-3)$	$-3x + 2x = -x$
$(1x-3)(2x+1)$	$x - 6x = -5x$

$2x^2 - 5x - 3 = (x-3)(2x+1)$

21. Positive

Factors of 2: 1, 2 Factors of −10: −1, +10; +1, −10; −2, +5; +2, −5

Trial Factors	*Middle Term*
$(1t-1)(2t+10)$	Common factor
$(1t+10)(2t-1)$	$-t+20t=19t$
$(1t+1)(2t-10)$	Common factor
$(1t-10)(2t+1)$	$t-20t=-19t$
$(1t-2)(2t+5)$	$5t-4t=t$
$(1t+5)(2t-2)$	Common factor
$(1t+2)(2t-5)$	$-5t+4t=-t$
$(1t-5)(2t+2)$	Common factor

$2t^2-t-10=(t+2)(2t-5)$

23. Positive Negative

Factors of 3: 1, 3 Factors of 5: −1, −5

Trial Factors	*Middle Term*
$(1p-1)(3p-5)$	$-5p-3p=-8p$
$(1p-5)(3p-1)$	$-p-15p=-16p$

$3p^2-16p+5=(p-5)(3p-1)$

25. Positive Negative

Factors of 12: 1, 12; 2, 6; 3, 4 Factors of 1: −1, −1

Trial Factors	*Middle Term*
$(1y-1)(12y-1)$	$-y-12y=-13y$
$(2y-1)(6y-1)$	$-2y-6y=-8y$
$(3y-1)(4y-1)$	$-3y-4y=-7y$

$12y^2-7y+1=(3y-1)(4y-1)$

27. Positive Negative

Factors of 6: 1, 6; 2, 3 Factors of 3: −1, −3

Trial Factors	*Middle Term*
$(1z-1)(6z-3)$	Common factor
$(1z-3)(6z-1)$	$-z-18z=-19z$
$(2z-1)(3z-3)$	Common factor
$(2z-3)(3z-1)$	$-2z-9z=-11z$

$6z^2-7z+3$ is nonfactorable over the integers.

29. Positive Negative

Factors of 6: 1, 6; 2, 3 Factors of 4: −1, −4; −2, −2

Trial Factors	*Middle Term*
$(1t-1)(6t-4)$	Common factor
$(1t-4)(6t-1)$	$-t-24t=-25t$
$(1t-2)(6t-2)$	Common factor
$(2t-1)(3t-4)$	$-8t-3t=-11t$
$(2t-4)(3t-1)$	Common factor
$(2t-2)(3t-2)$	Common factor

$6t^2-11t+4=(2t-1)(3t-4)$

31. Positive Positive

Factors of 8: 1, 8; 2, 4 Factors of 4: +1, +4; +2, +2

Trial Factors	*Middle Term*
$(1x+1)(8x+4)$	Common factor
$(1x+4)(8x+1)$	$x+32x=33x$
$(1x+2)(8x+2)$	Common factor
$(2x+1)(4x+4)$	Common factor
$(2x+4)(4x+1)$	Common factor
$(2x+2)(4x+2)$	Common factor

$8x^2+33x+4=(x+4)(8x+1)$

33. Positive

Factors of 5: 1, 5 Factors of −7: −1, +7; +1, −7

Trial Factors	*Middle Term*
$(1x-1)(5x+7)$	$7x-5x=2x$
$(1x+7)(5x-1)$	$-x+35x=34x$
$(1x+1)(5x-7)$	$-7x+5x=-2x$
$(1x-7)(5x+1)$	$x-35x=-34x$

$5x^2-62x-7$ is nonfactorable over the integers.

35. Positive Positive

Factors of 12: 1, 12; 2, 6; 3, 4 Factors of 5: +1, +5

Trial Factors	*Middle Term*
$(1y+1)(12y+5)$	$5y+12y=17y$
$(1y+5)(12y+1)$	$y+60y=61y$
$(2y+1)(6y+5)$	$10y+6y=16y$
$(2y+5)(6y+1)$	$2y+30y=32y$
$(3y+1)(4y+5)$	$15y+4y=19y$
$(3y+5)(4y+1)$	$3y+20y=23y$

$12y^2+19y+5=(3y+1)(4y+5)$

37. Positive

Factors of 2: 1, 2 Factors of -14: $-1, +14$; $+1, -14$; $-2, +7$; $+2, -7$

Trial Factors	*Middle Term*
$(1z-1)(2z+14)$	Common factor
$(1z+14)(2z-1)$	$-z+28z=27z$
$(1z+1)(2z-14)$	Common factor
$(1z-14)(2z+1)$	$z-28z=-27z$
$(1z-2)(2z+7)$	$7z-4z=3z$
$(1z+7)(2z-2)$	Common factor
$(1z+2)(2z-7)$	$-7z+4z=-3z$
$(1z-7)(2z+2)$	Common factor

$2z^2-27z-14=(z-14)(2z+1)$

39. Positive

Factors of 3: 1, 3 Factors of -16: $-1, +16$; $+1, -16$; $-2, +8$; $+2, -8$; $-4, +4$

Trial Factors	*Middle Term*
$(1p-1)(3p+16)$	$16p-3p=13p$
$(1p+16)(3p-1)$	$-p+48p=47p$
$(1p+1)(3p-16)$	$-16p+3p=-13p$
$(1p-16)(3p+1)$	$p-48p=-47p$
$(1p-2)(3p+8)$	$8p-6p=2p$
$(1p+8)(3p-2)$	$-2p+24p=22p$
$(1p+2)(3p-8)$	$-8p+6p=-2p$
$(1p-8)(3p+2)$	$2p-24p=-22p$
$(1p-4)(3p+4)$	$4p-12p=-8p$
$(1p+4)(3p-4)$	$-4p+12p=8p$

$3p^2+22p-16=(p+8)(3p-2)$

41. The GCF is 2.

$4x^2+6x+2=2(2x^2+3x+1)$

Factor the trinomial.

Positive Positive

Factors of 2: 1, 2 Factors of 1: $+1, +1$

Trial Factors	*Middle Term*
$(1x+1)(2x+1)$	$x+2x=3x$

$4x^2+6x+2=2(x+1)(2x+1)$

43. The GCF is 5.

$15y^2-50y+35=5(3y^2-10y+7)$

Factor the trinomial.

Positive Negative

Factors of 3: 1, 3 Factors of 7: $-1, -7$

Trial Factors	*Middle Term*
$(1y-1)(3y-7)$	$-7y-3y=-10y$
$(1y-7)(3y-1)$	$-y-21y=-22y$

$15y^2-50y+35=5(y-1)(3y-7)$

45. The GCF is x.

$2x^3-11x^2+5x=x(2x^2-11x+5)$

Factor the trinomial $2x^2-11x+5$.

Positive Negative

Factors of 2: 1, 2 Factors of 5: $-1, -5$

Trial Factors	*Middle Term*
$(1x-1)(2x-5)$	$-5x-2x=-7x$
$(1x-5)(2x-1)$	$-x-10x=-11x$

$2x^3-11x^2+5x=x(x-5)(2x-1)$

47. The GCF is b.

$3a^2b-16ab+16b=b(3a^2-16a+16)$

Factor the trinomial $3a^2-16a+16$.

Positive Negative

Factors of 3: 1, 3 Factors of 16: $-1, -16$; $-2, -8$; $-4, -4$

Trial Factors	*Middle Term*
$(1a-1)(3a-16)$	$-16a-3a=-19a$
$(1a-16)(3a-1)$	$-a-48a=-49a$
$(1a-2)(3a-8)$	$-8a-6a=-14a$
$(1a-8)(3a-2)$	$-2a-24a=-26a$
$(1a-4)(3a-4)$	$-4a-12a=-16a$

$3a^2b-16ab+16b=b(a-4)(3a-4)$

49. There is no common factor.

Factor the trinomial.

Positive Factors of 3:		Positive Factors of 10:	
	1, 3		+1, +10
			+2, +5

Trial Factors	*Middle Term*
$(1z+1)(3z+10)$	$10z+3z=13z$
$(1z+10)(3z+1)$	$z+30z=31z$
$(1z+2)(3z+5)$	$5z+6z=11z$
$(1z+5)(3z+2)$	$2z+8z=10z$

$3z^2+95z+10$ is nonfactorable over the integers

51. The GCF is $-3x$.

$36x-3x^2-3x^3=-3x(x^2+x-12)$

Factor the trinomial.

Positive Factors of 1:		Factors of -12:
	1, 1	$-1, +12$
		$+1, -12$
		$+2, -6$
		$-2, +6$
		$+3, -4$
		$-3, +4$

Trial Factors	*Middle Term*
$(1x-1)(1x+12)$	$12x-1x=11x$
$(1x+1)(1x-12)$	$-12x+1x=11x$
$(1x+2)(1x-6)$	$-6x+2x=-4x$
$(1x-2)(1x+6)$	$6x-2x=4x$
$(1x+3)(1x-4)$	$-4x+3x=-x$
$(1x-3)(1x+4)$	$4x-3x=x$

$36x-3x^2-3x^3=-3x(x-3)(x+4)$

53. The GCF is 4.

$80y^2-36y+4=4(20y^2-9y+1)$

Factor the trinomial.

Positive Factors of 20:		Negative Factors of 1:	
	1, 20		$-1, -1$
	2, 10		
	4, 5		

Trial Factors	*Middle Term*
$(1y-1)(20y-1)$	$-y-20y=-21y$
$(2y-1)(10y-1)$	$-2y-10y=-12y$
$(4y-1)(5y-1)$	$-4y-5y=-9y$

$80y^2-36y+4=4(4y-1)(5y-1)$

55. The GCF is z.

$8z^3+4z^2+3z=z(8z^2+14z+3)$

Factor the trinomial.

Positive Factors of 8:		Positive Factors of 3:	
	1, 8		$+1, +3$
	2, 4		

Trial Factors	*Middle Term*
$(1z+1)(8z+3)$	$3z+8z=11z$
$(1z+3)(8z+1)$	$z+24z=25z$
$(2z+1)(4z+3)$	$6z+4z=10z$
$(2z+3)(4z+1)$	$2z+12z=14z$

$8z^3+14z^2+3z=z(2z+3)(4z+1)$

57. The GCF is y.

$6x^2y-11xy-10y=y(6x^2-11x-10)$

Factor the trinomial.

Positive Factors of 6:		Factors of -10:	
	1, 6		$-1, +10$
	2, 3		$+1, -10$
			$-2, +5$
			$+2, -5$

Trial Factors	*Middle Term*
$(1x-1)(6x+10)$	Common factor
$(1x+10)(6x-1)$	$-x+60x=59x$
$(1x+1)(6x-10)$	Common factor
$(1x-10)(6x+1)$	$x-60x=-59x$
$(1x-2)(6x+5)$	$5x-12x=-7x$
$(1x+5)(6x-2)$	Common factor
$(1x+2)(6x-5)$	$-5x+12x=7x$
$(1x-5)(6x+2)$	Common factor
$(2x-1)(3x+10)$	$20x-3x=17x$
$(2x+10)(3x-1)$	Common factor
$(2x+1)(3x-10)$	$-20x+3x=-17x$
$(2x-10)(3x+1)$	Common factor
$(2x-2)(3x+5)$	Common factor
$(2x+5)(3x-2)$	$-4x+15x=11x$
$(2x+2)(3x-5)$	Common factor
$(2x-5)(3x+2)$	$4x-15x=-11x$

$6x^2y-11xy-10y=y(2x-5)(3x+2)$

59. The GCF is 5. $10t^2-5t-50=5(2t^2-t-10)$

Factor the trinomial.

Positive

Factors of 2: 1, 2 Factors of -10: $-1,\ +10$; $+1,\ -10$; $-2,\ +5$; $+2,\ -5$

Trial Factors	*Middle Term*
$(1t-1)(2t+10)$	Common factor
$(1t+10)(2t-1)$	$-t+20t=19t$
$(1t+1)(2t-10)$	Common factor
$(1t-10)(2t+1)$	$t-20t=-19t$
$(1t-2)(2t+5)$	$5t-4t=t$
$(1t+5)(2t-2)$	Common factor
$(1t+2)(2t-5)$	$-5t+4t=-t$
$(1t-5)(2t+2)$	Common factor

$10t^2-5t-50=5(t+2)(2t-5)$

61. The GCF is p.

$3p^3-16p^2+5p=p(3p^2-16p+5)$

Factor the trinomial.

Positive Negative

Factors of 3: 1, 3 Factors of 5: $-1,\ -5$

Trial Factors	*Middle Term*
$(1p-1)(3p-5)$	$-5p-3p=-8p$
$(1p-5)(3p-1)$	$-p-15p=-16p$

$3p^3-16p^2+5p=p(p-5)(3p-1)$

63. The GCF is 2.

$26z^2+98z-24=2(13z^2+49z-12)$

Factor the trinomial.

Positive

Factors of 13: 1, 13 Factors of -12: $-1,\ +12$; $+1,\ -12$; $-2,\ +6$; $+2,\ -6$; $-3,\ +4$; $+3,\ -4$

Trial Factors	*Middle Term*
$(1z-1)(13z+12)$	$2z-13z=-z$
$(1z+12)(13z-1)$	$-z+156z=155z$
$(1z+1)(13z-12)$	$-12z+13z=z$
$(1z-12)(13z+1)$	$z-156z=-155z$
$(1z-2)(13z+6)$	$6z-26z=-20z$
$(1z+6)(13z-2)$	$-2z+78z=76z$
$(1z+2)(13z-6)$	$-6z+26z=20z$
$(1z-6)(13z+2)$	$2z-78z=-76z$
$(1z-3)(13z+4)$	$4z-39z=-35z$
$(1z+4)(13z-3)$	$-3z+52z=49z$
$(1z+3)(13z-4)$	$-4z+39z=35z$
$(1z-4)(13z+3)$	$3z-52z=-49z$

$26z^2+98z-24=2(z+4)(13z-3)$

65. The GCF is $2y$.

$10y^3-44y^2+16y=2y(5y^2-22y+8)$

Factor the trinomial.

Positive Negative

Factors of 5: 1, 5 Factors of 8: $-1,\ -8$; $-2,\ -4$

Trial Factors	*Middle Term*
$(1y-1)(5y-8)$	$-8y-5y=-13y$
$(1y-8)(5y-1)$	$-y-40y=-41y$
$(1y-2)(5y-4)$	$-4y-10y=-14y$
$(1y-4)(5y-2)$	$-2y-20y=-22y$

$10y^3-44y^2+16y=2y(y-4)(5y-2)$

67. The GCF is yz.

$4yz^3+5yz^2-6yz=yz(4z^2+5z-6)$

Factor the trinomial.

Positive Factors of 4:		Factors of −6:
	1, 4	−1, +6
	2, 2	+1, −6
		−2, +3
		+2, −3

Trial Factors	*Middle Term*
$(1z-1)(4z+6)$	Common factor
$(1z+6)(4z-1)$	$-z+24z=23z$
$(1z+1)(4z-6)$	Common factor
$(1z-6)(4z+1)$	$z-24z=-23z$
$(1z-2)(4z+3)$	$3z-8z=-5z$
$(1z+3)(4z-2)$	Common factor
$(1z+2)(4z-3)$	$-3z+8z=5z$
$(1z-3)(4z+2)$	Common factor
$(2z-1)(2z+6)$	Common factor
$(2z+1)(2z-6)$	Common factor
$(2z-2)(2z+3)$	Common factor
$(2z+2)(2z-3)$	Common factor

$4yz^3+5yz^2-6yz=yz(z+2)(4z-3)$

69. The GCF is $3a$.

$42a^3+45a^2-27a=3a(14a^2+15a-9)$

Factor the trinomial.

Positive Factors of 14:		Factors of −9:
	1, 14	−1, +9
	2, 7	+1, −9
		−3, +3

Trial Factors	*Middle Term*
$(1a-1)(14a+9)$	$9a-14a=-5a$
$(1a+9)(14a-1)$	$-a+126a=125a$
$(1a+1)(14a-9)$	$-9a+14a=5a$
$(1a-9)(14a+1)$	$a-126a=-125a$
$(1a-3)(14a+3)$	$3a-42a=-39a$
$(1a+3)(14a-3)$	$-3a+42a=39a$
$(2a-1)(7a+9)$	$18a-7a=11a$
$(2a+9)(7a-1)$	$-2a+63a=61a$
$(2a+1)(7a-9)$	$-18a+7a=-11a$
$(2a-9)(7a+1)$	$2a-63a=-61a$
$(2a-3)(7a+3)$	$6a-21a=-15a$
$(2a+3)(7a-3)$	$-6a+21a=15a$

$42a^3+45a^2-27a=3a(2a+3)(7a-3)$

71. The GCF is y.

$9x^2y-30xy^2+25y^3=y(9x^2-30xy+25y^2)$

Factor the trinomial.

Positive		Negative	
Factors of 9:	1, 9	Factors of 25:	−1, −25
	3, 3		−5, −5

Trial Factors	*Middle Term*
$(1x-1y)(9x-25y)$	$-25xy-9xy=-34xy$
$(1x-25y)(9x-1y)$	$-xy-225xy=-226xy$
$(1x-5y)(9x-5y)$	$-5xy-45xy=-50xy$
$(3x-1y)(3x-25y)$	$-75xy-3xy=-78xy$
$(3x-5y)(3x-5y)$	$-15xy-15xy=-30xy$

$9x^2y-30xy^2+25y^3=y(3x-5y)(3x-5y)$

73. The GCF is xy.

$9x^3y-24x^2y^2+16xy^3=xy(9x^2-24xy+16y^2)$

Factor the trinomial.

Positive		Negative	
Factors of 9:	1, 9	Factors of 16:	−1, −16
	3, 3		−2, −8
			−4, −4

Trial Factors	*Middle Term*
$(1x-1y)(9x-16y)$	$-16xy-9xy=-25xy$
$(1x-16y)(9x-1y)$	$-xy-144xy=-145xy$
$(1x-2y)(9x-8y)$	$-8xy-18xy=-26xy$
$(1x-8y)(9x-2y)$	$-2xy-72xy=-74xy$
$(1x-4y)(9x-4y)$	$-4xy-36xy=-40xy$
$(3x-1y)(3x-16y)$	$-48xy-3xy=-51xy$
$(3x-2y)(3x-8y)$	$-24xy-6xy=-30xy$
$(3x-4y)(3x-4y)$	$-12xy-12xy=-24xy$

$9x^3y-24x^2y^2+16xy^3=xy(3x-4y)(3x-4y)$

75. p must be odd. If p were even, $(nx+p)$ would have a common factor of 2 since n is even.

Objective B Exercises

77. $6x^2-17x+12 \quad 6\cdot 12=72$

Factors of 72 whose sum is −17: −9 and −8

$$\begin{aligned}6x^2-17x+12&=6x^2-9x-8x+12\\&=(6x^2-9x)+(-8x+12)\\&=3x(2x-3)-4(2x-3)\\&=(2x-3)(3x-4)\end{aligned}$$

79. $5b^2+33b-14 \qquad 5(-14)=-70$

Factors of -70 whose sum is 33: 35 and –2

$$\begin{aligned}5b^2+33b-14 &= 5b^2+35b-2b-14\\ &= (5b^2+35b)+(-2b-14)\\ &= 5b(b+7)-2(b+7)\\ &= (b+7)(5b-2)\end{aligned}$$

81. $6a^2+7a-24 \qquad 6(-24)=-144$

Factors of -144 whose sum is 7: 16 and –9

$$\begin{aligned}6a^2+7a-24 &= 6a^2+16a-9a-24\\ &= (6a^2+16a)+(-9a-24)\\ &= 2a(3a+8)-3(3a+8)\\ &= (3a+8)(2a-3)\end{aligned}$$

83. $4z^2+11z+6 \qquad 4\cdot 6=24$

Factors of 24 whose sum is 11: 8 and 3

$$\begin{aligned}4z^2+11z+6 &= 4z^2+8z+3z+6\\ &= (4z^2+8z)+(3z+6)\\ &= 4z(z+2)+3(z+2)\\ &= (z+2)(4z+3)\end{aligned}$$

85. $22p^2+51p-10 \qquad 22(-10)=-220$

Factors of -220 whose sum is 51: 55 and –4

$$\begin{aligned}22p^2+51p-10 &= 22p^2+55p-4p-10\\ &= (22p^2+55p)+(-4p-10)\\ &= 11p(2p+5)-2(2p+5)\\ &= (2p+5)(11p-2)\end{aligned}$$

87. $8y^2+17y+9 \qquad 8\cdot 9=72$

Factors of 72 whose sum is 17: 9 and 8

$$\begin{aligned}8y^2+17y+9 &= 8y^2+8y+9y+9\\ &= (8y^2+8y)+(9y+9)\\ &= 8y(y+1)+9(y+1)\\ &= (y+1)(8y+9)\end{aligned}$$

89. $18t^2-9t-5 \qquad 18(-5)=-90$

Factors of -90 whose sum is -9: –15 and 6

$$\begin{aligned}18t^2-9t-5 &= 18t^2-15t+6t-5\\ &= (18t^2-15t)+(6t-5)\\ &= 3t(6t-5)+1(6t-5)\\ &= (6t-5)(3t+1)\end{aligned}$$

91. $6b^2+71b-12 \qquad 6(-12)=-72$

Factors of -72 whose sum is 71: 72 and –1

$$\begin{aligned}6b^2+71b-12 &= 6b^2+72b-b-12\\ &= (6b^2+72b)+(-b-12)\\ &= 6b(b+12)-1(b+12)\\ &= (b+12)(6b-1)\end{aligned}$$

93. $9x^2+12x+4 \qquad 9\cdot 4=36$

Factors of 36 whose sum is 12: 6 and 6

$$\begin{aligned}9x^2+12x+4 &= 9x^2+6x+6x+4\\ &= (9x^2+6x)+(6x+4)\\ &= 3x(3x+2)+2(3x+2)\\ &= (3x+2)(3x+2)\end{aligned}$$

95. $6b^2-13b+6 \qquad 6\cdot 6=36$

Factors of 36 whose sum is –13: –9 and –4

$$\begin{aligned}6b^2-13b+6 &= 6b^2-9b-4b+6\\ &= (6b^2-9b)+(-4b+6)\\ &= 3b(2b-3)-2(2b-3)\\ &= (2b-3)(3b-2)\end{aligned}$$

97. $33b^2+34b-35 \qquad 33(-35)=-1155$

Factors of -1155 whose sum is 34: 55 and –21

$$\begin{aligned}33b^2+34b-35 &= 33b^2+55b-21b-35\\ &= (33b^2+55b)+(-21b-35)\\ &= 11b(3b+5)-7(3b+5)\\ &= (3b+5)(11b-7)\end{aligned}$$

99. $18y^2-39y+20 \qquad 18\cdot 20=360$

Factors of 360 whose sum is -39: –24 and –15

$$\begin{aligned}18y^2-39y+20 &= 18y^2-24y-15y+20\\ &= (18y^2-24y)+(-15y+20)\\ &= 6y(3y-4)-5(3y-4)\\ &= (3y-4)(6y-5)\end{aligned}$$

101 $15a^2+26a-21 \qquad 15(-21)=-315$

Factors of -315 whose sum is 26: 35 and -9

$$\begin{aligned}15a^2+26a-21&=15a^2+35a-9a-21\\&=(15a^2+35a)+(-9a-21)\\&=5a(3a+7)-3(3a+7)\\&=(3a+7)(5a-3)\end{aligned}$$

103. $8y^2-26y+15 \qquad 8\cdot 15=120$

Factors of 120 whose sum is -26: -20 and -6

$$\begin{aligned}8y^2-26y+15&=8y^2-20y-6y+15\\&=(8y^2-20y)+(-6y+15)\\&=4y(2y-5)-3(2y-5)\\&=(2y-5)(4y-3)\end{aligned}$$

105. $8z^2+2z-15 \qquad 8(-15)=-120$

Factors of -120 whose sum is 2: 12 and -10

$$\begin{aligned}8z^2+2z-15&=8z^2+12z-10z-15\\&=(8z^2+12z)+(-10z-15)\\&=4z(2z+3)-5(2z+3)\\&=(2z+3)(4z-5)\end{aligned}$$

107. $15x^2-82x+24 \qquad 15\cdot 24=360$

Factors of 360 whose sum is -82: none

$15x^2-82x+24$ is nonfactorable over the integers.

109. $10z^2-29z+10 \qquad 10\cdot 10=100$

Factors of 100 whose sum is -29: -25 and -4

$$\begin{aligned}10z^2-29z+10&=10z^2-25z-4z+10\\&=(10z^2-25z)+(-4z+10)\\&=5z(2z-5)-2(2z-5)\\&=(2z-5)(5z-2)\end{aligned}$$

111. $36z^2+72z+35 \qquad 36\cdot 35=1260$

Factors of 1260 whose sum is 72: 30 and 42

$$\begin{aligned}36z^2+72z+35&=36z^2+30z+42z+35\\&=(36z^2+30z)+(42z+35)\\&=6z(6z+5)+7(6z+5)\\&=(6z+5)(6z+7)\end{aligned}$$

113. $3x^2+xy-2y^2 \qquad 3(-2)=-6$

Factors of -6 whose sum is 1: 3 and -2

$$\begin{aligned}3x^2+xy-2y^2&=3x^2+3xy-2xy-2y^2\\&=(3x^2+3xy)+(-2xy-2y^2)\\&=3x(x+y)-2y(x+y)\\&=(x+y)(3x-2y)\end{aligned}$$

115. $3a^2+5ab-2b^2 \qquad 3(-2)=-6$

Factors of -6 whose sum is 5: 6 and -1

$$\begin{aligned}3a^2+5ab-2b^2&=3a^2+6ab-ab-2b^2\\&=(3a^2+6ab)+(-ab-2b^2)\\&=3a(a+2b)-b(a+2b)\\&=(a+2b)(3a-b)\end{aligned}$$

117. $4y^2-11yz+6z^2 \qquad 4\cdot 6=24$

Factors of 24 whose sum is -11: -8 and -3

$$\begin{aligned}4y^2-11yz+6z^2&=4y^2-8yz-3yz+6z^2\\&=(4y^2-8yz)+(-3yz+6z^2)\\&=4y(y-2z)-3z(y-2z)\\&=(y-2z)(4y-3z)\end{aligned}$$

119. $28+3z-z^2 \qquad 28(-1)=-28$

Factors of -28 whose sum is 3: 7 and -4

$$\begin{aligned}28+3z-z^2&=-z^2+3z+28=-z^2+7z-4z+28\\&=(-z^2+7z)+(-4z+28)\\&=-z(z+4)+7(z+4)\\&=-(z-7)(z+4)\end{aligned}$$

121. $8-7x-x^2 \qquad 8(-1)=-8$

Factors of -8 whose sum is -7: -8 and 1

$$\begin{aligned}8-7x-x^2&=-x^2-7x+8=-x^2-8x+x+8\\&=(-x^2-8x)+(x+8)\\&=-x(x-1)-8(x-1)\\&=-(x-1)(x+8)\end{aligned}$$

123. $9x^2+33x-60$

Common factor 3: $3(3x^2+11x-20)$

$3(-20)=-60$

Factors of -60 whose sum is 11: 15 and -4

$$\begin{aligned}3(3x^2+11x-20)&=3(3x^2+15x-4x-20)\\&=3\left[(3x^2+15x)+(-4x-20)\right]\\&=3\left[3x(x+5)-4(x+5)\right]\\&=3(x+5)(3x-4)\end{aligned}$$

125. $24x^2-52x+24$

Common factor 4: $4(6x^2-13x+6)$

$6\cdot 6=36$

Factors of 36 whose sum is -13: -9 and -4

$$\begin{aligned}4(6x^2-13x+6)&=4(6x^2-9x-4x+6)\\&=4\left[(6x^2-9x)+(-4x+6)\right]\\&=4\left[3x(2x-3)-2(2x-3)\right]\\&=4(2x-3)(3x-2)\end{aligned}$$

127. $35a^4+9a^3-2a^2$

Common factor a^2: $a^2(35a^2+9a-2)$

$35(-2)=-70$

Factors of -70 whose sum is 9: 14 and -5

$$\begin{aligned}a^2(35a^2+9a-2)&=a^2(35a^2+14a-5a-2)\\&=a^2\left[(35a^2+14a)+(-5a-2)\right]\\&=a^2\left[7a(5a+2)-1(5a+2)\right]\\&=a^2(5a+2)(7a-1)\end{aligned}$$

129. $15b^2-115b+70$

Common factor 5: $5(3b^2-23b+14)$

$3\cdot 14=42$

Factors of 42 whose sum is -23: -21 and -2

$$\begin{aligned}5(3b^2-23b+14)&=5(3b^2-21b-2b+14)\\&=5\left[(3b^2-21b)+(-2b+14)\right]\\&=5\left[3b(b-7)-2(b-7)\right]\\&=5(b-7)(3b-2)\end{aligned}$$

131. $3x^2-26xy+35y^2$ $\quad 3\cdot 35=105$

Factors of 105 whose sum is -26: -21 and -5

$$\begin{aligned}3x^2-25xy+36y^2&=3x^2-21xy-5xy+35y^2\\&=(3x^2-21xy)+(-5xy+35y^2)\\&=3x(x-7y)-5y(x-7y)\\&=(x-7y)(3x-5y)\end{aligned}$$

133. $216y^2-3y-3$

Common factor 3: $3(72y^2-y-1)$

$72(-1)=-72$

Factors of -72 whose sum is -1: -9 and 8

$$\begin{aligned}3(72y^2-y-1)&=3(72y^2-9y+8y-1)\\&=3\left[(72y^2-9y)+(8y-1)\right]\\&=3\left[9y(8y-1)+1(8y-1)\right]\\&=3(8y-1)(9y+1)\end{aligned}$$

135. If c is negative, there must be 1 positive and 1 negative sign.

137. If c is negative, there must be 1 positive and 1 negative sign.

Critical Thinking 10.3

139. $(x+1)^2-(x+1)-6$ $\quad$ Let $a=x+1$

$$\begin{aligned}&=a^2-a-6\\&=(a-3)(a+2)\\&=(x+1-3)(x+1+2)\\&=(x-2)(x+3)\end{aligned}$$

141. $(y+3)^2-5(y+3)+6$ $\quad$ Let $a=y+3$

$$\begin{aligned}&=a^2-5a+6\\&=(a-2)(a-3)\\&=[y+3-2][y+3-3]\\&=(y+1)y \text{ or } y(y+1)\end{aligned}$$

143. $3(a+2)^2-(a+2)-4$ $\quad$ Let $x=a+2$

$$\begin{aligned}&=3x^2-x-4\\&=(3x-4)(x+1)\\&=\left[3(a+2)-4\right]\left[a+2+1\right]\\&=(3a+6-4)(a+3)\\&=(3a+2)(a+3)\end{aligned}$$

145. The GCF is $2y$.

$$6y+8y^3-26y^2=2y(3+4y^2-13y)$$
$$=2y(4y^2-13y+3)$$
$$=2y(4y-1)(y-3)$$

147. The GCF is ab.

$$a^3b-24ab-2a^2b=ab(a^2-24-2a)$$
$$=ab(a^2-2a-24)$$
$$=ab(a-6)(a+4)$$

Projects or Group Activities 10.3

149. $2x^2+kx+3$

$2 \times 3 = 6$

Factors	*Sum*
$+1, +6$	$+7$
$-1, -6$	-7
$+2, +3$	$+5$
$-2, -3$	-5

k can be 7, –7, 5, or –5.

151. $3x^2+kx+2$

$2 \times 3 = 6$

Factors	*Sum*
$+1, +6$	$+7$
$-1, -6$	-7
$+2, +3$	$+5$
$-2, -3$	-5

k can be 7, –7, 5, or –5.

153. $2x^2+kx+5$

$2 \times 5 = 10$

Factors	*Sum*
$+1, +10$	11
$-1, -10$	-11
$+2, +5$	7
$-2, -5$	-7

k can be 11, –11, 7, or –7.

155. $3x^2+x-2=(x+1)(3x-2)$

Length and width must both be positive.

$$3x-2>0$$
$$3x-2+2>0+2$$
$$3x>2$$
$$\frac{3x}{3}>\frac{2}{3}$$
$$x>\frac{2}{3}$$

The length must be greater than the width.

$3x - 2$ if the length if $3x - 2 > x + 1$.

$$3x-2>x+1$$
$$3x-x-2>x-x+1$$
$$2x-2>1$$
$$2x-2+2>1+2$$
$$2x>3$$
$$\frac{2x}{2}>\frac{3}{2}$$
$$x>\frac{3}{2}$$

$x + 1$ is the length is $x + 1 > 3x - 2$.

$3x - 2$ is the length if $\frac{2}{3}<x<\frac{3}{2}$ and $x + 1$ is the length if $x>\frac{3}{2}$.

Check Your Progress: Chapter 10

1. The GCF is 5. $20b+5=5(4b+1)$

2. The GCF is $(b + 7)$

$$2x(7+b)-y(b+7)=2x(b+7)-y(b+7)$$
$$=(b+7)(2x-y)$$

3. $x^2+20x+100$

Factors	*Sum*
$+1, +100$	$+101$
$+2, +50$	$+52$
$+4, +25$	$+29$
$+5, +20$	$+25$
$+10, +10$	$+20$

$x^2+20x+100=(x+10)(x+10)$

4. The GCF is y.

$x^2y - 2xy - 24y = y(x^2 - 2x - 24)$

Factors	*Sum*
$+1, -24$	-23
$-1, +24$	$+23$
$+2, -12$	-10
$-2, +12$	$+10$
$+3, -8$	-5
$-3, +8$	$+5$
$+4, -6$	-2
$-4, +6$	$+2$

$x^2 - 2x - 24 = (x+4)(x-6)$

$x^2y - 2xy - 24y = y(x+4)(x-6)$

5. Factor -1.

$35 + 2x - x^2 = -1(x^2 - 2x - 35)$

Factors	*Sum*
$+1, -35$	-34
$-1, +35$	$+34$
$+5, -7$	-2
$-5, +7$	$+2$

$x^2 - 2x - 35 = (x+5)(x-7)$

$35 + 2x - x^2 = -1(x+5)(x-7)$

6. $x^2 - 8x - 2$

Factors	*Sum*
$+1, -2$	-1
$-1, +2$	$+1$

$x^2 - 8x - 2$ is nonfactorable over the integers.

7. $21x^2 + 6xy - 49x - 14y$

$= (21x^2 + 6xy) + (-49x - 14y)$

$= 3x(7x + 2y) - 7(7x + 2y)$

$= (7x + 2y)(3x - 7)$

8. The GCF is $3a$. $6ab + 9a = 3a(2b + 3)$

9. Negative

Factors of 5: 1, 5 Factors of 8: $-1, -8$
$-2, -4$

Trial Factors	*Middle Term*
$(y-1)(5y-8)$	$-8y - 5y = -13y$
$(y-8)(5y-1)$	$-y - 40y = -41y$
$(y-2)(5y-4)$	$-4y - 10y = -14y$
$(y-4)(5y-2)$	$-2y - 20y = -22y$

$5y^2 - 22y + 8 = (y-4)(5y-2)$

10. $12x^2 + 31x + 9$ $12(9) = 108$

Factors of 108 whose sum is 31: 27 and 4

$12x^2 + 27x + 4x + 9$

$= (12x^2 + 27x) + (4x + 9)$

$= 3x(4x + 9) + (4x + 9)$

$= (4x + 9)(3x + 1)$

$12x^2 + 31x + 9 = (4x + 9)(3x + 1)$

11. The GCF is x. $9x - 5x^2 = x(9 - 5x)$

12. $2x^2 + x + 2xy + y$

$= (2x^2 + x) + (2xy + y)$

$= x(2x + 1) + y(2x + 1)$

$= (2x + 1)(x + y)$

13. $8a^2 - 2ab - 3b^2$ $8(-3) = -24$

Factors of -24 whose sum is -2: 4 and -6

$8a^2 + 4ab - 6ab - 3b^2$

$= (8a^2 + 4ab) + (-6ab - 3b^2)$

$= 4a(2a + b) - 3b(2a + b)$

$= (2a + b)(4a - 3b)$

14. $b^2 + 9b + 20$

Factors	*Sum*
$+1, +20$	$+21$
$+4, +5$	$+9$

$b^2 + 9b + 20 = (b + 4)(b + 5)$

15. The GCF is $2a$.

$2a^3+24a^2+54a=2a\left(a^2+12a+27\right)$

$a^2+12a+27$

Factors	*Sum*
$+1, +27$	$+28$
$+3, +9$	$+12$

$a^2+12a+27=(a+3)(a+9)$

$2a^2+24a^2+54a=2a(a+3)(a+9)$

16. Factors of 11: 1, 11 Factors of -5: $1, -5$; $-1, 5$

Trial Factors	*Middle Term*
$(a+1)(11a-5)$	$-5a+11a=-6a$
$(a-5)(11a+1)$	$a-55a=-54a$
$(a-1)(11a+5)$	$5a-11a=-6a$
$(a+5)(11a-1)$	$-a+55a=+54a$

$11a^2-54a-5=(a-5)(11a+1)$

17. The GCF is 4. $360y^2+4y-4=4\left(90y^2+y-1\right)$

$90y^2+y-1 \qquad 90(-1)=-90$

The factors of –90 whose sum is +1: +10 and –9

$90y^2-10y+9y-1$

$=\left(90y^2-10y\right)+(9y-1)$

$=10y(9y-1)+(9y-1)$

$=(9y-1)(10y+1)$

$90y^2+y-1=(9y-1)(10y+1)$

$360y^2+4y-4=4(9y-1)(10y+1)$

18. The GCF is y.

$14y^3+5y^2+11y=y\left(14y^2+5y+11\right)$

$14y^2+5y+11$

The factors of 14: $+1, +14$; $+2, +7$

The factors of 11: $+1, +11$

Trial Factors	Middle Term
$(x+1)(14x+11)$	$11x+14x=25x$
$(x+11)(14x+1)$	$154x+x=155x$
$(2x+1)(7x+11)$	$22x+7x=29x$
$(2x+11)(7x+1)$	$14x+77x=91x$

$14y^2+5y+11$ is nonfactorable over the integers.

$14y^3+5y^2+11y=y\left(14y^2+5y+11\right)$

19. Negative

$x^2-7x+10$

Factors	*Sum*
$-1, -10$	-11
$-2, -5$	-7

$x^2-7x+10=(x-2)(x-5)$

20. Positive

$x^2+8xy+9y^2$

Factors	*Sum*
$+1, +9$	$+10$
$+3, +3$	$+6$

$x^2+8xy+9y^2$ is nonfactorable over the integers.

21. Positive

$b^2+13b+40$

Factors	*Sum*
$+1, +40$	$+41$
$+2, +20$	$+22$
$+4, +10$	$+14$
$+5, +8$	$+13$

$b^2+13b+40=(b+5)(b+8)$

22. $2x^2-5x-6xy+15y$

$=2x^2-6xy-5x+15y$

$=\left(2x^2-6xy\right)+(-5x+15y)$

$=2x(x-3y)-5(x-3y)$

$=(x-3y)(2x-5)$

23. The GCF is xy.

$x^2y - xy^3 + x^3y = xy(x - y^2 + x^2)$

24. Positive

Factors of 3: +1, +3 Factors of 16: +1, +16; +2, +8; +4, +4

Trail Factors	Middle Term
$(b+1)(3b+16)$	$16b + 3b = 19b$
$(b+16)(3b+1)$	$b + 48b = 49b$
$(b+2)(3b+8)$	$8b + 6b = 14b$
$(b+8)(3b+2)$	$2b + 24b = 26b$
$(b+4)(3b+4)$	$4b + 12b = 16b$

$3b^2 + 16b + 16 = (b+4)(3b+4)$

25. $x^2 - 11x - 42$

Factors	*Sum*
+1, − 42	−41
−1, + 42	+41
+2, − 21	−19
−2, + 21	+19
+3, − 14	−11
−3, + 14	+11
+6, − 7	−1
−6, + 7	+1

$x^2 - 11x - 4 = (x+3)(x-14)$

Section 10.4

Concept Check

1. $4; 25x^6; 100x^4y^4$

3. i and iv

5. Sometimes true

7. Always true

9. $3x; 2$

11. Answers will vary. For example:

a. $x^2 - 16$

b. $(x+7)(x-7)$

c. $x^2 + 10x + 25$

d. $(x-3)^2$

e. $x^2 + 36$

Objective A Exercises

13. $x^2 - 4 = x^2 - 2^2 = (x+2)(x-2)$

15. $a^2 - 81 = a^2 - 9^2 = (a+9)(a-9)$

17. $y^2 + 2y + 1; \quad \sqrt{y^2} = y; \quad \sqrt{1} = 1; \quad 2(y \cdot 1) = 2y$

The trinomial is a perfect square.

$y^2 + 2y + 1 = (y+1)^2$

19. $a^2 - 2a + 1; \quad \sqrt{a^2} = a; \quad \sqrt{1} = 1; \quad -2(a \cdot 1) = -2a$

The trinomial is a perfect square.

$a^2 - 2a + 1 = (a-1)^2$

21. $4x^2 - 1 = (2x)^2 - 1^2 = (2x+1)(2x-1)$

23. $x^6 - 9 = (x^3)^2 - 3^2 = (x^3+3)(x^3-3)$

25. $x^2 + 8x - 16; \quad \sqrt{-16}$ is not a real number.

$x^2 + 8x - 16$ is non factorable over the integers.

27. $x^2 + 2xy + y^2; \sqrt{x^2} = x; \sqrt{y^2} = 7; 2(x \cdot y) = 2xy$

The trinomial is a perfect square.

$x^2 + 2xy + y^2 = (x+y)^2$

29. $4a^2 + 4a + 1; \sqrt{4a^2} = 2a; \sqrt{1} = 1; 2(4a \cdot 1) = 4a$

The trinomial is a perfect square.

$4a^2 + 4a + 1 = (2a+1)^2$

31. $9x^2 - 1 = (3x)^2 - 1^2 = (3x+1)(3x-1)$

33. $1 - 64x^2 = 1^2 - (8x)^2 = (1+8x)(1-8x)$

35. $x^2 + 64$ is nonfactorable over the integers.

37. $9a^2 + 6a + 1; \sqrt{9a^2} = 3a; \sqrt{1} = 1; 2(3a \cdot 1) = 6a$

The trinomial is a perfect square.

$9a^2 + 6a + 1 = (3a+1)^2$

39. $b^4 - 16a^2 = (b^2)^2 - (4a)^2 = (b^2 + 4a)(b^2 - 4a)$

41. $4a^2 - 20a + 25; \quad \sqrt{4a^2} = 2a; \quad \sqrt{25} = 5;$

$-2(2a \cdot 5) = -20a$

The trinomial is a perfect square.

$4a^2 - 20a + 25 = (2a-5)^2$

43. $9a^2-42a+49;\ \sqrt{9a^2}=3a;\ \sqrt{49}=7;$
$-2(3a\cdot 7)=-42a$
The trinomial is a perfect square.
$9a^2-42a+49=(3a-7)^2$

45. $25z^2-y^2=(5z)^2-y^2=(5z+y)(5x-y)$

47. $a^2b^2-25=(ab)^2-5^2=(ab+5)(ab-5)$

49. $25x^2-1=(5x)^2-1^2=(5x+1)(5x-1)$

51. $4a^2-12ab+9b^2;\ \sqrt{4a^2}=2a;\ \sqrt{9b^2}=3b;$
$-2(2a\cdot 3b)=-12ab$
The trinomial is a perfect square.
$4a^2-12ab+9b^2=(2a-3b)^2$

53. $4y^2-36xy+81z^2;\ \sqrt{4y^2}=2y;\ \sqrt{81z^2}=9z;$
$-2(2y\cdot 9z)=-36yz$
The trinomial is a perfect square.
$4y^2-36xy+81z^2=(2y-9z)^2$

55. **(i)** $a^2x^2-2abx+b^2;\ \sqrt{a^2x^2}=ax;\ \sqrt{b^2}=b;$
$-2(ax\cdot b)=-2abx$
The trinomial is a perfect square.
$a^2x^2-2ab+b^2=(ax-b)^2$ Yes

(ii) $a^2x^2-2abx-b^2;\ \sqrt{a^2x^2}=ax;\ \sqrt{-b^2}$ is not a real number.
The trinomial is nonfactorable over the integers. No

(iii) $a^2x^2+2abx+b^2;\ \sqrt{a^2x^2}=ax;\ \sqrt{b^2}=b;$
$2(ax\cdot b)=2abx$
The trinomial is a perfect square.
$a^2x^2+2ab+b^2=(ax+b)^2$ Yes

(iv) $a^2x^2+2abx-b^2;\ \sqrt{a^2x^2}=ax;\ \sqrt{-b^2}$ is not a real number.
The trinomial is nonfactorable over the integers. No

Critical Thinking 10.4

57. $x^2+6x+k \qquad \sqrt{x^2}=x;\ \sqrt{k}=\sqrt{k}$
$2(x\cdot\sqrt{k})=6x$
$2\sqrt{k}=6$
$\sqrt{k}=3$
$k=9$

59. $x^2-2x+k \qquad \sqrt{x^2}=x;\ \sqrt{k}=\sqrt{k}$
$-2(x\cdot\sqrt{k})=-2x$
$-2x\sqrt{k}=-2x$
$\sqrt{k}=1$
$k=1$

61. $x^2+10x+k \qquad \sqrt{x^2}=x;\ \sqrt{k}=\sqrt{k}$
$2(x\cdot\sqrt{k})=10x$
$2x\sqrt{k}=10x$
$\sqrt{k}=5$
$k=25$

Section 10.5

Concept Check

1. common

3. x^2-81

5. Determine if the polynomial is a perfect-square trinomial or try to "unFOIL" the trinomial.

7. To check factoring, multiply the factors to see if the product is the original polynomial.

Objective A Exercises

9. $12n^2-48$ The GCF is 12.
$12(n^2-4)=12[n^2-2^2]=12(n+2)(n-2)$

11. $4rs^2-4rs+r$ The GCF is r.
$4rs^2-4rs+r=r(4s^2-4s+1)$
$(4s^2-4s+1);\sqrt{4s^2}=2s;\sqrt{1}=1;-2(2s\cdot 1)=-4s$
$(4s^2-4s+1)=(2s-1)^2$
$4rs^2-4rs+r=r(2s-1)^2$

13. $81-t^4=9^2-(t^2)^2=(9+t^2)(9-t^2)$

$=(9+t^2)(3^2-t^2)$

$=(9+t^2)(3+t)(3-t)$

15. $x^2+10x+16$; $\sqrt{x^2}=x$; $\sqrt{16}=4$, $2(x\cdot 4)=8x$

Not a perfect square. $1\cdot 16=16$

Factors of 16 whose sum is 10: 8 and 2

$x^2+10x+16=x^2+8x+2x+16$

$=(x^2+8x)+(2x+16)$

$=x(x+8)+2(x+8)$

$=(x+8)(x+2)$

17. $36c^4-48c^3+16c^2$ The GCF is $4c^2$.

$36c^4-48c^3+16c^2=4c^2(9c^2-12c+4)$

$(9c^2-12c+4)$; $\sqrt{9c^2}=3c$; $\sqrt{4}=2$;

$-2(3c\cdot 2)=-12c$

$(9c^2-12c+4)=(3c-2)^2$

$36c^4-48c^3+16c^2=4c^2(3c-2)^2$

19. $32s^4-2=2(16s^4-1)=2\left[(4s^2)^2-1^2\right]$

$=2(4s^2+1)(4s^2-1)$

$=2(4s^2+1)\left[(2s)^2-1^2\right]$

$=2(4s^2+1)(2s+1)(2s-1)$

21. $9+24a+16a^2$; $\sqrt{9}=3$; $\sqrt{16a^2}=4a$;

$2(3\cdot 4a)=24a$

The trinomial is a perfect square.

$9+24a+16a^2=(3+4a)^2$

23. $(2x+5)^2-25=(2x+5)^2-5^2$

$=(2x+5+5)(2x+5-5)$

$=(2x+10)(2x)$

$=2(x+5)(2x)$

$=4x(x+5)$

25. $(4x^2+12x+9)-4y^2=(2x+3)^2-4y^2$

$=(2x+3)^2-(2y)^2$

$=(2x+3+2y)(2x+3-2y)$

27. $2x^2-18$ The GCF is 2.

$2x^2-18=2(x^2-9)$

Factor the difference of two squares.

$2(x^2-9)=2(x+3)(x-3)$

$2x^2-18=2(x+3)(x-3)$

29. y^3-10y^2+25y The GCF is y.

$y^3-10y^2+25y=y(y^2-10y+25)$

Factor the perfect-square trinomial.

$y(y^2-10y+25)=y(y-5)^2$

$y^3-10y^2+25=y(y-5)^2$

31. $a^4-11a^3+24a^2$ The GCF is a^2.

$a^4-11a^3+24a^2=a^2(a^2-11a+24)$

Factor the trinomial.

Factors	Sum
−1, −24	−25
−2, −12	−14
−3, −8	−11
−4, −6	−10

$a^2(a^2-11a+24)=a^2(a-3)(a-8)$

$a^4-11a^3+24a^2=a^2(a-3)(a-8)$

33. $6y^2-48y+72$ The GCF is 6.

$6y^2-48y+72=6(y^2-8y+12)$

Factor the trinomial.

$6(y^2-8y+12)=6(y-2)(y-6)$

$6y^2-48y+72=6(y-2)(y-6)$

35. There is no common factor.

Factor the trinomial $5a^2-30a+4$.

Factors of 3: 1, 5 Factors of 10: −1, −4
−2, −2

$5a^2-30a+4$ is nonfactorable over the integers.

37. $3a^2b+21ab-54b$ The GCF is $3b$.

$3a^2b+21ab-54b=3b\left(a^2+7a-18\right)$

Factor the trinomial.

Factors	Sum
−1,+18	+17
+1,−18	−17
−2, +9	+7
+2,−9	−7
−3, +9	+6
+3, −9	−6

$3b\left(a^2+7a-18\right)=3b(a+9)(a-2)$

$3a^2b+21ab-54b=3b(a+9)(a-2)$

39. b^3-8b^2-7b The GCF is b.

$b^3-8b^2-7b=b\left(b^2-8b-7\right)$

Factor the trinomial.

Factors	Sum
−1,+7	+6
+1, −7	−6

The trinomial is nonfactorable.

$b^3-8b^2-7b=b\left(b^2-8b-7\right)$

41. $3y^2-147$ The GCF is 3.

$3y^2-147=3\left(y^2-49\right)$

Factor the difference of squares.

$3\left(y^2-49\right)=3\left(y^2-7^2\right)=3(y+7)(y-7)$

$3y^2-147=3(y+7)(y-7)$

43. $12a^2-36a+27$ The GCF is 3.

$12a^2-36a+27=3\left(4a^2-12a+9\right)$

Factor the perfect-square trinomial.

$3\left(4a^2-12a+9\right)=3(2a-3)^2$

$12a^2-3a+17=3(2a-3)^2$

45. $a^2b^2+3a^2b-88a^2$ The GCF is a^2.

$a^2b^2+3a^2b-88a^2=a^2\left(b^2+3b-88\right)$

Factor the trinomial.

Factors	Sum
−1, +8	+87
+1, −88	−87
−2, +44	+42
+2, −44	−42
−4, +22	+18
+4, −22	−18
−8, +11	+3
+8, − 11	−3

$a^2\left(b^2+3b-88\right)=a^2(b+11)(b-8)$

$a^2b^2+3a^2b-88a^2=a^2(b+11)(b-8)$

47. $16x^2-32xy+12y^2$ The GCF is 4.

$16x^2-32xy+12y^2=4\left(4x^2-8xy+3y^2\right)$

Factor the trinomial.

Factors of 4	Factors of 3
1, 4	−1, −3
2, 2	

$4\left(4x^2-8xy+3y^2\right)=4(2x-y)(2x-3y)$

$16x^2-32xy+12y^2=4(2x-y)(2x-3y)$

49. $72-2x^2$ The GCF is −2.

$72-2x^2=-2\left(x^2-36\right)$

Factor the difference of two squares.

$-2\left(x^2-36\right)=-2\left(x^2-6^2\right)=-2(x+6)(x-6)$

$72-2x^2=-2(x+6)(x-6)$

51. $a^2b^2+6ab^2+9b^2$ The GCF is b^2.

$a^2b^2+6ab^2+9b^2=b^2\left(a^2+6a+9\right)$

Factor the perfect-square trinomial.

$b^2\left(a^2+6a+9\right)=b^2(a+3)^2$

$a^2b^2+6ab^2+9b^2=b^2(a+3)^2$

53. $2x^3y-7x^2y^2+6xy^3$ The GCF is xy.

$2x^3y-7x^2y^2+6xy^3=xy\left(2x^2-7xy+6y^2\right)$

Factor the trinomial.

Factors of 2	Factors of −3
1, 2	−1, −6
	−2, −3

$xy\left(2x^2-7xy+5y^2\right)=xy(x-2y)(2x-3y)$

$2x^3y-7x^2y^2+6xy^3=xy(x-2y)(2x-3y)$

55. $18a^3+24a^2+8a$ The GCF is $2a$.

$18a^3+24a^2+8a=2a\left(9a^2+12a+4\right)$

Factor the perfect-square trinomial.

$2a\left(9a^2+12a+4\right)=2a(3a+2)^2$

$18a^3+24a^2+8a=2a(3a+2)^2$

57. $75+27y^2$ The GCF is 3.

$75+27y^2=3\left(25+9y^2\right)$

$25+9y^2$ is nonfactorable over the integers.

$75+27y^2=3\left(25+9y^2\right)$

59. $24x^3-66x^2+15x$ The GCF is $3x$.

$24x^3-66x^2+15x=3x\left(8x^2-22x+5\right)$

Factor the trinomial.

Factors of 8	Factors of 5
1, 8	−1, −5
2, 4	

$3x\left(8x^2-22x+5x\right)=3x(2x-5)(4x-1)$

$24x^3-66x^2+15x=3x(2x-5)(4x-1)$

61. $2a^3-8a^2b+8ab^2$ The GCF is $2a$.

$2a^3-8a^2b+8ab^2=2a\left(a^2-4ab+4b^2\right)$

Factor the perfect-square trinomial.

$2a\left(a^2-4ab+4b^2\right)=2a(a-2b)^2$

$2a^3-8a^2b+8ab^2=2a(a-2b)^2$

63. $a^2b^2-6ab^2+9b^2$ The GCF is b^2.

$a^2b^2-6ab^2+9b^2=b^2\left(a^2-6a+9\right)$

Factor the perfect-square trinomial.

$b^2\left(a^2-6a+9\right)=b^2(a-3)^2$

$a^2b^2-6ab^2+9b^2=b^2(a-3)^2$

65. $21x^2-11x^2-2x^4$ The GCF is $-x^2$.

$21x^2-11x^3-2x^4=-x^2\left(2x^2+11x-21\right)$

Factor the trinomial.

Factors of 2	Factors of −21
1, 2	−1, +21
2, 4	+1, −21
	−3, +7
	+3, −7

$x^2\left(2x^2+11x-21\right)=-x^2(2x-3)(x+7)$

$21x^2-11x^2-2x^4=-x^2(2x-3)(x+7)$

67. $b^4-a^2b^2$ The GCF is b^2.

$b^4-a^2b^2=b^2\left(b^2-a^2\right)$

Factor the difference of two squares.

$b^2\left(b^2-a^2\right)=b^2(b+a)(b-a)$

$b^4-a^2b^2=b^2(b+a)(b-a)$

69. $32xy^2-48xy+18x$ The GCF is $2x$.

$32xy^2-48xy+18x=2x\left(16y^2-24y+9\right)$

Factor the perfect-square trinomial.

$2x\left(16y^2-24y+9\right)=2x(4y-3)^2$

$32xy^2-48xy+18x=2x(4y-3)^2$

71. $15y^2-2xy^2-x^2y^2$ The GCF is $-y^2$.

$15y^2-2xy^2-x^2y^2=-y^2\left(x^2+2x-15\right)$

Factor the trinomial.

Factors of −15	Sum
−1, +15	+14
+1, −15	−14
−3, +5	+2
+3, −5	−2

$-y^2\left(x^2+2x-15\right)=-y^2(x-3)(x+5)$

$15y^2-2xy^2-x^2y^2=-y^2(x-3)(x+5)$

73. $3x^2-27y^2$ The GCF is 3.

$3\left(x^2-9y^2\right)$

Factor the difference of two squares.

$3\left(x^2-9y^2\right)=3\left[x^2-(3y)^2\right]=3(x+3y)(x-3y)$

$3x^2-27y^2=3(x+3y)(x-3y)$

75. y^3-9y The GCF is y.

$y^3-9y^2=y\left(y^2-9\right)$

Factor the difference of two squares.

$y\left(y^2-9\right)=y\left(y^2-3^2\right)=y(y+3)(y+3)$

$y^3-9y=y(y+3)(y-3)$

77. $15x^4y^2 - 13x^3y^3 - 20x^2y^4$ The GCF is x^2y^2.

$15x^4y^2 - 13x^3y^3 - 20x^2y^4$

$= x^2y^2\left(15x^2 - 13xy - 20y^2\right)$

Factor the trinomial.

Factors of 15	Factors of −20
1, 15	−1, +20
3, 5	+1, −20
	−2, +10
	+2, −10
	−4, +5
	+4, −5

$x^2y^2\left(15x^2 - 13xy - 20y^2\right) = x^2y^2(5x+4y)(3x-5y)$

$15x^4y^2 - 13x^3y^3 - 20x^2y^4 = x^2y^2(5x+4y)(3x-5y)$

79. The common binomial factor is $2x - 2$.

$a(2x-2)+b(2x-2)=(2x-2)(a+b)$

Factor 2 from the binomial factor.

$(2x-2)(a+b)=2(x-1)(a+b)$

$a(2x-2)+b(2x-2)=2(x-1)(a+b)$

81. The common binomial factor is $x - 2$.

$x^2(x-2)-(x-2)=(x-2)\left(x^2-1\right)$

Factor the difference of two squares.

$(x-2)\left(x^2-1^2\right)=(x-2)(x+1)(x-1)$

$x^2(x-2)-(x-2)=(x-2)(x+1)(x-1)$

83. The common binomial factor is x^2-4.

$a\left(x^2-4\right)+b\left(x^2-4\right)=\left(x^2-4\right)(a+b)$

Factor the difference of two squares.

$\left(x^2-2^2\right)(a+b)=(x+2)(x-2)(a+b)$

$a\left(x^2-4\right)+b\left(x^2-4\right)=(x+2)(x-2)(a+b)$

85. The common binomial factor is $x - 5$.

$4(x-5)-x^2(x-5)=(x-5)\left(4-x^2\right)$

Factor the difference of two squares.

$(x-5)\left(2^2-x^2\right)=(x-5)(2+x)(2-x)$

$4(x-5)-x^2(x-5)=(x-5)(2+x)(2-x)$

Critical Thinking 10.5

87. $(4x-3)^2 - y^2 = [(4x-3)-y][(4x-3)+y]$

$= (4x-3-y)(4x-3+y)$

89. $\left(x^2-4x+4\right)-y^2=(x-2)^2-y^2$

$=(x-2-y)(x-2+y)$

91. 48

1·48	1 is a perfect square, 48 is not prime
2·24	no perfect squares
3·16	16 is a perfect square, 3 is prime
4·12	4 is a perfect square, 12 is not

3+16 = 19

The sum is 19.

Projects and Group Activities 10.5

93. For example: 7, 8, 9

Square of the middle number: $8^2 = 64$

Product of the first and third numbers: 7·9 = 63

The square of the middle number is one more than the product of the first and third number.

Let $n-1$, n, and $n+1$ represent three consecutive numbers.

Square of the middle number: n^2

Product of the first and the third number:

$(n-1)(n+1)=n^2-1$

The square of the middle number is one more than the product of the first and third number.

Section 10.6

Concept Check

1. **a.** Yes **b.** No **c.** Yes

3. **a.** Yes **b.** Yes **c.** No **d.** Yes **e.** No **f.** Yes

Objective A Exercises

5. $(y+3)(y+2)=0$

$y+3=0 \quad y+2=0$

$y=-3 \quad y=-2$

The solutions are −2 and −3.

7. $(z-7)(z-3)=0$

$z-7=0 \quad z-3=0$

$z=7 \quad z=3$

The solutions are 3 and 7.

9. $x(x-5)=0$

$x=0 \quad x-5=0$

$x=5$

The solutions are 0 and 5.

11. $a(a-9)=0$

$a=0 \quad a-9=0$

$a=9$

The solutions are 0 and 9.

13. $y(2y+3)=0$

$y=0 \quad 2y+3=0$

$2y=-3$

$y=-\frac{3}{2}$

The solutions are 0 and $-\frac{3}{2}$.

15. $2a(3a-2)=0$

$2a=0 \quad 3a-2=0$

$a=0 \quad 3a=2$

$a=\frac{2}{3}$

The solutions are 0 and $\frac{2}{3}$.

17. $(b+2)(b-5)=0$

$b+2=0 \quad b-5=0$

$b=-2 \quad b=5$

The solutions are −2 and 5.

19. $x^2-81=0$

$(x+9)(x-9)=0$

$x+9=0 \quad x-9=0$

$x=-9 \quad x=9$

The solutions are −9 and 9.

21. $4x^2-49=0$

$(2x+7)(2x-7)=0$

$2x+7=0 \quad 2x-7=0$

$2x=-7 \quad 2x=7$

$x=-\frac{7}{2} \quad x=\frac{7}{2}$

The solutions are $-\frac{7}{2}$ and $\frac{7}{2}$

23. $9x^2-1=0$

$(3x+1)(3x-1)=0$

$3x+1=0 \quad 3x-1=0$

$3x=-1 \quad 3x=1$

$x=-\frac{1}{3} \quad x=\frac{1}{3}$

The solutions are $-\frac{1}{3}$ and $\frac{1}{3}$.

25. $x^2+6x+8=0$

$(x+2)(x+4)=0$

$x+2=0 \quad x+4=0$

$x=-2 \quad x=-4$

The solutions are −2 and −4.

27. $z^2+5z-14=0$

$(z+7)(z-2)=0$

$z+7=0 \quad z-2=0$

$z=-7 \quad z=2$

The solutions are −7 and 2.

29. $2a^2-9a-5=0$

$(2a+1)(a-5)=0$

$2a+1=0 \quad a-5=0$

$2a=-1 \quad a=5$

$a=-\frac{1}{2}$

The solutions are $-\frac{1}{2}$ and 5.

31. $6z^2+5z+1=0$

$(3z+1)(2z+1)=0$

$3z+1=0 \quad 2z+1=0$

$3z=-1 \quad 2z=-1$

$z=-\frac{1}{3} \quad z=-\frac{1}{2}$

The solutions are $-\frac{1}{3}$ and $-\frac{1}{2}$.

33. $x^2-3x=0$

$x(x-3)=0$

$x=0 \quad x-3=0$

$x=3$

The solutions are 0 and 3.

35. $x^2-7x=0$

$x(x-7)=0$

$x=0 \quad x-7=0$

$x=7$

The solutions are 0 and 7.

37. $a^2+5a=-4$

$a^2+5a+4=0$

$(a+1)(a+4)=0$

$a+1=0 \quad a+4=0$

$a=-1 \quad a=-4$

The solutions are −1 and −4.

39.
$$y^2 - 5y = -6$$
$$y^2 - 5y + 6 = 0$$
$$(y-2)(y-3) = 0$$
$$y - 2 = 0 \quad y - 3 = 0$$
$$y = 2 \qquad y = 3$$
The solutions are 2 and 3.

41.
$$2t^2 + 7t = 4$$
$$2t^2 + 7t - 4 = 0$$
$$(2t - 1)(t + 4) = 0$$
$$2t - 1 = 0 \quad t + 4 = 0$$
$$2t = 1 \qquad t = -4$$
$$t = \frac{1}{2}$$
The solutions are $\frac{1}{2}$ and −4.

43.
$$3t^2 - 13t = -4$$
$$3t^2 - 13t + 4 = 0$$
$$(3t - 1)(t - 4) = 0$$
$$3t - 1 = 0 \quad t - 4 = 0$$
$$3t = 1 \qquad t = 4$$
$$t = \frac{1}{3}$$
The solutions are $\frac{1}{3}$ and 4.

45.
$$x(x - 12) = -27$$
$$x^2 - 12x = -27$$
$$x^2 - 12x + 27 = 0$$
$$(x - 3)(x - 9) = 0$$
$$x - 3 = 0 \quad x - 9 = 0$$
$$x = 3 \qquad x = 9$$
The solutions are 3 and 9.

47.
$$y(y - 7) = 18$$
$$y^2 - 7y = 18$$
$$y^2 - 7y - 18 = 0$$
$$(y + 2)(y - 9) = 0$$
$$y + 2 = 0 \quad y - 9 = 0$$
$$y = -2 \qquad y = 9$$
The solutions are −2 and 9.

49.
$$p(p + 3) = -2$$
$$p^2 + 3p = -2$$
$$p^2 + 3p + 2 = 0$$
$$(p + 1)(p + 2) = 0$$
$$p + 1 = 0 \quad p + 2 = 0$$
$$p = -1 \qquad p = -2$$
The solutions are −1 and −2.

51.
$$y(y + 4) = 45$$
$$y^2 + 4y = -45$$
$$y^2 + 4y - 45 = 0$$
$$(y + 9)(y - 5) = 0$$
$$y + 9 = 0 \quad y - 5 = 0$$
$$y = -9 \qquad y = 5$$
The solutions are −9 and 5.

53.
$$x(x+3) = 28$$
$$x^2 + 3x = 28$$
$$x^2 + 3x - 28 = 0$$
$$(x+7)(x-4) = 0$$
$$x + 7 = 0 \quad x - 4 = 0$$
$$x = -7 \qquad x = 4$$
The solutions are −7 and 4.

55.
$$(x+8)(x-3) = -30$$
$$x^2 + 5x - 24 = -30$$
$$x^2 + 5x + 6 = 0$$
$$(x+2)(x+3) = 0$$
$$x + 2 = 0 \quad x + 3 = 0$$
$$x = -2 \qquad x = -3$$
The solutions are −2 and −3.

57.
$$(z - 5)(z + 4) = 52$$
$$z^2 - z - 20 = 52$$
$$z^2 - z - 72 = 0$$
$$(z + 8)(z - 9) = 0$$
$$z + 8 = 0 \quad z - 9 = 0$$
$$z = -8 \qquad z = 9$$
The solutions are −8 and 9.

59.
$$(z - 6)(z + 1) = -10$$
$$z^2 - 5z - 6 = -10$$
$$z^2 - 5z + 4 = 0$$
$$(z - 1)(z - 4) = 0$$
$$z - 1 = 0 \quad z - 4 = 0$$
$$z = 1 \qquad z = 4$$
The solutions are 1 and 4.

61.
$$(a - 4)(a + 7) = -18$$
$$a^2 + 3a - 28 = -18$$
$$a^2 + 3a - 10 = 0$$
$$(a + 5)(a - 2) = 0$$
$$a + 5 = 0 \quad a - 2 = 0$$
$$a = -5 \qquad a = 2$$
The solutions are −5 and 2.

63. To have one positive solution and one negative solution, the factors must have one negative sign and one positive sign, so c must be less than zero.

Objective B Exercises

65. Strategy:
The positive number: x
The square of the positive number is six more than five times the positive number.

Solution:

$$x^2 = 5x + 6$$
$$x^2 - 5x - 6 = 0$$
$$(x - 6)(x + 1) = 0$$
$$x - 6 = 0 \quad x + 1 = 0$$
$$x = 6 \qquad x = -1$$

Because –1 is not a positive number, it is not a solution. The number is 6.

67. Strategy:
One of the numbers: x
The other number: $6 - x$
The sum of the squares of the numbers is twenty.

Solution:

$$x^2 + (-x + 6)^2 = 20$$
$$2x^2 - 12x + 36 = 20$$
$$2x^2 - 12x + 16 = 0$$
$$(2x - 4)(x - 4) = 0$$
$$2x - 4 = 0 \quad x - 4 = 0$$
$$x = 2 \qquad x = 4$$

The numbers are 2 and 4.

69. Strategy:
First positive integer: x
Next positive integer: $x + 1$
The sum of the squares of two consecutive positive integers is 113.

Solution:

$x^2 + (x + 1)^2 = 113$ is (ii).

70. −8 should be eliminated because the problem asked for only *positive* integers that satisfy the given condition.

71. Strategy:
The first positive integer: x
The next positive integer: $x + 1$
The sum of the squares of two consecutive positive integers is forty-one.

Solution:

$$x^2 + (x + 1)^2 = 41$$
$$2x^2 + 2x + 1 = 41$$
$$2x^2 + 2x - 40 = 0$$
$$(2x + 10)(x - 4) = 0$$
$$2x + 10 = 0 \quad x - 4 = 0$$
$$x = -5 \qquad x = 4$$

Because −5 is not a positive number, it is not a solution. The numbers are 4 and 5.

73. Strategy:
The first positive integer: x
The next positive integer: $x + 1$
The product of two consecutive positive integers is two hundred forty.

Solution:

$$x(x + 1) = 240$$
$$x^2 + x - 240 = 0$$
$$(x - 15)(x + 16) = 240$$
$$x - 15 = 0 \quad x + 16 = 0$$
$$x = 15 \qquad x = -16$$

Because −16 is not a positive number, it is not a solution. The numbers are 15 and 16.

75. Strategy:
Height of the triangle: x
Base of the triangle: $3x$
The area of the rectangle is 54 ft^2.
The equation for the area of a triangle is $A=\frac{1}{2}bh$. Substitute in the equation and solve for x.

Solution:

$$A=\frac{1}{2}bh$$
$$54=\frac{1}{2}3x\cdot x$$
$$54=\frac{3}{2}x^2$$
$$\frac{2}{3}\cdot 54=\frac{2}{3}\cdot\frac{3}{2}x^2$$
$$36=x^2$$
$$0=x^2-36$$
$$0=(x-6)(x+6)$$

$x-6=0 \quad x+6=0$
$x=6 \quad x=-6$

Because -6 is not positive, it can not be a solution.
$3\cdot 6=18$
The height is 6 ft.
The base is 18 ft.

77. Strategy:
Width of the rectangle: x
Length of the rectangle: $2x+2$
The area of the rectangle is144 ft^2.
The equation for the area of a rectangle is $A=LW$. Substitute in the equation and solve for x.

Solution:
$$A=LW$$
$$144=(2x+2)(x)$$
$$144=2x^2+2x$$
$$0=2x^2+2x-144$$
$$0=2(x^2+x-72)$$
$$0=2(x+9)(x-8)$$

$x+9=0 \quad x-8=0$
$x=-9 \quad x=8$

Because -9 is not positive, it can not be a solution.
$2x+2=2(8)+2=18$

The length is 8 ft.
The width is 18 ft.

79. Strategy:
Side of the original square: x
Side of the larger square: $x+4$
The area of the larger square is 64 cm^2.
The equation for the area of a square is $A=s^2$. Substitute in the equation and solve for x.

Solution:
$$A=s^2$$
$$64=(x+4)^2$$
$$64=x^2+8x+16$$
$$0=x^2+8x-48$$
$$0=(x+12)(x-4)$$

$x+12=0 \quad x-4=0$
$x=-12 \quad x=4$

Because the length of a side of a square cannot be a negative number, -12 is not a solution.
The length of a side of the original square is 4 m.

81. Strategy:
Radius of the original circle: x
Radius of the larger circle: $x + 3$
The equation for the area of a circle is $A = \pi r^2$. The area of the new circle is 100 in^2 more than the area of the original circle.

Solution:

$$\pi x^2 + 100 = \pi(x+3)^2$$
$$\pi x^2 + 100 = \pi\left(x^2 + 6x + 9\right)$$
$$\pi x^2 + 100 = \pi x^2 + 6\pi x + 9\pi$$
$$\pi x^2 - \pi x^2 + 100 = \pi x^2 + 6\pi x + 9\pi$$
$$100 - 9\pi = 6\pi x + 9\pi - 9\pi$$
$$100 - 9\pi = 6\pi x$$
$$\frac{100 - 9\pi}{6\pi} = \frac{6\pi x}{6\pi}$$
$$3.81 \approx x$$

The radius of the original circle was 3.81 in.

83. Strategy:
Length of the border : x
Width of type area: $6 - 2x$
Length of type area: $9 - 2x$
The area of the type area is 28 in^2.
The equation for the area of a square is $A = lw$. Substitute in the equation and solve for x.

Solution:

$$28 = (6-2x)(9-2x)$$
$$28 = 54 - 30x + 4x^2$$
$$0 = 4x^2 - 30x + 26$$
$$0 = 2\left(2x^2 - 15x + 13\right)$$
$$0 = 2(2x-13)(x-1)$$
$$2x - 13 = 0 \qquad x - 1 = 0$$
$$x = \frac{13}{2} \qquad x = 1$$

Since $6 - 2x = 6 - 2\left(\frac{13}{2}\right) = 6 - 13 = -7$, $\frac{13}{2}$ cannot be the border.
Width: $6 - 2(1) = 4$ in
Length: $9 - 2(1) = 7$ in

85. Strategy:
Width of the rectangle: x
Length of the rectangle: $x + 3$
The area of the rectangle is 304 ft^2.
The equation for the area of a rectangle is $A = LW$. Substitute in the equation and solve for x.

Solution:

$$A = LW$$
$$304 = (x+3)(x)$$
$$304 = x^2 + 3x$$
$$0 = x^2 + 3x - 304$$
$$0 = (x+19)(x-16)$$
$$x + 19 = 0 \qquad x - 16 = 0$$
$$x = -19 \qquad x = 16$$

Because -19 is not positive, it can not be a solution.
The width is 16 ft.

87. Strategy:
Known: $d = 320$; $v = 16$
Unknown: t

Solution:

$$d = vt + 16t^2$$
$$320 = 16t + 16t^2$$
$$0 = 16t^2 + 16t - 320$$
$$0 = 16\left(t^2 + t - 20\right)$$
$$0 = 16(t-4)(t+5)$$
$$t + 5 = 0 \qquad t - 4 = 0$$
$$t = -5 \qquad t = 4$$

Time cannot be negative, so -5 cannot be a solution. In 4 s, the object will hit the ground.

89. Strategy:
Known: $S = 120$
Unknown: n

Solution:

$$S = \frac{n^2 + n}{2}$$
$$120 = \frac{n^2 + n}{2}$$
$$240 = n^2 + n$$
$$n^2 + n - 240 = 0$$
$$(n+16)(n-15) = 0$$

$n + 16 = 0 \qquad n - 15 = 0$

$n = -16 \qquad n = 15$

There cannot be a negative number of numbers, so −16 is not a solution.
Fifteen consecutive numbers beginning with 1 will give a sum of 120.

91. Strategy:
Known: $N = 45$
Unknown: t

Solution:

$$N = \frac{t^2 - t}{2}$$
$$45 = \frac{t^2 - t}{2}$$
$$90 = t^2 - t$$
$$t^2 - t - 90 = 0$$
$$(t+9)(t-10) = 0$$

$t + 9 = 0 \qquad t - 10 = 0$

$t = -9 \qquad t = 10$

There cannot be a negative number of teams, so −9 is not a solution. There are 10 teams in the league.

93. Strategy:
Known: $h = 0$; $v = 48$
Unknown: t

Solution:

$$h = vt - 16t^2$$
$$0 = 48t - 16t^2$$
$$0 = -16t(t-3)$$

$t = 0 \qquad t - 3 = 0$

$t = 3$

The ball will return to the ground in 3 s.

Critical Thinking 10.6

95.
$$2y(y+4) = 3(y+4)$$
$$2y^2 + 8y = 3y + 12$$
$$2y^2 + 5y - 12 = 0$$
$$(2y-3)(y+4) = 0$$

$2y - 3 = 0 \qquad y - 4 = 0$

$2y = 3 \qquad y = 4$

$y = \frac{3}{2}$

The solutions are $\frac{3}{2}$ and −4.

97.
$$(b+5)^2 = 16$$
$$b^2 + 10b + 25 = 16$$
$$b^2 + 10b + 9 = 0$$
$$(b+1)(b+9) = 0$$

$b + 1 = 0 \qquad b + 9 = 0$

$b = -1 \qquad b = -9$

The solutions are −1 and −9.

99.
$$p^3 = 7p^2$$
$$p^3 - 7p^2 = 0$$
$$p^2(p-7) = 0$$

$p^2 = 0 \qquad p - 7 = 0$

$p = 0 \qquad p = 7$

The solutions are 0 and 7.

101.
$$(x+3)(2x-1) = (3-x)(5-3x)$$
$$2x^2 + 5x - 3 = 15 - 14x + 3x^2$$
$$-x^2 + 19x - 18 = 0$$
$$x^2 - 19x + 18 = 0$$
$$(x-1)(x-18) = 0$$

$x - 1 = 0 \qquad x - 18 = 0$

$x = 1 \qquad x = 18$

The solutions are 1 and 18.

103.
$$n(n+3) = 4$$
$$n^2 + 3n - 4 = 0$$
$$(n+4)(n-1) = 0$$

$n + 4 = 0 \qquad n - 1 = 0$

$n = -4 \qquad n = 1$

$2n^2 = 2(-4)^3 \qquad 2n^2 = 2(1)^3$

$= 2(-64) \qquad = 2(1)$

$= -128 \qquad = 2$

$2n^3$ equals −128 or 2.

105. The error of accidentally dividing by zero takes place in the solution

$$x^2 = x$$
$$\frac{x^2}{x} = \frac{x}{x}$$
$$x = 1$$

Since one of the solutions of $x^2 = x$ is $x = 0$, an error occurs when the quotients in the equation $\frac{x^2}{x} = \frac{x}{x}$ are introduced. Here is the correct solution.

$$x^2 = x$$
$$x^2 - x = 0$$
$$x(x-1) = 0$$
$$x = 0 \quad x - 1 = 0$$
$$x = 1$$

The solutions are 0 and 1.

Projects and Group Activities 10.6

107. Strategy:
Width: x
Length: $x + 10$
After 2 in is cut off both ends.
New width: $x - 4$
New length: $x + 10 - 4 = x + 6$
Height: 2

Solution:
$V = lwh$.

$$2(x+6)(x-4) = 192$$
$$2(x^2 + 2x - 24) = 192$$
$$2x^2 + 4x - 48 = 192$$
$$2x^2 + 4x - 240 = 0$$
$$2(x^2 + 2x - 120) = 0$$
$$2(x-10)(x+12) = 0$$
$$x - 10 = 0 \qquad x + 12 = 0$$
$$x = 10 \qquad x = -12$$

The width cannot be –12.
Width: 10 cm
Length: 10 + 10 = 20 cm

Chapter Review

1.
$$\begin{aligned} b^2 - 13b + 30 &= b - 10b - 3b + 30 \\ &= (b - 10b) + (-3b + 30) \\ &= b(b-10) - 3(b-10) \\ &= (b-10)(b-3) \end{aligned}$$

2.
$$\begin{aligned} 4x(x-3) - 5(3-x) &= 4x(x-3) + 5(x-3) \\ &= (x-3)(4x+5) \end{aligned}$$

3. $2x^2 - 5x + 6$

Factors of 2	*Negative Factors of* 6
1 and 2	−1 and −6
	−2 and −3

Trial Factors	*Middle Term*
$(x-1)(2x-6)$	Common factor
$(x-2)(2x-3)$	$-3x - 4x = -7x$
$(2x-1)(x-6)$	$-12x - x = -13x$
$(2x-2)(x-3)$	Common factor

$2x^2 - 5x + 6$ is nonfactorable over the integers.

4. $5x^3 + 10x^2 + 35x$

The GCF is $5x$: $5x(x^2 + 2x + 7)$

5. $14y^9 - 49y^6 + 7y^3$

The GCF is $7y^3$: $7y^3(2y^6 - 7y^3 + 1)$

6. $y^2 + 5y - 36$

Factors of −36 whose sum is 5: 9 and −4

$$\begin{aligned} y^2 + 9y - 4y - 36 &= y(y+9) - 4(y+9) \\ &= (y+9)(y-4) \end{aligned}$$

7. $6x^2 - 29x + 28$

Factors of 6	*Negative Factors of* 28
1 and 6	−1 and −28
2 and 3	−2 and −14
	−4 and −7

Trial Factors	*Middle Term*
$(x-1)(6x-28)$	Common factor
$(x-4)(6x-7)$	$-7x - 24x = -31x$
$(2x-1)(3x-28)$	$-56x - 6x = -62x$
$(2x-4)(3x-7)$	Common factor
$(2x-7)(3x-4)$	$-8x - 21x = -29x$

$6x^2 - 29x + 28 = (2x-7)(3x-4)$

8. $12a^2b + 3ab^2$

The GCF is $3ab$: $3ab(4a + b)$

9. $a^6 - 100 = (a^3)^2 - 10^2 = (a^3 + 10)(a^3 - 10)$

10. $n^4 - 2n^3 - 3n^2$

The GCF is n^2: $n^2(n^2 - 2n - 3)$

Factors of –3 whose sum is –2: –3 and 1

$$n^2(n^2 - 3n + n - 3) = n^2[n(n-3) + 1(n-3)]$$
$$= n^2(n-3)(n+1)$$

11. $12y^2 + 16y - 3$

Factors of 12	*Factors of* –3
1 and 12	1 and –3
2 and 6	–1 and 3
3 and 4	

Trial Factors	*Middle Term*
$(3y+1)(4y-3)$	$-9y + 4y = -5y$
$(3y-1)(4y+3)$	$9y - 4y = 5y$
$(2y+1)(6y-3)$	$-6y + 6y = 0$
$(6y-1)(2y+3)$	$18y - 2y = 16y$

$12y^2 + 16y - 3 = (6y-1)(2y+3)$

12. $12b^3 - 58b^2 + 56b$

The GCF is $2b$: $2b(6b^2 - 29b + 28)$

$6 \cdot 28 = 168$

Factors of 168 whose sum is –29: –21 and –8

$$2b(6b^2 - 21b - 8b + 28)$$
$$2b[3b(2b-7) - 4(2b-7)]$$
$$2b(2b-7)(3b-4)$$

$12b^3 - 58b^2 + 56b = 2b(2b-7)(3b-4)$

13. $$9y^4 - 25z^2 = (3y^2)^2 - (5z)^2$$
$$= (3y^2 + 5z)(3y^2 - 5z)$$

14. $c^2 + 8c + 12$

Factors of 12 whose sum is 8: 6 and 2

$$c^2 + 6c + 2c + 12 = (c^2 + 6c) + (2c + 12)$$
$$= c(c+6) + 2(c+6)$$
$$= (c+6)(c+2)$$

15. $18a^2 - 3a - 10$

$18(-10) = -180$

Factors of –180 whose sum is –3: –15 and 12

$$18a^2 - 15a + 12a - 10 = 3a(6a-5) + 2(6a-5)$$
$$= (6a-5)(3a+2)$$
$$18a^2 - 3a - 10 = (6a-5)(3a+2)$$

16. $$4x^2 + 27x = 7$$
$$4x^2 + 27x - 7 = 0$$
$$(4x-1)(x+7) = 0$$

$4x - 1 = 0 \quad x + 7 = 0$

$4x = 1 \quad x = -7$

$x = \frac{1}{4}$

The solutions are $\frac{1}{4}$ and –7.

17. $4x^3 - 20x^2 - 24x$

The GCF is 4x: $4x(x^2 - 5x - 6)$

Factors of –6 whose sum is –5: –6 and 1

$$4x(x^2 - 5x - 6) = 4x(x^2 - 6x + x - 6)$$
$$= 4x[(x^2 - 6x) + (x - 6)]$$
$$= 4x[x(x-6) + 1(x-6)]$$
$$= 4x(x-6)(x+1)$$

18. $3a^2 - 15a - 42$

The GCF is 3: $3(a^2 - 5a - 14)$

Factors of –14	Sum
–1, 14	+13
+1, –14	–13
–2, + 7	+5
+2, –7	–5

$3(a^2 - 5a - 14) = 3(a+2)(a-7)$

$3a^2 - 15a - 42 = 3(a+2)(a-7)$

19. $2a^2 - 19a - 60$

$2(-60) = -120$

Factors of –120 whose sum is –19: –24 and 5

$$2a^2 - 24a + 5a - 60$$
$$= (2a^2 - 24a) + (5a - 60)$$
$$= 2a(a-12) + 5(a-12)$$
$$= (a-12)(2a+5)$$
$$2a^2 - 19a - 60 = (a-12)(2a+5)$$

20. $$(x+1)(x-5) = 16$$
$$x^2 - 4x - 5 = 16$$
$$x^2 - 4x - 21 = 0$$
$$(x-7)(x+3) = 0$$

$x - 7 = 0 \quad x + 3 = 0$

$x = 7 \quad x = -3$

The solutions are 7 and –3.

21. $21ax - 35bx - 10by + 6ay$

$= (21ax - 35bx) + (-10by + 6ay)$

$= 7x(3a - 5b) + 2y(-5b + 3a)$

$= 7x(3a - 5b) + 2y(3a - 5b)$

$= (3a - 5b)(7x + 2y)$

22. $a^2b^2 - 1 = (ab)^2 - 1^2 = (ab + 1)(ab - 1)$

23. $10x^2 + 25x + 4xy + 10y$

$= (10x^2 + 25x) + (4xy + 10y)$

$= 5x(2x + 5) + 2y(2x + 5)$

$= (2x + 5)(5x + 2y)$

24. $5x^2 - 5x - 30$

The GCF is 5: $5(x^2 - x - 6)$

$1(-6) = -6$ Factors of −6 whose sum is −1: −3 and 2

$5(x^2 - 3x + 2x - 6)$

$= 5\left[(x^2 - 3x) + (2x - 6)\right]$

$= 5[x(x - 3) + 2(x - 3)] = 5(x - 3)(x + 2)$

$5x^2 - 5x - 30 = 5(x - 3)(x + 2)$

25. $3x^2 + 36x + 108$

The GCF is 3: $3(x^2 + 12x + 36)$

$\sqrt{x^2} = x$ $2(x \cdot 6) = 12x$

$\sqrt{36} = 6$ The trinomial is a perfect square.

$3x^2 + 36x + 108 = 3(x + 6)^2$

26. $3x^2 - 17x + 10$

$3 \cdot 10 = 30$

Factors of 30 whose sum is −17: −15 and −2

$3x^2 - 15x - 2x + 10$

$= (3x^2 - 15x) + (2x + 10)$

$= 3x(x - 5) - 2(x - 5) = (x - 5)(3x - 2)$

$3x^2 - 17x + 10 = (x - 5)(3x - 2)$

27. Strategy:

→Width: x

→Length: $2x - 20$

→Use the equation for the area of a rectangle:

$A = LW$

Solution:

$LW = A$

$x(2x - 20) = 6000$

$2x^2 - 20x - 6000 = 0$

$2(x^2 - 10x - 3000) = 0$

$2(x - 60)(x + 50) = 0$

$x - 60 = 0$ $x + 50 = 0$

$x = 60$ $x = -50$

Because the width of a rectangle cannot be a negative number, −50 is not a solution.

$2x - 20 = 120 - 20 = 100$

The width is 60 yd and the length is 100 yd.

28. Strategy:

Known: $S = 400$

Unknown: d

Solution:

$S = d^2$

$400 = d^2$

$0 = d^2 - 400$

$0 = (d + 20)(d - 20)$

$d + 20 = 0$ $d - 20 = 0$

$d = -20$ $d = 20$

Since the distance cannot be negative, −20 is not a solution.

The distance is 20 ft.

29. Strategy:

→Width of the picture frame: x

New width: $12+2x$

New length: $15+2x$

→Use the equation for the area of a rectangle.

Solution:

$$LW = A$$
$$(12+2x)(15+2x) = 270$$
$$180+24x+30x+4x^2 = 270$$
$$4x^2+54x-90 = 0$$
$$2(2x^2+27x-45) = 0$$
$$2(2x-3)(x+15) = 0$$

$2x-3=0 \qquad x+15=0$

$2x=3 \qquad x=-15$

$x=1.5$

Because the width of a picture frame cannot be a negative number, –15 is not a solution. The width of the frame is 1.5 in.

30. Strategy:

→First consecutive positive integer: n

→Second consecutive positive integer: $n+1$

→The sum of the squares is 41.

Solution:

$$n^2+(n+1)^2 = 41$$
$$n^2+n^2+2n+1 = 41$$
$$2n^2+2n-40 = 0$$
$$2(n^2+n-20) = 0$$
$$2(n+5)(n-4) = 0$$

$x+5=0 \qquad x-4=0$

$x=-5 \qquad x=4$

Since –5 is not positive, –5 is not a solution. The integers are 4 and 5.

Chapter Test

1. $ab+6a-3b-18$

$=(ab+6a)+(-3b-18)$

$=a(b+6)-3(b+6)=(b+6)(a-3)$

2. $2y^4-14y^3-16y^2$

$=2y^2(y^2-7y-8)$

$=2y^2[y^2-8y+y-8]$

$=2y^2[(y^2-8y)+(y-8)]$

$=2y^2[y(y-8)+(y-8)]$

$=2y^2(y+1)(y-8)$

3. $8x^2+20x-48$

$8(-48)=-384$

Factors of –384 whose sum is 20: 32 and –12

$8x^2+20x-48$

$=8x^2-12x+32x-48$

$=(8x^2-12x)+(32x-48)$

$=4x(2x-3)+16(2x-3)$

$=(4x+16)(2x-3)=4(x+4)(2x-3)$

4. There is no common factor.

Factor the trinomial.

Positive Factors of 6	*Positive Factors of* 8
1 and 6	1 and 8
2 and 3	2 and 4

Trial Factors	*Middle Term*
$(x+1)(6x+8)$	$8x+6x=14x$
$(x+8)(6x+1)$	$x+48x=49x$
$(2x+1)(3x+8)$	$16x+3x=19x$
$(2x+8)(3x+1)$	$2x+24x=26x$

$6x^2+19x+8=(2x+1)(3x+8)$

5.

Factors	*Sum*
$-1, -48$	-49
$-2, -24$	-26
$-3, -16$	-19
$-4, -12$	-16
$-6, -8$	-14

$a^2-19a+48=(a-3)(a-16)$

6. $6x^3-8x^2+10x=2x(3x^2-4x+5)$

7.

Factors	*Sum*
$-1, +15$	$+14$
$+1, -15$	-14
$-3, +5$	$+2$
$+3, -5$	-2

$x^2+2x-15=(x+5)(x-3)$

8.
$$4x^2 - 1 = 0$$
$$(2x - 1)(2x + 1) = 0$$

$$2x - 1 = 0 \quad 2x + 1 = 0$$
$$x = \frac{1}{2} \quad x = -\frac{1}{2}$$

The solutions are $-\frac{1}{2}$ and $\frac{1}{2}$.

9. $5x^2 - 45x - 15 = 5(x^2 - 9x - 3)$

10. $p^2 + 12p + 36 = (p + 6)^2$

11.
$$x(x - 8) = -15$$
$$x^2 - 8x = -15$$
$$x^2 - 8x + 15 = 0$$
$$(x - 3)(x - 5) = 0$$
$$x - 3 = 0 \quad x - 5 = 0$$
$$x = 3 \quad x = 5$$

The solutions are 3 and 5.

12.
$$3x^2 + 12xy + 12y^2 = 3(x^2 + 4xy + 4y^2)$$
$$3(x + 2y)(x + 2y) = 3(x + 2y)^2$$

13. $b^2 - 16 = (b + 4)(b - 4)$

14. $6x^2y^2 + 9xy^2 + 3y^2$

Common Factor $3y^2$:

$3y^2(2x^2 + 3x + 1)$

$2 \cdot 2 = 2$

Factors of 2 whose sum is 3: 2, 1

$$3y^2(2x^2 + 3x + 1) = 3y^2(2x^2 + 2x + x + 1)$$
$$= 3y^2\left[(2x^2 + 2x) + (x + 1)\right]$$
$$= 3y^2\left[2x(x + 1) + (x + 1)\right]$$
$$= 3y^2(2x + 1)(x + 1)$$

15.
$$p^2 + 5p + 6 = p + 2p + 3p + 6$$
$$= (p + 2p) + (3p + 6)$$
$$= p(p + 2) + 3(p + 2)$$
$$= (p + 2)(p + 3)$$

16. $a(x - 2) + b(x - 2) = (x - 2)(a + b)$

17. $x(p + 1) - (p + 1) - (p + 1)(x - 1)$

18.
$$3a^2 - 75 = 3(a^2 - 25)$$
$$= 3(a + 5)(a - 5)$$

19. There is no common factor.

Factor the trinomial.

Factors of 2	*Factors of* −5
1 and 2	−1 and 5
	1 and −5

Trial Factors	*Middle Term*
$(x - 1)(2x + 5)$	$5x - 2x = 3x$
$(x + 1)(2x - 5)$	$-5x + 2x = -3x$

$2x^2 + 4x - 5$ is nonfactorable over the integers.

20.

Factors	*Sum*
−1, +36	+35
+1, −36	−35
−2, +18	+16
+2, −18	−16
−3, +12	+9
+3, −12	−9
−4, +9	+5
+4, −9	−5
−6, +6	0

$x^2 - 9x - 36 = (x + 3)(x - 12)$

21. $4a^2 - 12ab + 9b^2 = (2a - 3b)^2$

22. $4x^2 - 49y^2 = (2x + 7y)(2x - 7y)$

23.
$$(2a - 3)(a + 7) = 0$$
$$2a - 3 = 0 \quad a + 7 = 0$$
$$a = \frac{3}{2} \quad a = -7$$

The solutions are $\frac{3}{2}$ and −7.

24. Strategy:

The first number: x

The second number: $10 - x$

The sum of the squares of the two numbers is fifty-eight.

Solution:

$$x^2 + (-x + 10)^2 = 58$$
$$2x^2 - 20x + 100 = 58$$
$$2x^2 - 20x + 42 = 0$$
$$(2x - 6)(x - 7) = 0$$
$$2x - 6 = 0 \quad x - 7 = 0$$
$$x = 3 \quad x = 7$$

The numbers are 3 and 7.

25. Strategy:
The width of the rectangle: x
The length of a rectangle is 3 cm more than twice its width.
The length of the triangle: $2w + 3$
The area of the rectangle is 90 cm^2.

Solution:

$$(2w+3)w = 90$$
$$2w^2 + 3w = 90$$
$$2w^2 + 3w - 90 = 0$$
$$(2w+15)(w-6) = 0$$
$$2w+15=0 \qquad w-6=0$$
$$w=-7.5 \qquad w=6$$

Because −7.5 is not positive, it cannot be a solution. Therefore the width is 6 cm and the length is 15 cm.

Cumulative Review Exercises

1. $$-2-(-3)-5-(-11) = -2+3-5+11 = 1-5+11 = -4+11 = 7$$

2. $$\begin{aligned}(3-7)^2 \div (-2) - 3(-4) &= (-4)^2 \div (-2) - 3(-4)\\ &= 16 \div (-2) - (-12)\\ &= -8+12 = 4\end{aligned}$$

3. $$\begin{aligned}&-2a^2 \div (2b) - c\\ &-2(-4)^2 \div [2(2)] - (-1)\\ &= -2(16) \div 4 + 1\\ &= -32 \div 4 + 1 = -8+1 = -7\end{aligned}$$

4. $$-\frac{3}{4}(-20x^2) = \frac{3 \cdot 20x^2}{4} = 15x^2$$

5. $$\begin{aligned}-2[4x-2(3-2x)-8x] &= -2[4x-6+4x-8x]\\ &= -2[-6] = 12\end{aligned}$$

6. $$\begin{aligned}-\frac{5}{7}x &= -\frac{10}{21}\\ -\frac{7}{5}\left(-\frac{5}{7}x\right) &= -\frac{7}{5}\left(-\frac{10}{21}\right)\\ x &= \frac{2}{3}\end{aligned}$$

The solution is $\frac{2}{3}$.

7. $$\begin{aligned}3x-2 &= 12-5x\\ 3x+5x &= 12+2\\ 8x &= 14\\ x &= \frac{14}{8}\\ x &= \frac{7}{4}\end{aligned}$$

The solution is $\frac{7}{4}$.

8. $$\begin{aligned}-2+4[3x-2(4-x)-3] &= 4x+2\\ -2+4[3x-8+2x-3] &= 4x+2\\ -2+4[5x-11] &= 4x+2\\ -2+20x-44 &= 4x+2\\ 20x-46 &= 4x+2\\ 16x &= 48\\ x &= 3\end{aligned}$$

The solution is 3.

9. $$\begin{aligned}P \cdot B &= A\\ 120\% \cdot B &= 54\\ 1.2B &= 54\\ B &= \frac{54}{1.2}\\ B &= 45\end{aligned}$$

The number is 45.

10. $$(-3a^3b^2)^2 = (-3)^2a^6b^4 = 9a^6b^4$$

11.
$$\begin{array}{rrrr} & x^2 & -\ 5x & +\ 4\\ & & x & +\ 2\\ \hline & 2x^2 & -\ 10x & +\ 8\\ x^3 & -\ 5x^2 & +\ 4x & \\ \hline x^3 & -\ 3x^2 & -\ 6x & +\ 8\end{array}$$

12.
$$\begin{array}{r} 4x\ +8 \\ 2x-3\overline{)8x^2+\ 4x-3} \\ \underline{8x^2-12x} \\ 16x-\ 3 \\ \underline{16x-24} \\ 21 \end{array}$$

$$(8x^2+4x-3) \div (2x-3) = 4x+8+\frac{21}{2x-3}$$

13. $$(x^{-4}y^3)^2 = x^{-8}y^6 = \frac{y^6}{x^8}$$

14. $$\begin{aligned}3a-3b-ax+bx &= (3a-3b)+(-ax+bx)\\ &= 3(a-b)-x(a-b)\\ &= (a-b)(3-x)\end{aligned}$$

15. $15xy^2 - 20xy^4$

The GCF is $5xy^2$: $5xy^2(3-4y^2)$

16. $x^2 - 5xy - 14y^2$

Factors of –14 whose sum is –5: –7 and 2

$$\begin{aligned} x^2 - 7xy + 2xy - 14y^2 &= (x^2 - 7xy) + (2xy - 14y^2) \\ &= x(x-7y) + 2y(x-7y) \\ &= (x-7y)(x+2y) \end{aligned}$$

17. $$\begin{aligned} p^2 - 9p - 10 &= p^2 - 10p + p - 10 \\ &= (p^2 - 10p) + (p-10) \\ &= p(p-10) + (p-10) \\ &= (p-10)(p+1) \end{aligned}$$

18. $$\begin{aligned} 18a^3 + 57a^2 + 30a &= 3a(6a^2 + 19a + 10) \\ &= 3a(3a+2)(2a+5) \end{aligned}$$

19. $$\begin{aligned} 36a^2 - 49b^2 &= (6a)^2 - (7b)^2 \\ &= (6a+7b)(6a-7b) \end{aligned}$$

20. $4x^2 + 28xy + 49y^2$ $\quad \sqrt{4x^2} = 2x; \quad \sqrt{49y^2} = 7y$

$2(2x \cdot 7y) = 28xy$

The trinomial is a perfect square.

$4x^2 + 28xy + 49y^2 = (2x+7y)^2$

21. $9x^2 + 15x - 14$

$9(-14) = -126$

Factors of –126 whose sum is 15: 21 and –6

$$\begin{aligned} 9x^2 + 15x - 14 &= 9x^2 + 21x - 6x - 14 \\ &= (9x^2 + 21x) + (-6x - 14) \\ &= 3x(3x+7) - 2(3x+7) \\ &= (3x+7)(3x-2) \end{aligned}$$

22. $18x^2 - 48xy + 32y^2$ The GCF is 2

$2(9x^2 - 24xy + 16y^2)$

$\sqrt{9x^2} = 3x; \quad \sqrt{16y^2} = 4y \quad -2(3x \cdot 4y) = -24xy$

The trinomial is a perfect square.

$18x^2 - 48xy + 32y^2 = 2(3x-4y)^2$

23. $3y(x-3) - 2(x-3) = (x-3)(3y-2)$

24. $3x^2 + 19x - 14 = 0$

$(3x-2)(x+7) = 0$

$3x - 2 = 0 \quad x + 7 = 0$

$3x = 2 \qquad x = -7$

$x = \dfrac{2}{3}$

The solutions are $\dfrac{2}{3}$ and –7.

25. Strategy:

→Shorter piece: x

Longer piece: $10 - x$

→Four times the length of the shorter piece is 2 ft less than three times the length of the longer piece.

Solution:

$$\begin{aligned} 4x &= 3(10-x) - 2 \\ 4x &= 30 - 3x - 2 \\ 4x &= 28 - 3x \\ 7x &= 28 \\ x &= 4 \end{aligned}$$

$10 - x = 10 - 4 = 6$

The shorter piece is 4 ft long and the longer piece is 6 ft long.

26. Strategy:

To find the percent, use the basic percent equation.

Percent = n, base = 165, amount = 99

Solution:

Percent · base = amount

$n \cdot 165 = 99$

$n = \dfrac{99}{165}$

$n = 0.6 = 60\%$

The sale price is 60% of the regular price.

27. Strategy:

→To find the measure of $\angle a$, use the fact that $\angle a$ and the 72° angle are the alternate interior angles of parallel lines.

→To find the measure of b, use the fact that $\angle a$ and $\angle b$ are supplementary angles.

Solution:

$m\angle a = 72^\circ$

$m\angle a + m\angle b = 180^\circ$

$72^\circ + m\angle b = 180^\circ$

$m\angle b = 108^\circ$

$m\angle a = 72^\circ$

$m\angle b = 108^\circ$

28. Strategy:

→Time driving to resort: x

	Rate	Time	Distance
To resort	42	x	$42x$
From resort	56	$7-x$	$56(7-x)$

→The distances are the same.

Solution:

$42x = 56(7-x)$

$42x = 392 - 56x$

$98x = 392$

$x = 4$

$42x = 42(4) = 168$

The distance to the resort is 168 mi.

29. Strategy:

→First integer: n, Middle integer: $n + 2$

Third integer: $n + 4$

→Five times the middle integer is twelve more than twice the sum of the first and third.

Solution:

$5(n+2) = 2(n+n+4)+12$

$5n+10 = 2(2n+4)+12$

$5n+10 = 4n+8+12$

$5n+10 = 4n+20$

$n = 10$

$n+2 = 12$

$n+4 = 14$

The integers are 10, 12, and 14.

30. Strategy:

→Height: x

Base: $3x$

→Use the equation for the area of a triangle.

Solution:

$\frac{1}{2}bh = A$

$\frac{1}{2}(3x)(x) = 24$

$\frac{3}{2}x^2 = 24$

$\frac{2}{3}\left(\frac{3}{2}x^2\right) = 24 \cdot \frac{2}{3}$

$x^2 = 16$

$x = 4$

$3x = 3(4) = 12$

The base of the triangle is 12 in.

Chapter 11: Rational Expressions

Prep Test

1. $12 = 2 \cdot 2 \cdot 3$
   $18 = 2 \cdot 3 \cdot 3$
   $\text{LCM} = 2 \cdot 2 \cdot 3 \cdot 3 = 36$

2. $\dfrac{9x^3y^4}{3x^2y^7} = \dfrac{9x^{3-2}}{3y^{7-4}} = \dfrac{3x}{y^3}$

3. $\dfrac{3}{4} - \dfrac{8}{9} = \dfrac{3 \cdot 9}{4 \cdot 9} - \dfrac{8 \cdot 4}{9 \cdot 4} = \dfrac{27}{36} - \dfrac{32}{36} = -\dfrac{5}{36}$

4. $\left(-\dfrac{8}{11}\right) \div \dfrac{4}{5} = \left(-\dfrac{8}{11}\right)\dfrac{5}{4} = -\dfrac{10}{11}$

5. No. $\dfrac{0}{a} = 0$, but $\dfrac{a}{0}$ is undefined.

6. $$\frac{2}{3}x - \frac{3}{4} = \frac{5}{6}$$
   $$12\left(\frac{2}{3}x - \frac{3}{4}\right) = \left(\frac{5}{6}\right) \cdot 12$$
   $$12 \cdot \frac{2}{3}x - 12 \cdot \frac{3}{4} = \frac{5}{6} \cdot 12$$
   $$8x - 9 = 10$$
   $$8x = 19$$
   $$x = \frac{19}{8}$$
   The solution is $\dfrac{19}{8}$.

7. $x^2 - 4x - 12 = (x-6)(x+2)$

8. $2x^2 - x - 3 = (2x-3)(x+1)$

9. Strategy:
   →Anthony's time running: t
   →Jean's time running: $t - 10$

	Rate	Time	Distance
Anthony	9	t	$9t$
Jean	12	$t - 10$	$12(t - 10)$

   →The distance traveled is the same.

   Solution:
   $$9t = 12(t-10)$$
   $$9t = 12t - 120$$
   $$-3t = -120$$
   $$t = 40$$
   Jean will catch up with Anthony at 9:40 A.M.

Section 11.1

Concept Check

1. A rational expression is a fraction whose numerator and denominator are polynomials. For example: $\dfrac{x^2-2}{3x}$

3. If x were equal to 4, then the denominator $x - 4$ would equal to zero and division by zero is undefined.

Objective A Exercises

5. $\dfrac{9x^3}{12x^4} = \dfrac{9}{12x^{4-3}} = \dfrac{3}{4x}$

7. $\dfrac{(x+3)^2}{(x+3)^3} = \dfrac{1}{(x+3)^{3-2}} = \dfrac{1}{x+3}$

9. $\dfrac{3n-4}{4-3n} = \dfrac{-\overset{1}{\cancel{(4-3n)}}}{\underset{1}{\cancel{4-3n}}} = -1$

11. $\dfrac{6y(y+2)}{9y^2(y+2)} = \dfrac{6\overset{1}{\cancel{(y+2)}}}{9y^{2-1}\underset{1}{\cancel{(y+2)}}} = \dfrac{2}{3y}$

13. $\dfrac{6x(x-5)}{8x^2(5-x)}=\dfrac{6\cancel{(x-5)}}{8x^{2-1}(-1)\cancel{(x-5)}}=-\dfrac{3}{4x}$

15. $\dfrac{\alpha^2+4a}{ab+4b}=\dfrac{a\cancel{(a+4)}}{b\cancel{(a+4)}}=\dfrac{a}{b}$

17. $\dfrac{4-6x}{3x^2-2x}=\dfrac{-1(2)\cancel{(3x-2)}}{x\cancel{(3x-2)}}=-\dfrac{2}{x}$

19. $\dfrac{y^2-3y+2}{y^2-4y+3}=\dfrac{\cancel{(y-1)}(y-2)}{\cancel{(y-1)}(y-3)}=\dfrac{y-2}{y-3}$

21. $\dfrac{x^2+3x-10}{x^2+2x-8}=\dfrac{(x+5)\cancel{(x-2)}}{(x+4)\cancel{(x-2)}}=\dfrac{x+5}{x+4}$

23. $\dfrac{x^2+x-12}{x^2-6x+9}=\dfrac{(x+4)\cancel{(x-3)}}{(x-3)\cancel{(x-3)}}=\dfrac{x+4}{x-3}$

25. $\dfrac{x^2-3x-10}{25-x^2}=\dfrac{(x+2)\cancel{(x-5)}}{-1(x+5)\cancel{(x-5)}}=-\dfrac{x+2}{x+5}$

27. $\dfrac{2x^3+2x^2-4x}{x^3+2x^2-3x}$

$=\dfrac{2x(x^2+x-2)}{x(x^2+2x-3)}=\dfrac{2x(x+2)\cancel{(x-1)}}{x(x+3)\cancel{(x-1)}}=\dfrac{2(x+2)}{x+3}$

29. $\dfrac{6x^2-7x+2}{6x^2+5x-6}=\dfrac{\cancel{(3x-2)}(2x-1)}{\cancel{(3x-2)}(2x+3)}=\dfrac{2x-1}{2x+3}$

31. $\dfrac{x^2+3x-28}{24-2x-x^2}=\dfrac{(x+7)\cancel{(x-4)}}{-1(x+6)\cancel{(x-4)}}=-\dfrac{x+7}{x+6}$

Objective B Exercises

33. $\dfrac{8x^2}{9y^3}\cdot\dfrac{3y^2}{4x^3}=\dfrac{8\cdot 3}{9y^{3-2}4x^{3-2}}=\dfrac{2}{3xy}$

35. $\dfrac{12x^3y^4}{7a^2b^3}\cdot\dfrac{14a^3b^4}{9x^2y^2}=\dfrac{12x^{3-2}y^{4-2}\cdot 14a^{3-2}b^{4-3}}{7\cdot 9}=\dfrac{8abxy^2}{3}$

37. $\dfrac{3x-6}{5x-20}\cdot\dfrac{10x-40}{27x-54}=\dfrac{3(x-2)}{5(x-4)}\cdot\dfrac{10(x-4)}{27(x-2)}$

$=\dfrac{3\cancel{(x-2)}10\cancel{(x-4)}}{5\cancel{(x-4)}\cdot 27\cancel{(x-2)}}=\dfrac{2}{9}$

39. $\dfrac{3x^2+2x}{2xy-3y}\cdot\dfrac{2xy^3-3y^3}{3x^3+2x^2}$

$=\dfrac{x(3x+2)}{y(2x-3)}\cdot\dfrac{y^3(2x-3)}{x^2(3x+2)}$

$=\dfrac{\cancel{(3x+2)}}{\cancel{(2x-3)}}\cdot\dfrac{y^{3-2}\cancel{(2x-3)}}{x^{2-1}\cancel{(3x+2)}}=\dfrac{y^2}{x}$

41. $\dfrac{x^2+5x+4}{x^3y^2}\cdot\dfrac{x^2y^3}{x^2+2x+1}$

$=\dfrac{(x+4)\cancel{(x+1)}}{x^{3-2}}\cdot\dfrac{y^{3-2}}{\cancel{(x+1)}(x+1)}=\dfrac{y(x+4)}{x(x+1)}$

43. $\dfrac{x^4y^2}{x^2+3x-28}\cdot\dfrac{x^2-49}{xy^4}$

$=\dfrac{x^{4-1}}{\cancel{(x+7)}(x-4)}\cdot\dfrac{(x-7)\cancel{(x+7)}}{y^{4-2}}=\dfrac{x^3(x-7)}{y^2(x-4)}$

45. $\dfrac{2x^2-5x}{2xy+y}\cdot\dfrac{2xy^2+y^2}{5x^2-2x^3}$

$=\dfrac{x(2x-5)}{y(2x+1)}\cdot\dfrac{y^2(2x+1)}{-x^2(2x-5)}$

$=\dfrac{\cancel{(2x-5)}}{\cancel{(2x+1)}}\cdot\dfrac{y^{2-1}\cancel{(2x+1)}}{-1x^{2-1}\cancel{(2x-5)}}=-\dfrac{y}{x}$

47. $\frac{x^2-2x-24}{x^2-5x-6}\cdot\frac{x^2+5x+6}{x^2+6x+8}$

$=\frac{\cancel{(x+4)}\cancel{(x-6)}}{(x+1)\cancel{(x-6)}}\cdot\frac{\cancel{(x+2)}(x+3)}{\cancel{(x+2)}\cancel{(x+4)}}=\frac{x+3}{x+1}$

49. $\frac{x^2+2x-35}{x^2+4x-21}\cdot\frac{x^2+3x-18}{x^2+9x+18}$

$=\frac{\cancel{(x+7)}(x-5)}{\cancel{(x+7)}\cancel{(x-3)}}\cdot\frac{\cancel{(x+6)}\cancel{(x-3)}}{(x+3)\cancel{(x+6)}}=\frac{x-5}{x+3}$

51. $\frac{x^2-3x-4}{x^2+6x+5}\cdot\frac{x^2+5x+6}{8+2x-x^2}$

$=\frac{\cancel{(x-4)}\cancel{(x+1)}}{(x+5)\cancel{(x+1)}}\cdot\frac{\cancel{(x+2)}(x+3)}{\cancel{(4-x)}\cancel{(2+x)}}=-\frac{x+3}{x+5}$

53. $\frac{16+6x-x^2}{x^2-10x-24}\cdot\frac{x^2-6x-27}{x^2-17x+72}$

$=\frac{\cancel{(8-x)}\cancel{(2+x)}}{\cancel{(x+2)}(x-12)}\cdot\frac{(x+3)\cancel{(x-9)}}{\cancel{(x-8)}\cancel{(x-9)}}=-\frac{x+3}{x-12}$

55. $\frac{2x^2+5x+2}{2x^2+7x+3}\cdot\frac{x^2-7x-30}{x^2-6x-40}$

$=\frac{\cancel{(2x+1)}(x+2)}{\cancel{(2x+1)}\cancel{(x+3)}}\cdot\frac{\cancel{(x+3)}\cancel{(x-10)}}{(x+4)\cancel{(x-10)}}=\frac{x+2}{x+4}$

57. Since $a > d$, the x will be in the numerator, and since $c > b$, the y will also be in the numerator. The denominator will be 1.

59. Since $a < d$, the x will be in the denominator, and since $c = b$, there will be no y variable. The numerator will be 1.

Objective C Exercises

61. $\frac{9x^3y^4}{16a^4b^2}\div\frac{45x^4y^2}{14a^7b}$

$=\frac{9x^3y^4}{16a^4b^2}\cdot\frac{14a^7b}{45x^4y^2}=\frac{9y^{4-2}\cdot 14a^{7-4}}{16b^{2-1}\cdot 45x^{4-3}}=\frac{7a^3y^2}{40bx}$

63. $\frac{28x+14}{45x-30}\div\frac{14x+7}{30x-20}$

$=\frac{14(2x+1)}{15(3x-2)}\cdot\frac{10(3x-2)}{7(2x+1)}$

$=\frac{14\cancel{(2x+1)}\cdot 10\cancel{(3x-2)}}{15\cancel{(3x-2)}\cdot 7\cancel{(2x+1)}}=\frac{4}{3}$

65. $\frac{5a^2y+3a^2}{2x^3+5x^2}\div\frac{10ay+6a}{6x^3+15x^2}$

$=\frac{a^2(5y+3)}{x^2(2x+5)}\cdot\frac{3x^2(2x+5)}{2a(5y+3)}$

$=\frac{a^{2-1}\cancel{(5y+3)}}{x^2\cancel{(2x+5)}}\cdot\frac{3x^2\cancel{(2x+5)}}{2\cancel{(5y+3)}}=\frac{3a}{2}$

67. $\frac{x^3y^2}{x^2-3x-10}\div\frac{xy^4}{x^2-x-20}$

$=\frac{x^3y^2}{(x-5)(x+2)}\cdot\frac{(x-5)(x+4)}{xy^4}$

$=\frac{x^{3-1}}{\cancel{(x-5)}(x+2)}\cdot\frac{\cancel{(x-5)}(x+4)}{y^{4-2}}=\frac{x^2(x+4)}{y^2(x+2)}$

69. $\frac{x^2y^5}{x^2-11x+30}\div\frac{xy^6}{x^2-7x+10}$

$=\frac{x^{2-1}}{\cancel{(x-5)}(x-6)}\cdot\frac{(x-2)\cancel{(x-5)}}{y^{6-5}}=\frac{x(x-2)}{y(x-6)}$

71. $\frac{3x^2y-9xy}{a^2b}\div\frac{3x^2-x^3}{ab^2}$

$=\frac{3xy(x-3)}{a^2b}\cdot\frac{ab^2}{x^2(3-x)}$

$=\frac{3y\cancel{(x-3)}}{a^{2-1}}\cdot\frac{b^{2-1}}{x^{2-1}\cancel{(3-x)}}=-\frac{3by}{ax}$

73. $\frac{x^2+3x-40}{x^2+2x-35}\div\frac{x^2+2x-48}{x^2+3x-18}$

$=\frac{\cancel{(x+8)}\cancel{(x-5)}}{(x+7)\cancel{(x-5)}}\cdot\frac{(x+6)(x-3)}{\cancel{(x+8)}(x-6)}=\frac{(x+6)(x-3)}{(x+7)(x-6)}$

75. $\dfrac{y^2-y-56}{y^2+8y+7} \div \dfrac{y^2-13y+40}{y^2-4y-5}$

$= \dfrac{\overset{1}{\cancel{(y+7)}}\,\overset{1}{\cancel{(y-8)}}}{\underset{1}{\cancel{(y+1)}}\,\underset{1}{\cancel{(y+7)}}} \cdot \dfrac{\overset{1}{\cancel{(y+1)}}\,\overset{1}{\cancel{(y-5)}}}{\underset{1}{\cancel{(y-5)}}\,\underset{1}{\cancel{(y-8)}}} = 1$

77. $\dfrac{x^2-x-2}{x^2-7x+10} \div \dfrac{x^2-3x-4}{40-3x-x^2}$

$= \dfrac{\overset{1}{\cancel{(x+1)}}\,\overset{1}{\cancel{(x-2)}}}{\underset{1}{\cancel{(x-2)}}\,\underset{1}{\cancel{(x-5)}}} \cdot \dfrac{(8+x)\overset{-1}{\cancel{(5-x)}}}{\underset{1}{\cancel{(x+1)}}(x-4)} = -\dfrac{x+8}{x-4}$

79. $\dfrac{6n^2+13n+6}{4n^2-9} \div \dfrac{6n^2+n-2}{4n^2-1}$

$= \dfrac{\overset{1}{\cancel{(2n+3)}}\,\overset{1}{\cancel{(3n+2)}}}{\underset{1}{\cancel{(2n+3)}}(2n-3)} \cdot \dfrac{(2n+1)\overset{1}{\cancel{(2n-1)}}}{\underset{1}{\cancel{(3n+2)}}\,\underset{1}{\cancel{(2n-1)}}} = \dfrac{2n+1}{2n-3}$

81. $\dfrac{x+1}{x+6} \div \dfrac{x-1}{x-4} = \dfrac{x+1}{x+6} \cdot \dfrac{x-4}{x-1}$

$= \dfrac{(x+1)(x-4)}{(x+6)(x-1)} = \dfrac{x^2-3x-4}{x^2+5x-6}$ Yes

83. $\dfrac{x-1}{x+1} \div \dfrac{x-4}{x+6} = \dfrac{x-1}{x+1} \cdot \dfrac{x+6}{x-4}$

$= \dfrac{(x-1)(x+6)}{(x+1)(x-4)} = \dfrac{x^2+5x-6}{x^2-3x-4}$ No

Critical Thinking 11.1

85. Strategy:
To find where the expression is undefined, find where the denominator x^2-4x-5 is equal to zero.
Solution:

$x^2-4x-5=0$

$(x-5)(x+1)=0$

$x-5=0 \qquad x+1=0$

$x=5 \qquad x=-1$

When x is 5 or -1, the expression is undefined.

87. $\dfrac{\text{shaded area}}{\text{total area}} = \dfrac{\pi(2x)^2}{\pi(5x)^2} = \dfrac{4x^2\pi}{25x^2\pi} = \dfrac{4}{25}$

89. $\dfrac{2x^2+7x-4}{3x^2-8x-3} = \dfrac{(2x-1)(x+4)}{(3x+1)(x-3)} = \dfrac{2x-1}{3x+1} \cdot \dfrac{x+4}{x-3}$

or $\dfrac{2x-1}{x-3} \cdot \dfrac{x+4}{3x+1}$

Projects or Group Activities 11.1

91. Yes. For example: $x = 3.000000001$

$\dfrac{1}{y-3}$

$\dfrac{1}{3.000000001-3} = \dfrac{1}{0.000000001}$

$= 1{,}000{,}000{,}000 > 10{,}000{,}000$

Section 11.2

Concept Check

1. True

3. False

Objective A Exercises

5. $8x^3y = 2\cdot2\cdot2\cdot x\cdot x\cdot x\cdot y$

$12xy^2 = 2\cdot2\cdot3\cdot x\cdot y\cdot y$

$\text{LCM} = 2\cdot2\cdot2\cdot3\cdot x\cdot x\cdot x\cdot y\cdot y = 24x^3y^2$

7. $2x^2y = 2\cdot x\cdot x\cdot y$

$3x^2+12x = 3x(x+4)$

$\text{LCM} = 2\cdot3\cdot x\cdot x\cdot y(x+4) = 6x^2y(x+4)$

9. $9x(x+2) = 3\cdot3(x+2)$

$12(x+2)^2 = 2\cdot2\cdot3(x+2)(x+2)$

$\text{LCM} = 2\cdot2\cdot3\cdot3\cdot x(x+2)(x+2) = 36x(x+2)^2$

11. $3x+3 = 3(x+1)$

$2x^2+4x+2 = 2(x^2+2x+1) = 2(x+1)(x+1)$

$\text{LCM} = 2\cdot3\cdot(x+1)(x+1) = 6(x+1)^2$

13. $(x-1)(x+2)$

$(x-1)(x+3)$

$\text{LCM} = (x-1)(x+2)(x+3)$

15. $(2x+3)^2 = (2x+3)(2x+3)$

$(2x+3)(x-5)$

$\text{LCM} = (2x+3)(2x+3)(x-5)$

$= (2x+3)^2(x-5)$

17. $(x-1)$

$(x-2)$

$(x-1)(x-2)$

$\text{LCM} = (x-1)(x-2)$

19. $x^2-x-6=(x+2)(x-3)$

$x^2+x-12=(x-3)(x+4)$

$\text{LCM} = (x+2)(x-3)(x+4)$

21. $x^2+5x+4=(x+4)(x+1)$

$x^2-3x-28=(x+4)(x-7)$

$\text{LCM} = (x+1)(x+4)(x-7)$

23. $x^2-2x-24=(x-6)(x+4)$

$x^2-36=(x-6)(x+6)$

$\text{LCM} = (x-6)(x+4)(x+6)$

25. $2x^2-7x+3=(2x-1)(x-3)$

$2x^2+x-1=(2x-1)(x+1)$

$\text{LCM} = (2x-1)(x-3)(x+1)$

27. $6+x-x^2=-1(x^2-x-6)=-1(x-3)(x+2)$

$x+2$

$x-3$

$\text{LCM} = (x+2)(x-3)$

Objective B Exercises

29. The LCM is x^2.

$\frac{4}{x}=\frac{4}{x}\cdot\frac{x}{x}=\frac{4x}{x^2}$

$\frac{3}{x^2}$

31. The LCM is $12y^2$.

$\frac{x}{3y^2}=\frac{x}{3y^2}\cdot\frac{4}{4}=\frac{4x}{12y^2}$

$\frac{z}{4y}\cdot\frac{3y}{3y}=\frac{3yz}{12y^2}$

33. The LCM is $x^2(x-3)$.

$\frac{y}{x(x-3)}=\frac{y}{x(x-3)}\cdot\frac{x}{x}=\frac{xy}{x^2(x-3)}$

$\frac{6}{x^2}=\frac{6}{x^2}\cdot\frac{(x-3)}{(x-3)}=\frac{6x-18}{x^2(x-3)}$

35. The LCM is $x(x-1)^2$.

$\frac{9}{(x-1)^2}=\frac{9}{(x-1)^2}\cdot\frac{x}{x}=\frac{9x}{x(x-1)^2}$

$\frac{6}{x(x-1)}=\frac{6}{x(x-1)}\cdot\frac{(x-1)}{(x-1)}=\frac{6x-6}{x(x-1)^2}$

$\frac{5}{x(3-x)}=\frac{5}{-x(x-3)}=-\frac{5}{x(x-3)}$

37. The LCM is $x(x-3)$.

$\frac{3}{x-3}=\frac{3}{x-3}\cdot\frac{x}{x}=\frac{3x}{x(x-3)}$

$\frac{5}{x(3-x)}=-\frac{5}{x(x-3)}$

39. The LCM is $(x+3)(x-4)$.

$\frac{x-2}{x+3}=\frac{x-2}{x+3}\cdot\frac{x-4}{x-4}=\frac{x^2-6x+8}{(x+3)(x-4)}$

$\frac{x}{x-4}=\frac{x}{x-4}\cdot\frac{x+3}{x+3}=\frac{x^2+3x}{(x-4)(x+3)}$

41. The LCM is $(x+2)(x-1)$.

$\frac{3}{x^2+x-2}=\frac{3}{(x+2)(x-1)}$

$\frac{x}{x+2}=\frac{x}{x+2}\cdot\frac{x-1}{x-1}=\frac{x^2-x}{(x+2)(x-1)}$

43. The LCM is $(x+3)(x-2)(x-3)$.

$$\frac{x}{x^2+x-6}=\frac{x}{(x+3)(x-2)}\cdot\frac{x-3}{x-3}$$
$$=\frac{x^2-3x}{(x+3)(x-2)(x-3)}$$
$$\frac{2x}{x^2-9}=\frac{2x}{(x+3)(x-3)}\cdot\frac{x-2}{x-2}$$
$$=\frac{2x^2-4x}{(x+3)(x-2)(x-3)}$$

Critical Thinking 11.2

45. The LCM is 10^4.

$$\frac{3}{10^2}=\frac{3}{10^2}\cdot\frac{10^2}{10^2}=\frac{300}{10^4}$$
$$\frac{5}{10^4}$$

47. The LCM is b.

$$b=\frac{b}{1}\cdot\frac{b}{b}=\frac{b^2}{b}$$
$$\frac{5}{b}$$

49. The LCM of two polynomials is equal to the product of the two polynomials when the two polynomials have no common factors other than 1.

Section 11.3

Concept Check

1. False

3. False

Objective A Exercises

5. $\frac{3}{y^2}+\frac{8}{y^2}=\frac{11}{y^2}$

7. $\frac{3}{x+4}-\frac{10}{x+4}=\frac{-7}{x+4}=-\frac{7}{x+4}$

9. $\frac{3x}{2x+3}+\frac{5x}{2x+3}=\frac{8x}{2x+3}$

11. $\frac{2x+1}{x-3}+\frac{3x+6}{x-3}=\frac{5x+7}{x-3}$

13. $\frac{5x-1}{x+9}-\frac{3x+4}{x+9}=\frac{5x-1-(3x+4)}{x+9}$

$$=\frac{5x-1-3x-4}{x+9}=\frac{2x-5}{x+9}$$

15. $\frac{x-7}{2x+7}-\frac{4x-3}{2x+7}=\frac{x-7-(4x-3)}{2x+7}$

$$=\frac{x-7-4x+3}{2x+7}=\frac{-3x-4}{2x+7}$$

17. $\frac{x}{x^2+2x-15}-\frac{3}{x^2+2x-15}=\frac{x-3}{x^2+2x-15}$

$$=\frac{x-3}{(x+5)(x-3)}=\frac{1}{x+5}$$

19. $\frac{2x+3}{x^2-x-30}-\frac{x-2}{x^2-x-30}$

$$=\frac{2x+3-(x-2)}{x^2-x-30}=\frac{2x+3-x+2}{x^2-x-30}$$
$$=\frac{\overset{1}{\cancel{(x+5)}}}{\underset{1}{\cancel{(x+5)}}(x-6)}=\frac{1}{x-6}$$

21. $\frac{4y+7}{2y^2+7y-4}-\frac{y-5}{2y^2+7y-4}$

$$=\frac{4y+7-(y-5)}{2y^2+7y-4}$$
$$=\frac{4y+7-y+5}{y^2+7y-4}=\frac{3y+12}{(y+4)(2y-1)}$$
$$=\frac{3\overset{1}{\cancel{(y+4)}}}{\underset{1}{\cancel{(y+4)}}(2y-1)}=\frac{3}{2y-1}$$

23. $\frac{2x^2+3x}{x^2-9x+20}+\frac{2x^2-3}{x^2-9x+20}-\frac{4x^2+2x+1}{x^2-9x+20}$

$$=\frac{2x^2+3x+(2x^2-3)-(4x^2+2x+1)}{x^2-9x+20}$$
$$=\frac{2x^2+3x+2x^2-3-4x^2-2x-1}{x^2-9x+20}$$
$$=\frac{\overset{1}{\cancel{(x-4)}}}{(x-5)\underset{1}{\cancel{(x-4)}}}=\frac{1}{x-5}$$

25. $\frac{3}{y-5}-\frac{y-2}{y-5}=\frac{3-(y-2)}{y-5}$

$=\frac{-y+5}{y-5}=\frac{\overset{-1}{\cancel{(-y+5)}}}{\underset{1}{\cancel{(y-5)}}}=-1$

i. $\frac{5-y}{y-5}=\frac{\overset{-1}{\cancel{(5-y)}}}{\underset{1}{\cancel{(y-5)}}}=-1$ Yes

ii. $\frac{1-y}{y-5}$ No

iii. $\frac{5-y}{2y-10}=\frac{\overset{-1}{\cancel{(5-y)}}}{2\underset{1}{\cancel{(y-5)}}}=-\frac{1}{2}$ No

iv. -1 Yes

v. $\frac{1-y}{-10}$ No

Objective B Exercises

27. The LCM is xy.

$\frac{4}{x}+\frac{5}{y}=\frac{4}{x}\cdot\frac{y}{y}+\frac{5}{y}\cdot\frac{x}{x}$

$=\frac{4y}{xy}+\frac{5x}{xy}=\frac{4y+5x}{xy}$

29. The LCM is $2x$.

$\frac{12}{x}-\frac{5}{2x}=\frac{12}{x}.\frac{2}{2}-\frac{5}{2x}=\frac{24}{2x}-\frac{5}{2x}=\frac{19}{2x}$

31. The LCM is $12x$.

$\frac{1}{2x}-\frac{5}{4x}+\frac{7}{6x}=\frac{1}{2x}\cdot\frac{6}{6}-\frac{5}{4x}\cdot\frac{3}{3}+\frac{7}{6x}\cdot\frac{2}{2}$

$=\frac{6}{12x}-\frac{15}{12x}+\frac{14}{12x}$

$=\frac{6-15+14}{12x}=\frac{5}{12x}$

33. The LCM is $6x^2$.

$\frac{5}{3x}-\frac{2}{x^2}+\frac{3}{2x}=\frac{5}{3x}\cdot\frac{2x}{2x}-\frac{2}{x^2}\cdot\frac{6}{6}+\frac{3}{2x}\cdot\frac{3x}{3x}$

$=\frac{10x}{6x^2}-\frac{12}{6x^2}+\frac{9x}{6x^2}$

$=\frac{10x-12+9x}{6x^2}=\frac{19x-12}{6x^2}$

35. The LCM is $20xy$.

$\frac{2}{x}-\frac{3}{2y}+\frac{3}{5x}-\frac{1}{4y}$

$=\frac{2}{x}\cdot\frac{20y}{20y}-\frac{3}{2y}\cdot\frac{10x}{10x}+\frac{3}{5x}\cdot\frac{4y}{4y}-\frac{1}{4y}\cdot\frac{5x}{5x}$

$=\frac{40y}{20xy}-\frac{30x}{20xy}+\frac{12y}{20xy}-\frac{5x}{20xy}$

$=\frac{40y-30x+12y-5x}{20xy}=\frac{52y-35x}{20xy}$

37. The LCM is $15x$.

$\frac{2x+1}{3x}+\frac{x-1}{5x}=\frac{2x+1}{3x}\cdot\frac{5}{5}+\frac{x-1}{5x}\cdot\frac{3}{3}$

$=\frac{10x+5}{15x}+\frac{3x-3}{15x}$

$=\frac{10x+5+3x-3}{15x}=\frac{13x+2}{15x}$

39. The LCM is $24x$.

$\frac{x-3}{6x}+\frac{x+4}{8x}=\frac{x-3}{6x}\cdot\frac{4}{4}+\frac{x+4}{8x}\cdot\frac{3}{3}$

$=\frac{4x-12}{24x}+\frac{3x+12}{24x}$

$=\frac{4x-12+3x+12}{24x}$

$=\frac{7x}{24x}=\frac{7}{24}$

41. The LCM is $45x$.

$\frac{2x+9}{9x}-\frac{x-5}{5x}=\frac{2x+9}{9x}\cdot\frac{5}{5}-\frac{x-5}{5x}\cdot\frac{9}{9}$

$=\frac{10x+45}{45x}-\frac{9x-45}{45x}$

$=\frac{10x+45-(9x-45)}{45x}$

$=\frac{10x+45-9x+45}{45x}=\frac{x+90}{45x}$

43. The LCM is $2x^2$.

$\frac{x+4}{2x}-\frac{x-1}{x^2}=\frac{x+4}{2x}\cdot\frac{x}{x}-\frac{x-1}{x^2}\cdot\frac{2}{2}$

$=\frac{x^2+4x}{2x^2}-\frac{2x-2}{2x^2}$

$=\frac{x^2+4x-(2x-2)}{2x^2}$

$=\frac{x^2+4x-2x+2}{2x^2}=\frac{x^2+2x+2}{2x^2}$

45. The LCM is $4x^2$.

$$\frac{x-10}{4x^2}+\frac{x+1}{2x}=\frac{x-10}{4x^2}+\frac{x+1}{2x}\cdot\frac{2x}{2x}$$
$$=\frac{x-10}{4x^2}+\frac{2x^2+2x}{4x^2}$$
$$=\frac{x-10+2x^2+2x}{4x^2}$$
$$=\frac{2x^2+3x-10}{4x^2}$$

47. The LCM is $x+4$.

$$\frac{4}{x+4}+x=\frac{4}{x+4}-\frac{x}{1}\cdot\frac{x+4}{x+4}$$
$$=\frac{4}{x+4}-\frac{x^2+4x}{x+4}$$
$$=\frac{4-(x^2+4x)}{x+4}$$
$$=\frac{-x^2-4x+4}{x+4}$$

49. The LCM is $x+1$.

$$5-\frac{x-2}{x+1}=\frac{5}{1}\cdot\frac{x+1}{x+1}-\frac{x-2}{x+1}=\frac{5x+5}{x+1}-\frac{x-2}{x+1}$$
$$=\frac{5x+5-(x-2)}{x+1}$$
$$=\frac{5x+5-x+2}{x+1}=\frac{4x+7}{x+1}$$

51. The LCM is $24x^2$.

$$\frac{x+3}{6x}-\frac{x-3}{8x^2}=\frac{x+3}{6x}\cdot\frac{4x}{4x}-\frac{x-3}{8x^2}\cdot\frac{3}{3}$$
$$=\frac{4x^2+12x}{24x^2}-\frac{3x-9}{24x^2}$$
$$=\frac{4x^2+12x-(3x-9)}{24x^2}$$
$$=\frac{4x^2+12x-3x+9}{24x^2}$$
$$=\frac{4x^2+9x+9}{24x^2}$$

53. The LCM is xy^2.

$$\frac{3x-1}{xy^2}-\frac{2x+3}{xy}=\frac{3x-1}{xy^2}-\frac{2x+3}{xy}\cdot\frac{y}{y}$$
$$=\frac{3x-1}{xy^2}-\frac{2xy+3y}{xy^2}$$
$$=\frac{3x-1-(2xy+3y)}{xy^2}$$
$$=\frac{3x-1-2xy-3y}{xy^2}$$

55. The LCM is $24x^2y^2$.

$$\frac{5x+7}{6xy^2}-\frac{4x-3}{8x^2y}=\frac{5x+7}{6xy^2}\cdot\frac{4x}{4x}-\frac{4x-3}{8x^2y}\cdot\frac{3y}{3y}$$
$$=\frac{20x^2+28x}{24x^2y^2}-\frac{12xy-9y}{24x^2y^2}$$
$$=\frac{20x^2+28x-(12xy-9y)}{24x^2y^2}$$
$$=\frac{20x^2+28x-12xy+9y}{24x^2y^2}$$

57. The LCM is $18xy^2$.

$$\frac{3x-1}{6y^2}-\frac{x+5}{9xy}=\frac{3x-1}{6y^2}\cdot\frac{3x}{3x}-\frac{x+5}{9xy}\cdot\frac{2y}{2y}$$
$$=\frac{9x^2-3x}{18xy^2}-\frac{2xy+10y}{18xy^2}$$
$$=\frac{9x^2-3x-(2xy+10y)}{18xy^2}$$
$$=\frac{9x^2-3x-2xy-10y}{18xy^2}$$

59. The LCM is $(x-3)(x-4)$.

$$\frac{2}{x-3}+\frac{5}{x-4}=\frac{2}{x-3}\cdot\frac{x-4}{x-4}+\frac{5}{x-4}\cdot\frac{x-3}{x-3}$$
$$=\frac{2x-8}{(x-3)(x-4)}+\frac{5x-15}{(x-3)(x-4)}$$
$$=\frac{7x-23}{(x-3)(x-4)}$$

61. The LCM is $(y+6)(y-3)$.

$$\frac{3}{y+6}-\frac{4}{y-3}=\frac{3}{y+6}\cdot\frac{y-3}{y-3}-\frac{4}{y-3}\cdot\frac{y+6}{y+6}$$
$$=\frac{3y-9}{(y+6)(y-3)}-\frac{4y+24}{(y+6)(y-3)}$$
$$=\frac{3y-9-(4y+24)}{(y+6)(y-3)}$$
$$=\frac{3y-9-4y-24}{(y+6)(y-3)}$$
$$=\frac{-y-33}{(y+6)(y-3)}$$

63. The LCM is $(x-4)(x+6)$.

$$\frac{3x}{x-4}+\frac{2}{x+6}=\frac{3x}{x-4}\cdot\frac{x+6}{x+6}+\frac{2}{x+6}\cdot\frac{x-4}{x-4}$$
$$=\frac{3x^2+18x}{(x-4)(x+6)}+\frac{2x-8}{(x-4)(x+6)}$$
$$=\frac{3x^2+20x-8}{(x-4)(x+6)}$$

65. The LCM is $(x+5)(2x+3)$.

$$\frac{6x}{x+5}-\frac{3}{2x+3}=\frac{6x}{x+5}\cdot\frac{2x+3}{2x+3}-\frac{3}{2x+3}\cdot\frac{x+5}{x+5}$$
$$=\frac{12x^2+18x}{(x+5)(2x+3)}-\frac{3x+15}{(2x+3)(x+5)}$$
$$=\frac{12x^2+18x-(3x+15)}{(x+5)(2x+3)}$$
$$=\frac{12x^2+18x-3x-15}{(x+5)(2x+3)}$$
$$=\frac{12x^2+15x-15}{(x+5)(2x+3)}=\frac{3(4x^2+5x-5)}{(x+5)(2x+3)}$$

67. The LCM is $x-6$.

$$\frac{4x}{6-x}+\frac{5}{x-6}=\frac{4x}{-1(x-6)}+\frac{5}{x-6}$$
$$=-\frac{4x}{x-6}+\frac{5}{x-6}=\frac{-4x+5}{x-6}$$

69. The LCM is $(y+4)(y-4)$.

$$\frac{y}{y^2-16}+\frac{1}{y-4}$$
$$=\frac{y}{(y-4)(y+4)}+\frac{1}{y-4}\cdot\frac{y+4}{y+4}$$
$$=\frac{y}{(y-4)(y+4)}+\frac{y+4}{(y-4)(y+4)}$$
$$=\frac{2y+4}{(y-4)(y+4)}=\frac{2(y+2)}{(y-4)(y+4)}$$

71. The LCM is $(x+1)^2$.

$$\frac{(x-1)^2}{(x+1)^2}-1$$
$$=\frac{(x-1)(x-1)}{(x+1)^2}-\frac{1}{1}\cdot\frac{(x+1)^2}{(x+1)^2}$$
$$=\frac{x^2-2x+1}{(x+1)^2}-\frac{x^2+2x+1}{(x+1)^2}$$
$$=\frac{x^2-2x+1-(x^2+2x+1)}{(x+1)^2}$$
$$=\frac{x^2-2x+1-x^2-2x-1}{(x+1)^2}=\frac{-4x}{(x+1)^2}=-\frac{4x}{(x+1)^2}$$

73. The LCM is $(1-x)(1+x)$.

$$\frac{x}{1-x^2}-1+\frac{x}{1+x}$$
$$=\frac{x}{(1-x)(1+x)}-\frac{1}{1}\cdot\frac{(1-x)(1+x)}{(1-x)(1+x)}+\frac{x}{1+x}\cdot\frac{1-x}{1-x}$$
$$=\frac{x}{(1-x)(1+x)}-\frac{1-x^2}{(1-x)(1+x)}+\frac{x-x^2}{(1-x)(1+x)}$$
$$=\frac{x-(1-x^2)+x-x^2}{(1-x)(1+x)}=\frac{x-1+x^2+x-x^2}{(1-x)(1+x)}$$
$$=\frac{2x-1}{(1-x)(1+x)}$$

75. The LCM is $(x-5)(x-5)$.

$$\frac{3x-1}{x^2-10x+25}-\frac{3}{x-5}$$
$$=\frac{3x-1}{(x-5)(x-5)}-\frac{3}{x-5}\cdot\frac{x-5}{x-5}$$
$$=\frac{3x-1}{x^2-10x+25}-\frac{3x-15}{x^2-10x+25}$$
$$=\frac{3x-1-(3x-15)}{x^2-10x+25}=\frac{3x-1-3x+15}{(x-5)(x-5)}$$
$$=\frac{14}{(x-5)^2}$$

77. The LCM is $(x+6)(x-7)$.

$$\frac{x+4}{x^2-x-42}+\frac{3}{7-x}$$
$$=\frac{x+4}{(x-7)(x+6)}+\frac{3}{-1(x-7)}\cdot\frac{x+6}{x+6}$$
$$=\frac{x+4}{(x-7)(x+6)}-\frac{3x+18}{(x-7)(x+6)}$$
$$=\frac{x+4-(3x+18)}{(x-7)(x+6)}=\frac{x+4-3x-18}{(x-7)(x+6)}$$
$$=\frac{-2x-14}{(x-7)(x+6)}=\frac{-2(x+7)}{(x-7)(x+6)}$$

79. The LCM is $(x+1)(x-6)$.

$$\frac{1}{x+1}+\frac{x}{x-6}-\frac{5x-2}{x^2-5x-6}$$
$$=\frac{1}{x+1}\cdot\frac{x-6}{x-6}+\frac{x}{x-6}\cdot\frac{x+1}{x+1}-\frac{5x-2}{(x+1)(x-6)}$$
$$=\frac{x-6}{(x+1)(x-6)}+\frac{x^2+x}{(x+1)(x-6)}-\frac{5x-2}{(x+1)(x-6)}$$
$$=\frac{x-6+x^2+x-(5x-2)}{(x+1)(x-6)}$$
$$=\frac{x-6+x^2+x-5x+2}{(x+1)(x-6)}=\frac{x^2-3x-4}{(x+1)(x-6)}$$
$$=\frac{(x-4)\overset{1}{\cancel{(x+1)}}}{\underset{1}{\cancel{(x+1)}}(x-6)}=\frac{x-4}{x-6}$$

81. The LCM is $(x-1)(x-3)$.

$$\frac{3x+1}{x-1}-\frac{x-1}{x-3}+\frac{x+1}{x^2-4x+3}$$
$$=\frac{3x+1}{x-1}\cdot\frac{x-3}{x-3}-\frac{x-1}{x-3}\cdot\frac{x-1}{x-1}+\frac{x+1}{(x-1)(x-3)}$$
$$=\frac{3x^2-8x-3}{(x-1)(x-3)}-\frac{x^2-2x+1}{(x-1)(x-3)}+\frac{x+1}{(x-1)(x-3)}$$
$$=\frac{3x^2-8x-3-(x^2-2x+1)+x+1}{(x-1)(x-3)}$$
$$=\frac{3x^2-8x-3-x^2+2x-1+x+1}{(x-1)(x-3)}$$
$$=\frac{2x^2-5x-3}{(x-1)(x-3)}=\frac{(2x+1)\overset{1}{\cancel{(x-3)}}}{(x-1)\underset{1}{\cancel{(x-3)}}}=\frac{2x+1}{x-1}$$

83. The LCM is $(x-3)(x+7)$.

$$\frac{2x+9}{3-x}+\frac{x+5}{x+7}-\frac{2x^2+3x-3}{x^2+4x-21}$$
$$=\frac{2x+9}{-1(x-3)}\cdot\frac{x+7}{x+7}+\frac{x+5}{x+7}\cdot\frac{x-3}{x-3}-\frac{2x^2+3x-3}{(x-3)(x+7)}$$
$$=-\frac{2x^2+23x+63}{(x-3)(x+7)}+\frac{x^2+2x-15}{(x-3)(x+7)}-\frac{2x^2+3x-3}{(x-3)(x+7)}$$
$$=\frac{-(2x^2+23x+63)+x^2+2x-15-(2x^2+3x-3)}{(x-3)(x+7)}$$
$$=\frac{-2x^2-23x-63+x^2+2x-15-2x^2-3x+3}{(x-3)(x+7)}$$
$$=\frac{-3x^2-24x-75}{(x-3)(x+7)}=\frac{-3(x^2+8x+25)}{(x-3)(x+7)}$$

Critical Thinking 11.3

85. $\frac{5b+4a}{ab}=\frac{5b}{ab}+\frac{4a}{ab}=\frac{5}{a}+\frac{4}{b}$

87. $\frac{3x^2+4xy}{x^2y^2}=\frac{3x^2}{x^2y^2}+\frac{4xy}{x^2y^2}=\frac{3}{y^2}+\frac{4}{xy}$

Projects or Group Activities 11.3

89. a.

Strategy:

x = number of miles per gallon

→Determine the number of gallons used per year.

→Multiply the number of gallons by price per gallon.

Solution:

Number of gallons = miles traveled per years ÷ miles per gallon

$=12{,}000 \div x = \dfrac{12{,}000}{x}$

Amount spent on gasoline per year

$= \dfrac{12{,}000}{x} \cdot 3.70 = \dfrac{44{,}400}{x}$ dollars

b.

Strategy:

→Determine how much you would spend each year if you increase gas mileage by 5 mi/gal.

→Subtract the expression found below from the expression found in part a.

Solution:

$x + 5$ increase in miles per gallon

Number of gallons =

$12{,}000 \div (x+5) = \dfrac{12{,}000}{x+5}$

Amount spent on gasoline per year

$= \dfrac{12{,}000}{x+5} \cdot 3.70 = \dfrac{44{,}400}{x+5}$

$$\frac{44{,}400}{x} - \frac{44{,}400}{x+5} = \frac{44{,}400}{x} \cdot \frac{x+5}{x+5} - \frac{44{,}400}{x+5} \cdot \frac{x}{x}$$

$$= \frac{44{,}400x + 222{,}000}{x(x+5)} - \frac{44{,}400x}{x(x+5)}$$

$$= \frac{222{,}000}{x(x+5)} \text{ dollars}$$

c.

Strategy:

Evaluate the expression found in part b by using 25 for x.

Solution:

$$\frac{222{,}000}{x(x+5)} = \frac{222{,}000}{25(25+5)} = \frac{222{,}000}{25(30)} = 296$$

You would save $296 a year.

Section 11.4

Concept Check

1. $\dfrac{2}{2}$

3. $\dfrac{(x-1)(x+4)}{(x-1)(x+4)}$

5. True

Objective A Exercises

7. The LCM is x^2.

$$\frac{1+\frac{3}{x}}{1-\frac{9}{x^2}} = \frac{1+\frac{3}{x}}{1-\frac{9}{x^2}} \cdot \frac{x^2}{x^2} = \frac{1 \cdot x^2 + \frac{3}{x} \cdot x^2}{1 \cdot x^2 - \frac{9}{x^2} \cdot x^2}$$

$$= \frac{x^2+3x}{x^2-9} = \frac{x\overset{1}{\cancel{(x+3)}}}{(x-3)\underset{1}{\cancel{(x+3)}}} = \frac{x}{x-3}$$

9. The LCM is $x+4$.

$$\frac{2-\frac{8}{x+4}}{3-\frac{12}{x+4}} = \frac{2-\frac{8}{x+4}}{3-\frac{12}{x+4}} \cdot \frac{x+4}{x+4}$$

$$= \frac{2(x+4) - \frac{8}{x+4} \cdot (x+4)}{3(x+4) - \frac{12}{x+4} \cdot (x+4)}$$

$$= \frac{2x+8-8}{3x+12-12} = \frac{2x}{3x} = \frac{2}{3}$$

11. The LCM is $y-2$.

$$\frac{1+\frac{5}{y-2}}{1-\frac{2}{y-2}}=\frac{1+\frac{5}{y-2}}{1-\frac{2}{y-2}}\cdot\frac{y-2}{y-2}$$

$$=\frac{1\cdot(y-2)+\frac{5}{y-2}\cdot(y-2)}{1\cdot(y-2)-\frac{2}{y-2}\cdot(y-2)}$$

$$=\frac{y-2+5}{y-2-2}=\frac{y+3}{y-4}$$

13. The LCM is $x+7$.

$$\frac{4-\frac{2}{x+7}}{5+\frac{1}{x+7}}=\frac{4-\frac{2}{x+7}}{5+\frac{1}{x+7}}\cdot\frac{x+7}{x+7}$$

$$=\frac{4\cdot(x+7)-\frac{2}{x+7}\cdot(x+7)}{5\cdot(x+7)+\frac{1}{x+7}\cdot(x+7)}$$

$$=\frac{4x+28-2}{5x+35+1}=\frac{4x+26}{5x+36}=\frac{2(2x+13)}{5x+36}$$

15. The LCM is x^2.

$$\frac{1-\frac{1}{x}-\frac{6}{x^2}}{1-\frac{9}{x^2}}=\frac{1-\frac{1}{x}-\frac{6}{x^2}}{1-\frac{9}{x^2}}\cdot\frac{x^2}{x^2}$$

$$=\frac{1\cdot x^2-\frac{1}{x}\cdot x^2-\frac{6}{x^2}\cdot x^2}{1\cdot x^2-\frac{9}{x^2}\cdot x^2}$$

$$=\frac{x^2-x-6}{x^2-9}=\frac{\overset{1}{\cancel{(x-3)}}(x+2)}{\underset{1}{\cancel{(x-3)}}(x+3)}=\frac{x+2}{x+3}$$

17. The LCM is x^2.

$$\frac{1-\frac{5}{x}-\frac{6}{x^2}}{1+\frac{6}{x}+\frac{5}{x^2}}=\frac{1-\frac{5}{x}-\frac{6}{x^2}}{1+\frac{6}{x}+\frac{5}{x^2}}\cdot\frac{x^2}{x^2}$$

$$=\frac{1\cdot x^2-\frac{5}{x}\cdot x^2-\frac{6}{x^2}\cdot x^2}{1\cdot x^2+\frac{6}{x}\cdot x^2+\frac{5}{x^2}\cdot x^2}$$

$$=\frac{x^2-5x-6}{x^2+6x+5}=\frac{(x-6)\overset{1}{\cancel{(x+1)}}}{(x+5)\underset{1}{\cancel{(x+1)}}}=\frac{x-6}{x+5}$$

19. The LCM is x^2.

$$\frac{1-\frac{6}{x}+\frac{8}{x^2}}{\frac{4}{x^2}+\frac{3}{x}-1}=\frac{1-\frac{6}{x}+\frac{8}{x^2}}{\frac{4}{x^2}+\frac{3}{x}-1}\cdot\frac{x^2}{x^2}$$

$$=\frac{1\cdot x^2-\frac{6}{x}\cdot x^2+\frac{8}{x^2}\cdot x^2}{\frac{4}{x^2}\cdot x^2+\frac{3}{x}\cdot x^2-1\cdot x^2}$$

$$=\frac{x^2-6x+8}{4+3x-x^2}=\frac{(x-2)\overset{-1}{\cancel{(x-4)}}}{(1+x)\underset{1}{\cancel{(4-x)}}}$$

$$=\frac{-x+2}{x+1}$$

21. The LCM is $x+3$.

$$\frac{x-\frac{4}{x+3}}{1+\frac{1}{x+3}}=\frac{x-\frac{4}{x+3}}{1+\frac{1}{x+3}}\cdot\frac{x+3}{x+3}$$

$$=\frac{x\cdot(x+3)-\frac{4}{x+3}\cdot(x+3)}{1\cdot(x+3)+\frac{1}{x+3}\cdot(x+3)}$$

$$=\frac{x^2+3x-4}{x+3+1}=\frac{(x-1)\overset{1}{\cancel{(x+4)}}}{\underset{1}{\cancel{(x+4)}}}$$

$$=x-1$$

23. The LCM is $2x + 1$.

$$\frac{1-\frac{x}{2x+1}}{x-\frac{1}{2x+1}} = \frac{1-\frac{x}{2x+1}}{x-\frac{1}{2x+1}} \cdot \frac{2x+1}{2x+1}$$

$$= \frac{1(2x+1)-\frac{x}{2x+1}(2x+1)}{x\cdot(2x+1)-\frac{1}{2x+1}\cdot(2x+1)}$$

$$= \frac{2x+1-x}{2x^2+x-1} = \frac{\overset{1}{\cancel{(x+1)}}}{(2x-1)\underset{1}{\cancel{(x+1)}}}$$

$$= \frac{1}{2x-1}$$

25. The LCM is $x + 4$.

$$\frac{x-5+\frac{14}{x+4}}{x+3-\frac{2}{x+4}}$$

$$= \frac{x-5+\frac{14}{x+4}}{x+3-\frac{2}{x+4}} \cdot \frac{x+4}{x+4}$$

$$= \frac{x\cdot(x+4)-5\cdot(x+4)+\frac{14}{x+4}\cdot(x+4)}{x\cdot(x+4)+3\cdot(x+4)-\frac{2}{x+4}\cdot(x+4)}$$

$$= \frac{x^2+4x-5x-20+14}{x^2+4x+3x+12-2} = \frac{x^2-x-6}{x^2+7x+10}$$

$$= \frac{(x-3)\overset{1}{\cancel{(x+2)}}}{(x+5)\underset{1}{\cancel{(x+2)}}} = \frac{x-3}{x+5}$$

27. The LCM is $x - 6$.

$$\frac{x+3-\frac{10}{x-6}}{x+2-\frac{20}{x-6}}$$

$$= \frac{x+3-\frac{10}{x-6}}{x+2-\frac{20}{x-6}} \cdot \frac{x-6}{x-6}$$

$$= \frac{x\cdot(x-6)+3\cdot(x-6)-\frac{10}{x-6}\cdot(x-6)}{x\cdot(x-6)+2\cdot(x-6)-\frac{20}{x-6}\cdot(x-6)}$$

$$= \frac{x^2-6x+3x-18-10}{x^2-6x+2x-12-20} = \frac{x^2-3x-28}{x^2-4x-32}$$

$$= \frac{(x-7)\overset{1}{\cancel{(x+4)}}}{(x-8)\underset{1}{\cancel{(x+4)}}} = \frac{x-7}{x-8}$$

29. The LCM is $2y + 3$.

$$\frac{y-6+\frac{22}{2y+3}}{y-5+\frac{11}{2y+3}}$$

$$= \frac{y-6+\frac{22}{2y+3}}{y-5+\frac{11}{2y+3}} \cdot \frac{2y+3}{2y+3}$$

$$= \frac{y\cdot(2y+3)-6\cdot(2y+3)+\frac{22}{2y+3}\cdot(2y+3)}{y\cdot(2y+3)-5\cdot(2y+3)+\frac{11}{2y+3}\cdot(2y+3)}$$

$$= \frac{2y^2+3y-12y-18+22}{2y^2+3y-10y-15+11}$$

$$= \frac{2y^2-9y+4}{2y^2-7y-4} = \frac{(2y-1)\overset{1}{\cancel{(y-4)}}}{(2y+1)\underset{1}{\cancel{(y-4)}}} = \frac{2y-1}{2y+1}$$

31. The LCM is $2x-3$.

$$\frac{x-\frac{2}{2x-3}}{2x-1-\frac{8}{2x-3}}$$

$$=\frac{x-\frac{2}{2x-3}}{2x-1-\frac{8}{2x-3}}\cdot\frac{2x-3}{2x-3}$$

$$=\frac{x\cdot(2x-3)-\frac{2}{2x-3}\cdot(2x-3)}{2x\cdot(2x-3)-1\cdot(2x-3)-\frac{8}{2x-3}\cdot(2x-3)}$$

$$=\frac{2x^2-3x-2}{4x^2-6x-2x+3-8}=\frac{2x^2-3x-2}{4x^2-8x-5}$$

$$=\frac{\overset{1}{\cancel{(2x+1)}}(x-2)}{\underset{1}{\cancel{(2x+1)}}(2x-5)}=\frac{x-2}{2x-5}$$

33. The LCM is $x(x-1)$.

$$\frac{\frac{1}{x}-\frac{2}{x-1}}{\frac{3}{x}+\frac{1}{x-1}}=\frac{\frac{1}{x}-\frac{2}{x-1}}{\frac{3}{x}+\frac{1}{x-1}}\cdot\frac{x(x-1)}{x(x-1)}$$

$$=\frac{\frac{1}{x}\cdot x(x-1)-\frac{2}{x-1}\cdot x(x-1)}{\frac{3}{x}\cdot x(x-1)+\frac{1}{x-1}\cdot x(x-1)}$$

$$=\frac{x-1-2x}{3x-3+x}=\frac{-x-1}{4x-3}$$

35. The LCM is $x(2x-1)$.

$$\frac{\frac{3}{2x-1}-\frac{1}{x}}{\frac{4}{x}+\frac{2}{2x-1}}=\frac{\frac{3}{2x-1}-\frac{1}{x}}{\frac{4}{x}+\frac{2}{2x-1}}\cdot\frac{x(2x-1)}{x(2x-1)}$$

$$=\frac{\frac{3}{2x-1}\cdot x(2x-1)-\frac{1}{x}\cdot x(2x-1)}{\frac{4}{x}\cdot x(2x-1)+\frac{2}{2x-1}\cdot x(2x-1)}$$

$$=\frac{3x-2x+1}{8x-4+2x}=\frac{x+1}{10x-4}=\frac{x+1}{2(5x-2)}$$

37. True.

$$\frac{\frac{x}{y}}{\frac{y}{x}}=\frac{\frac{x}{y}}{\frac{y}{x}}\cdot\frac{xy}{xy}=\frac{x^2}{y^2}$$

Critical Thinking 11.4

39.

$$1+\frac{1}{1+\frac{1}{1+\frac{1}{2}}}=1+\frac{1}{1+\frac{1}{1+\frac{1}{2}}\cdot\frac{2}{2}}=1+\frac{1}{1+\frac{2}{2+1}}$$

$$=1+\frac{1}{1+\frac{2}{3}}=1+\frac{1}{1+\frac{2}{3}}\cdot\frac{3}{3}=1+\frac{3}{3+2}$$

$$=1+\frac{3}{5}=\frac{5}{5}+\frac{3}{5}=\frac{8}{5}$$

41.

$$1-\frac{1}{1-\frac{1}{y+1}}=1-\frac{1}{1-\frac{1}{y+1}}\cdot\frac{y+1}{y+1}=1-\frac{y+1}{y+1-1}$$

$$=1-\frac{y+1}{y}=\frac{y}{y}-\frac{y+1}{y}=-\frac{1}{y}$$

43.

$$\frac{x^{-2}-y^{-2}}{x^{-2}y^{-2}}=\frac{\frac{1}{x^2}-\frac{1}{y^2}}{\frac{1}{x^2y^2}}=\frac{\frac{1}{x^2}-\frac{1}{y^2}}{\frac{1}{x^2y^2}}\cdot\frac{x^2y^2}{x^2y^2}=y^2-x^2$$

Projects or Group Activities 11.4

45.

$$\frac{1}{\frac{1}{r_1}+\frac{1}{r_2}}$$

$$\frac{1}{\frac{1}{2}+\frac{1}{3}}=\frac{1}{\frac{1}{2}+\frac{1}{3}}\cdot\frac{6}{6}=\frac{6}{\frac{1}{2}\cdot 6+\frac{1}{3}\cdot 6}=\frac{6}{3+2}$$

$$=\frac{6}{5}\text{ ohms}$$

$$\frac{r_1r_2}{r_2+r_1}$$

$$\frac{2\cdot 3}{3+2}=\frac{6}{5}\text{ ohms}$$

47. The simplified form without complex fractions was easier to work with because you had to simplify the complex fraction each time in the original form.

Check Your Progress: Chapter 11

1. $\dfrac{x^2-4}{x^2+3x-10}=\dfrac{\cancel{(x-2)}(x+2)}{(x+5)\cancel{(x-2)}}=\dfrac{x+2}{x+5}$

2. $\dfrac{10x^2-50x}{12x+24}\cdot\dfrac{2x+4}{x^2-5x}=\dfrac{10\cancel{x}\cancel{(x-5)}}{12\cancel{(x+2)}}\cdot\dfrac{2\cancel{(x+2)}}{\cancel{x}\cancel{(x-5)}}$

$=\dfrac{20}{12}=\dfrac{5}{3}$

3. $\dfrac{6x^3y^2}{18a^4b}\div\dfrac{3xy}{9a^2b^5}=\dfrac{\cancel{6}x^3y^2}{\cancel{18}a^4b}\cdot\dfrac{\cancel{9}a^2b^5}{\cancel{3}xy}=\dfrac{b^4x^2y}{a^2}$

4. $\dfrac{a^3b}{a^2-5a-14}\div\dfrac{ab^6}{a^2-3a-28}$

$=\dfrac{a^3b}{\cancel{(a-7)}(a+2)}\cdot\dfrac{\cancel{(a-7)}(a+4)}{ab^6}=\dfrac{a^2(a+4)}{b^5(a+2)}$

5. $10x^4y^2=2\cdot5x^4y^2$

$15x^3y=3\cdot5x^3y$

$\text{LCM}=2\cdot3\cdot5x^4y^2=30x^4y^2$

6. $8x^2=2\cdot2\cdot2x^2$

$4x^2+8x=4x(x+2)=2\cdot2x(x+2)$

$\text{LCM}=2\cdot2\cdot2x^2(x+2)=8x^2(x+2)$

7. $\text{LCM}=x^2(x+2)$

$\dfrac{3}{x^2+2x}=\dfrac{3}{x(x+2)}\cdot\dfrac{x}{x}=\dfrac{3x}{x^2(x+2)}$

$\dfrac{4}{x^2}=\dfrac{4}{x^2}\cdot\dfrac{(x+2)}{(x+2)}=\dfrac{4x+8}{x^2(x+2)}$

8. $\dfrac{8a}{3a-1}-\dfrac{10a}{3a-1}=\dfrac{-2a}{3a-1}=-\dfrac{2a}{3a-1}$

9. $\dfrac{a-1}{a-2}-\dfrac{3a+1}{a^2+3a-10}$

$=\dfrac{a-1}{a-2}-\dfrac{3a+1}{(a+5)(a-2)}$

$=\dfrac{a-1}{a-2}\cdot\dfrac{a+5}{a+5}+\dfrac{-3a-1}{(a-2)(a+5)}$

$=\dfrac{a^2+4a-5}{(a-2)(a+5)}+\dfrac{-3a-1}{(a-2)(a+5)}$

$=\dfrac{a^2+a-6}{(a-2)(a+5)}=\dfrac{(a+3)\cancel{(a-2)}}{\cancel{(a-2)}(a+5)}$

$=\dfrac{a+3}{a+5}$

10. $\dfrac{\dfrac{7}{x-3}-\dfrac{2}{3x}}{\dfrac{5}{3x}+\dfrac{1}{x-3}}=\dfrac{\dfrac{7}{x-3}-\dfrac{2}{3x}}{\dfrac{5}{3x}+\dfrac{1}{x-3}}\cdot\dfrac{3x(x-3)}{3x(x-3)}$

$=\dfrac{\dfrac{7}{\cancel{(x-3)}}\cdot\dfrac{3x\cancel{(x-3)}}{1}-\dfrac{2}{\cancel{3x}}\cdot\dfrac{\cancel{3x}(x-3)}{1}}{\dfrac{5}{\cancel{3x}}\cdot\dfrac{\cancel{3x}(x-3)}{1}+\dfrac{1}{\cancel{(x-3)}}\cdot\dfrac{3x\cancel{(x-3)}}{1}}$

$=\dfrac{21x-2x+6}{5x-15+3x}=\dfrac{19x+6}{8x-15}$

Section 11.5

Concept Check

1. Multiplication Property of Equations

3. We can clear denominators in an *equation* as in part (a), but not in an *expression* as in part (b).

Objective A Exercises

5. The following values would make a denominator 0.

$x+1=0$ and $x-2=0$

$x=-1$ $\qquad$ $x=2$

7. The following values would make a denominator 0.

$x^2 - 9x = x(x-9)$

$x = 0$ and $x - 9 = 0$

$x = 9$

9. The LCM is 12.

$$\frac{x}{3} - \frac{1}{4} = \frac{1}{12}$$

$$\frac{12}{1}\left(\frac{x}{3} - \frac{1}{4}\right) = \frac{\cancel{12}}{1}\left(\frac{1}{\cancel{12}}\right)$$

$$\frac{\overset{4}{\cancel{12}}}{1} \cdot \frac{x}{\underset{1}{\cancel{3}}} - \frac{\overset{3}{\cancel{12}}}{1} \cdot \frac{1}{\underset{1}{\cancel{4}}} = 1$$

$$4x - 3 = 1$$

$$4x = 4$$

$$x = 1$$

1 checks as a solution. The solution is 1.

11. The LCM is 18.

$$\frac{2y}{9} - \frac{1}{6} = \frac{y}{9} + \frac{1}{6}$$

$$\frac{18}{1}\left(\frac{2y}{9} - \frac{1}{6}\right) = \frac{18}{1}\left(\frac{y}{9} + \frac{1}{6}\right)$$

$$\frac{\overset{2}{\cancel{18}}}{1} \cdot \frac{2y}{\underset{1}{\cancel{9}}} - \frac{\overset{3}{\cancel{18}}}{1} \cdot \frac{1}{\underset{1}{\cancel{6}}} = \frac{\overset{2}{\cancel{18}}}{1} \cdot \frac{y}{\underset{1}{\cancel{9}}} + \frac{\overset{3}{\cancel{18}}}{1} \cdot \frac{1}{\underset{1}{\cancel{6}}}$$

$$4y - 3 = 2y + 3$$

$$2y - 3 = 3$$

$$2y = 6$$

$$y = 3$$

3 checks as a solution. The solution is 3.

13. The LCM is 12.

$$\frac{3x+4}{12} - \frac{1}{3} = \frac{5x+2}{12} - \frac{1}{2}$$

$$\frac{12}{1}\left(\frac{3x+4}{12} - \frac{1}{3}\right) = \frac{12}{1}\left(\frac{5x+2}{12} - \frac{1}{2}\right)$$

$$\frac{\overset{1}{\cancel{12}}}{1} \cdot \frac{3x+4}{\underset{1}{\cancel{12}}} - \frac{\overset{4}{\cancel{12}}}{1} \cdot \frac{1}{\underset{1}{\cancel{3}}} = \frac{\overset{1}{\cancel{12}}}{1} \cdot \frac{5x+2}{\underset{1}{\cancel{12}}} - \frac{\overset{6}{\cancel{12}}}{1} \cdot \frac{1}{\underset{1}{\cancel{2}}}$$

$$3x + 4 - 4 = 5x + 2 - 6$$

$$3x = 5x - 4$$

$$-2x = -4$$

$$x = 2$$

2 checks as a solution. The solution is 2.

15. The LCM is $3x - 2$.

$$\frac{12}{3x-2} = 3$$

$$\frac{\overset{1}{\cancel{3x-2}}}{1} \cdot \frac{12}{\underset{1}{\cancel{3x-2}}} = \frac{3x-2}{1} \cdot \frac{3}{1}$$

$$12 = 9x - 6$$

$$18 = 9x$$

$$2 = x$$

2 checks as a solution. The solution is 2.

17. The LCM is $4 - 3x$.

$$\frac{6}{4-3x} = 3$$

$$\frac{\overset{1}{\cancel{4-3x}}}{1} \cdot \frac{6}{\underset{1}{\cancel{4-3x}}} = \frac{4-3x}{1} \cdot \frac{3}{1}$$

$$6 = 12 - 9x$$

$$-6 = -9x$$

$$\frac{-6}{-9} = x$$

$$\frac{2}{3} = x$$

$\frac{2}{3}$ checks as a solution. The solution is $\frac{2}{3}$.

19. The LCM is n.

$$3+\frac{8}{n}=5$$
$$\frac{n}{1}\left(3+\frac{8}{n}\right)=\frac{n}{1}\cdot 5$$
$$\frac{n}{1}\cdot 3+\frac{n}{1}\cdot\frac{8}{n}=5n$$
$$3n+8=5n$$
$$8=2n$$
$$4=n$$

4 checks as a solution. The solution is 4.

21. The LCM is x.

$$3-\frac{12}{x}=7$$
$$\frac{x}{1}\left(3-\frac{12}{x}\right)=\frac{x}{1}\cdot 7$$
$$\frac{x}{1}.3-\frac{x}{1}\cdot\frac{12}{x}=7x$$
$$3x-12=7x$$
$$-12=4x$$
$$-3=x$$

-3 checks as a solution. The solution is -3.

23. The LCM is x.

$$\frac{6}{x}+3=11$$
$$\frac{x}{1}\left(\frac{6}{x}+3\right)=\frac{x}{1}\cdot 11$$
$$\frac{x}{1}\cdot\frac{6}{x}+\frac{x}{1}\cdot 3=11x$$
$$6+3x=11x$$
$$6=8x$$
$$\frac{6}{8}=x$$
$$\frac{3}{4}=x$$

$\frac{3}{4}$ checks as a solution. The solution is $\frac{3}{4}$.

25. The LCM is $(x+3)(x-1)$.

$$\frac{5}{x+3}=\frac{3}{x-1}$$
$$\frac{\overset{1}{\cancel{(x+3)}}(x-1)}{1}\cdot\frac{5}{\underset{1}{\cancel{(x+3)}}}=\frac{(x+3)\overset{1}{\cancel{(x-1)}}}{1}\cdot\frac{3}{\underset{1}{\cancel{(x-1)}}}$$
$$5x-5=3x+9$$
$$2x=14$$
$$x=7$$

7 checks as a solution. The solution is 7.

27. The LCM is $(3x-4)(1-2x)$.

$$\frac{5}{3x-4}=\frac{-3}{1-2x}$$
$$\frac{\overset{1}{\cancel{(3x-4)}}(1-2x)}{1}\cdot\frac{5}{\underset{1}{\cancel{3x-4}}}=\frac{(3x-4)\overset{1}{\cancel{(1-2x)}}}{1}\cdot\frac{-3}{\underset{1}{\cancel{1-2x}}}$$
$$5-10x=-9x+12$$
$$-x=7$$
$$x=-7$$

-7 checks as a solution. The solution is -7.

29. The LCM is $(5y-1(2y-1)$.

$$\frac{4}{5y-1}=\frac{2}{2y-1}$$

$$\frac{\cancel{(5y-1)}^{1}(2y-1)}{1}\cdot\frac{4}{\cancel{5y-1}_{1}}=\frac{(5y-1)\cancel{(2y-1)}^{1}}{1}\cdot\frac{2}{\cancel{2y-1}_{1}}$$

$$8y-4=10y-2$$
$$-2y=2$$
$$y=-1$$

-1 checks as a solution. The solution is -1.

31. The LCM is $x+2$.

$$\frac{2x}{x+2}-5=\frac{7x}{x+2}$$

$$\frac{x+2}{1}\left(\frac{2x}{x+2}-5\right)=\frac{\cancel{x+2}^{1}}{1}\cdot\frac{7x}{\cancel{x+2}_{1}}$$

$$\frac{\cancel{x+2}^{1}}{1}\cdot\frac{2x}{\cancel{x+2}_{1}}-\frac{x+2}{1}\cdot 5=7x$$

$$2x-5x-10=7x$$
$$-3x-10=7x$$
$$-10x=10$$
$$x=-1$$

-1 checks as a solution. The solution is -1.

33. The LCM is $x+4$.

$$\frac{x}{x+4}=3-\frac{4}{x+4}$$

$$\frac{\cancel{x+4}^{1}}{1}\cdot\frac{x}{\cancel{x+4}_{1}}=\frac{x+4}{1}\left(3-\frac{4}{x+4}\right)$$

$$x=\frac{x+4}{1}\cdot 3-\frac{\cancel{x+4}^{1}}{1}\cdot\frac{4}{\cancel{x+4}_{1}}$$

$$x=3x+12-4$$
$$x=3x+8$$
$$-2x=8$$
$$x=-4$$

-4 does not check as a solution. The equation has no solution.

35. The LCM is $(x+12)(x+5)$.

$$\frac{x}{x+12}=\frac{1}{x+5}$$

$$\frac{\cancel{(x+12)}^{1}(x+5)}{1}\cdot\frac{x}{\cancel{x+12}_{1}}=\frac{(x+12)\cancel{(x+5)}^{1}}{1}\cdot\frac{1}{\cancel{x+5}_{1}}$$

$$x^2+5x=x+12$$
$$x^2+4x-12=0$$
$$(x+6)(x-2)=0$$

$$x+6=0 \qquad x-2=0$$
$$x=-6 \qquad x=2$$

Both -6 and 2 check as solutions. The solutions are -6 and 2.

37. The LCM is $(3n-8)(n+2)$.

$$\frac{5}{3n-8}=\frac{n}{n+2}$$

$$\frac{\cancel{(3n-8)}^{1}(n+2)}{1}\cdot\frac{5}{\cancel{3n-8}_{1}}=\frac{(3n-8)\cancel{(n+2)}^{1}}{1}\cdot\frac{n}{\cancel{n+2}_{1}}$$

$$5n+10=3n^2-8n$$
$$0=3n^2-13n-10$$
$$0=(3n+2)(n-5)$$

$$3n+2=0 \qquad n-5=0$$
$$3n=-2 \qquad n=5$$
$$n=-\frac{2}{3}$$

Both $-\frac{2}{3}$ and 5 check as solutions. The solutions are $-\frac{2}{3}$ and 5.

39. The LCM is $x-3$.

$$x-\frac{6}{x-3}=\frac{2x}{x-3}$$

$$\frac{x-3}{1}\cdot\left(x-\frac{6}{x-3}\right)=\frac{\cancel{x-3}^{1}}{1}\cdot\frac{2x}{\cancel{x-3}_{1}}$$

$$\frac{x-3}{1}\cdot x-\frac{\cancel{x-3}^{1}}{1}\cdot\frac{6}{\cancel{x-3}_{1}}=2x$$

$$x^2-3x-6=2x$$

$$x^2-5x-6=0$$

$$(x-6)(x+1)=0$$

$$x-6=0 \qquad x+1=0$$

$$x=6 \qquad x=-1$$

Both 6 and -1 check as solutions. The solutions are 6 and -1.

Critical Thinking 11.5

41. The LCM is 15.

$$\frac{3}{5}y-\frac{1}{3}(1-y)=\frac{2y-5}{15}$$

$$\frac{\cancel{15}^{3}}{1}\cdot\frac{3}{\cancel{5}}y-\frac{\cancel{15}^{5}}{1}\cdot\frac{1}{\cancel{3}_{1}}(1-y)=\frac{\cancel{15}^{1}}{1}\cdot\frac{2y-5}{\cancel{15}_{1}}$$

$$9y-5+5y=2y-5$$

$$14y-5=2y-5$$

$$12y=0$$

$$y=0$$

0 checks as a solution. The solution is 0.

43. The LCM is $(x+2)(x+1)(x-1)$.

$$\frac{x+1}{x^2+x-2}=\frac{x+2}{x^2-1}+\frac{3}{x+2}$$

$$\frac{x+1}{(x+2)(x-1)}=\frac{x+2}{(x-1)(x+1)}+\frac{3}{x+2}$$

$$\frac{\cancel{(x+2)}^{1}(x+1)\cancel{(x-1)}^{1}}{1}\cdot\frac{x+1}{\cancel{(x+2)}_{1}\cancel{(x-1)}_{1}}$$

$$=\frac{(x+2)\cancel{(x+1)}^{1}\cancel{(x-1)}^{1}}{1}\cdot\frac{x+2}{\cancel{(x-1)}_{1}\cancel{(x+1)}_{1}}$$

$$+\frac{\cancel{(x+2)}^{1}(x+1)(x-1)}{1}\cdot\frac{3}{\cancel{x+2}_{1}}$$

$$(x+1)(x+1)=(x+2)(x+2)+(x+1)(x-1)3$$

$$x^2+2x+1=x^2+4x+4+(x^2-1)3$$

$$x^2+2x+1=x^2+4x+4+3x^2-3$$

$$x^2+2x+1=4x^2+4x+1$$

$$0=3x^2+2x$$

$$0=x(3x+2)$$

$$x=0 \qquad 3x+2=0$$

$$3x=-2$$

$$x=-\frac{2}{3}$$

Both 0 and $-\frac{2}{3}$ check as solutions. The solutions are 0 and $-\frac{2}{3}$.

Projects and Group Activities 11.5

45. Strategy:

To find the illumination, solve $I = \frac{s}{r^2}$ for I when $r = 5$ and $s = 100$.

Solution:

$$I = \frac{s}{r^2}$$
$$I = \frac{100}{5^2}$$
$$I = \frac{100}{25}$$
$$I = 4$$

The illumination on the desk is 4 lm.

47. Strategy:

To find the candela, solve $I = \frac{s}{r^2}$ for s when $r = 4$ and $I = 20$.

Solution:

$$I = \frac{s}{r^2}$$
$$20 = \frac{s}{4^2}$$
$$20 = \frac{s}{16}$$
$$16 \cdot 20 = 16 \cdot \frac{s}{16}$$
$$320 = s$$

The lamp should have 320 candela.

49. Strategy:

To find the distance, solve $I = \frac{s}{r^2}$ for r when $s = 40$ and $I = 10$.

Solution:

$$I = \frac{s}{r^2}$$
$$10 = \frac{40}{s^2}$$
$$s^2 \cdot 10 = s^2 \cdot \frac{40}{s^2}$$
$$10s^2 = 40$$
$$\frac{10s^2}{10} = \frac{40}{10}$$
$$s^2 = 4$$
$$x = 2$$

The lamp should be placed 2 m above the desk.

Section 11.6

Concept Check

1. True

3. R

Objective A Exercises

5. $d = rt$

$$\frac{d}{r} = \frac{rt}{r}$$
$$\frac{d}{r} = t$$

7. $PV = nRT$

$$\frac{PV}{nR} = \frac{nRT}{nR}$$
$$\frac{PV}{nR} = T$$

9. $P = 2l + 2w$

$$P - 2w = 2l + 2w - 2w$$
$$P - 2w = 2l$$
$$\frac{P - 2w}{2} = \frac{2l}{2}$$
$$\frac{P - 2w}{2} = l$$

11. $A = \frac{1}{2}h(b_1 + b_2)$

$$2 \cdot A = 2 \cdot \frac{1}{2}h(b_1 + b_2)$$
$$2A = h(b_1 + b_2)$$
$$2A = hb_1 + hb_2$$
$$2A - hb_2 = hb_1 + hb_2 - hb_2$$
$$2A - hb_2 = hb_1$$
$$\frac{2A - hb_2}{h} = \frac{hb_1}{h}$$
$$\frac{2A - hb_2}{h} = b_1$$

13. $V = \frac{1}{3}Ah$

$$3 \cdot V = 3 \cdot \frac{1}{3}Ah$$
$$3V = Ah$$
$$\frac{3V}{A} = \frac{Ah}{A}$$
$$\frac{3V}{A} = h$$

15.
$$\begin{aligned} R &= \frac{C-S}{t} \\ t\cdot R &= t\cdot\frac{C-S}{t} \\ Rt &= C-S \\ Rt-C &= C-C-S \\ Rt-C &= -S \\ -1(Rt-C) &= -1(-S) \\ C-Rt &= S \end{aligned}$$

17.
$$\begin{aligned} A &= P+Prt \\ A &= P(1+rt) \\ \frac{A}{1+rt} &= \frac{P(1+rt)}{1+rt} \\ \frac{A}{1+rt} &= P \end{aligned}$$

19.
$$\begin{aligned} A &= Sw+w \\ A &= w(S+1) \\ \frac{A}{S+1} &= \frac{w(S+1)}{S+1} \\ \frac{A}{S+1} &= w \end{aligned}$$

21.
$$\begin{aligned} 3x+y &= 10 \\ 3x-3x+y &= -3x+10 \\ y &= -3x+10 \end{aligned}$$

23.
$$\begin{aligned} 4x-y &= 3 \\ 4x-4x-y &= -4x+3 \\ -y &= -4x+3 \\ -1(-y) &= -1(-4x+3) \\ y &= 4x-3 \end{aligned}$$

25.
$$\begin{aligned} 3x+2y &= 6 \\ 3x-3x+2y &= -3x+6 \\ 2y &= -3x+6 \\ \frac{2y}{2} &= \frac{-3x+6}{2} \\ y &= -\frac{3}{2}x+3 \end{aligned}$$

27.
$$\begin{aligned} 2x-5y &= 10 \\ 2x-2x-5y &= -2x+10 \\ -5y &= -2x+10 \\ \frac{-5y}{-5} &= \frac{-2x+10}{-5} \\ y &= \frac{2}{5}x-2 \end{aligned}$$

29.
$$\begin{aligned} 2x+7y &= 14 \\ 2x-2x+7y &= -2x+14 \\ 7y &= -2x+14 \\ \frac{7y}{7} &= \frac{-2x+14}{7} \\ y &= -\frac{2}{7}x+2 \end{aligned}$$

31.
$$\begin{aligned} x+3y &= 6 \\ x-x+3y &= -x+6 \\ 3y &= -x+6 \\ \frac{3y}{3} &= \frac{-x+6}{3} \\ y &= -\frac{1}{3}x+2 \end{aligned}$$

33.
$$\begin{aligned} x+3y &= 6 \\ x+3y-3y &= -3y+6 \\ x &= -3y+6 \end{aligned}$$

35.
$$\begin{aligned} 3x-y &= 3 \\ 3x-y+y &= y+3 \\ 3x &= y+3 \\ \frac{3x}{3} &= \frac{y+3}{3} \\ x &= \frac{1}{3}y+1 \end{aligned}$$

37.
$$\begin{aligned} 2x+5y &= 10 \\ 2x+5y-5y &= -5y+10 \\ 2x &= -5y+10 \\ \frac{2x}{2} &= \frac{-5y+10}{2} \\ x &= -\frac{5}{2}y+5 \end{aligned}$$

39.
$$\begin{aligned} x-2y+1 &= 0 \\ x-2y+1-1 &= 0-1 \\ x-2y &= -1 \\ x-2y+2y &= 2y-1 \\ x &= 2y-1 \end{aligned}$$

41. a.

$$A = P(1+i)$$
$$A = P + Pi$$
$$A - P = P - P + Pi$$
$$A - P = Pi$$
$$\frac{A-P}{P} = \frac{Pi}{P}$$
$$\frac{A-P}{P} = i$$

$$A = P(1+i)$$
$$\frac{A}{P} = \frac{P(1+i)}{P}$$
$$\frac{A}{P} = 1+i$$
$$\frac{A}{P} - 1 = 1 - 1 + i$$
$$\frac{A}{P} - 1 = i$$

Yes

b.

$$A = P(1+i)$$
$$A = P + Pi$$
$$A - A - Pi = P + Pi - Pi - A$$
$$-Pi = P - A$$
$$\frac{-Pi}{-P} = \frac{P-A}{-P}$$
$$i = -\frac{P-A}{P}$$

$$A = P(1+i)$$
$$A = P + Pi$$
$$A - P = P - P + Pi$$
$$A - P = Pi$$
$$\frac{A-P}{P} = \frac{Pi}{P}$$
$$\frac{A-P}{P} = i$$

Yes

Critical Thinking 11.6

43.

$$\frac{1}{R_1} + \frac{1}{R_2} = \frac{1}{R}$$
$$R_1 \cdot R_2 \cdot R \cdot \frac{1}{R_1} + R_1 \cdot R_2 \cdot R \cdot \frac{1}{R_2} = R_1 \cdot R_2 \cdot R \cdot \frac{1}{R}$$
$$R_2R + R_1R = R_1R_2$$
$$R_2R - R_2R + R_1R = R_1R_2 - R_2R$$
$$R_1R = R_2(R_1 - R)$$
$$\frac{RR_1}{R_1 - R} = \frac{R_2(R_1-R)}{(R_1-R)}$$
$$\frac{RR_1}{R_1 - R} = R_2$$

Section 11.7

Objective A Exercises

1. The rate of work is the amount of a task that is competed per unit of time.

3. $\frac{1}{x}$

5. Jen

7. a. $\frac{1}{x}$ of the job per hour

b. $\frac{3}{x}$

9. $\frac{5}{h}$

11. a. 8 + 4 = 12 mph

b. 8 – 4 = 4 mph

Objective A Exercises

13.

	Rate	•	Time	=	Part of Job
Elect.	$\frac{1}{10}$	•	t	=	$\frac{t}{10}$
Assist	$\frac{1}{12}$	•	t	=	$\frac{t}{12}$

15. Strategy:

→Time to fill the fountain together: t

	Rate	Time	Part
First	$\frac{1}{3}$	t	$\frac{t}{3}$
Second	$\frac{1}{6}$	t	$\frac{t}{6}$

Solution:

$$\frac{t}{3} + \frac{t}{6} = 1$$
$$6 \cdot \left(\frac{t}{3} + \frac{t}{6}\right) = 6 \cdot 1$$
$$2t + t = 6$$
$$3t = 6$$
$$t = 2$$

It would take 2 h to fill the fountain room.

17. Strategy:

→ Time to remove the earth with both skiploaders working together: t

	Rate	Time	Part
First	$\frac{1}{12}$	t	$\frac{t}{12}$
Second	$\frac{1}{4}$	t	$\frac{t}{4}$

→The sum of the parts of the task completed by each must equal 1.

Solution:

$$\frac{t}{12}+\frac{t}{4}=1$$
$$12\left(\frac{t}{12}+\frac{t}{4}\right)=12\cdot 1$$
$$t+3t=12$$
$$4t=12$$
$$t=3$$

With both skiploaders, it would take 3 h to remove the earth.

19. Strategy:

→Time to solve with both computers working: t

	Rate	Time	Part
Small	$\frac{1}{75}$	t	$\frac{t}{75}$
Large	$\frac{1}{50}$	t	$\frac{t}{50}$

→The sum of the parts completed by each air conditioner must equal 1.

Solution:

$$\frac{t}{75}+\frac{t}{50}=1$$
$$150\left(\frac{t}{75}+\frac{t}{50}\right)=150\cdot 1$$
$$2t+3t=150$$
$$5t=150$$
$$t=30$$

It would take 30 h to solve the problem with both computers working.

21. Strategy:

→Time cool the room with both air conditioners: t

	Rate	Time	Part
Smaller	$\frac{1}{60}$	t	$\frac{t}{60}$
Larger	$\frac{1}{40}$	t	$\frac{t}{40}$

→The sum of the parts printed by each press must equal 1.

Solution:

$$\frac{t}{60}+\frac{t}{40}=1$$
$$240\left(\frac{t}{60}+\frac{t}{40}\right)=240\cdot 1$$
$$4t+6t=240$$
$$10t=240$$
$$t=24$$

It would take 24 min for both air conditioners.

23. Strategy:

→Time for the second welder to compete the job: t

	Rate	Time	Part
First welder	$\frac{1}{10}$	6	$\frac{6}{10}$
Second welder	$\frac{1}{t}$	6	$\frac{6}{t}$

→The sum of the parts printed by each press must equal 1.

Solution:

$$\frac{6}{10}+\frac{6}{t}=1$$
$$10t\left(\frac{6}{10}+\frac{6}{t}\right)=10t\cdot 1$$
$$6t+60=10t$$
$$6t-6t+60=10t-6t$$
$$60=4t$$
$$15=t$$

It would take 15 h for the second welder working alone.

25. Strategy:

→Time for the second pipeline to fill the tank: t

	Rate	Time	Part
First pipeline	$\frac{1}{45}$	30	$\frac{30}{45}$
Second pipeline	$\frac{1}{t}$	30	$\frac{30}{t}$

→The sum of the parts completed by both machines must equal 1.

Solution:

$$\frac{30}{45}+\frac{30}{t}=1$$
$$45t\left(\frac{30}{45}+\frac{30}{t}\right)=45t\cdot 1$$
$$30t+1350=45t$$
$$1350=15t$$
$$90=t$$

It would take the second pipeline 90 min to fill the tank.

27. Strategy:

→Time for the old machine working alone to complete the wall: t

	Rate	Time	Part
New reaper	$\frac{1}{1.5}$	1	$\frac{1}{1.5}$
Old reaper	$\frac{1}{t}$	1	$\frac{1}{t}$

→The sum of the parts completed by both must equal 1.

Solution:

$$\frac{1}{1.5}+\frac{1}{t}=1$$
$$1.5t\left(\frac{1}{1.5}+\frac{1}{t}\right)=1.5t\cdot 1$$
$$t+1.5=1.5t$$
$$t-t+1.5=1.5t-t$$
$$1.5=0.5t$$
$$3=t$$

It would take the old reaper 3 h alone.

29. Strategy:

→Time for the second technician to complete the task: t

	Rate	Time	Part
First tech	$\frac{1}{4}$	2	$\frac{2}{4}$
Second tech	$\frac{1}{6}$	t	$\frac{t}{6}$

→The sum completed by each must equal 1.

Solution:

$$\frac{2}{4}+\frac{t}{6}=1$$
$$12\left(\frac{2}{4}+\frac{t}{6}\right)=12\cdot 1$$
$$6+2t=12$$
$$2t=6$$
$$t=3$$

It would take the second technician 3 h to complete the wiring.

31. Strategy:

→Time for the smaller heating unit: t

	Rate	Time	Part
Large heating unit	$\frac{1}{8}$	2	$\frac{2}{8}$
Second heating unit	$\frac{1}{t}$	11	$\frac{11}{t}$

→The sum of the parts of the task must equal 1.

Solution:

$$\frac{2}{8}+\frac{11}{t}=1$$
$$8t\left(\frac{2}{8}+\frac{11}{t}\right)=8t\cdot 1$$
$$2t+88=8t$$
$$88=6t$$
$$14\frac{2}{3}=t$$

The small heating unit would take $14\frac{2}{3}$ h to heat the pool.

33. Strategy:

→Time for the apprentice to complete the repairs: t

	Rate	Time	Part
Mechanic	$\frac{1}{2}$	1	$\frac{1}{2}$
Apprentice	$\frac{1}{6}$	t	$\frac{t}{6}$

→The sum of the parts completed by each worker must equal 1.

Solution:

$$\frac{1}{2}+\frac{t}{6}=1$$
$$6\left(\frac{1}{2}+\frac{t}{6}\right)=6\cdot 1$$
$$3+t=6$$
$$t=3$$

It would take 3 h to complete the repairs.

35. Less than k

Objective B Exercises

37. **a.**

a.	*Dist.*	÷	*Rate*	=	*Time*
Against	1440	÷	$380-r$	=	$\frac{1440}{380-r}$
With	1600	÷	$380+r$	=	$\frac{1600}{380+r}$

b. $\frac{1440}{3880-r}=\frac{1600}{380+r}$

39. Strategy:

→Rate walking: r

→Rate driving: $10r$

	Distance	Rate	Time
Driving	80	$10r$	$\frac{80}{10r}$
Hiking	4	r	$\frac{4}{r}$

→The total time of the trip was 3 h.

Solution:

$$\frac{80}{10r}+\frac{4}{r}=3$$
$$10r\left(\frac{80}{10r}+\frac{4}{r}\right)=10r(3)$$
$$80+40=30r$$
$$120=30r$$
$$4=r$$

The hiking was 4 mph.

41. Strategy:

→Rate of helicopter: r

→Rate of jet: $4r$

	Distance	Rate	Time
Helicopter	180	r	$\frac{180}{r}$
Jet	1080	$4r$	$\frac{1080}{4r}$

→The total time of the trip was 5 h.

Solution:

$$\frac{180}{r}+\frac{1080}{4r}=5$$
$$4r\left(\frac{180}{r}+\frac{1080}{4r}\right)=4r\cdot 5$$
$$720+1080=20r$$
$$1800=20r$$
$$90=r$$
$$4r=90\cdot 4=360$$

The rate of the jet was 360 mph.

43. Strategy

→First rate: r

→Second rate: $r + 2$

	Distance	Rate	Time
First rate	15	r	$\frac{15}{r}$
Second rate	19	$r+2$	$\frac{19}{r+2}$

→The total time of the trip was 4 h.

Solution:

$$\frac{15}{r}+\frac{19}{r+2}=4$$
$$r(r+2)\left(\frac{15}{r}+\frac{19}{r+2}\right)=r(r+2)(4)$$
$$15r+30+19r=4r^2+8r$$
$$0=4r^2-26r-30$$
$$0=2(2r^2-13r-15)$$
$$0=2(2r-15)(r+1)$$

$2r-15=0 \qquad r+1=0$

$2r=15 \qquad r=-1$

$r=\frac{15}{2}$

Since the rate cannot be –1, the rate for the first 15 mi was 7.5 mph.

45. Strategy:

→Rate of the technician through congested traffic: r

	Distance	Rate	Time
Congested traffic	10	r	$\frac{10}{r}$
Expressway	20	$r+20$	$\frac{20}{r+20}$

→The total time of the trip was 1 h.

Solution:

$$\frac{10}{r}+\frac{20}{r+20}=1$$
$$r(r+20)\left(\frac{10}{r}+\frac{20}{r+20}\right)=r(r+20)\cdot 1$$
$$10(r+20)+20r=r(r+20)$$
$$10r+200+20r=r^2+20r$$
$$30r+200=r^2+20r$$
$$0=r^2-10r-200$$
$$0=(r-20)(r+10)$$

$r-20=0 \qquad r+10=0$

$r=20 \qquad r=-10$

The solution −10 is not possible because the rate cannot be negative. The rate of travel in congested traffic was 20 mph.

47. Strategy:

→Distance on way: d

	Distance	Rate	Time
With current	d	$15+3$	$\frac{d}{18}$
Against current	d	$15-3$	$\frac{d}{12}$

→The total time traveled is 3 h.

Solution:

$$\frac{d}{18}+\frac{d}{12}=3$$
$$36\left(\frac{d}{18}+\frac{d}{12}\right)=36(3)$$
$$2d+3d=108$$
$$5d=108$$
$$d=21.6$$

The family can travel 21.6 mi downstream.

49. Strategy:

→Rate of the current: r

	Distance	Rate	Time
With current	75	$20+r$	$\frac{75}{20+r}$
Against current	45	$20-r$	$\frac{45}{20-r}$

→The time traveled with the current equals the time traveled against the current.

Solution:

$$\frac{75}{20+r}=\frac{45}{20-r}$$
$$(20+r)(20-r)\frac{75}{20+r}=(20+r)(20-r)\frac{45}{20-r}$$
$$1500-75r=900+45r$$
$$600=120r$$
$$5=r$$

The rate of the current is 5 mph.

51. Strategy:

→Rate of the freight train: r

→Rate of the express train: $r+20$

	Distance	Rate	Time
Freight train	360	r	$\frac{360}{r}$
Express train	600	$r+20$	$\frac{600}{r+20}$

→The time of the freight train equals the time of the express train.

Solution:

$$\frac{360}{r}=\frac{600}{r+20}$$
$$r(r+20)\cdot\frac{360}{r}=r(r+20)\cdot\frac{600}{r+20}$$
$$360(r+20)=600r$$
$$360r+7200=600r$$
$$7200=240r$$
$$30=r$$
$$50=r+20$$

The rate of the freight train is 30 mph.
The rate of the express train is 50 mph.

53. Strategy:

→Rate of the current: r

	Distance	Rate	Time
With current	24	$6+r$	$\frac{24}{6+r}$
Against gulf current	12	$6-r$	$\frac{12}{6-r}$

→The time traveled with the current equals the time traveled against the gulf current.

Solution:

$$\frac{24}{6+r}=\frac{12}{6-r}$$
$$(6+r)(6-r)\left(\frac{24}{6+r}\right)=(6+r)(6-r)\left(\frac{12}{6-r}\right)$$
$$144-24r=72+12r$$
$$72=36r$$
$$2=r$$

The rate of the current is 2 mph.

55. Strategy:

→Rate of the jet stream: r

	Distance	Rate	Time
With jet stream	2400	$550+r$	$\frac{2400}{550+r}$
Against jet stream	2000	$550-r$	$\frac{2000}{550-r}$

→The time traveled with the jet stream equals the time traveled against the jet stream.

Solution:

$$\frac{2400}{550+r}=\frac{2000}{550-r}$$
$$\overset{1}{\cancel{(550+r)}}(550-r)\frac{2400}{\underset{1}{\cancel{550+r}}}=(550+r)\overset{1}{\cancel{(550-r)}}\frac{2000}{\underset{1}{\cancel{550-r}}}$$
$$(550-r)2400=2000(550+r)$$
$$1{,}320{,}000-2400r=1{,}100{,}000+2000r$$
$$220{,}000=4400r$$
$$50=r$$

The rate of the jet stream is 50 mph.

57. Strategy:

→Rate of the current: r

	Distance	Rate	Time
With current	25	$20+r$	$\frac{25}{20+r}$
Against current	15	$20-r$	$\frac{15}{20-r}$

→The time traveled with the current equals the time traveled against the current.

Solution:

$$\frac{25}{20+r}=\frac{15}{20-r}$$
$$(20+r)(20-r)\frac{25}{20+r}=(20+r)(20-r)\frac{15}{20-r}$$
$$500-25r=300+15r$$
$$200=40r$$
$$5=r$$

The rate of the current is 5 mph.

Critical Thinking 11.7

59. Strategy:

→Time to fill the tank with all three pipes working: t

	Rate	Time	Part
First pipe	$\frac{1}{2}$	t	$\frac{t}{2}$
Second pipe	$\frac{1}{4}$	t	$\frac{t}{4}$
Third pipe	$\frac{1}{5}$	t	$\frac{t}{5}$

→The sum of the parts completed by each pipe must equal 1.

Solution:

$$\frac{t}{2}+\frac{t}{4}+\frac{t}{5}=1$$
$$20\left(\frac{t}{2}+\frac{t}{4}+\frac{t}{5}\right)=20\cdot 1$$
$$10t+5t+4t=20$$
$$19t=20$$
$$t=\frac{20}{19}=1\frac{1}{19}$$

It would take $1\frac{1}{19}$ h to fill the tank with all three pipes working.

61. Strategy:

To find the canoeing time

→Hiking rate: r

→Canoe rate: $3r$

	Distance	Rate	Time
Canoe	18	$3r$	$\frac{18}{3r}$
Hike	3	r	$\frac{3}{r}$

→The time hiking was 1 h less than the time canoeing.

→Find the time canoeing.

Solution:

$$\frac{3}{r}=\frac{18}{3r}-1$$
$$3r\left(\frac{3}{r}\right)=3r\left(\frac{18}{3r}-1\right)$$
$$9=18-3r$$
$$-9=-3r$$
$$3=r$$

Time canoeing: $\frac{18}{3r}=\frac{18}{3(3)}=\frac{18}{9}=2$

The time canoeing was 2 h.

Projects or Group Activities 11.7

63. Strategy

→Usual speed: r

→Speed in bad weather: $r - 10$

	Distance	Rate	Time
Usual	150	r	$\frac{150}{r}$
Bad weather	150	$r - 10$	$\frac{150}{r-10}$

→The time during bad weather is $\frac{1}{2}$ h more than the usual time.

Solution:

$$\frac{150}{r-10} = \frac{150}{r} + \frac{1}{2}$$
$$2r(r-10)\cdot\frac{150}{r-10} = 2r(r-10)\left(\frac{150}{r}+\frac{1}{2}\right)$$
$$300r = 300(r-10) + r(r-10)$$
$$300r = 300r - 3000 + r^2 - 10r$$
$$0 = r^2 - 10r - 3000$$
$$0 = (r-60)(r+50)$$

$r - 60 = 0 \qquad r + 50 = 0$

$r = 60 \qquad r = -50$

The solution −50 is not possible because the rate cannot be negative. The bus usually travels 60 mph.

Chapter Review Exercises

1. $\frac{6a^2b^7}{25x^3y} \div \frac{12a^3b^4}{5x^2y^2} = \frac{6a^2b^7}{25x^3y}\cdot\frac{5x^2y^2}{12a^3b^4}$

$= \frac{\overset{1}{\cancel{6}}\cdot\overset{1}{\cancel{5}}a^2b^7x^2y^2}{\underset{5}{\cancel{25}}\cdot\underset{2}{\cancel{12}}a^3b^4x^3y} = \frac{b^3y}{10ax}$

2. $\frac{x+7}{15x} + \frac{x-2}{20x} = \frac{4}{4}\left(\frac{x+7}{15x}\right) + \frac{3}{3}\left(\frac{x-2}{20x}\right)$

$= \frac{4x+28}{60x} + \frac{3x-6}{60x} = \frac{7x+22}{60x}$

3. $\frac{3x^3+9x^2}{6xy^2-18y^2}\cdot\frac{4xy^3-12y^3}{5x^2+15x}$

$\frac{3x^2\cancel{(x+3)}}{6y^2\cancel{(x-3)}}\cdot\frac{4y^3\cancel{(x-3)}}{5x\cancel{(x+3)}} = \frac{\cancel{3}x^{2-1}\cancel{2}\cdot 2y^{3-2}}{\cancel{2}\cdot\cancel{3}\cdot 5}$

$= \frac{2xy}{5}$

4. $\frac{2x(x-y)}{x^2y(x+y)} \div \frac{3(x-y)}{x^2y^2} = \frac{2x\cancel{(x-y)}}{x^2y(x+y)}\cdot\frac{x^2y^2}{3\cancel{(x-y)}}$

$= \frac{2xy}{3(x+y)}$

5. $\dfrac{x-\frac{16}{5x-2}}{3x-4-\frac{88}{5x-2}} = \dfrac{x-\frac{16}{5x-2}}{3x-4-\frac{88}{5x-2}}\cdot\frac{5x-2}{5x-2}$

$= \dfrac{x(5x-2)-\frac{16}{\cancel{5x-2}}\cdot\cancel{(5x-2)}}{3x(5x-2)-4(5x-2)-\frac{88}{\cancel{5x-2}}\cdot\cancel{(5x-2)}}$

$= \frac{5x^2-2x-16}{15x^2-6x-20x+8-88} = \frac{5x^2-2x-16}{15x^2-26x-80}$

$= \frac{\cancel{(5x+8)}(x-2)}{\cancel{(5x+8)}(3x-10)} = \frac{x-2}{3x-10}$

6. $\frac{x^2+x-30}{15+2x-x^2} = \frac{(x+6)\overset{1}{\cancel{(x-5)}}}{(3+x)\underset{-1}{\cancel{(5-x)}}} = -\frac{x+6}{x+3}$

7. $\frac{16x^5y^3}{24xy^{10}} = \frac{\cancel{2}\cdot\cancel{2}\cdot\cancel{2}\cdot 2x^{5-1}}{\cancel{2}\cdot\cancel{2}\cdot\cancel{2}\cdot 3y^{10-3}} = \frac{2x^4}{3y^7}$

8. $\frac{20}{x+2} = \frac{5}{16}$

$16\cancel{(x+2)}\cdot\frac{20}{\cancel{x+2}} = \cancel{16}(x+2)\cdot\frac{5}{\cancel{16}}$

$320 = 5x + 10$

$310 = 5x$

$62 = x$

62 checks as a solution. 62 is the solution.

9. $\dfrac{10-23x+12y^2}{6y^2-y-5}\div\dfrac{4y^2-13y+10}{18y^2+3y-10}$

$=\dfrac{12y^2-23y+10}{6y^2-y-5}\cdot\dfrac{18y^2+3y-10}{4y^2-13y+10}$

$=\dfrac{(3y-2)\overset{1}{\cancel{(4y-5)}}}{\underset{1}{\cancel{(6y+5)}}(y-1)}\cdot\dfrac{(3y-2)\overset{1}{\cancel{(6y+5)}}}{\underset{1}{\cancel{(4y-5)}}(y-2)}$

$=\dfrac{(3y-2)^2}{(y-1)(y-2)}$

10. $3ax-x=5$

$x(3a-1)=5$

$\dfrac{x(3a-1)}{3a-1}=\dfrac{5}{3a-1}$

$x=\dfrac{5}{3a-1}$

11. $\dfrac{2}{x}+\dfrac{3}{4}=1$

$4x\cdot\dfrac{2}{x}+4x\cdot\dfrac{3}{4}=4x\cdot 1$

$8+3x=4x$

$8=x$

8 checks as a solution. The solution is 8.

12. $\dfrac{x}{y}+\dfrac{3}{x}=\dfrac{x}{y}\cdot\dfrac{x}{x}+\dfrac{3}{x}\cdot\dfrac{y}{y}$

$=\dfrac{x^2}{xy}+\dfrac{3y}{xy}=\dfrac{x^2+3y}{xy}$

13. $5x+4y=20$

$5x-5x+4y=-5x+20$

$4y=-5x+20$

$\dfrac{4y}{4}=\dfrac{-5x+20}{4}$

$y=-\dfrac{5}{4}x+5$

14. $\dfrac{8ab^2}{15x^3y}\cdot\dfrac{5xy^4}{16a^2b}$

$\dfrac{8\cdot 5ab^2xy^4}{15\cdot 16a^2bx^3y}=\dfrac{by^3}{6ax^2}$

15. $\dfrac{1-\frac{1}{x}}{1-\frac{8x-7}{x^2}}=\dfrac{1-\frac{1}{x}}{1-\frac{8x-7}{x^2}}\cdot\dfrac{x^2}{x^2}$

$=\dfrac{1\cdot x^2-\frac{1}{x}\cdot x^2}{1\cdot x^2-\frac{8x-7}{x^2}\cdot x^2}=\dfrac{x^2-x}{x^2-8x+7}$

$=\dfrac{x\overset{1}{\cancel{(x-1)}}}{\underset{1}{\cancel{(x-1)}}(x-7)}=\dfrac{x}{x-7}$

16. $\dfrac{x}{12x^2+16x-3}$, $\dfrac{4x^2}{6x^2+7x-3}$

$\dfrac{x}{(6x-1)(2x+3)}$, $\dfrac{4x^2}{(2x+3)(3x-1)}$

The LCM is $(6x-1)(2x+3)(3x-1)$.

$\dfrac{x}{(6x-1)(2x+3)}\cdot\dfrac{3x-1}{3x-1}=\dfrac{3x^2-x}{(6x-1)(2x+3)(3x-1)}$

$\dfrac{4x^2}{(2x+3)(3x-1)}\cdot\dfrac{6x-1}{6x-1}=\dfrac{24x^3-4x^2}{(6x-1)(2x+3)(3x-1)}$

17. $T=2(ab+bc+ca)$

$T=2ab+2bc+2ca$

$T-2bc=2a(b+c)$

$\dfrac{T-2bc}{2(b+c)}=a$

$\dfrac{T-2bc}{2b+2C}=a$

18. $\dfrac{5}{7}+\dfrac{x}{2}=2-\dfrac{x}{7}$

$14\left(\dfrac{5}{7}+\dfrac{x}{2}\right)=14\left(2-\dfrac{x}{7}\right)$

$\overset{2}{\cancel{14}}\cdot\dfrac{5}{\underset{1}{\cancel{7}}}+\overset{7}{\cancel{14}}\cdot\dfrac{x}{\underset{2}{\cancel{2}}}=14\cdot 2-\overset{2}{\cancel{14}}\cdot\dfrac{x}{\underset{1}{\cancel{7}}}$

$10+7x=28-2x$

$9x=18$

$x=2$

2 checks as a solution. The solution is 2.

19. $\dfrac{2+\frac{1}{x}}{3-\frac{2}{x}}=\dfrac{2+\frac{1}{x}}{3-\frac{2}{x}}\cdot\dfrac{x}{x}$

$=\dfrac{2\cdot x+\frac{1}{x}\cdot x}{3\cdot x-\frac{2}{x}\cdot x}=\dfrac{2x+1}{3x-2}$

20. $\dfrac{2x}{x-5}-\dfrac{x+1}{x-2}=\dfrac{2x}{x-5}\cdot\dfrac{x-2}{x-2}-\dfrac{x+1}{x-2}\cdot\dfrac{x-5}{x-5}$

$=\dfrac{2x^2-4x}{(x-5)(x-2)}-\dfrac{x^2-4x-5}{(x-5)(x-2)}$

$=\dfrac{2x^2-4x-(x^2-4x-5)}{(x-5)(x-2)}=\dfrac{2x^2-4x-x^2+4x+5}{(x-5)(x-2)}$

$=\dfrac{x^2+5}{(x-5)(x-2)}$

21. $i=\dfrac{100m}{c}$

$c(i)=\dfrac{100m}{c}(c)$

$ci=100m$

$c=\dfrac{100m}{i}$

22. $\dfrac{x+8}{x+4}=1+\dfrac{5}{x+4}$

$(x+4)\left(\dfrac{x+8}{x+4}\right)=(x+4)\cdot 1+(x+4)\left(\dfrac{5}{x+4}\right)$

$x+8=x+4+5$

$0=1$

There is no solution.

23. $\dfrac{20x^2-45x}{6x^3+4x^2}\div\dfrac{40x^3-90x^2}{12x^2+8x}$

$=\dfrac{20x^2-45x}{6x^3+4x^2}\cdot\dfrac{12x^2+8x}{40x^3-90x^2}$

$=\dfrac{5x(4x-9)}{2x^2(3x+2)}\cdot\dfrac{4x(3x+2)}{10x^2(4x-9)}$

$=\dfrac{5x(4x-9)}{2x^2(3x+2)}\cdot\dfrac{2\cdot 2\cdot x(3x+2)}{5\cdot 2x^2(4x-9)}=\dfrac{1}{x^2}$

24. $\dfrac{2y}{5y-7}+\dfrac{3}{7-5y}=\dfrac{2y}{5y-7}+\dfrac{3}{-1(5y-7)}$

$=\dfrac{2y}{5y-7}-\dfrac{3}{5y-7}=\dfrac{2y-3}{5y-7}$

25. $\dfrac{5x+3}{2x^2+5x-3}-\dfrac{3x+4}{2x^2+5x-3}$

$=\dfrac{5x+3-(3x+4)}{2x^2+5x-3}=\dfrac{5x+3-3x-4}{2x^2+5x-3}$

$=\dfrac{2x-1}{(2x-1)(x+3)}=\dfrac{1}{x+3}$

26. $10x^2-11x+3=(5x-3)(2x-1)$

$20x^2-17x+3=(5x-3)(4x-1)$

$\text{LCM}=(5x-3)(2x-1)(4x-1)$

27. $4x+9y=18$

$4x-4x+9y=-4x+18$

$9y=-4x+18$

$\dfrac{9y}{9}=\dfrac{-4x+18}{9}$

$y=-\dfrac{4}{9}x+2$

28. $\dfrac{2x^2-5x-3}{3x^2-7x-6}\cdot\dfrac{3x^2+8x+4}{x^2+4x+4}$

$=\dfrac{(2x+1)(x-3)}{(3x+2)(x-3)}\cdot\dfrac{(3x+2)(x+2)}{(x+2)(x+2)}=\dfrac{2x+1}{x+2}$

29. $\dfrac{20}{2x+3}=\dfrac{17x}{2x+3}-5$

$(2x+3)\cdot\dfrac{20}{2x+3}=(2x+3)\cdot\dfrac{17x}{2x+3}-5(2x+3)$

$20=17x-10x-15$

$20=7x-15$

$35=7x$

$5=x$

5 checks as a solution. The solution is 5.

30. $\dfrac{x-1}{x+2}+\dfrac{3x-2}{5-x}+\dfrac{5x^2+15x-11}{x^2-3x-10}$

LCM $=(x-5)(x+2)$

$$\frac{x-1}{x+2}+\frac{3x-2}{5-x}+\frac{5x^2+15x-11}{x^2-3x-10}$$
$$=\frac{x-1}{x+2}-\frac{3x-2}{x-5}+\frac{5x^2+15x-11}{(x+2)(x-5)}$$
$$=\frac{x-1}{x+2}\cdot\frac{x-5}{x-5}-\frac{3x-2}{x-5}\cdot\frac{x+2}{x+2}+\frac{5x^2+15x-11}{(x+2)(x-5)}$$
$$=\frac{x^2-6x+5}{(x+2)(x-5)}-\frac{3x^2+4x-4}{(x+2)(x-5)}+\frac{5x^2+15x-11}{(x+2)(x-5)}$$
$$=\frac{x^2-6x+5-(3x^2+4x-4)+5x^2+15x-11}{(x+2)(x-5)}$$
$$=\frac{x^2-6x+5-3x^2-4x+4+5x^2+15x-11}{(x+2)(x-5)}$$
$$=\frac{3x^2+5x-2}{(x+2)(x-5)}=\frac{(3x-1)\overset{1}{\cancel{(x+2)}}}{\underset{1}{\cancel{(x+2)}}(x-5)}=\frac{3x-1}{x-5}$$

31.
$$\frac{6}{x-7}=\frac{8}{x-6}$$
$$(x-6)\overset{1}{\cancel{(x-7)}}\frac{6}{\underset{1}{\cancel{x-7}}}=\overset{1}{\cancel{(x-6)}}(x-7)\frac{8}{\underset{1}{\cancel{x-6}}}$$
$$6x-36=8x-56$$
$$20=2x$$
$$10=x$$

10 checks as a solution. The solution is 10.

32.
$$\frac{3}{20}=\frac{x}{80}$$
$$80\cdot\frac{3}{20}=\frac{x}{80}\cdot 80$$
$$12=x$$

12 checks as a solution. The solution is 12.

33. Strategy:

→Time to fill the pool using both hoses: t

	Rate	Time	Part
First hose	$\frac{1}{15}$	t	$\frac{t}{15}$
Second hose	$\frac{1}{10}$	t	$\frac{t}{10}$

→The sum of the parts of the task completed by each hose must equal 1.

Solution:
$$\frac{t}{15}+\frac{t}{10}=1$$
$$\overset{2}{\cancel{30}}\cdot\frac{t}{\underset{1}{\cancel{15}}}+\overset{3}{\cancel{30}}\cdot\frac{t}{\underset{1}{\cancel{10}}}=30\cdot 1$$
$$2t+3t=30$$
$$5t=30$$
$$t=6$$

It would take 6 h to fill the pool.

34. Strategy:

→Rate of the bus: r

→Rate of the car: $r+10$

	Distance	Rate	Time
Bus	245	r	$\frac{245}{r}$
Car	315	$r+10$	$\frac{315}{r+10}$

→The time of the bus equals the time of the car.

Solution:
$$\frac{315}{r+10}=\frac{245}{r}$$
$$r\overset{1}{\cancel{(r+10)}}\frac{315}{\underset{1}{\cancel{r+10}}}=\overset{1}{\cancel{r}}(r+10)\frac{245}{\underset{1}{\cancel{r}}}$$
$$315r=245r+2450$$
$$70r=2450$$
$$r=35$$
$$r+10=45$$

The rate of the car is 45 mph.

35. Strategy:

→Rate of the wind: r

	Distance	Rate	Time
With the wind	2100	400 + r	$\frac{2100}{400+r}$
Against the wind	1900	400 − r	$\frac{1900}{400-r}$

→The time traveled with the wind equals the time traveled against the wind.

Solution:

$$\frac{2100}{400+r}=\frac{1900}{400-r}$$

$$(400+r)(400-r)\frac{2100}{400+r}=(400+r)(400-r)\frac{1900}{400-r}$$

$$840{,}000-2100r=760{,}000+1900r$$

$$80{,}000=4000r$$

$$20=r$$

The rate of the wind is 20 mph.

Chapter Test

1. $\frac{x}{x+3}-\frac{2x-5}{x^2+x-6}$

$\frac{x}{x+3}-\frac{2x-5}{(x+3)(x-2)}$

LCM $=(x+3)(x-2)$

$\frac{x}{x+3}\cdot\frac{x-2}{x-2}-\frac{2x-5}{(x+3)(x-2)}$

$=\frac{x^2-2x}{(x+3)(x-2)}-\frac{2x-5}{(x+3)(x-2)}$

$=\frac{x^2-2x-(2x-5)}{(x+3)(x-2)}=\frac{x^2-2x-2x+5}{(x+3)(x-2)}$

$=\frac{x^2-4x+5}{(x+3)(x-2)}$

2. $\frac{3}{x+4}=\frac{5}{x+6}$

$(x+6)(x+4)\cdot\frac{3}{x+4}=(x+6)(x+4)\frac{5}{x+6}$

$3x+18=5x+20$

$-2=2x$

$-1=x$

−1 checks as a solution. The solution is −1.

3. $\frac{x^2+2x-3}{x^2+6x9}\cdot\frac{2x^2-11x+5}{2x^2+3x-5}$

$\frac{(x+3)(x-1)}{(x+3)(x+3)}\cdot\frac{(2x-1)(x-5)}{(2x+5)(x-1)}=\frac{(2x-1)(x-5)}{(x+3)(2x+5)}$

4. $\frac{16x^5y}{24x^2y^4}=\frac{16x^5y}{24x^2y^4}=\frac{2x^3}{3y^3}$

5. $d=s+rt$

$d-s=rt$

$\frac{d-s}{r}=\frac{rt}{r}$

$\frac{d-s}{r}=t$

6. $\frac{6}{x}-2=1$

$x\cdot\frac{6}{x}-2\cdot x=1\cdot x$

$6-2x=x$

$6=3x$

$2=x$

2 checks as a solution. The solution is 2.

7. $\frac{x^2+4x-5}{1-x^2}=\frac{(x+5)(x-1)}{(1+x)(1-x)}=-\frac{x+5}{x+1}$

8. $6x-3=3(2x-1)$

$2x^2+x-1=(2x-1)(x+1)$

LCM $=3(2x-1)(x+1)$

9. $\dfrac{2}{2x-1}-\dfrac{3}{3x+1}=\dfrac{2}{2x-1}\cdot\dfrac{3x+1}{3x+1}-\dfrac{3}{3x+1}\cdot\dfrac{2x-1}{2x-1}$

$$=\frac{6x+2}{(2x-1)(3x+1)}-\frac{6x-3}{(2x-1)(3x+1)}$$

$$=\frac{6x+2-(6x-3)}{(2x-1)(3x+1)}$$

$$=\frac{6x+2-6x+3}{(2x-1)(3x+1)}=\frac{5}{(2x-1)(3x+1)}$$

10. $\dfrac{x^2+3x+2}{x^2+5x+4}\div\dfrac{x^2-x-6}{x^2+2x-15}$

$$=\frac{\cancel{(x+2)}\cancel{(x+1)}}{\cancel{(x+1)}(x+4)}\cdot\frac{(x+5)\cancel{(x-3)}}{\cancel{(x-3)}\cancel{(x+2)}}=\frac{x+5}{x+4}$$

11. $\dfrac{1+\frac{1}{x}-\frac{12}{x^2}}{1+\frac{2}{x}-\frac{8}{x^2}}=\dfrac{1+\frac{1}{x}-\frac{12}{x^2}}{1+\frac{2}{x}-\frac{8}{x^2}}\cdot\dfrac{x^2}{x^2}$

$$=\frac{x^2\cdot 1+x^2\cdot\frac{1}{x}-x^2\cdot\frac{12}{x^2}}{x^2\cdot 1+x^2\cdot\frac{2}{x}-x^2\cdot\frac{8}{x^2}}=\frac{x^2+x-12}{x^2+2x-8}$$

$$=\frac{\cancel{(x+4)}(x-3)}{\cancel{(x+4)}(x-2)}=\frac{x-3}{x-2}$$

12. $\dfrac{3}{x^2-2x}=\dfrac{3}{x(x-2)}$ $\qquad\dfrac{x}{x^2-4}=\dfrac{x}{(x-2)(x+2)}$

LCM $=x(x-2)(x+2)$

$$\frac{3}{x(x-2)}\cdot\frac{x+2}{x+2}=\frac{3x+6}{x(x-2)(x+2)}$$

$$\frac{x}{(x-2)(x+2)}\cdot\frac{x}{x}=\frac{x^2}{x(x-2)(x+2)}$$

13. $\dfrac{2x}{x^2+3x-10}-\dfrac{4}{x^2+3x-10}=\dfrac{2x-4}{x^2+3x-10}$

$$=\frac{2\cancel{(x-2)}}{(x+5)\cancel{(x-2)}}=\frac{2}{x+5}$$

14. $3x-8y=16$

$$3x-3x-8y=-3x+16$$
$$-8y=-3x+16$$
$$\frac{-8y}{-8}=\frac{-3x+16}{-8}$$
$$y=\frac{3}{8}x-2$$

15. $\dfrac{2x}{x+1}-3=\dfrac{-2}{x+1}$

$$\cancel{(x+1)}\left(\frac{2x}{\cancel{x+1}}\right)-3(x+1)=\cancel{(x+1)}\left(\frac{-2}{\cancel{x+1}}\right)$$
$$2x-3x-3=-2$$
$$-x=1$$
$$x=-1$$

-1 does not check as a solution. There is no solution.

16. $\dfrac{x^3y^4}{x^2-4x+4}\cdot\dfrac{x^2-x-2}{x^6y^4}$

$$=\frac{x^3y^4}{(x-2)\cancel{(x-2)}}\cdot\frac{\cancel{(x-2)}(x+1)}{x^6y^4}=\frac{x+1}{x^3(x-2)}$$

17. $\dfrac{8a^2b^5}{3xy^4}\div\dfrac{4a^3b}{9x^2y}=\dfrac{8a^2b^5}{3xy^4}\cdot\dfrac{9x^2y}{4a^3b}=\dfrac{6b^4x}{ay^3}$

18. $\dfrac{4}{5x^2y}+\dfrac{1}{5x^2y}=\dfrac{4+1}{5x^2y}=\dfrac{5}{5x^2y}=\dfrac{1}{x^2y}$

19. Strategy:

→Time to fill the pool with both pipes: t

	Rate	Time	Part
First pipe	$\frac{1}{6}$	t	$\frac{t}{6}$
Second pipe	$\frac{1}{12}$	t	$\frac{t}{12}$

→The sum of the parts completed by each pipe must equal 1.

Solution:

$$\frac{t}{6}+\frac{t}{12}=1$$
$$12\cdot\frac{t}{6}+12\cdot\frac{t}{12}=12\cdot 1$$
$$2t+t=12$$
$$3t=12$$
$$t=4$$

It would take both pipes 4 h to fill the pool.

20. Strategy:

→Rate of the wind: r

	Distance	Rate	Time
With the wind	260	$110+r$	$\frac{260}{110+r}$
Against the wind	180	$110-r$	$\frac{180}{110-r}$

→The time traveled with the wind equals the time traveled against the wind.

Solution:

$$\frac{260}{110+r}=\frac{180}{110-r}$$
$$(110-r)(110+r)\frac{260}{110+r}=(110-r)(110+r)\frac{180}{110-r}$$
$$28{,}600-260r=19{,}800+180r$$
$$8800=440r$$
$$20=r$$

The rate of the wind is 20 mph.

21. Strategy:

→Time to make enough snow using man-made snow and natural: t

	Rate	Time	Part
Man-made	$\frac{1}{4}$	t	$\frac{t}{4}$
Natural	$\frac{1}{12}$	t	$\frac{t}{12}$

→The sum of the parts of the task completed by each snow maker must equal 1.

Solution:

$$\frac{t}{4}+\frac{t}{12}=1$$
$$12\left(\frac{t}{4}+\frac{t}{12}\right)=12\cdot 1$$
$$3t+t=12$$
$$4t=12$$
$$t=3$$

Using man-made and natural snow, it would take 3 h to open the ski run.

22. Strategy:

→Rate of the current: r

	Distance	Rate	Time
With current	4	$35+r$	$\frac{4}{35+r}$
Against	3	$35-r$	$\frac{3}{35-r}$

→The time with the current equals the time against the current.

Solution:

$$\frac{4}{35+r}=\frac{3}{35-r}$$
$$4(35-r)=3(35+r)$$
$$140-4r=105+3r$$
$$35=7r$$
$$5=r$$

The rate of the current is 5 mph.

Cumulative Review Exercises

1. $\left(\frac{2}{3}\right)^2 \div \left(\frac{3}{2}-\frac{2}{3}\right)+\frac{1}{2}=\left(\frac{2}{3}\right)^2 \div \left(\frac{9}{6}-\frac{4}{6}\right)+\frac{1}{2}$

$=\frac{4}{9}\div\left(\frac{5}{6}\right)+\frac{1}{2}=\frac{4}{9}\cdot\frac{6}{5}+\frac{1}{2}$

$=\frac{2\cdot 2\cdot 2\cdot \cancel{3}}{\cancel{3}\cdot 3\cdot 5}+\frac{1}{2}=\frac{8}{15}+\frac{1}{2}$

$=\frac{16}{30}+\frac{15}{30}=\frac{31}{30}$

2. $-a^2+(a-b)^2$

$-(-2)^2+(-2-3)^2=-4+(-5)^2=-4+25=21$

3. $-2x-(-3y)+7x-5y=-2x+3y+7x-5y$

$=-2x+7x+3y-5y$

$=5x-2y$

4. $2[3x-7(x-3)-8]=2[3x-7x+21-8]$

$=2[-4x+13]=-8x+26$

5. $4-\frac{2}{3}x=7$

$-\frac{2}{3}x=3$

$-\frac{3}{2}\left(-\frac{2}{3}x\right)=3\left(-\frac{3}{2}\right)$

$x=-\frac{9}{2}$

The solution is $-\frac{9}{2}$.

6. $3[x-2(x-3)]=2(3-2x)$

$3[x-2x+6]=6-4x$

$3[-x+6]=6-4x$

$-3x+18=6-4x$

$x=-12$

The solution is −12.

7. $P\cdot B=A$

$16\frac{2}{3}\%\cdot 60=A$

$\frac{1}{6}\cdot 60=A$

$10=A$

8. $(a^2b^5)(ab^2)=a^{2+1}b^{5+2}=a^3b^7$

9. $(a-3b)(a+4b)=a^2+4ab-3ab-12b^2$

$=a^2+ab-12b^2$

10. $\frac{15b^4-5b^2+10b}{5b}=\frac{\overset{3}{\cancel{15}}b^4}{\underset{1}{\cancel{5}}b}-\frac{\overset{1}{\cancel{5}}b^2}{\underset{1}{\cancel{5}}b}+\frac{\overset{2}{\cancel{10}}b}{\underset{1}{\cancel{5}}b}$

$=3b^3-b+2$

11.
$$\begin{array}{r}
x^2+2x+4\\
x-2\overline{)x^3+0x^2+0x-8}\\
\underline{x^3-2x^2}\qquad\quad\\
2x^2+0x\quad\\
\underline{2x^2-4x}\quad\\
4x-8\\
\underline{4x-8}\\
0
\end{array}$$

$(x^3-8)\div(x-2)=x^2+2x+4$

12. $12x^2-x-1$

$12\cdot(-1)=-12$

Factors of −12 whose sum is −1: −4 and 3

$12x^2-3x+3x-1$

$4x(3x-1)+1(3x-1)=(3x-1)(4x+1)$

13. y^2-7y+6

$1(6)=6$

Factors of 6 whose sum is −7: −6 and −1

$y^2-6y-y+6$

$y(y-6)-1(y-6)=(y-6)(y-1)$

14. $2a^3+7a^2-15a$

The GCF is *a*: $a(2a^2+7a-15)$

$2\cdot(-15)=-30$

Factors of −30 whose sum is 7: 10 and −3

$a(2a^2+10a-3a-15)$

$a[2a(a+5)-3(a+5)]=a(a+5)(2a-3)$

15. $4b^2-100$

The GCF is 4: $4(b^2-25)$

$4[b^2-(5)^2]=4(b+5)(b-5)$

16. $(x+3)(2x-5)=0$

$$x+3=0 \qquad 2x-5=0$$
$$x=-3 \qquad 2x=5$$
$$x=\frac{5}{2}$$

The solutions are −3 and $\frac{5}{2}$.

17. $$\frac{12x^4y^2}{18xy^7}=\frac{\cancel{2}\cdot 2\cdot \cancel{3}x^{4-1}}{\cancel{2}\cdot\cancel{3}\cdot 3y^{7-2}}=\frac{2x^3}{3y^5}$$

18. $$\frac{x^2-7x+10}{25-x^2}=\frac{\cancel{(x-5)}(x-2)}{(5+x)\cancel{(5-x)}}=-\frac{x-2}{x+5}$$

19. $$\frac{x^2-x-56}{x^2+8x+7}\div\frac{x^2-13x+40}{x^2-4x-5}$$
$$=\frac{\cancel{(x-8)}\cancel{(x+7)}}{\cancel{(x+7)}\cancel{(x+1)}}\cdot\frac{\cancel{(x-5)}\cancel{(x+1)}}{\cancel{(x-8)}\cancel{(x-5)}}=1$$

20. The GCF is $(2x-1)(x+1)$.

$$\frac{2}{2x-1}-\frac{1}{x+1}=\frac{2}{2x-1}\cdot\frac{x+1}{x+1}-\frac{1}{x+1}\cdot\frac{2x-1}{2x-1}$$
$$=\frac{2x+2}{(2x-1)(x+1)}-\frac{2x-1}{(2x-1)(x+1)}$$
$$=\frac{2x+2-(2x-1)}{(2x-1)(x+1)}$$
$$=\frac{2x+2-2x+1}{(2x-1)(x+1)}=\frac{3}{(2x-1)(x+1)}$$

21. $$\frac{1-\frac{2}{x}-\frac{15}{x^2}}{1-\frac{25}{x^2}}=\frac{1-\frac{2}{x}-\frac{15}{x^2}}{1-\frac{25}{x^2}}\cdot\frac{x^2}{x^2}$$
$$=\frac{1\cdot x^2-\frac{2}{x}\cdot x^2-\frac{15}{x^2}\cdot x^2}{1\cdot x^2-\frac{25}{x^2}\cdot x^2}$$
$$=\frac{x^2-2x-15}{x^2-25}$$
$$=\frac{\cancel{(x-5)}(x+3)}{\cancel{(x-5)}(x+5)}=\frac{x+3}{x+5}$$

22. $$\frac{3x}{x-3}-2=\frac{10}{x-3}$$
$$\cancel{(x-3)}\left(\frac{3x}{\cancel{x-3}}\right)-2(x-3)=\cancel{(x-3)}\left(\frac{10}{\cancel{x-3}}\right)$$
$$3x-2(x-3)=10$$
$$3x-2x+6=10$$
$$x+6=10$$
$$x=4$$

4 checks as a solution. The solution is 4.

23. $$\frac{2}{x-2}=\frac{12}{x+3}$$
$$(x+3)\cancel{(x-2)}\frac{2}{\cancel{x-2}}=\cancel{(x+3)}(x-2)\frac{12}{\cancel{x+3}}$$
$$2x+6=12x-24$$
$$30=10x$$
$$3=x$$

3 checks as a solution. The solution is 3.

24. $$f=v+at$$
$$f-v=v-v+at$$
$$f-v=at$$
$$\frac{f-v}{a}=\frac{at}{a}$$
$$\frac{f-v}{a}=t$$

25. Strategy:

→The unknown number: x

Solution:
$$5x-13=-8$$
$$5x-13+13=-8+13$$
$$5x=5$$
$$\frac{5x}{5}=\frac{5}{5}$$
$$x=1$$

The number is 1.

26. Strategy:

Change 2.2% to a decimal and use the basic percent equation.

Solution:
$$\frac{1.1}{0.022}=50$$

There are 50 million school-age students in the U.S.

27. Strategy:

→Height: x

→Base: $2x - 2$

→Use the equation for the area of a triangle.

Solution:

$$\frac{1}{2}bh = A$$
$$\frac{1}{2}(2x-2)x = 30$$
$$(x-1)x = 30$$
$$x^2 - x - 30 = 0$$
$$(x-6)(x+5) = 0$$
$$x-6=0 \qquad x+5=0$$
$$x=6 \qquad x=-5$$

The height cannot be −5, so the height is 6 in.

The base is $2x - 2 = 2(6) - 2 = 12 - 2 = 10$ in.

28. Strategy:

To find the cost of a \$5000 policy, write and solve a proportion using x to represent the cost for a \$5000 policy.

Solution:

$$\frac{16}{1000} = \frac{x}{5000}$$
$$\overset{5}{\cancel{5000}} \cdot \frac{16}{\underset{1}{\cancel{1000}}} = \overset{1}{\cancel{5000}} \cdot \frac{x}{\underset{1}{\cancel{5000}}}$$
$$80 = x$$

The cost of a \$5000 policy is \$80.

29. Strategy:

→Time to fill the pool with both pipes: t

	Rate	Time	Part
First pipe	$\frac{1}{9}$	t	$\frac{t}{9}$
Second pipe	$\frac{1}{18}$	t	$\frac{t}{18}$

→The sum of the parts completed by each pipe must equal 1.

Solution:

$$\frac{t}{9} + \frac{t}{18} = 1$$
$$\overset{2}{\cancel{18}} \cdot \frac{t}{\underset{1}{\cancel{9}}} + \overset{1}{\cancel{18}} \cdot \frac{t}{\underset{1}{\cancel{18}}} = 18 \cdot 1$$
$$2t + t = 18$$
$$3t = 18$$
$$t = 6$$

It would take both pipes 6 min to fill the tank.

30. Strategy:

→Rate of the current: r

	Distance	Rate	Time
With current	14	$5+r$	$\frac{14}{5+r}$
Against current	6	$5-r$	$\frac{6}{5-r}$

→The time traveled with the current equals the time traveled against the current.

Solution:

$$\frac{14}{5+r}=\frac{6}{5-r}$$

$$(5+r)(5-r)\frac{14}{5+r}=(5+r)(5-r)\frac{6}{5-r}$$

$$70-14r=30+6r$$

$$-20r=-40$$

$$r=2$$

The rate of the current is 2 mph.

Chapter 12: Linear Equations in Two Variables

Prep Test

1. $-\frac{5-(-7)}{4-8}=-\frac{5+7}{-4}=\frac{12}{4}=3$

2. $\frac{a-b}{c-d}$

$\frac{3-(-2)}{-3-2}=\frac{3+2}{-3-2}=\frac{5}{-5}=-1$

3. $-3(x-4)=-3x+12$

4. $3x+6=0$

$3x=-6$

$x=-2$

The solution is −2.

5. $4x+5y=20$

$4x+5(0)=20$

$4x=20$

$x=5$

The solution is 5.

6. $3x-7y=11$

$3(-1)-7y=11$

$-3-7y=11$

$-7y=14$

$y=-2$

The solution is −2.

7. $\frac{12x-15}{-3}=\frac{-12x}{-3}+\frac{-15}{3}=-4x+5$

8. $\frac{2x+1}{3}=\frac{3x}{4}$

$12\cdot\frac{2x+1}{3}=12\cdot\frac{3x}{4}$

$8x+4=9x$

$4=x$

The solution is 4.

9. $3x-5y=15$

$-5y=-3x+15$

$y=\frac{-3x+15}{-5}$

$y=\frac{3}{5}x-3$

10. $y+3=-\frac{1}{2}(x+4)$

$y+3=-\frac{1}{2}x-2$

$y=-\frac{1}{2}x-5$

Section 12.1

Concept Check

1. Quadrant II

3. y-axis

5. Answers will vary. For example, (–3, 2) and (5, 2).

7. Right; down

9. (6, –5)

11. x; y

13. Ordered pairs; domain; range

Objective A Exercises

15.

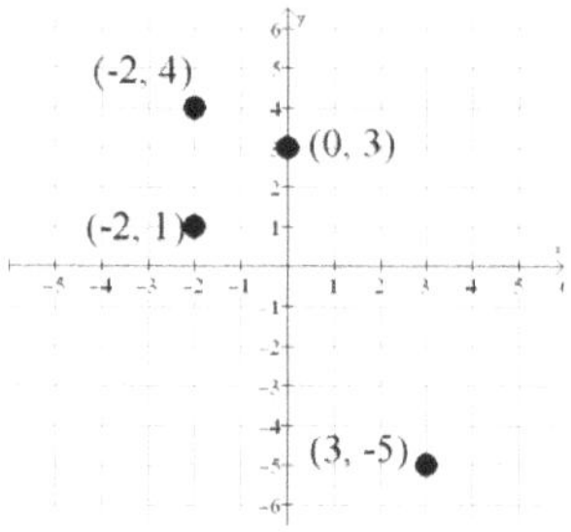

17.

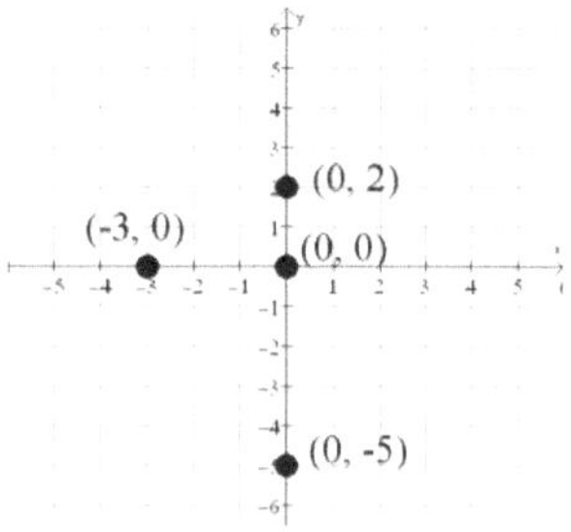

19.

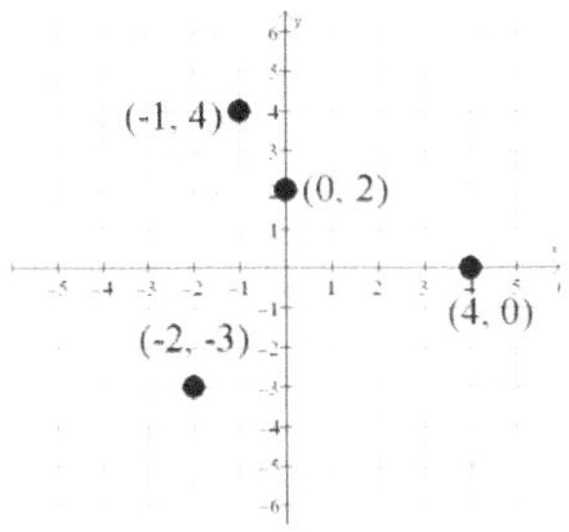

21. $A\ (2, 3)$
$B\ (4, 0)$
$C\ (-4, 1)$
$D\ (-2, -2)$

23. $A\ (-2, 5)$
$B\ (3, 4)$
$C\ (0, 0)$
$D\ (-3, -2)$

25. **a.** Abscissa of point A: 2
Abscissa of point C: -4

b. Ordinate of point B: 1
Ordinate of point D: -3

27. **a.** I
b. II
c. IV
d. III

Objective B Exercises

29.

$$\begin{array}{c|c} \multicolumn{2}{c}{y=-x+7} \\ \hline 4 & -(3)+7 \\ & -3+7 \\ & 4 \end{array}$$
$$4=4$$

Yes, (3, 4) is a solution of $y=-x+7$.

31.

$$\begin{array}{c|c} \multicolumn{2}{c}{y=\frac{1}{2}x-1} \\ \hline 2 & \frac{1}{2}(-1)-1 \\ & -\frac{1}{2}-1 \\ & -\frac{3}{2} \end{array}$$
$$2\neq -\frac{3}{2}$$

No, (2, −3) is not a solution of $y=\frac{1}{2}x-1$.

33.

$$\begin{array}{c|c} \multicolumn{2}{c}{2x-5y=4} \\ \hline 2(4)-5(1) & 4 \\ 8-5 & \\ 3 & \end{array}$$
$$3\neq 4$$

No, (4, 1) is a not solution of $2x-5y=4$.

35. If $x>2$, $-3x<-6$, then $-3x+6<0$.
y is negative.

37.

x	$y=2x$	(x, y)
-2	$2(-2)=-4$	$(-2, -4)$
-1	$2(-1)=-2$	$(-1, -2)$
0	$2(0)=0$	$(0, 0)$
2	$2(2)=4$	$(2, 4)$

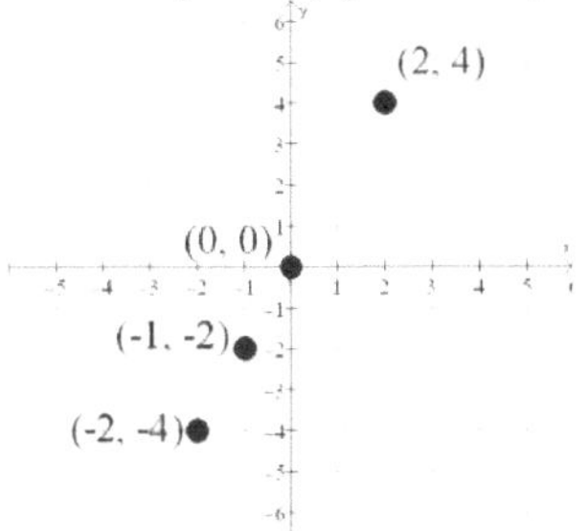

39.

x	$y=\frac{2}{3}x+1$	(x, y)
-3	$\frac{2}{3}(-3)+1=-1$	$(-3, -1)$
0	$\frac{2}{3}(0)+1=1$	$(0, 1)$
3	$\frac{2}{3}(3)+1=3$	$(3, 3)$

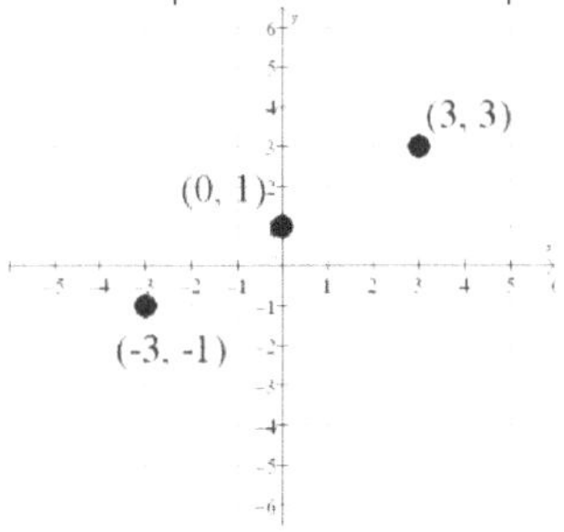

41. Solve $2x+3y=6$ for y.

$$2x+3y=6$$
$$3y=-2x+6$$
$$y=-\frac{2}{3}x+2$$

x	$y=-\frac{2}{3}x+2$	(x, y)
-3	$-\frac{2}{3}(-3)+2=4$	$(-3, 4)$
0	$-\frac{2}{3}(0)+2=2$	$(0, 2)$
3	$-\frac{2}{3}(3)+2=0$	$(3, 0)$

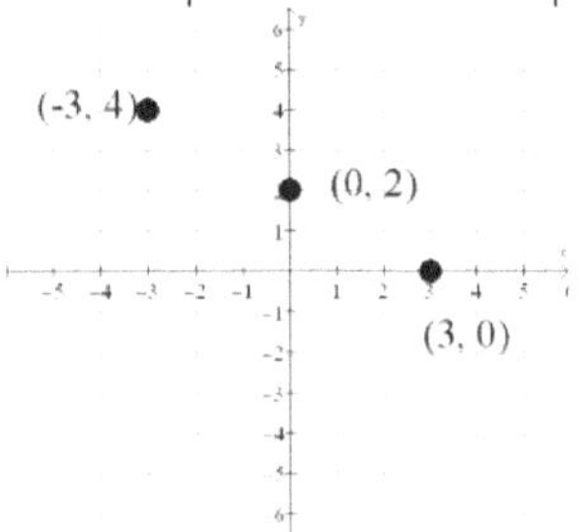

Objective C Exercises

43. No, a date could have more than one person born on that date, so two ordered pairs could have the same first coordinate but different second coordinates.

45. The relation is {(35, 7.50), (45, 7.58), (38, 7.63), (24, 7.78), (47, 7.80), (51, 7.86), (35, 7.89), (48, 7.92)}
No, the relation is not a function. The two ordered pairs (35, 7.50) and (35, 7.89) have the same first coordinate but different second coordinates.

47. The relation is {(1600, 18), (2000, 20), (2200, 24), (2500, 25), (2800, 31)}.
No two ordered pairs have the same first coordinate, so, yes, the relation is a function.

49. D: {0, 2, 4, 6} R: {0}
No two ordered pairs have the same first coordinate, so, yes, the relation is a function.

51. D: {2} R: {2, 4, 6, 8}
No, the relation is not a function. All ordered pairs have the same first coordinate but different second coordinates.

53.

x	$y=2x+3$	(x, y)
-2	$y=2(-2)+3=-1$	$(-2, -1)$
-1	$y=2(-1)+3=1$	$(-1, 1)$
1	$y=2(1)+3=5$	$(1, 5)$
4	$y=2(4)+3=11$	$(4, 11)$

No two ordered pairs have the same first coordinate, so, yes, the relation is a function.

55.

x	$y=x^2$	(x, y)
-2	$y=(-2)^2=4$	$(-2, 4)$
-1	$y=(-1)^2=1$	$(-1, 1)$
0	$y=(0)^2=0$	$(0, 0)$
1	$y=(1)^2=1$	$(1, 1)$
2	$y=(2)^2=4$	$(2, 4)$

No two ordered pairs have the same first coordinate, so, yes, the relation is a function.

Objective D Exercises

57. $f(x)=5x+1$
$f(2)=5(2)+1=10+1=11$

59. $f(x)=x^2-1$
$f(1)=1^2-1=1-1=0$

61. $H(x)=x^2-x$
$H(-2)=(-2)^2-(-2)=4+2=6$

63. $P(x)=\dfrac{4}{2x+1}$
$P(-2)=\dfrac{4}{2(-2)+1}=\dfrac{4}{-4+1}=-\dfrac{4}{3}$

65. $Q(r)=4r^2-r-3$
$Q(2)=4(2)^2-2-3=4(4)-2-3$
$=16-2-3=11$

67. $v(t)=\dfrac{2t}{2t+1}$
$v(3)=\dfrac{2(3)}{2(3)+1}=\dfrac{6}{6+1}=\dfrac{6}{7}$

69. Either

71. Positive

Critical Thinking 12.1

73.

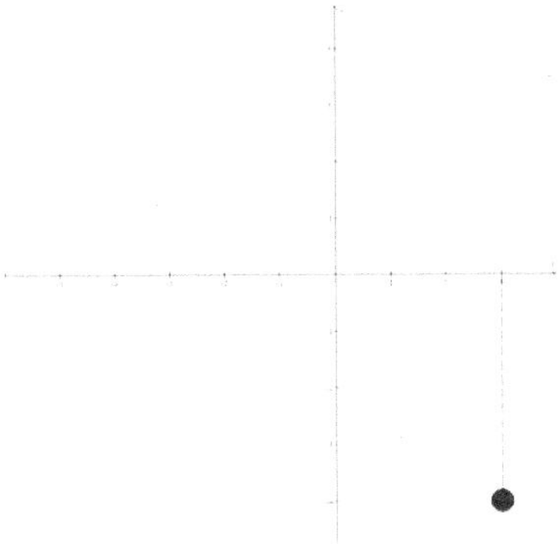

4 units

75.

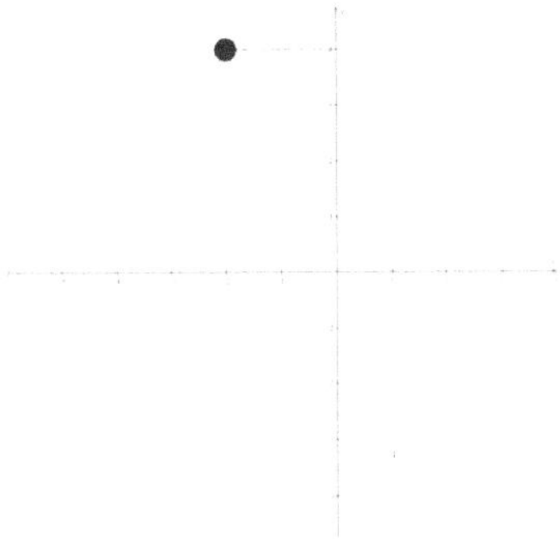

2 units

77.

5 units

79. To plot points in a rectangular coordinate system, move to the right or left the number of units indicated by the first number in the coordinates (right for a positive value, left for a negative value). Then move up or down the number of units indicated by the second number in the coordinates (up for a positive value, down for a negative value). Then place a point at the location.

Projects or Group Activities 12.1

81. Answers will vary.

Section 12.2

1. **(i)** linear equation

(ii) linear equation

(iii) $y = -x^2 + 4$ is not a linear equation because it has a term with a variable squared.

(iv) $y^2 = x - 6$ is not a linear equation because it has a term with a variable squared.

3. **a.** The graph of $y = x^2 + 1$ is not a straight line because it has a term with a variable squared.

b. straight line

c. The graph of $y = \frac{1}{x}$ is not a straight line because there is a variable in the denominator.

d. straight line

e. The graph of $y = \sqrt{x} - 1$ is not a straight line because there is a variable under a radical.

5. Multiples of 3 such as –3, 0, 3, and 6.

Objective A Exercises

7. $y = 2x - 3$

x	y
0	−3
1	−1
2	1

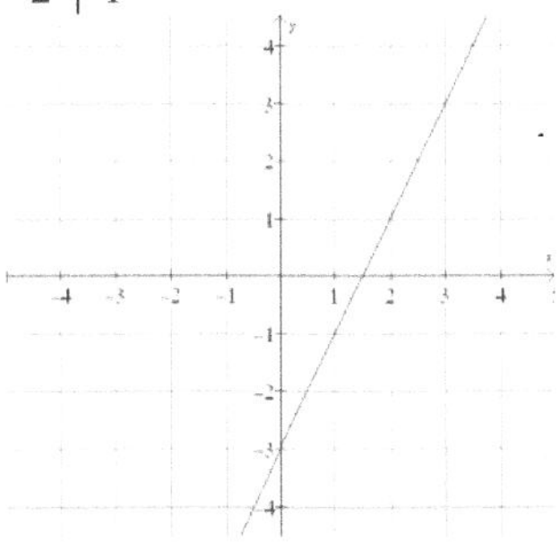

9. $y = \frac{1}{3}x$

x	y
−3	−1
0	0
3	1

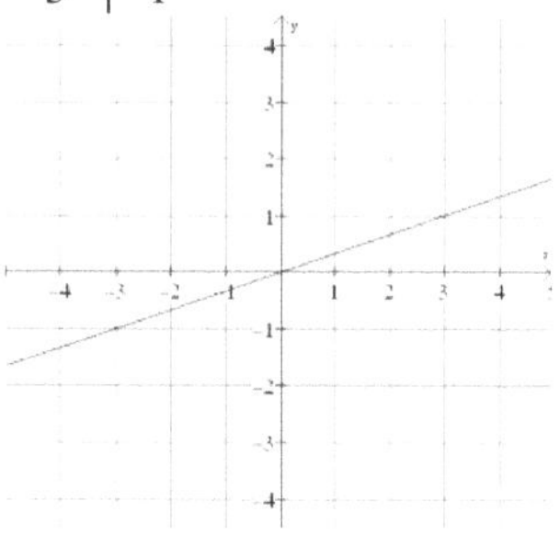

11. $y = \frac{2}{3}x - 1$

x	y
−3	−3
0	−1
3	1

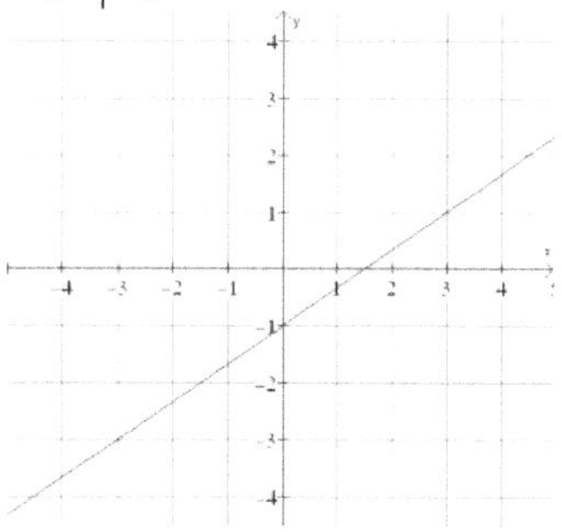

13. $y = -\frac{1}{4}x + 2$

x	y
−4	3
0	2
4	1

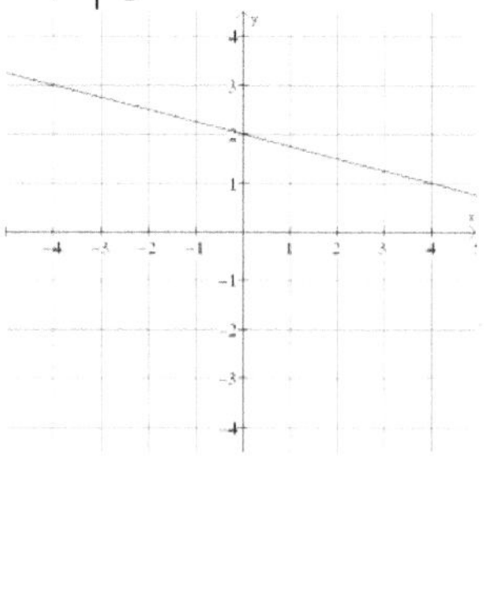

15. $y = -\frac{2}{5}x + 1$

x	y
−5	3
0	1
5	−1

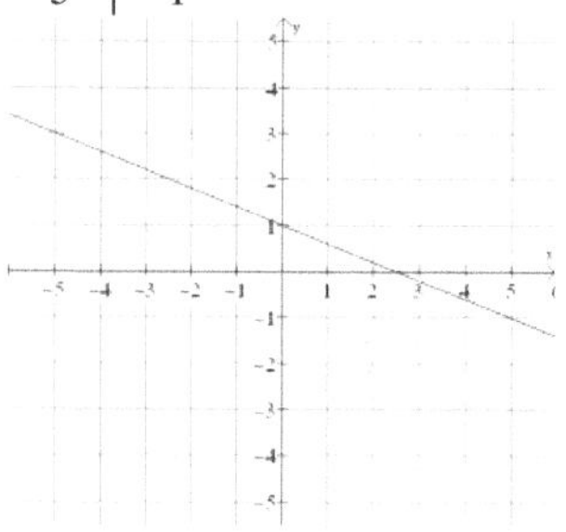

17. $y = 2x - 4$

x	y
0	−4
1	−2
2	0

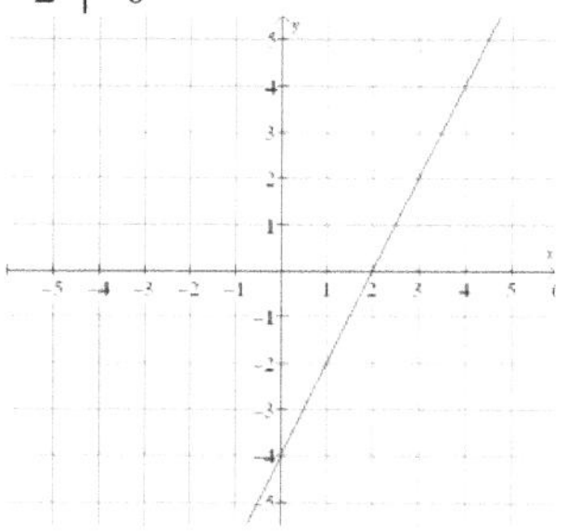

19. $y = x - 3$

x	y
0	−3
1	−2
3	0

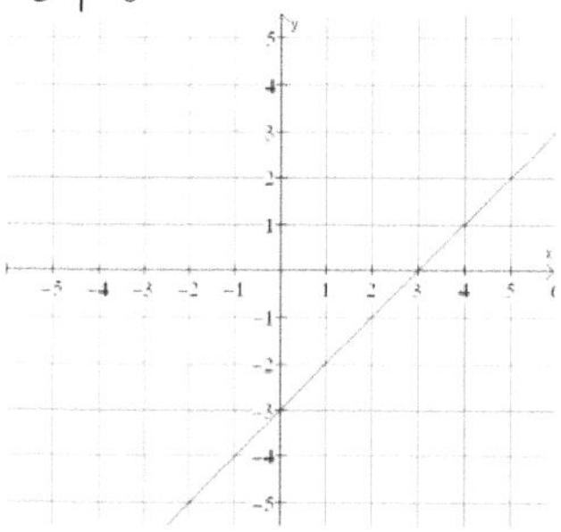

21. $y = -x + 2$

x	y
−2	4
0	2
2	0

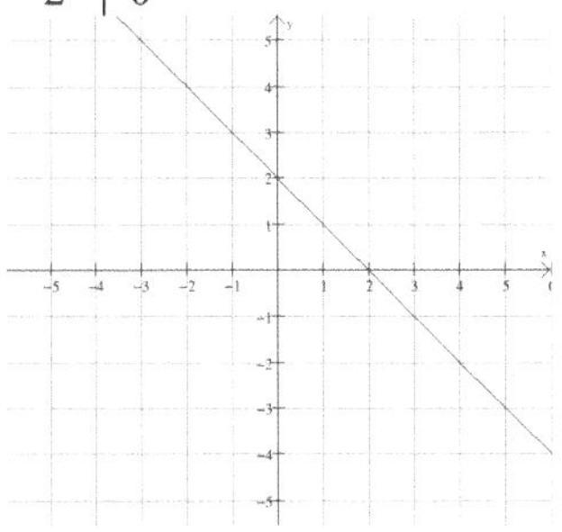

23. $y = -\frac{2}{3}x + 1$

x	y
−3	3
0	1
3	−1

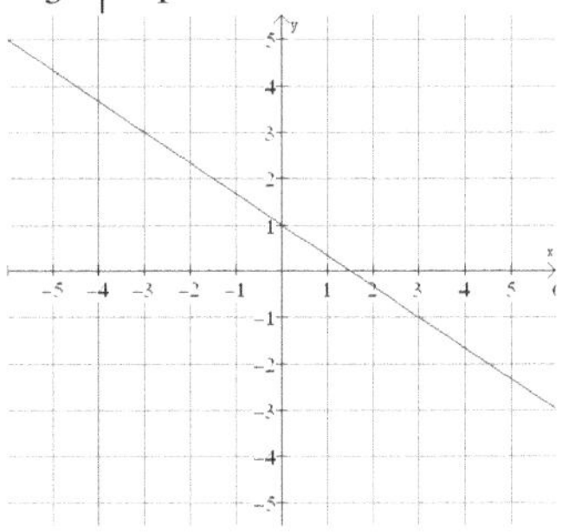

25. The origin is the point (0, 0).
If the graph passes through the point (0, 0), we can replace x with 0 and y with 0 and solve for b.

$y = mx + b$

$0 = m(0) + b$

$0 = b$

b is equal to 0.

Objective B Exercises

27. $2x + y = 5$

$y = -2x + 5$

29. $5x - y = 7$

$-y = -5x + 7$

$y = 5x - 7$

31. $2x + 3y = 9$

$3y = -2x + 9$

$y = -\frac{2}{3}x + 3$

33. $5x - 2y = 4$

$-2y = -5x + 4$

$y = \frac{5}{2}x - 2$

35. $6x - 5y = 10$

$-5y = -6x + 10$

$y = \frac{6}{5}x - 2$

37. $x - 4y = 12$

$-4y = -x + 12$

$y = \frac{1}{4}x - 3$

39. $2x + y = 4$

$y = -2x + 4$

x	y
0	4
1	2
2	0

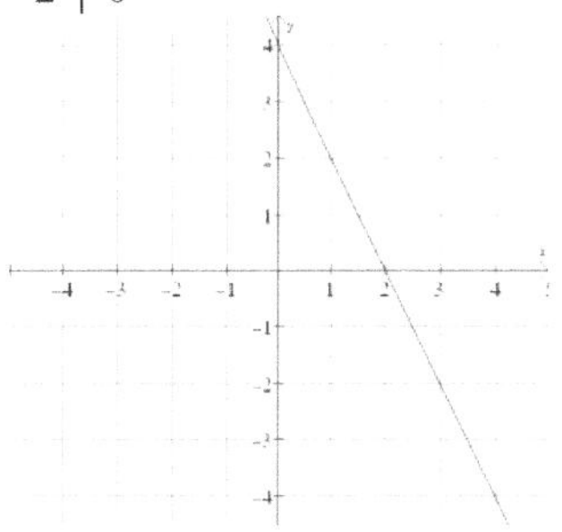

41. $3x + 2y = 4$

$2y = -3x + 4$

$y = -\frac{3}{2}x + 2$

x	y
−2	5
0	2
2	−1

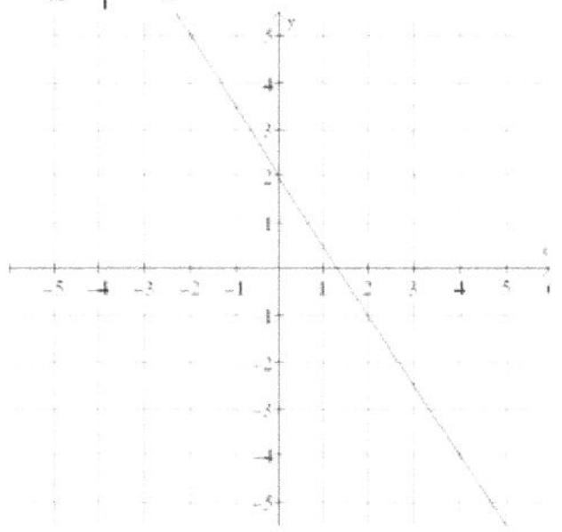

43. $x - 3y = 6$

$$-3y = -x + 6$$

$$y = \frac{1}{3}x - 2$$

x	y
−3	−3
0	−2
3	−1

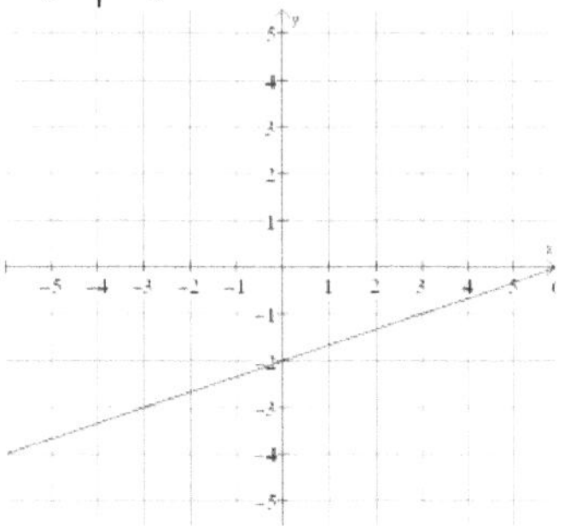

45. $3x - 2y = 8$

$$-2y = -3x + 8$$

$$y = \frac{3}{2}x - 4$$

x	y
0	−4
2	−1
4	2

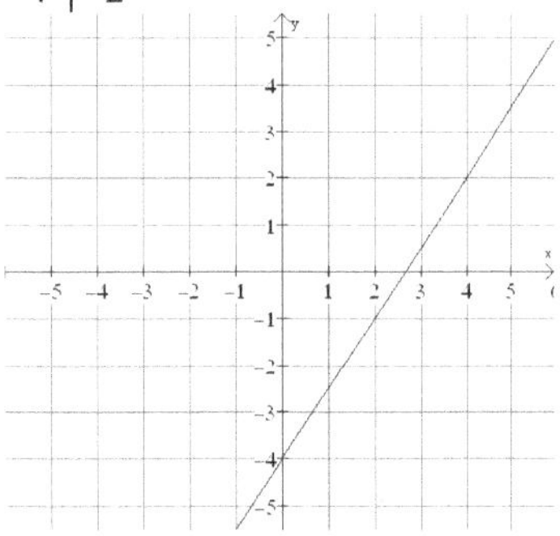

47. $3x + 4y = 12$

$$4y = -3x + 12$$

$$y = -\frac{3}{4}x + 3$$

x	y
−4	6
0	3
4	0

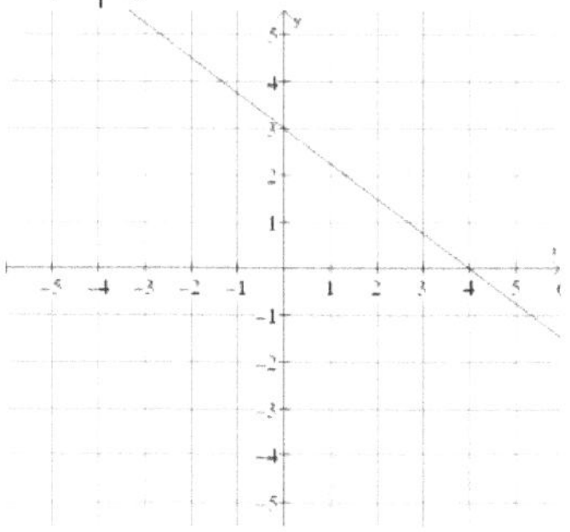

49. $y = -4$

The graph of an equation of the form $y = b$ is a horizontal line passing through the point (0, b).

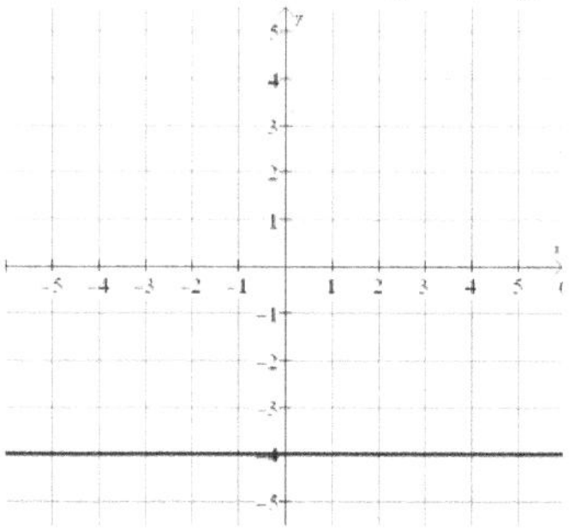

51. $4x - 3y = 12$

$$-3y = -4x + 12$$

$$y = \frac{4}{3}x - 4$$

x	y
0	−4
3	0
6	4

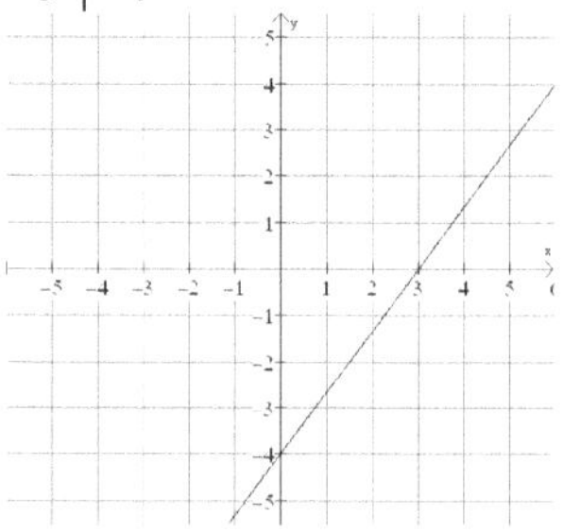

53. *A*

55. To the right

Objective C Exercises

57. $y = 10.5x + 25$

x	y
0	25
5	77.5
20	235

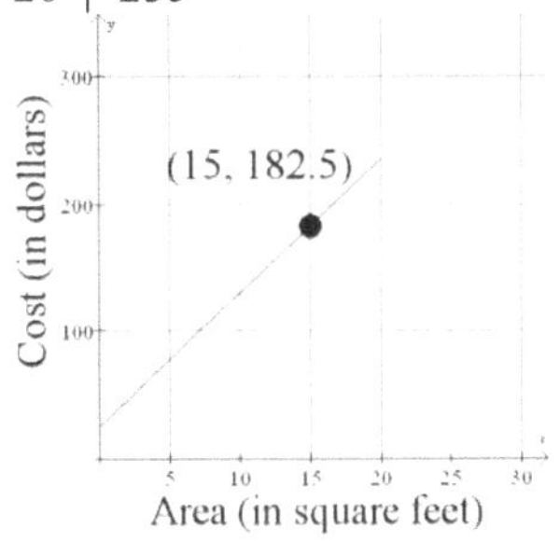

The ordered pair (15, 182.5) means that it costs \$182.50 for a custom sign 15 ft^2 in area.

59. $H = 4x + 16$

x	H
2	24
4	32
21	100

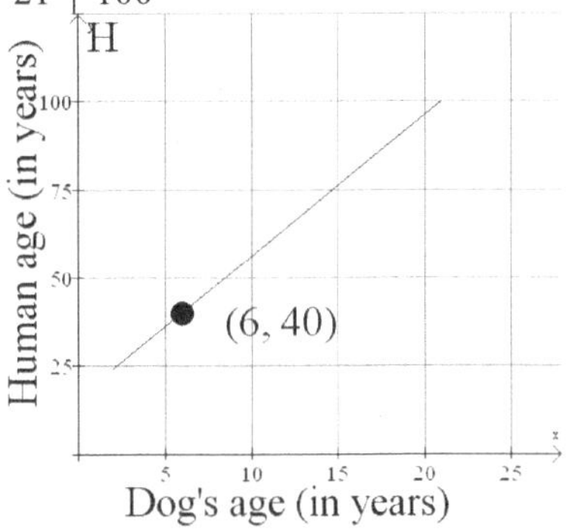

The ordered pair (6, 40) means that a dog 6 years old is equivalent in age to a human 40 years old.

Critical Thinking 12.2

61. Answers will vary. For example, $y = 2x$.

Projects and Group Activities 12.2

63. $y = 2x - 2$ $y = 2x + 3$ $y = 2x$

x	y
0	−2
1	−1
2	2

x	y
0	3
1	5
−1	1

x	y
0	0
1	2
2	4

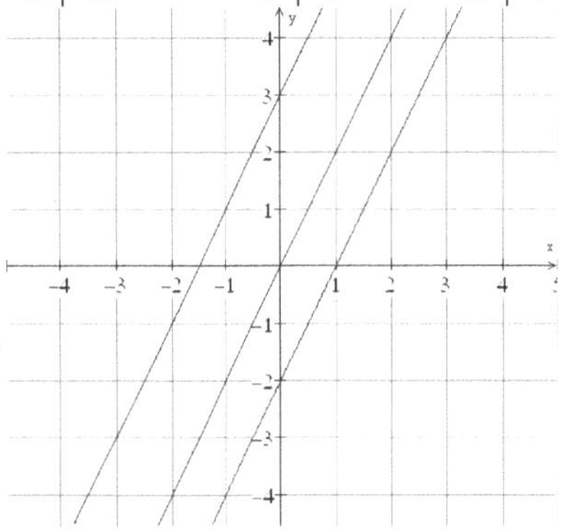

All the graphs are parallel.

Check Your Progress: Chapter 12

1.

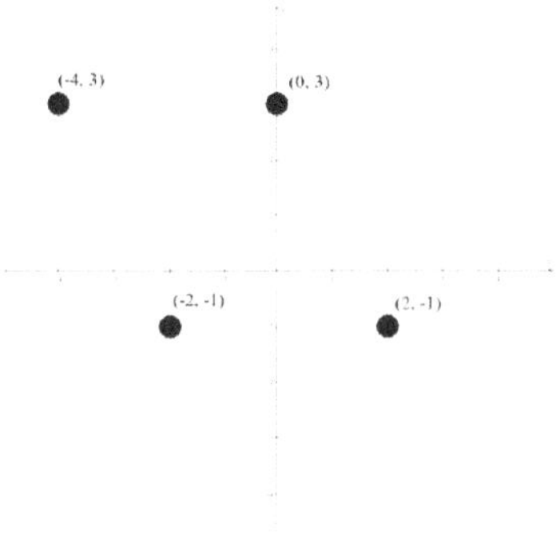

2. $y = -\frac{4}{5}x + 4$

x	y
−5	0
0	4
−5	8

3.

$$3x - 4y = 12$$
$$-4y = -3x + 12$$
$$y = \frac{3}{4}x - 3$$

x	y
0	–3
4	0
–4	–6

4.

x	$y = -3x + 2$	(x, y)
–1	$y = -3(-1) + 2 = 5$	(–1, 5)
0	$y = -3(0) + 2 = 2$	(0, 2)
1	$y = -3(1) + 2 = -1$	(1, –1)
2	$y = -3(2) + 2 = -4$	(2, –4)

5. A vertical line passing through (–5, 0).

6. D: {–3, –2, –1} R: {–2, –1, 0}
No two ordered pairs have the same first coordinate, so, yes, the relation is a function.

7.

$$h(s) = s^2 + 3s$$
$$h(-3) = (-3)^2 + 3(-3)$$
$$h(-3) = 9 - 9$$
$$h(-3) = 0$$

8.

y	$= -\frac{1}{3}x - 1$
0	$-\frac{1}{3}(-3) - 1$
0	$1 - 1$
0	0

$$0 = 0$$

Yes, (–3, 0) is a solution of $y = -\frac{1}{3}x - 1$.

9.

x	$y = x^2 - 4$	(x, y)
–3	$y = (-3)^2 - 4 = 5$	(–3, 5)
–1	$y = (-1)^2 - 4 = -3$	(–1, –3)
0	$y = (0)^2 - 4 = -4$	(0, –4)
1	$y = (1)^2 - 4 = -3$	(1, –3)
3	$y = (3)^2 - 4 = 5$	(3, 5)

The relation is {(–3, 4), (–1, –3), (0, –4), (1, –3), (3, 5)}.
No two ordered pairs have the same first coordinate, so, yes, the relation is a function.

Section 12.3

Concept Check

1. m

3. **a.** positive
b. negative
c. zero
d. undefined

5. x-coordinate

7. 2; –4; 3; 1

9. Vertical

11. $\frac{6}{5}$

Objective A Exercises

13. x-intercept

$$x - y = 3$$
$$x - 0 = 3$$
$$x = 0$$

(3, 0)

y-intercept

$$x - y = 3$$
$$0 - y = 3$$
$$-y = 3$$
$$y = -3$$

(0, –3)

15. x-intercept

$$3x - y = 6$$
$$3x - 0 = 6$$
$$3x = 6$$
$$x = 2$$

(2, 0)

y-intercept

$$3x - y = 6$$
$$3(0) - y = 6$$
$$-y = 6$$
$$y = -6$$

(0, –6)

17. x-intercept

$$x - 5y = 10$$
$$x - 5(0) = 10$$
$$x = 10$$
$(10, 0)$

y-intercept

$$x - 5y = 10$$
$$0 - 5y = 10$$
$$-5y = 10$$
$$y = -2$$
$(0, -2)$

19. x-intercept

$$3x - y = -12$$
$$3x - 0 = -12$$
$$3x = -12$$
$$x = -4$$
$(-4, 0)$

y-intercept

$$3x - y = -12$$
$$3(0) - y = -12$$
$$-y = -12$$
$$y = 12$$
$(0, 12)$

21. x-intercept

$$2x - 3y = 0$$
$$2x - 3(0) = 0$$
$$2x = 0$$
$$x = 0$$
$(0, 0)$

y-intercept

$$2x - 3y = 0$$
$$2(0) - 3y = 0$$
$$-3y = 0$$
$$y = 0$$
$(0, 0)$

23. x-intercept

$$x + 2y = 6$$
$$x + 2(0) = 6$$
$$x = 6$$
$(6, 0)$

y-intercept

$$x + 2y = 6$$
$$0 + 2y = 6$$
$$2y = 6$$
$$y = 3$$
$(0, 3)$

25. x-intercept

$$5x + 2y = 10$$
$$5x + 2(0) = 10$$
$$5x = 10$$
$$x = 2$$
$(2, 0)$

y-intercept

$$5x + 2y = 10$$
$$5(0) + 2y = 10$$
$$2y = 10$$
$$y = 5$$
$(0, 5)$

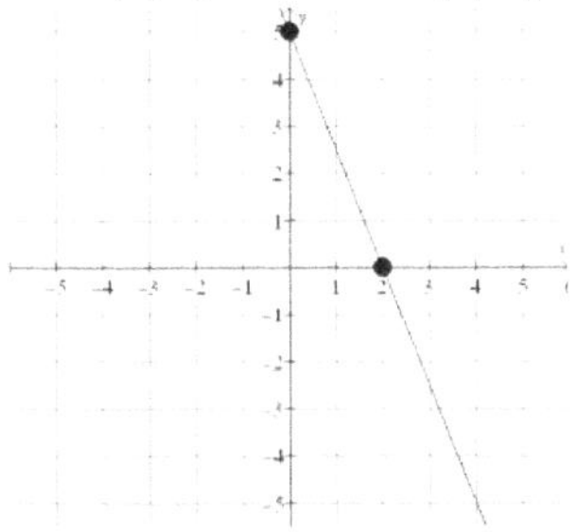

27. x-intercept

$$3x - 4y = 12$$
$$3x - 4(0) = 12$$
$$3x = 12$$
$$x = 4$$
$(4, 0)$

y-intercept

$$3x - 4y = 12$$
$$3(0) - 4y = 12$$
$$-4y = 12$$
$$y = -3$$
$(-3, 0)$

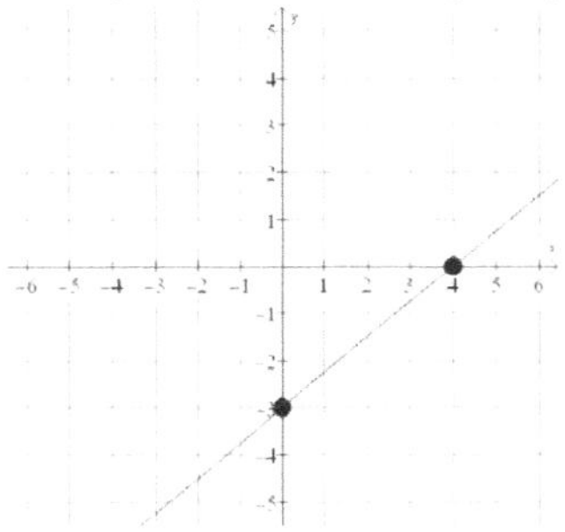

29. x-intercept

$$5y - 3x = 15$$
$$5(0) - 3x = 15$$
$$-3x = 15$$
$$x = -5$$
$(-5, 0)$

y-intercept

$$5y - 3x = 15$$
$$5y - 3(0) = 15$$
$$5y = 15$$
$$y = 3$$
$(0, 3)$

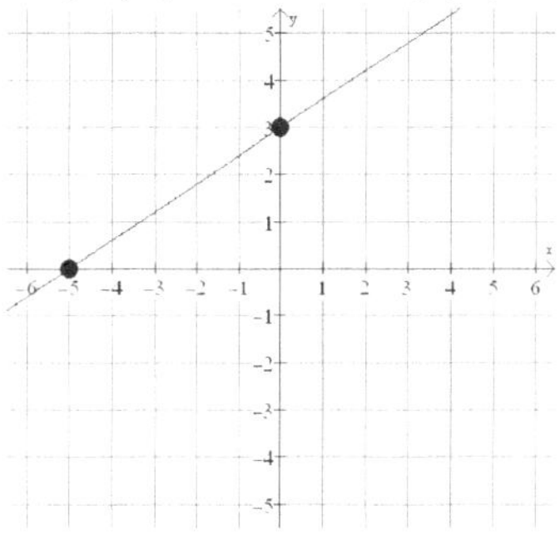

31. Find the y-intercept.

$$Ax + By = C$$
$$A(0) + By = C$$
$$By = C$$
$$y = \frac{C}{B}$$

If $B > 0$ and $C > 0$, then $y > 0$. The y-intercept is above the x-axis.

Objective B Exercises

33. $P_1(4,2),\ P_2(3,4)$

$$m = \frac{y_2 - y_1}{x_2 - x_1} = \frac{4-2}{3-4} = \frac{2}{-1} = -2$$

The slope is -2.

35. $P_1(-1,3),\ P_2(2,4)$

$$m=\frac{y_2-y_1}{x_2-x_1}=\frac{4-3}{2-(-1)}=\frac{1}{3}=\frac{1}{3}$$

The slope is $\frac{1}{3}$.

37. $P_1(2,4),\ P_2(4,-1)$

$$m=\frac{y_2-y_1}{x_2-x_1}=\frac{-1-4}{4-2}=\frac{-5}{2}=-\frac{5}{2}$$

The slope is $-\frac{5}{2}$.

39. $P_1(3,-4),\ P_2(3,5)$

$$m=\frac{y_2-y_1}{x_2-x_1}=\frac{5-(-4)}{3-3}=\frac{9}{0}$$

The slope is undefined.

41. $P_1(4,-2),\ P_2(3,-2)$

$$m=\frac{y_2-y_1}{x_2-x_1}=\frac{-2-(-2)}{3-4}=\frac{0}{-1}=0$$

The slope is 0.

43. $P_1(0,-1),\ P_2(3,-2)$

$$m=\frac{y_2-y_1}{x_2-x_1}=\frac{-2-(-1)}{3-0}=\frac{-1}{3}=-\frac{1}{3}$$

The slope is $-\frac{1}{3}$.

45. A line that has zero slope is a horizontal line; a line whose slope is undefined is a vertical line.

47. For the slope to be zero, the numerator must be zero and the denominator must not be zero.

$$m=\frac{y_2-y_1}{x_2-x_1}$$

$$m=\frac{d-b}{c-a}$$

$$d-b=0$$

$$d=b$$

So $b=d$, but $a\neq c$.

49. No, the slopes are not equal.

51. No, the product of the slopes is not -1.

53. $P_1(4,-5),\ P_2(6,-9)$

$$m=\frac{-9-(-5)}{6-4}=\frac{-4}{2}=-2$$

$Q_1(5,-4),\ \ Q_2(1,4)$

$$m=\frac{4-(-4)}{1-5}=\frac{-8}{-4}=-2$$

Parallel. The slopes are equal.

55. $P_1(5,1),\ P_2(3,-2)$

$$m=\frac{-2-1}{3-5}=\frac{-3}{-2}=\frac{3}{2}$$

$Q_1(0,-2),\ \ Q_2(3,-4)$

$$m=\frac{-4-(-2)}{3-0}=\frac{-2}{3}=-\frac{2}{3}$$

Perpendicular. The product of the slopes is -1.

57. $P_1(1,-1),\ P_2(3,-2)$

$$m=\frac{-2-(-1)}{3-1}=\frac{-1}{2}=-\frac{1}{2}$$

$Q_1(-4,1),\ \ Q_2(2,-5)$

$$m=\frac{-5-1}{2-(-4)}=\frac{-6}{6}=-1$$

Neither. The slopes are not equal nor is their product -1.

59. $P_1(5,-2),\ P_2(-1,3)$

$$m=\frac{3-(-2)}{-1-5}=\frac{5}{-6}=-\frac{5}{6}$$

$Q_1(3,4),\ \ Q_2(-2,-2)$

$$m=\frac{-2-4}{-2-3}=\frac{-6}{-5}=\frac{6}{5}$$

Perpendicular. The product of the slopes is -1.

61. The points are (2, 3.7) and (5, 1.8).

$$m=\frac{1.8-3.7}{5-2}=\frac{-1.9}{3}=-\frac{1.9}{3}\cdot\frac{10}{10}=-\frac{19}{30}$$

The slope is $-\frac{19}{30}=-0.6\overline{3}$. Each minute, the water in the lock decreases by $0.6\overline{3}$ million gallons.

63. The points are (2, 140) and (4, 280).

$$m=\frac{280-140}{4-2}=\frac{140}{2}=70$$

Walking burns 70 calories per minute.

65. The points are (2001, 73) and (2005, 82).

$$m = \frac{82-73}{2005-2001} = \frac{9}{4} = 2.25$$

The slope is 2.25. The percent of people using seat belts has increased by 2.25% per year.

Objective C Exercises

67.
$$y = -\frac{3}{8}x + 5$$

The slope is $-\frac{3}{8}$ and the y-intercept is (0, 5).

69.
$$2x - 3y = 6$$
$$-3y = -2x + 6$$
$$y = \frac{2}{3}x - 2$$

The slope is $\frac{2}{3}$ and the y-intercept is (0, −2).

71.
$$2x + 5y = 10$$
$$5y = -2x + 10$$
$$y = -\frac{2}{5}x + 2$$

The slope is $-\frac{2}{5}$ and the y-intercept is (0, 2).

73.
$$x - 4y = 0$$
$$-4y = -x$$
$$y = \frac{1}{4}x$$

The slope is $\frac{1}{4}$ and the y-intercept is (0, 0).

75.
$$y = 3x + 1$$
$$y\text{-intercept} = (0, b) = (0, 1)$$
$$m = 3 = \frac{3}{1} = \frac{\text{change in } y}{\text{change in } x}$$

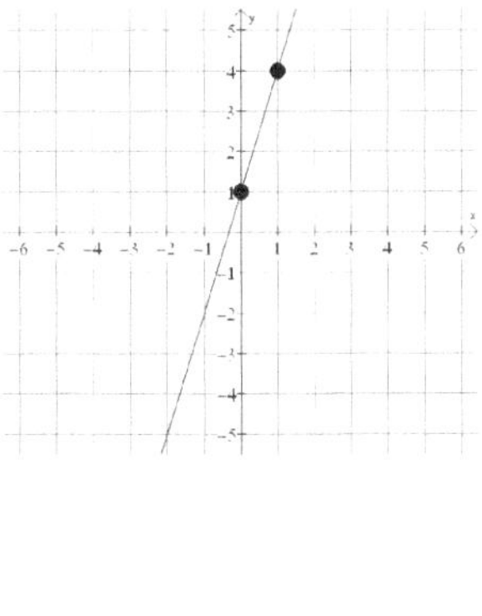

77.
$$y = \frac{2}{5}x - 2$$
$$y\text{-intercept} = (0, b) = (0, -2)$$
$$m = \frac{2}{5} = \frac{\text{change in } y}{\text{change in } x}$$

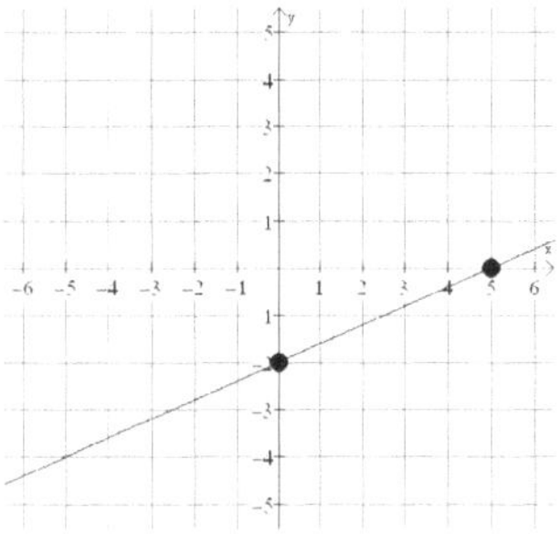

79.
$$2x + y = 3$$
$$y = -2x + 3$$
$$y\text{-intercept} = (0, b) = (0, 3)$$
$$m = -2 = \frac{-2}{1} = \frac{\text{change in } y}{\text{change in } x}$$

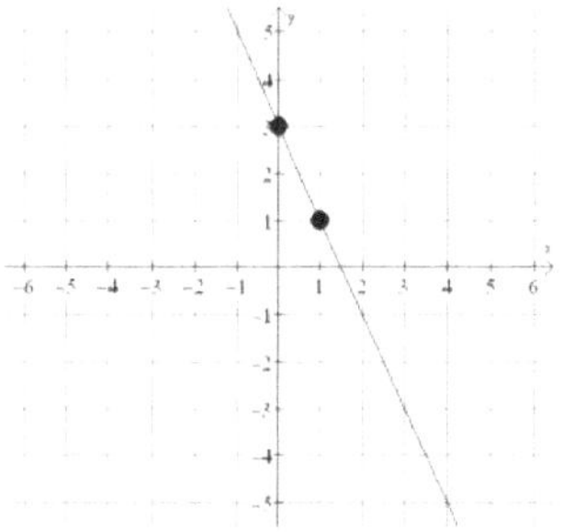

81.
$$x - 2y = 4$$
$$-2y = -x + 4$$
$$y = \frac{1}{2}x - 2$$
$$y\text{-intercept} = (0, b) = (0, -2)$$
$$m = \frac{1}{2} = \frac{\text{change in } y}{\text{change in } x}$$

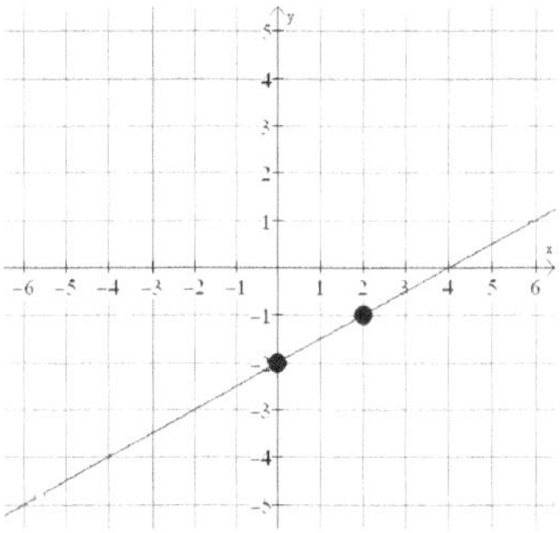

83. $y = \frac{2}{3}x$

$y\text{-intercept} = (0,b) = (0,0)$

$m = \frac{2}{3} = \frac{\text{change in } y}{\text{change in } x}$

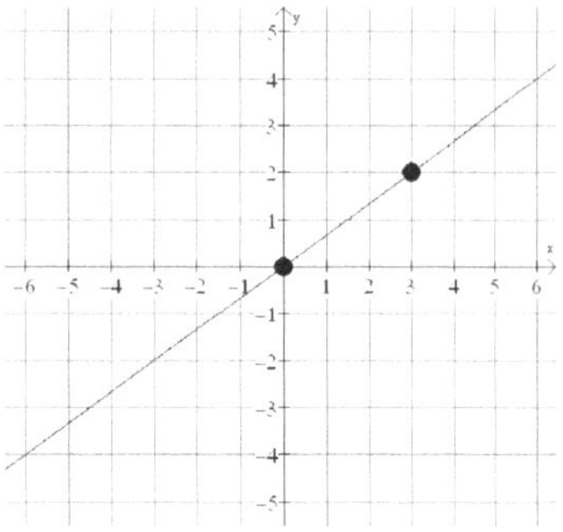

85. $y = -x + 1$

$y\text{-intercept} = (0,b) = (0,1)$

$m = -1 = \frac{-1}{1} = \frac{\text{change in } y}{\text{change in } x}$

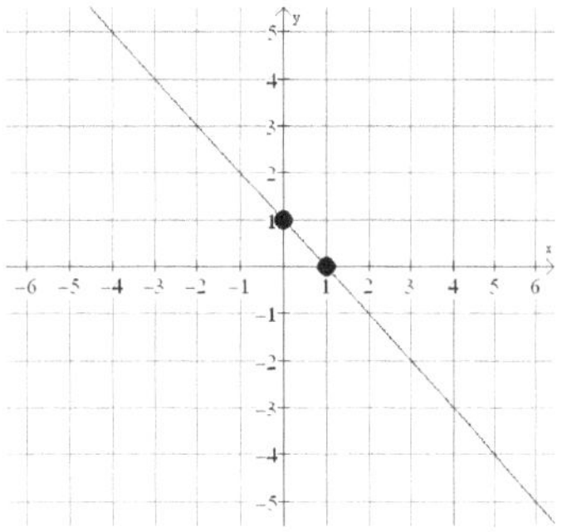

87. $3x - 4y = 12$

$-4y = -3x + 12$

$y = \frac{3}{4}x - 3$

$y\text{-intercept} = (0,b) = (0,-3)$

$m = \frac{3}{4} = \frac{\text{change in } y}{\text{change in } x}$

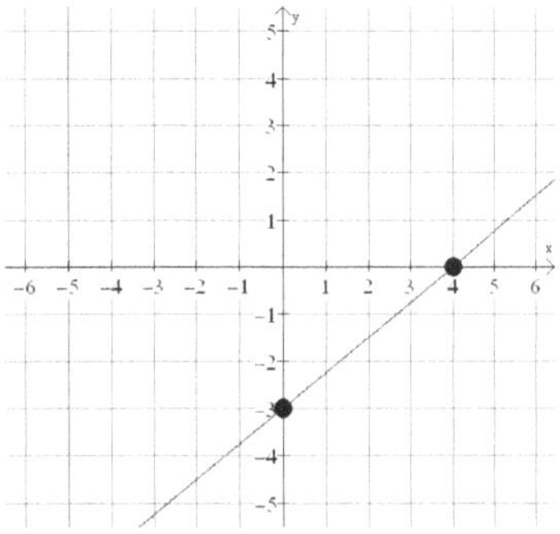

89. $y = -4x + 2$

$y\text{-intercept} = (0,b) = (0,2)$

$m = -4 = \frac{-4}{1} = \frac{\text{change in } y}{\text{change in } x}$

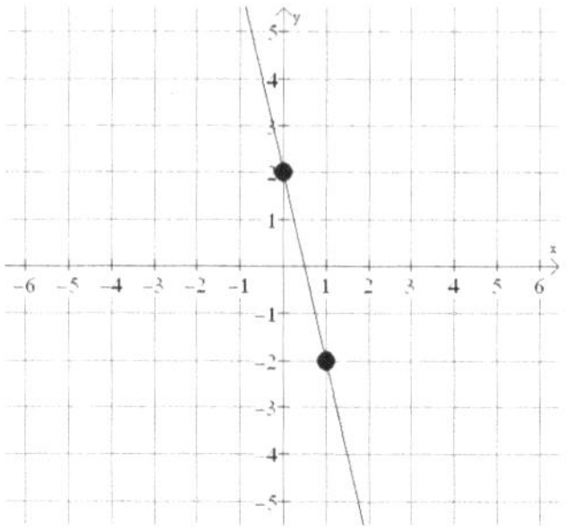

91. To find the y-intercept, let $x = 0$:

$Ax + By = C$

$A(0) + By = C$

$By = C$

$y = \frac{C}{B}$

As both values are negative, y is positive, so it is above the x-axis.

To determine which direction the line slants, find the slope.

$Ax + By = C$

$By = -Ax + C$

$y = -\frac{A}{B}x + \frac{C}{B}$

The slope is $-\frac{A}{B}$. If $A < 0$ and $B > 0$, then the slope is positive. The graph slants upward to the right.

Critical Thinking 12.3

93. Decreases the slope.

95. Decreases the y-coordinate of the x-intercept.

97. No, not all straight lines have a y-intercept. Any line that is parallel to the y-axis does not have a y-intercept. An example is the graph of $x = 2$.

Projects and Group Activities 12.3

99. Line a slants upward to the right. Line b slants downward to the right.

101. Line a has a y-intercept of (0, 5). Line b has a y-intercept of (0, –5).

103. Line a is a horizontal line. Line b is a vertical line.

105. $y = 2x + 1$

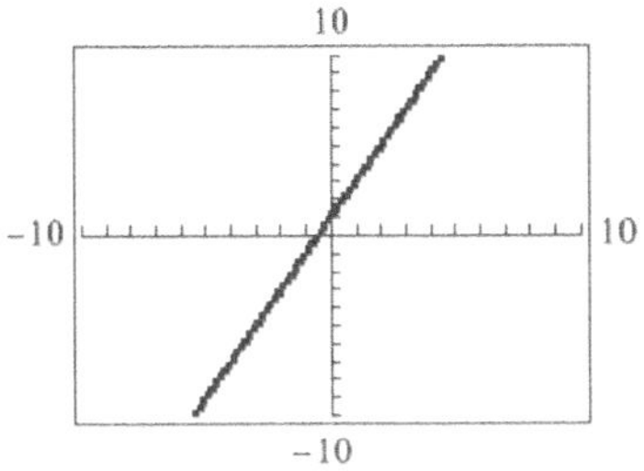

107. $3x + 2y = 6$

$2y = -3x + 6$

$y = -\frac{3}{2}x + 3$

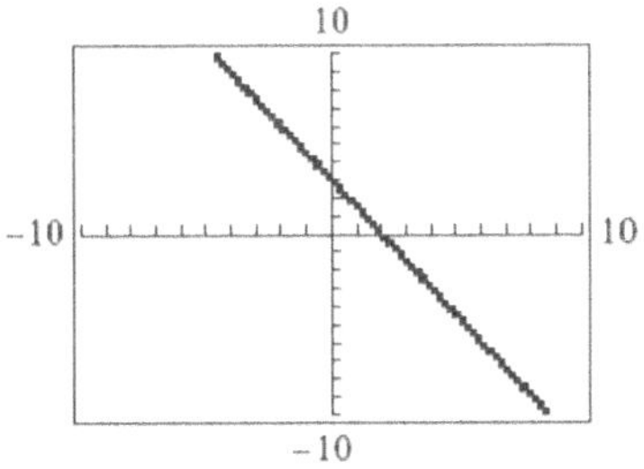

Section 12.4

Concept Check

1. 5; 7

3. 0; 2

5. 3; 1; $y = 3x + 1$

Objective A Exercises

7. $m = 2,\ b = 2$

$y = mx + b$

$y = 2x + 2$

The equation of the line is $y = 2x + 2$.

9. $m = -3,\ (x_1, y_1) = (-1, 2)$

$y = mx + b$

$2 = -3(-1) + b$

$2 = 3 + b$

$-1 = b$

$y - -3x - 1$

The equation of the line is $y = -3x - 1$.

11. $m = \frac{1}{3},\ (x_1, y_1) = (3, 1)$

$y = mx + b$

$1 = \frac{1}{3}(3) + b$

$1 = 1 + b$

$0 = b$

$y = \frac{1}{3}x$

The equation of the line is $y = \frac{1}{3}x$.

13. $m = \frac{3}{4},\ (x_1, y_1) = (4, -2)$

$y = mx + b$

$-2 = \frac{3}{4}(4) + b$

$-2 = 3 + b$

$-5 = b$

$y = \frac{3}{4}x - 5$

The equation of the line is $y = \frac{3}{4}x - 5$.

15. $m = -\frac{3}{5},\ (x_1, y_1) = (5, -3)$

$y = mx + b$

$-3 = -\frac{3}{5}(5) + b$

$-3 = -3 + b$

$0 = b$

$y = -\frac{3}{5}x$

The equation of the line is $y = -\frac{3}{5}x$.

17. $m = \frac{1}{4},\ (x_1, y_1) = (2, 3)$

$y = mx + b$

$3 = \frac{1}{4}(2) + b$

$3 = \frac{1}{2} + b$

$\frac{5}{2} = b$

$y = \frac{1}{4}x + \frac{5}{2}$

The equation of the line is $y = \frac{1}{4}x + \frac{5}{2}$.

19. $m=-\frac{2}{3};\quad (x_1,y_1)=(-3,-5)$

$$
\begin{aligned}
y &= mx+b\\
-5 &= -\frac{2}{3}(-3)+b\\
-5 &= 2+b\\
-7 &= b\\
y &= -\frac{2}{3}x-7
\end{aligned}
$$

The equation of the line is $y=-\frac{2}{3}x-7$.

Objective B Exercises

21. $m=2,\ (x_1,y_1)=(1,\ -1)$

$$
\begin{aligned}
y-y_1 &= m(x-x_1)\\
y-(-1) &= 2(x-1)\\
y+1 &= 2x-2\\
y &= 2x-3
\end{aligned}
$$

The equation of the line is $y=2x-3$.

23. $m=-2,\ (x_1,y_1)=(-2,\ 1)$

$$
\begin{aligned}
y-y_1 &= m(x-x_1)\\
y-1 &= -2(x-(-2))\\
y-1 &= -2(x+2)\\
y-1 &= -2x-4\\
y &= -2x-3
\end{aligned}
$$

The equation of the line is $y=-2x-3$.

25. $m=\frac{2}{3},\ (x_1,y_1)=(0,\ 0)$

$$
\begin{aligned}
y-y_1 &= m(x-x_1)\\
y-0 &= \frac{2}{3}(x-0)\\
y &= \frac{2}{3}x
\end{aligned}
$$

The equation of the line is $y=\frac{2}{3}x$.

27. $m=\frac{1}{2},\ (x_1,y_1)=(2,\ 3)$

$$
\begin{aligned}
y-y_1 &= m(x-x_1)\\
y-3 &= \frac{1}{2}(x-2)\\
y-3 &= \frac{1}{2}x-1\\
y &= \frac{1}{2}x+2
\end{aligned}
$$

The equation of the line is $y=\frac{1}{2}x+2$.

29. $m=-\frac{3}{4},\ (x_1,y_1)=(-4,\ 1)$

$$
\begin{aligned}
y-y_1 &= m(x-x_1)\\
y-1 &= -\frac{3}{4}(x-(-4))\\
y-1 &= -\frac{3}{4}(x+4)\\
y-1 &= -\frac{3}{4}x-3\\
y &= -\frac{3}{4}x-2
\end{aligned}
$$

The equation of the line is $y=-\frac{3}{4}x-2$.

31. $m=\frac{3}{4},\ (x_1,y_1)=(-2,\ 1)$

$$
\begin{aligned}
y-y_1 &= m(x-x_1)\\
y-1 &= \frac{3}{4}(x-(-2))\\
y-1 &= \frac{3}{4}(x+2)\\
y-1 &= \frac{3}{4}x+\frac{3}{2}\\
y &= \frac{3}{4}x+\frac{5}{2}
\end{aligned}
$$

The equation of the line is $y=\frac{3}{4}x+\frac{5}{2}$.

33. $m=-\frac{4}{3},\ (x_1,y_1)=(-3,-5)$

$$
\begin{aligned}
y-y_1 &= m(x-x_1)\\
y-(-5) &= -\frac{4}{3}(x-(-3))\\
y+5 &= -\frac{4}{3}x-4\\
y &= -\frac{4}{3}x-9
\end{aligned}
$$

The equation of the line is $y=-\frac{4}{3}x-9$..

35. $m, (0, b)$

$$y - y_1 = m(x - x_1)$$
$$y - b = m(x - 0)$$
$$y - b = mx$$
$$y = mx + b$$

The equation of the line is $y = mx + b$.

Objective C Exercises

37. $m = \dfrac{y_2 - y_1}{x_2 - x_1} = \dfrac{7 - (-2)}{1 - (-2)} = \dfrac{9}{3} = 3$

$m = 3, \qquad (x_1, y_1) = (-2, -2)$

$$y - (-2) = 3(x - (-2))$$
$$y + 2 = 3x + 6$$
$$y = 3x + 4$$

The equation of the line is $y = 3x + 4$.

39. $m = \dfrac{y_2 - y_1}{x_2 - x_1} = \dfrac{-6 - 1}{2 - (-5)} = \dfrac{-7}{7} = -1$

$m = -1, \quad (x_1, y_1) = (-5, 1)$

$$y - y_1 = m(x - x_1)$$
$$y - 1 = -1(x - (-5))$$
$$y - 1 = -x - 5$$
$$y = -x - 4$$

The equation of the line is $y = -x - 4$.

41. $m = \dfrac{y_2 - y_1}{x_2 - x_1} = \dfrac{11 - (-1)}{-5 - 5} = \dfrac{12}{-10} = -\dfrac{6}{5}$

$m = -\dfrac{6}{5}, \quad (x_1, y_1) = (5, -1)$

$$y - y_1 = m(x - x_1)$$
$$y - (-1) = -\frac{6}{5}(x - 5)$$
$$y + 1 = -\frac{6}{5}x + 6$$
$$y = -\frac{6}{5}x + 5$$

The equation of the line is $y = -\dfrac{6}{5}x + 5$.

43. $m = \dfrac{y_2 - y_1}{x_2 - x_1} = \dfrac{-9 - (-3)}{5 - (-10)} = \dfrac{-6}{15} = -\dfrac{2}{5}$

$m = -\dfrac{2}{5}, \quad (x_1, y_1) = (-10, -3)$

$$y - y_1 = m(x - x_1)$$
$$y - (-3) = -\frac{2}{5}(x - (-10))$$
$$y + 3 = -\frac{2}{5}x - 4$$
$$y = -\frac{2}{5}x - 7$$

The equation of the line is $y = -\dfrac{2}{5}x - 7$.

45. $m = \dfrac{y_2 - y_1}{x_2 - x_1} = \dfrac{5 - 5}{-6 - 1} = \dfrac{0}{-7} = 0$

$m = 0, \quad (x_1, y_1) = (1, 5)$

$$y - y_1 = m(x - x_1)$$
$$y - 5 = 0(x - 1)$$
$$y - 5 = 0$$
$$y = 5$$

The equation of the line is $y = 5$.

47. $m = \dfrac{y_2 - y_1}{x_2 - x_1} = \dfrac{-7 - (-1)}{5 - 5} = \dfrac{-6}{0} = \text{undefined}$

If the slope is undefined, the graph of the line is vertical and the equation is $x = a$.
The equation of the line is $x = 5$.

49. $m = \dfrac{y_2 - y_1}{x_2 - x_1} = \dfrac{12 - (-8)}{5 - (-20)} = \dfrac{20}{25} = \dfrac{4}{5}$

$m = \dfrac{4}{5}, \quad (x_1, y_1) = (-20, -8)$

$$y - y_1 = m(x - x_1)$$
$$y - (-8) = \frac{4}{5}(x - (-20))$$
$$y + 8 = \frac{4}{5}x + 16$$
$$y = \frac{4}{5}x + 8$$

The equation of the line is $y = \dfrac{4}{5}x + 8$.

51.
$$m=\frac{y_2-y_1}{x_2-x_1}=\frac{1-(-2)}{-6-0}=\frac{3}{-6}=-\frac{1}{2}$$
$$m=-\frac{1}{2},\ (x_1, y_1)=(0, -2)$$
$$y-y_1=m(x-x_1)$$
$$y-(-2)=-\frac{1}{2}(x-0)$$
$$y+2=-\frac{1}{2}x$$
$$y=-\frac{1}{2}x-2$$
The equation of the line is $y=-\frac{1}{2}x-2.$

53.
$$m=\frac{y_2-y_1}{x_2-x_1}=\frac{1-(-11)}{-3-6}=\frac{12}{-9}=-\frac{4}{3}$$
$$m=-\frac{4}{3},\ (x_1, y_1)=(6, -11)$$
$$y-y_1=m(x-x_1)$$
$$y-(-11)=-\frac{4}{3}(x-6)$$
$$y+11=-\frac{4}{3}x+8$$
$$y=-\frac{4}{3}x-3$$
The equation of the line is $y=-\frac{4}{3}x-3.$

55.
$$m=\frac{y_2-y_1}{x_2-x_1}=\frac{-3-6}{0-3}=\frac{-9}{-3}=3$$
$$m=3,\ (x_1, y_1)=(3, 6)$$
$$y-y_1=m(x-x_1)$$
$$y-6=3(x-3)$$
$$y-6=3x-9$$
$$y=3x-3$$
The equation of the line is $y=3x-3.$

57.
$$m=\frac{y_2-y_1}{x_2-x_1}=\frac{6-(-3)}{2-(-1)}=\frac{9}{3}=3$$
$$m=3,\ (x_1, y_1)=(-1, -3)$$
$$y-y_1=m(x-x_1)$$
$$y-(-3)=3(x-(-1))$$
$$y+3=3x+3$$
$$y=3x$$
The equation of the line is $y=3x.$

59.
$$m=\frac{y_2-y_1}{x_2-x_1}=\frac{1-(-5)}{3-3}=\frac{6}{0}=\text{undefined}$$
If the slope is undefined, the graph of the line is vertical and the equation is $x=a$.
The equation of the line is $x=3$.

61. 2

Critical Thinking 12.4

63. Strategy:
Find the slope between the first two points and the slope between the second and third points. If the slopes are the same, the points are on a straight line. Use the slope and any point to find the equation of the line.

Solution:
$$m_1=\frac{-3-(-4)}{0-(-2)}=\frac{1}{2}=\frac{1}{2}$$
$$m_2=\frac{-1-(-3)}{4-0}=\frac{2}{4}=\frac{1}{2}$$
Since the slopes are the same, the three points are on a straight line.
$$m=\frac{1}{2}\ (x_1, y_1)=(-2, -4)$$
$$y-y_1=m(x-x_1)$$
$$y-(-4)=\frac{1}{2}(x-(-2))$$
$$y+4=\frac{1}{2}x+1$$
$$y=\frac{1}{2}x-3$$
The equation of the line is $y=\frac{1}{2}x-3.$

65. Strategy:
Find the slope between the first two points and the slope between the second and third points. If the slopes are the same, the points are on a straight line. Use the slope and any point to find the equation of the line.

Solution:

$$m_1 = \frac{-4-(-1)}{12-3} = \frac{-3}{9} = -\frac{1}{3}$$
$$m_2 = \frac{2-(-4)}{-6-12} = \frac{6}{-18} = -\frac{1}{3}$$

Since the slopes are the same, the three points are on a straight line.

$$m = -\frac{1}{3} \quad (x_1, y_1) = (3, -1)$$
$$y - y_1 = m(x - x_1)$$
$$y - (-1) = -\frac{1}{3}(x-3)$$
$$y + 1 = -\frac{1}{3}x + 1$$
$$y = -\frac{1}{3}x$$

The equation of the line is $y = -\frac{1}{3}x$.

67. Strategy:
To solve for n:
→Find the slope of the line using the first two points.
→Find an expression for the slope using the second and third points.
→The slopes must be equal.

Solution:

$$m_1 = \frac{5-2}{-1-2} = \frac{3}{-3} = -1$$
$$m_2 = \frac{n-5}{3-(-1)} = \frac{n-5}{4}$$
$$\frac{n-5}{4} = -1$$
$$4\left(\frac{n-5}{4}\right) = 4(-1)$$
$$n - 5 = -4$$
$$n = 1$$

69. Strategy:
To solve for n:
→Find the slope of the line using the first two points.
→Find an expression for the slope using the second and third points.
→The slopes must be equal.

Solution:

$$m_1 = \frac{4-(-2)}{-2-1} = \frac{6}{-3} = -2$$
$$m_2 = \frac{n-4}{4-(-2)} = \frac{n-4}{6}$$
$$\frac{n-4}{6} = -2$$
$$6\left(\frac{n-4}{6}\right) = 6(-2)$$
$$n - 4 = -12$$
$$n = -8$$

Projects and Group Activities 12.4

71. x-intercept: (3, 0)
y-intercept: (0, 2)

$$m = -\frac{2}{3}$$
$$y = mx + b$$
$$y = -\frac{2}{3}x + 2$$

The equation of the line is $y = -\frac{2}{3}x + 2$.

73. x-intercept: (–1, 0)
y-intercept: (0, –2)

$$m = -2$$
$$y = mx + b$$
$$y = -2x - 2$$

The equation of the line is $y = -2x - 2$.

Chapter Review Exercises

1. **a.**

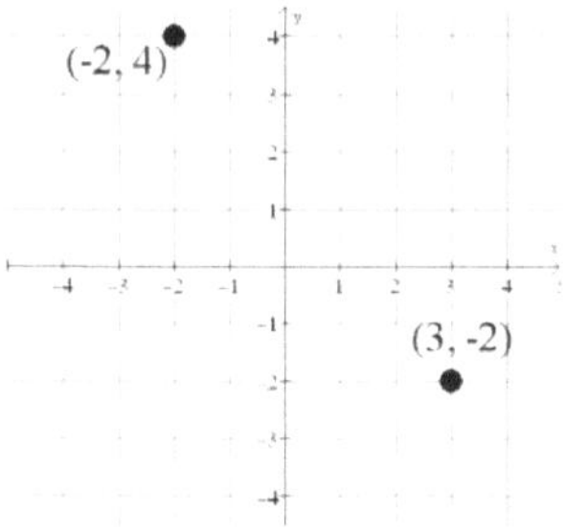

b. Abscissa of point A: -2

c. Ordinate of point B: -4

2.

x	$y=-\frac{1}{2}x-2$	(x, y)
-4	$-\frac{1}{2}(-4)-2=0$	$(-4, 0)$
-2	$-\frac{1}{2}(-2)-2=-1$	$(-2, -1)$
0	$-\frac{1}{2}(0)-2=-2$	$(0, -2)$
2	$-\frac{1}{2}(2)-2=-3$	$(2, -3)$

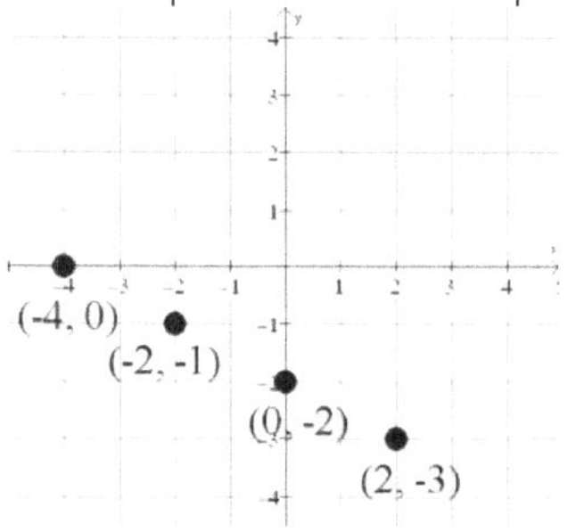

3.

$$m=\frac{y_2-y_1}{x_2-x_1}=\frac{-5-3}{2-(-1)}=\frac{-8}{3}=-\frac{8}{3}$$

$$m=-\frac{8}{3},\ (x_1, y_1)=(-1, 3)$$

$$y-y_1=m(x-x_1)$$

$$y-3=-\frac{8}{3}(x+1)$$

$$y-3=-\frac{8}{3}x-\frac{8}{3}$$

$$y=-\frac{8}{3}x+\frac{1}{3}$$

The equation of the line is $y=-\frac{8}{3}x+\frac{1}{3}$.

4.

$$m=-\frac{5}{2},\ (x_1, y_1)=(6, 1)$$

$$y-y_1=m(x-x_1)$$

$$y-1=-\frac{5}{2}(x-6)$$

$$y-1=-\frac{5}{2}x+15$$

$$y=-\frac{5}{2}x+16$$

The equation of the line is $y=-\frac{5}{2}x+16$.

5.

$$y=\frac{1}{4}x+3$$

$$y-\text{intercept}=(0,b)=(0,3)$$

$$m=\frac{1}{4}=\frac{\text{change in } y}{\text{change in } x}$$

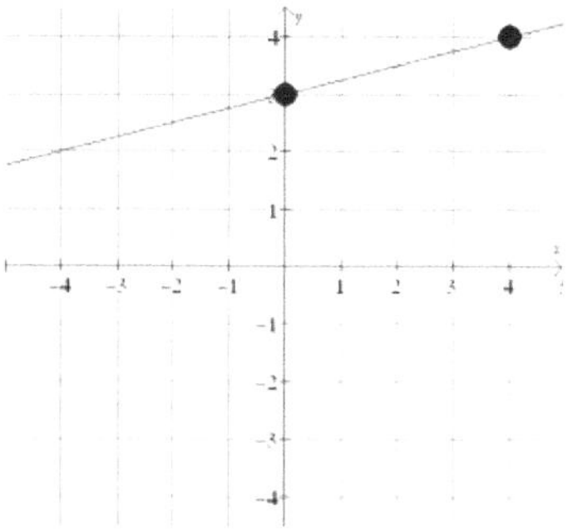

6.

$$5x+3y=15$$

$$3y=-5x+15$$

$$y=-\frac{5}{3}x+5$$

$$y-\text{intercept}=(0,b)=(0,5)$$

$$m=-\frac{5}{3}=\frac{-5}{3}=\frac{\text{change in } y}{\text{change in } x}$$

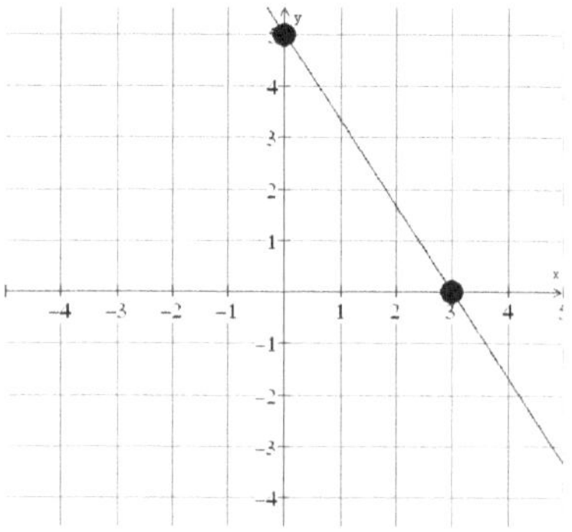

7. $P_1(7,-5),\ P_2(6,-1)$

$$m=\frac{-1-(-5)}{6-7}=\frac{4}{-1}=-4$$

$Q_1(4,5),\quad Q_2(2,-3)$

$$m=\frac{-3-5}{2-4}=\frac{-8}{-2}=4$$

Neither. The slopes are not equal nor is their product -1.

8.

$$f(x)=x^2-2$$

$$f(-1)=(-1)^2-2=1-2=-1$$

9. No two ordered pairs have the same first coordinates, so yes, y is a function of x.

10.

$$m=\frac{y_2-y_1}{x_2-x_1}=\frac{1-8}{-2-9}=\frac{-7}{-11}=\frac{7}{11}$$

The slope is $\frac{7}{11}$.

11. $3x-2y=24$

y-intercept	x-intercept
$3(0)-2y=24$	$3x-2(0)=24$
$-2y=24$	$3x=24$
$y=-12$	$x=8$
$(0,-12)$	$(8,0)$

12.

$$m=\frac{y_2-y_1}{x_2-x_1}=\frac{-3-(-3)}{-2-4}=\frac{0}{-6}=0$$

The line has zero slope.

13. y-intercept = (0, −1)

$$m=\frac{1}{2}=\frac{\text{change in } y}{\text{change in } x}$$

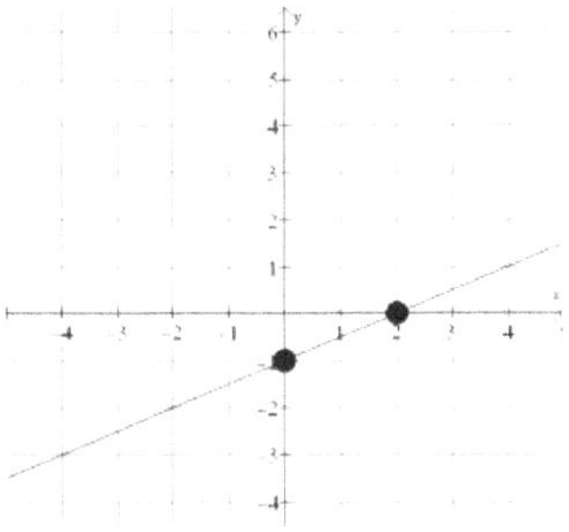

14. $x=-3$

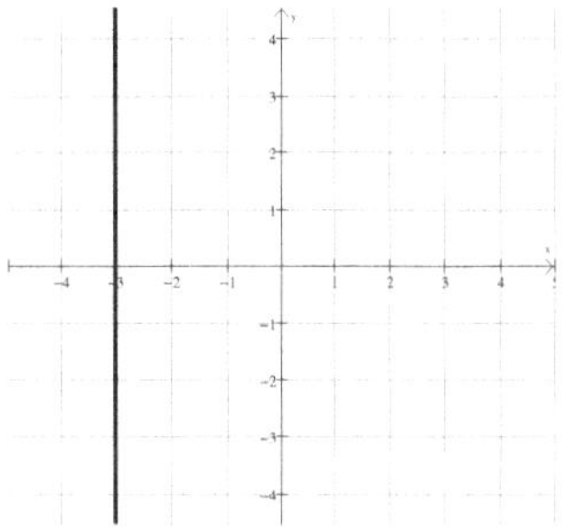

15. y-intercept = (0, 2)

$$m=-\frac{2}{3}=\frac{-2}{3}=\frac{\text{change in } y}{\text{change in } x}$$

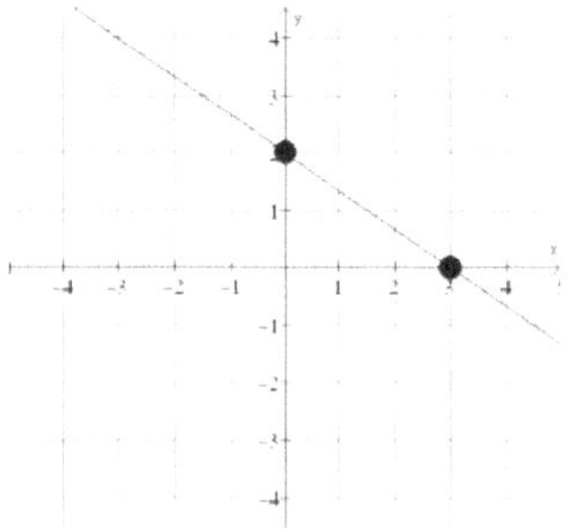

16. $y=-2x-1$

x	y
0	−1
1	−3
−1	1

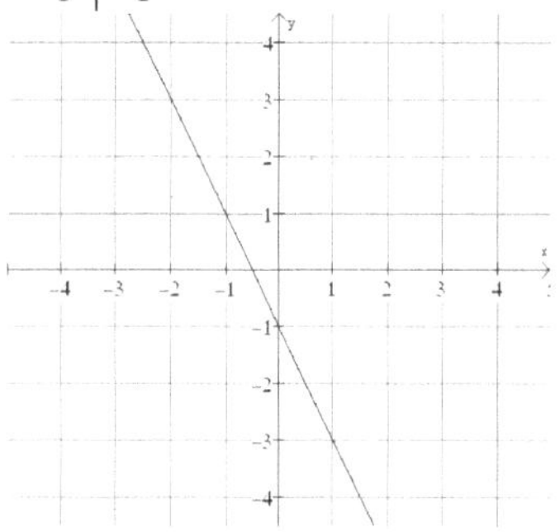

17. y-intercept = (0, −4)

$$m=2=\frac{2}{1}=\frac{\text{change in } y}{\text{change in } x}$$

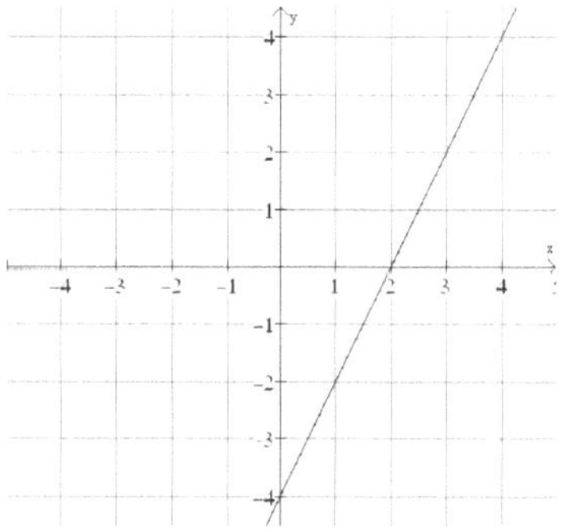

18. $3x - 2y = -6$

$$-2y = -3x - 6$$

$$y = \frac{3}{2}x + 3$$

x	y
0	3
−2	0
−4	−3

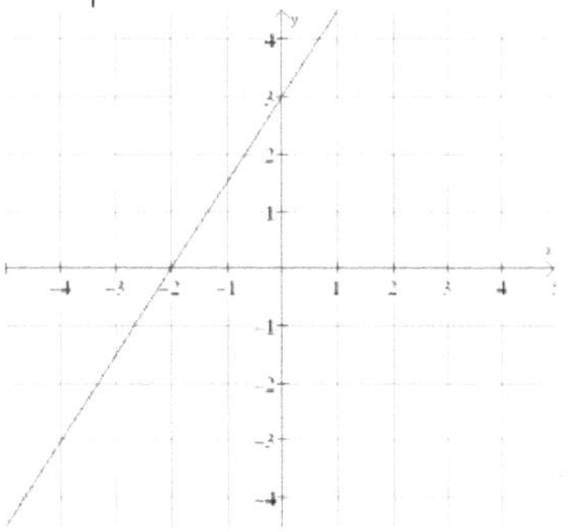

19. {(55, 95), (57, 101), (53, 94), (57, 98), (60, 100), (61, 105), (58, 97), (54, 64)}
No, the relation is not a function. The two ordered pairs (57, 101) and (57, 98) have the same first component but different second components.

20.

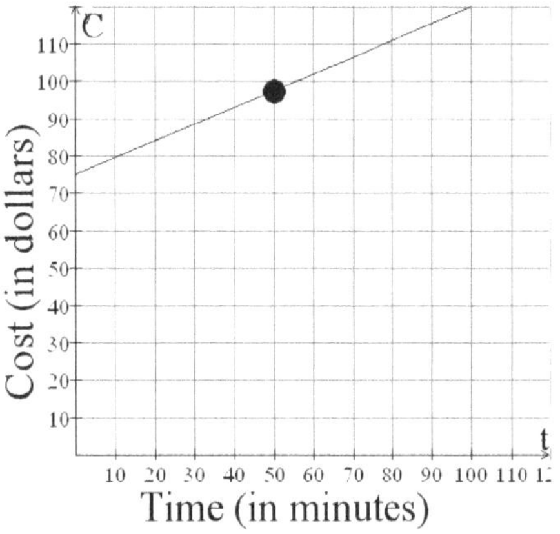

The cost of 50 min of access time for one month is \$97.50.

Chapter Test

1. $2x - 3y = 15$

$$2(3) - 3y = 15$$

$$6 - 3y = 15$$

$$-3y = 9$$

$$y = -3$$

The ordered pair solution is (3, −3).

2.

x	$y = -\frac{3}{2}x + 1$	(x, y)
−2	$-\frac{3}{2}(-2) + 1 = 4$	(−2, 4)
0	$-\frac{3}{2}(0) + 1 = 1$	(0, 1)
4	$-\frac{3}{2}(4) + 1 = -5$	(4, −5)

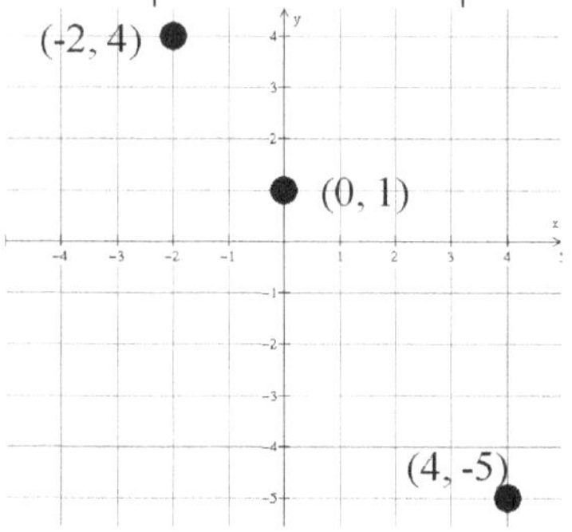

3. The relation is {(−2, −4), (0, −3), (4, −1)}. No two ordered pairs have the same first coordinate, so, yes, y is a function of x.

4. $f(t) = t^2 + t$

$$f(2) = (2)^2 + 2 = 4 + 2 = 6$$

5. $f(x) = x^2 - 2x$

$$f(-1) = (-1)^2 - 2(-1) = 1 + 2 = 3$$

6. {(3.5, 25), (4.0, 30), (5.2, 45), (5.0, 38), (4.0, 42), (6.3, 12), (5.4, 34)}
No, the relation is not a function. The two ordered pairs (4.0, 30) and (4.0, 42) have the same first coordinate but different second coordinates.

7. $y = 3x + 1$

$y\text{-intercept} = (0,1)$

$$m = 3 = \frac{3}{1} = \frac{\text{change in } y}{\text{change in } x}$$

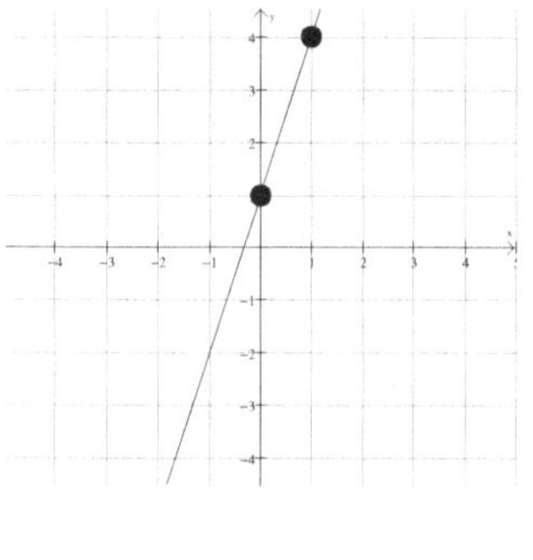

8. $y = -\frac{3}{4}x + 3$

y-intercept $= (0,3)$

$m = -\frac{3}{4} = \frac{-3}{4} = \frac{\text{change in } y}{\text{change in } x}$

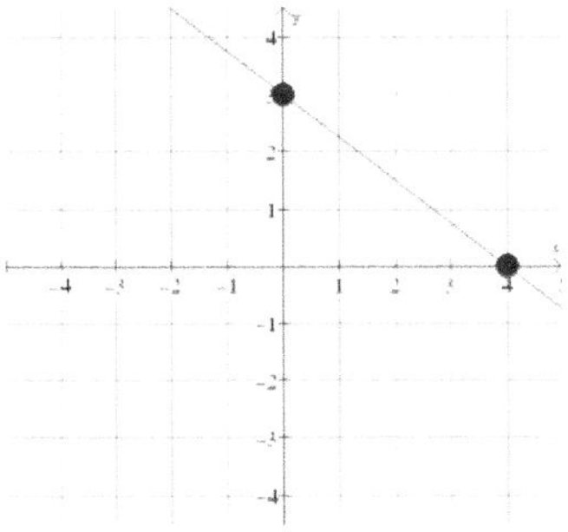

9. $3x - 2y = 6$

$-2y = -3y + 6$

$y = \frac{3}{2}x - 3$

y-intercept $= (0,-3)$

$m = \frac{3}{2} = \frac{\text{change in } y}{\text{change in } x}$

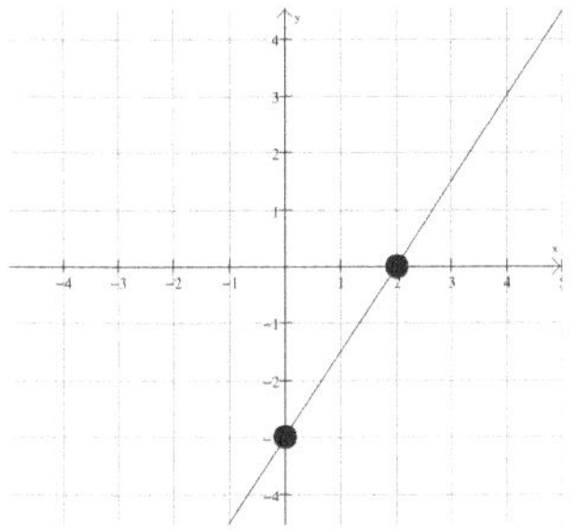

10. $x + 3 = 0$

$x = -3$

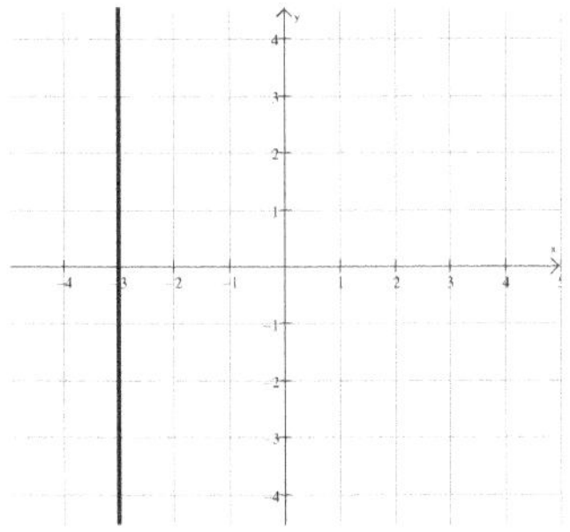

11. y-intercept $= (0,4)$

$m = -\frac{2}{3} = \frac{-2}{3} = \frac{\text{change in } y}{\text{change in } x}$

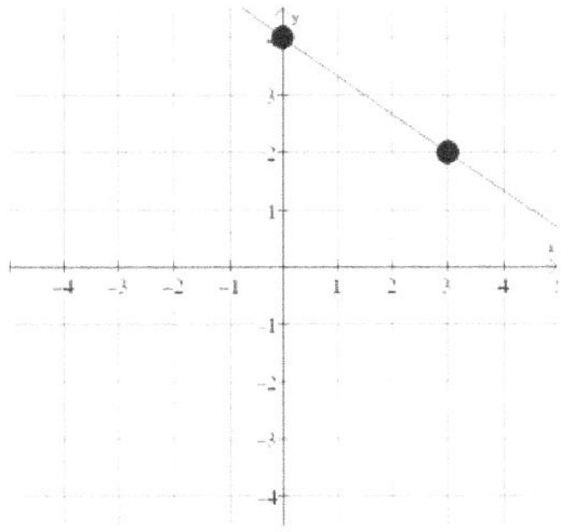

12. y-intercept $= (0,-2)$

$m = 2 = \frac{2}{1} = \frac{\text{change in } y}{\text{change in } x}$

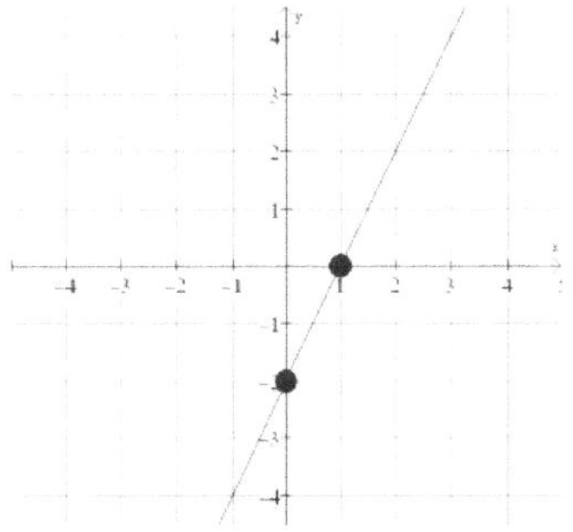

13.

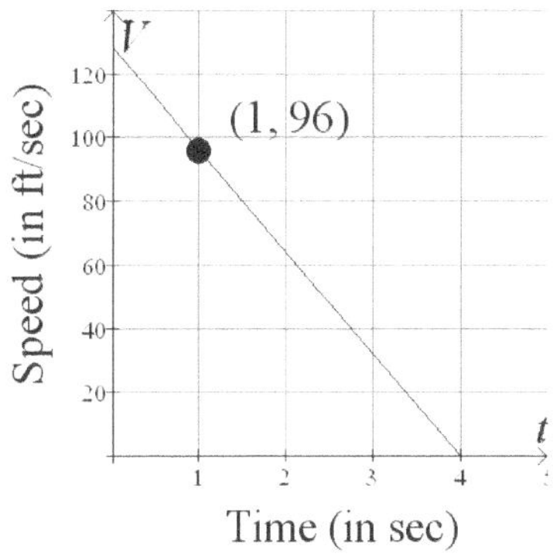

The ordered pair (1, 96) means that after 1 s, the speed of the ball is 96 ft/s.

14. $6x - 4y = 12$

y-intercept	x-intercept
$6(0) - 4y = 12$	$6x - 4(0) = 12$
$-4y = 12$	$6x = 12$
$y = -3$	$x = 2$
(0, −3)	(2, 0)

15. $y=\frac{1}{2}x+1$

y-intercept

$y=\frac{1}{2}(0)+1$

$y=1$

(0, 1)

x-intercept

$0=\frac{1}{2}x+1$

$-1=\frac{1}{2}x$

$2(-1)=2\left(\frac{1}{2}x\right)$

$-2=x$

(−2, 0)

16. $m=\frac{y_2-y_1}{x_2-x_1}=\frac{1-(-3)}{4-2}=\frac{4}{2}=2$

The slope is 2.

17. $P_1(2,5),\ P_2(-1,1)$

$m=\frac{1-5}{-1-2}=\frac{-4}{-3}=\frac{4}{3}$

$Q_1(-2,3),\ \ Q_2(4,11)$

$m=\frac{11-3}{4-(-2)}=\frac{8}{6}=\frac{4}{3}$

The slopes are the same. The lines are parallel.

18. $m=\frac{y_2-y_1}{x_2-x_1}=\frac{7-2}{-5-(-5)}=\frac{5}{0}$

The slope is undefined.

19. $2x+3y=6$

$3y=-2x+6$

$y=-\frac{2}{3}x+2$

$m=-\frac{2}{3}$

20. $m=3$, $(0,-1)$ so $b=-1$.

$y=3x-1$

21. $m=\frac{2}{3}\quad (x_1,y_1)=(-3,1)$

$y-y_1=m(x-x_1)$

$y-1=\frac{2}{3}(x+3)$

$y-1=\frac{2}{3}x+2$

$y=\frac{2}{3}x+3$

22. $(x_1,y_1)=(5,-4)\qquad (x_2,y_2)=(-3,1)$

$m=\frac{y_2-y_1}{x_2-x_1}=\frac{1-(-4)}{-3-5}=\frac{5}{-8}=-\frac{5}{8}$

$m=-\frac{5}{8}\qquad (x_1,y_1)=(5,-4)$

$y-(-4)=-\frac{5}{8}(x-5)$

$y+4=-\frac{5}{8}x+\frac{25}{8}$

$y=-\frac{5}{8}x-\frac{7}{8}$

23. $(x_1,y_1)=(-2,0)\qquad (x_2,y_2)=(5,-2)$

$m=\frac{y_2-y_1}{x_2-x_1}=\frac{-2-0}{5-(-2)}=\frac{-2}{7}=-\frac{2}{7}$

$m=-\frac{2}{7}\qquad (x_1,y_1)=(-2,0)$

$y-0=-\frac{2}{7}(x-(-2))$

$y=-\frac{2}{7}(x+2)$

$y=-\frac{2}{7}x-\frac{4}{7}$

Cumulative Review Exercises

1. $12-18\div 3\cdot(-2)^2=12-18\div 3\cdot 4$

$=12-6\cdot 4$

$=12-24$

$=-12$

2. $\frac{a-b}{a^2-c}$

$\frac{-2-(3)}{(-2)^2-(-4)}=\frac{-5}{4+4}=-\frac{5}{8}$

3. $f(x)=\frac{2}{x-1}$

$f(-2)=\frac{2}{-2-1}=\frac{2}{-3}=-\frac{2}{3}$

4. $2x-\frac{2}{3}=\frac{7}{3}$

$2x=\frac{9}{3}$

$2x=3$

$x=\frac{3}{2}$

The solution is $\frac{3}{2}$.

5.

$$\begin{aligned} 3x-2[x-3(2-3x)] &= x-7 \\ 3x-2[x-6+9x] &= x-7 \\ 3x-2[10x-6] &= x-7 \\ 3x-20x+12 &= x-7 \\ -17x+12 &= x-7 \\ -18x &= -19 \\ x &= \frac{-19}{-18} \\ x &= \frac{19}{18} \end{aligned}$$

The solution is $\frac{19}{18}$.

6. $6\frac{2}{3}\% = \frac{6\frac{2}{3}}{100} = \frac{\frac{20}{3}}{100} = \frac{\overset{1}{\cancel{20}}}{3}\cdot\frac{1}{\underset{5}{\cancel{100}}} = \frac{1}{15}$

7. $(-2x^2y)^3(2xy^2)^2 = (-8x^6y^3)(4x^2y^4) = -32x^8y^7$

8. $\frac{-15x^7}{5x^5} = -3x^2$

9.

$$\begin{array}{r} x+3 \\ x-7\overline{)x^2-4x-21} \\ \underline{x^2-7x} \\ 3x-21 \\ \underline{3x-21} \end{array}$$

$(x^2-4x-21)\div(x-7) = x+3$

10.

$$\begin{aligned} &15x^2+15x+10 \\ &= 5(x^2+3x+2) \\ &= 5(x+2)(x+1) \end{aligned}$$

11. $x(a+2)+y(a+2) = (a+2)(x+y)$

12.

$$\begin{aligned} x(x-2) &= 8 \\ x^2-2x &= 8 \\ x^2-2x-8 &= 0 \\ (x-4)(x+2) &= 0 \end{aligned}$$

$x-4=0 \qquad x+2=0$

$x=4 \qquad x=-2$

The solutions are 4 and −2.

13.

$$\frac{x^5y^3}{x^2-x-6}\cdot\frac{x^2-9}{x^2y^4}$$

$$= \frac{x^5y^3}{\underset{1}{\cancel{(x-3)}}(x+2)}\cdot\frac{(x+3)\overset{1}{\cancel{(x-3)}}}{x^2y^4}$$

$$= \frac{x^3(x+3)}{y(x+2)}$$

14.

$$\frac{3x}{x^2+5x-24}-\frac{9}{x^2+5x-24}$$

$$= \frac{3x-9}{x^2+5x-24} = \frac{3\overset{1}{\cancel{(x-3)}}}{(x+8)\underset{1}{\cancel{(x-3)}}}$$

$$= \frac{3}{x+8}$$

15.

$$\begin{aligned} 3-\frac{1}{x} &= \frac{5}{x} \\ x\left(3-\frac{1}{x}\right) &= x\left(\frac{5}{x}\right) \\ 3x-1 &= 5 \\ 3x &= 6 \\ x &= 2 \end{aligned}$$

The solution is 2.

16.

$$\begin{aligned} 4x-5y &= 15 \\ -5y &= -4x+15 \\ y &= \frac{4}{5}x-3 \end{aligned}$$

17.

$$\begin{aligned} y &= 2x-1 \\ y &= 2(-2)-1 \\ y &= -4-1 \\ y &= -5 \end{aligned}$$

The ordered-pair solution is (−2, −5).

18. $m = \frac{y_2-y_1}{x_2-x_1} = \frac{3-3}{-2-2} = \frac{0}{-4} = 0$

The line has zero slope.

19. $m=\frac{1}{2}$ $(x_1,y_1)=(2,-1)$

$$y-y_1=m(x-x_1)$$
$$y-(-1)=\frac{1}{2}(x-2)$$
$$y+1=\frac{1}{2}x-1$$
$$y=\frac{1}{2}x-2$$

The equation of the line is $y=\frac{1}{2}x-2$.

20. $m=-3$ $(x_1,y_1)=(0,2)$ so $b=2$

$$y=mx+b$$
$$y=-3x+2$$

The equation of the line is $y=-3x+2$.

21. $m=2$ $(x_1,y_1)=(-1,0)$

$$y-y_1=m(x-x_1)$$
$$y-0=2(x-(-1))$$
$$y=2(x+1)$$
$$y=2x+2$$

The equation of the line is $y=2x+2$.

22. $m=\frac{2}{3}$ $(x_1,y_1)=(6,1)$

$$y-y_1=m(x-x_1)$$
$$y-1=\frac{2}{3}(x-6)$$
$$y-1=\frac{2}{3}x-4$$
$$y=\frac{2}{3}x-3$$

The equation of the line is $y=\frac{2}{3}x-3$.

23. Strategy:
To calculate the probability:
→Count the number of possible outcomes.
→Count the number of favorable outcomes.
→Use the probability formula.

Solution:
There are 9 possible outcomes (4 blue + 2 green + 3 red).
There are 4 + 2 = 6 favorable outcomes.

$$\text{Probability}=\frac{6}{9}=\frac{2}{3}$$

The probability is $\frac{2}{3}$ that the marble chosen is not red.

24. Strategy:
Measure of first angle: $x+3$
Measure of second angle: x
Measure of third angle: $2x+5$
Use the equation $A+B+C=180^\circ$.

Solution:

$$x+(x+3)+(2x+5)=180$$
$$4x+8=180$$
$$4x=172$$
$$x=43$$
$$x+3=43+3=46$$
$$2x+5=2(43)+3=91$$

The measures of the angles are 43°, 46°, and 91°.

25. Strategy:
Write and solve a proportion using x to represent the value of the home.

Solution:

$$\frac{6250}{500{,}000}=\frac{13{,}750}{x}$$
$$6250x=6{,}875{,}000{,}000$$
$$x=1{,}100{,}000$$

The value of the home is $1,100,000.

26. Strategy:

→Time working together: t

	Rate	Time	Part
Electrician	$\frac{1}{6}$	t	$\frac{t}{6}$
Apprentice	$\frac{1}{10}$	t	$\frac{t}{10}$

→The sum of the parts of the task completed by each person must equal 1.

Solution:

$$\frac{t}{6}+\frac{t}{10}=1$$
$$30\left(\frac{t}{6}+\frac{t}{10}\right)=30\cdot 1$$
$$5t+3t=30$$
$$8t=30$$
$$t=3\frac{3}{4}$$

It would take $3\frac{3}{4}$ h for both working together to wire the garage.

27.
$$y=\frac{1}{2}x-1$$
$$m=\frac{1}{2}=\frac{\text{change in } y}{\text{change in } x}$$
$$b=-1$$

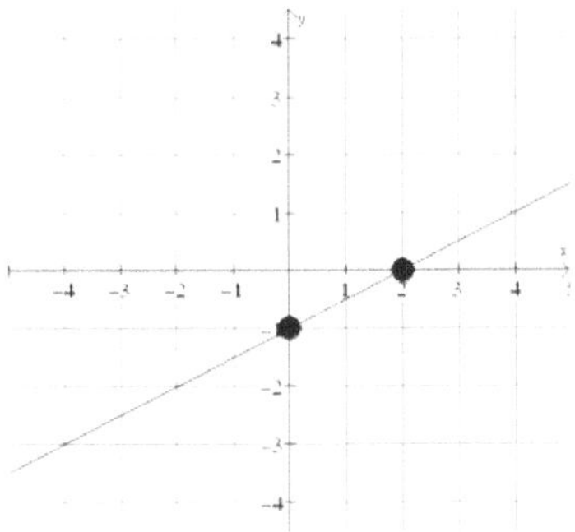

28. y – intercept $=(0,2)$

$$m=-\frac{2}{3}=\frac{-2}{3}=\frac{\text{change in } y}{\text{change in } x}$$

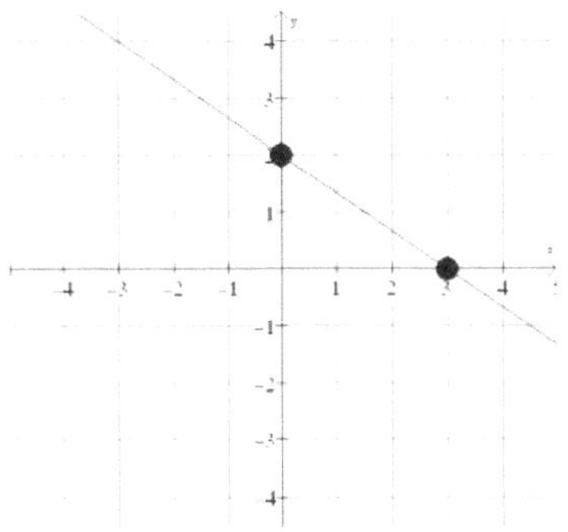

Chapter 13: Systems of Linear Equations

Prep Test

1.
$$3x-4y=24$$
$$-4y=-3x+24$$
$$y=\frac{3}{4}x-6$$

2.
$$50+0.07x=0.05(x+1400)$$
$$50+0.07x=0.05x+70$$
$$0.02x=20$$
$$x=1000$$
The solution is 1000.

3.
$$-3(2x-7y)+3(2x+4y)=-6x+21y+6x+12y$$
$$=33y$$

4.
$$4x+2(3x-5)=4x+6x-10$$
$$=10x-10$$

5.
$$3x-5y=-22$$

$3(-4)-5(2)$	-22
$-12-10$	-22
-22	-22

Yes, (−4, 2) is a solution of $3x-5y=-22$.

6.
$$3x-4y=12$$

x-intercept	y-intercept
$3x-4(0)=12$	$3(0)-4y=12$
$3x=12$	$-4y=12$
$x=4$	$y=-3$
(4, 0)	(0, −3)

7. To determine if the graphs of the equations are parallel, write each equation in slope-intercept form and then compare slopes.
$$3x+y=6$$
$$y=-3x+6$$
$$y=-3x-4$$
The slopes are equal, so, yes, the lines are parallel.

8.
$$y=\frac{5}{4}x-2$$

x	y
0	−2
4	3
−4	−7

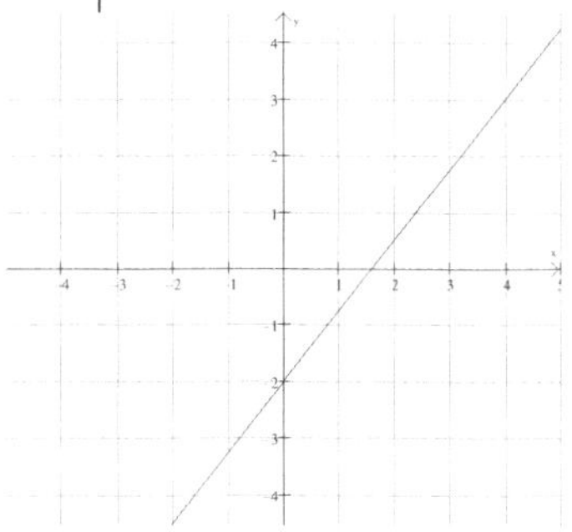

10. Strategy:
→Time second hiker spends hiking: t
→Time first hiker spends hiking: $t+0.5$

	Distance	Rate	Time
Hiker 1	$3(t+0.5)$	3 mph	$t+0.5$
Hiker 2	$4t$	4 mph	t

→The hikers travel the same distance.

Solution:
$$3(t+0.5)=4t$$
$$3t+1.5=4t$$
$$1.5=t$$
The hikers are side-by-side after 1.5 h.

Section 13.1

Concept Check

1. Always true

3. Never true

Objective A Exercises

5.

$$\begin{array}{r|l} \multicolumn{2}{c}{3x+4y=18} \\ \hline 3(2)+4(3) & 18 \\ 6+12 & 18 \\ \multicolumn{2}{c}{18=18} \end{array} \qquad \begin{array}{r|l} \multicolumn{2}{c}{2x-y=1} \\ \hline 2(2)-3 & 1 \\ 4-3 & 1 \\ \multicolumn{2}{c}{1=1} \end{array}$$

Yes, (2, 3) is a solution of the system of equations.

7.

$$\begin{array}{r|l} \multicolumn{2}{c}{5x-2y=14} \\ \hline 5(4)-2(3) & 14 \\ 20-6 & 14 \\ \multicolumn{2}{c}{14=14} \end{array} \qquad \begin{array}{r|l} \multicolumn{2}{c}{x+y=8} \\ \hline 4+3 & 8 \\ & \\ \multicolumn{2}{c}{7\neq 8} \end{array}$$

No, (4, 3) is not a solution of the system of equations.

9.

$$\begin{array}{r|l} \multicolumn{2}{c}{y=2x-7} \\ \hline -3 & 2(2)-7 \\ -3 & 4-7 \\ \multicolumn{2}{c}{-3=-3} \end{array} \qquad \begin{array}{r|l} \multicolumn{2}{c}{3x-y=9} \\ \hline 3(2)-(-3) & 9 \\ 6+3 & 9 \\ \multicolumn{2}{c}{9=9} \end{array}$$

Yes, (2, −3) is a solution of the system of equations.

11.

$$\begin{array}{r|l} \multicolumn{2}{c}{y=x} \\ \hline 0 & 0 \\ & \\ \multicolumn{2}{c}{0=0} \end{array} \qquad \begin{array}{r|l} \multicolumn{2}{c}{3x+4y=0} \\ \hline 3(0)+4(0) & 0 \\ 0+0 & 0 \\ \multicolumn{2}{c}{0=0} \end{array}$$

Yes, (0, 0) is a solution of the system of equations.

13. The lines are parallel, so the system is (iii) inconsistent.

15. The lines intersect, so the system is (i) independent.

17. To find the solution of the system, identify the point where the graphs intersect.

The point (2, −1) is the solution to the system of equations.

19. To find the solution of the system, identify the point where the graphs intersect.

The lines are the same. So, the ordered-pair solutions of $y=-\frac{3}{2}x+1$ are solutions of the system of equations.

21. To find the solution of the system, identify the point where the graphs intersect.

The lines do not intersect. So, there is no solution.

23.

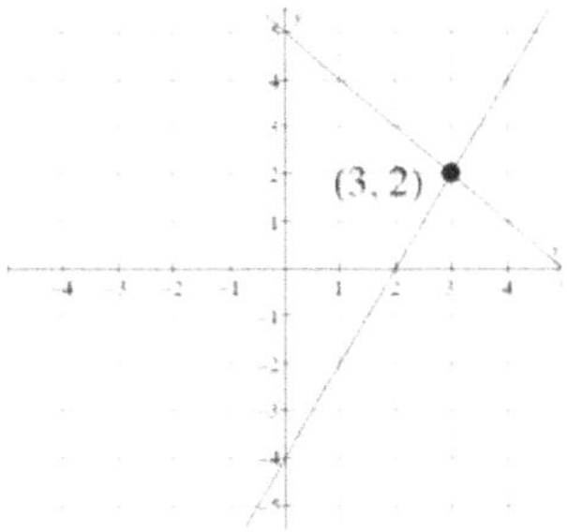

The solution is (3, 2).

25.

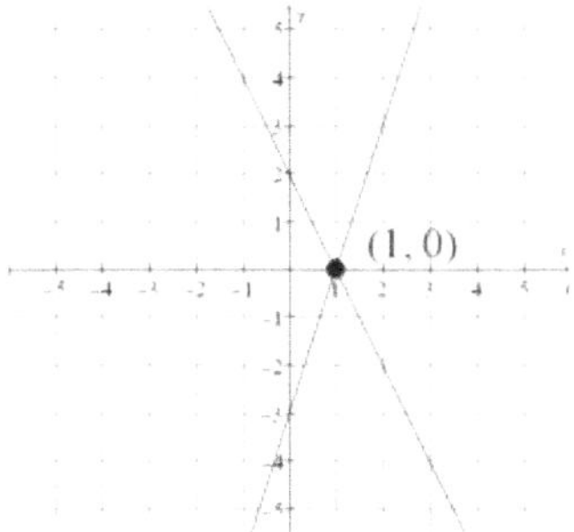

The solution is (1, 0).

27.

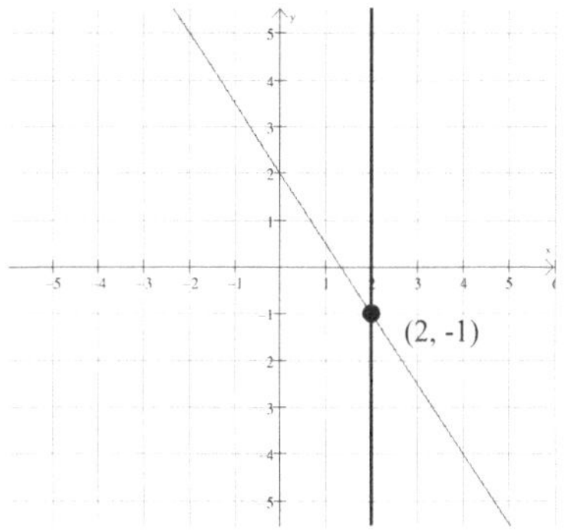

The solution is (2, −1).

29.

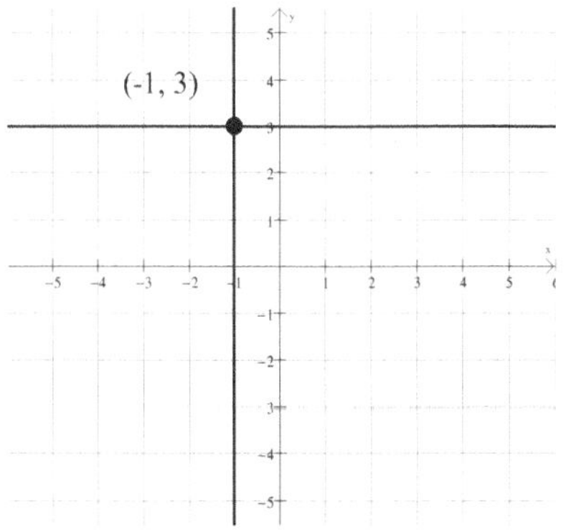

The solution is (−1, 3).

31.

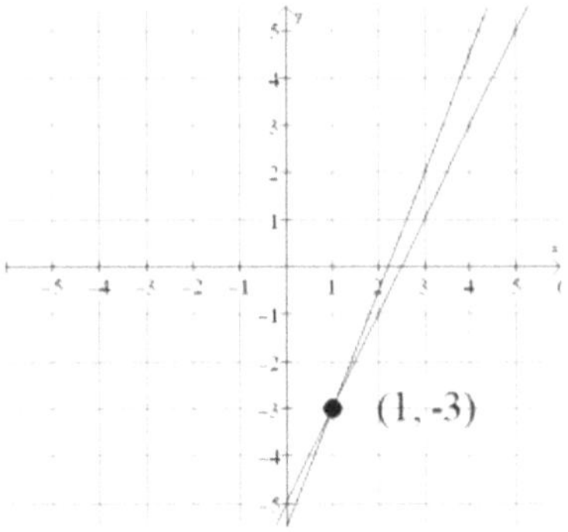

The solution is $(1, -3)$.

33.

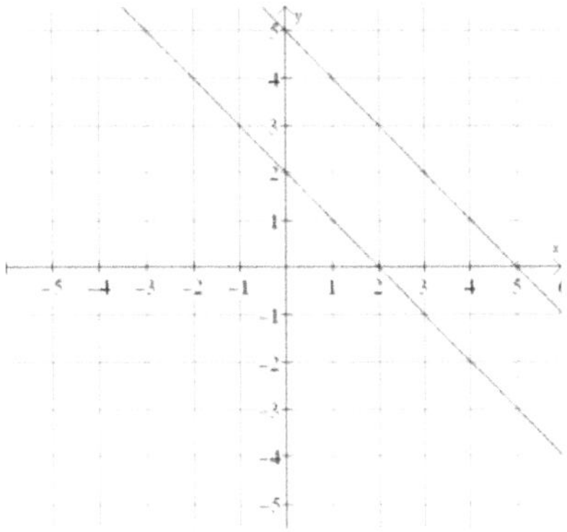

No solution

35.

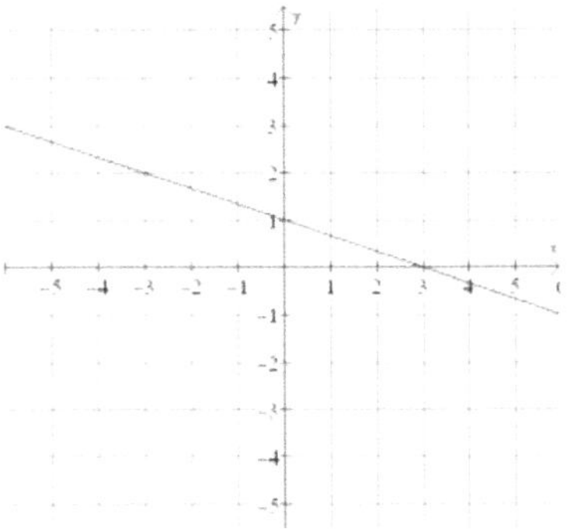

Dependent. The solutions are the ordered-pair solutions of $y = -\frac{1}{3}x + 1$.

37.

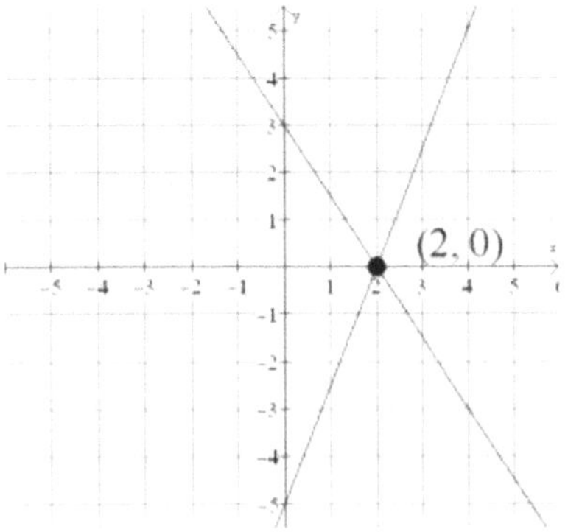

The solution is $(2, 0)$.

39.

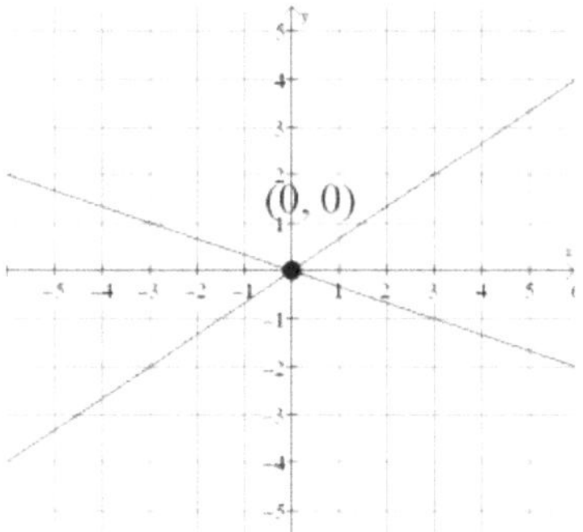

The solution is $(0, 0)$.

41.

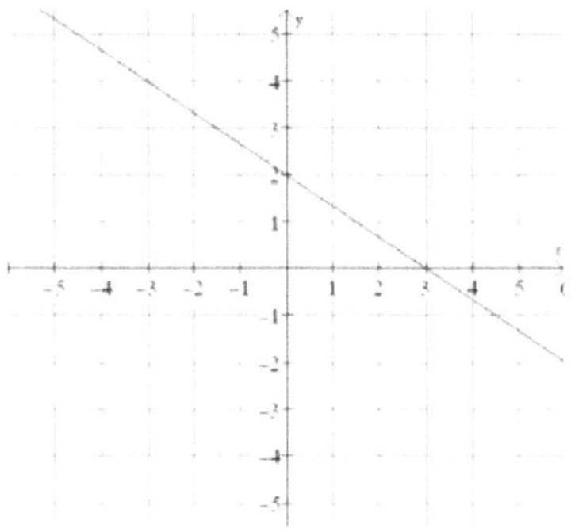

Dependent. The solutions are the ordered-pair solutions of $4x + 6y = 12$.

43. The lines will intersect so the system is independent.

Critical Thinking 13.1

45. $x = 3$
$y = x - 1$

47. $y = 2x - 2$
$y = \frac{2}{3}x + \frac{2}{3}$

Projects or Group Activities 13.1

49. Answers will vary. For example:

a. $y = x + 8$
$y = -x - 2$

b. $y = 2x - 3$
$y = 2x + 4$

c. $y = 3x - 5$
$3x - y = 5$

Section 13.2

Concept Check

1. The system of equations is dependent if the equation reduces to a true statement, such as 4 = 4. In this case (as in the case of a inconsistent system of equations), the variable drops out and we are left with only the constants.

The system of equations is inconsistent if the equation reduces to a false statement, such as 0 = 2. In this case (as in the case of a dependent system of equations), the variable drops out and we are left with only the constants.

3. True

5. False

Objective A Exercises

7. (1) $2x+3y=7$

(2) $x=2$

Substitute 2 for x in Equation (1).

$$2x+3y=7$$
$$2(2)+3y=7$$
$$4+3y=7$$
$$3y=3$$
$$y=1$$

The solution is (2, 1).

9. (1) $y=x-3$

(2) $x+y=5$

Substitute $x-3$ for y in Equation (2).

$$x+y=5$$
$$x+(x-3)=5$$
$$2x-3=5$$
$$2x=8$$
$$x=4$$

Substitute 4 for x in Equation (1).

$$y=x-3$$
$$y=4-3$$
$$y=1$$

The solution is (4, 1).

11. (1) $x=y-2$

(2) $x+3y=2$

Substitute $y-2$ for x in Equation (2).

$$x+3y=2$$
$$(y-2)+3y=2$$
$$4y-2=2$$
$$4y=4$$
$$y=1$$

Substitute 1 for y in Equation (1).

$$x=y-2$$
$$x=1-2$$
$$x=-1$$

The solution is (−1, 1).

13. (1) $y=4-3x$

(2) $3x+y=5$

Substitute $4-3x$ for y in Equation (2).

$$3x+y=5$$
$$3x+(4-3x)=5$$
$$4=5$$

The system is inconsistent and has no solution.

15. (1) $x=3y+3$

(2) $2x-6y=12$

Substitute $3y+3$ for x in Equation (2).

$$2x-6y=12$$
$$2(3y+3)-6y=12$$
$$6y+6-6y=12$$
$$6=12$$

The system is inconsistent and has no solution.

17. (1) $3x+5y=-6$

(2) $x=5y+3$

Substitute $5y+3$ for x in Equation (1).

$$3x+5y=-6$$
$$3(5y+3)+5y=-6$$
$$15y+9+5y=-6$$
$$20y+9=-6$$
$$20y=-15$$
$$y=-\frac{3}{4}$$

Substitute $-\frac{3}{4}$ for y in Equation (2).

$$x=5y+3$$
$$x=5\left(-\frac{3}{4}\right)+3$$
$$x=-\frac{15}{4}+3$$
$$x=-\frac{3}{4}$$

The solution is $\left(-\frac{3}{4},-\frac{3}{4}\right)$.

19. (1) $3x+y=4$

(2) $4x-3y=1$

Solve Equation (1) for y.

$$3x+y=4$$
$$y=-3x+4$$

Substitute $-3x+4$ for y in Equation (2).

$$4x-3y=1$$
$$4x-3(-3x+4)=1$$
$$4x+9x-12=1$$
$$13x-12=1$$
$$13x=13$$
$$x=1$$

Substitute 1 for x in Equation (1).

$$3x+y=4$$
$$3(1)+y=4$$
$$3+y=4$$
$$y=1$$

The solution is (1, 1).

21. (1) $3x-y=6$

(2) $x+3y=2$

Solve Equation (2) for x.

$$x+3y=2$$
$$x=-3y+2$$

Substitute $-3y+2$ for x in Equation (1).

$$3x-y=6$$
$$3(-3y+2)-y=6$$
$$-9y+6-y=6$$
$$-10y+6=6$$
$$-10y=0$$
$$y=0$$

Substitute 0 for y in Equation (2).

$$x+3y=2$$
$$x+3(0)=2$$
$$x=2$$

The solution is (2, 0).

23. (1) $3x-y=5$

(2) $2x+5y=-8$

Solve Equation (1) for y.

$$3x-y=5$$
$$-y=-3x+5$$
$$y=3x-5$$

Substitute $3x-5$ for y in Equation (2).

$$2x+5y=-8$$
$$2x+5(3x-5)=-8$$
$$2x+15x-25=-8$$
$$17x=17$$
$$x=1$$

Substitute 1 for x in Equation (1).

$$3x-y=5$$
$$3(1)-y=5$$
$$3-y=5$$
$$-y=2$$
$$y=-2$$

The solution is (1, −2).

25. (1) $4x+3y=0$

(2) $2x-y=0$

Solve Equation (2) for y.

$2x-y=0$

$-y=-2x$

$y=2x$

Substitute $2x$ for y in Equation (1).

$4x+3y=0$

$4x+3(2x)=0$

$4x+6x=0$

$10x=0$

$x=0$

Substitute 0 for x in Equation (2).

$2x-y=0$

$2(0)-y=0$

$-y=0$

$y=0$

The solution is (0, 0).

27. (1) $2x-y=2$

(2) $6x-3y=6$

Solve Equation (1) for y.

$2x-y=2$

$-y=-2x+2$

$y=2x-2$

Substitute $2x-2$ for y in Equation (2).

$6x-3y=6$

$6x-3(2x-2)=6$

$6x-6x+6=6$

$6=6$

The system is dependent. The solutions are the ordered-pairs that satisfy the equation $2x-y=2$.

29. (1) $x=3y+2$

(2) $y=2x+6$

Substitute $3y+2$ for x in Equation (2).

$y=2x+6$

$y=2(3y+2)+6$

$y=6y+4+6$

$y=6y+10$

$-5y=10$

$y=-2$

Substitute -2 for y in Equation (1).

$x=3y+2$

$x=3(-2)+2$

$x=-6+2$

$x=-4$

The solution is (−4, −2).

31. (1) $y=2x+11$

(2) $y=5x-19$

Substitute $2x+11$ for y in Equation (2).

$y=5x-19$

$2x+11=5x-19$

$-3x+11=-19$

$-3x=-30$

$x=10$

Substitute 10 for x in Equation (1).

$y=2x+11$

$y=2(10)+11$

$y=20+11$

$y=31$

The solution is (10, 31).

33. (1) $y=-4x+2$

(2) $y=-3x-1$

Substitute $-4y+2$ for y in Equation (2).

$y=-3x-1$

$-4x+2=-3x-1$

$-x+2=-1$

$-x=-3$

$x=3$

Substitute 3 for x in Equation (1).

$y=-4x+2$

$y=-4(3)+2$

$y=-12+2$

$y=-10$

The solution is (3, −10).

35. (1) $x = 4y - 2$

(2) $x = 6y + 8$

Substitute $4y - 2$ for x in Equation (2).

$$x = 6y + 8$$
$$4y - 2 = 6y + 8$$
$$-2y - 2 = 8$$
$$-2y = 10$$
$$y = -5$$

Substitute -5 for y in Equation (1).

$$x = 4y - 2$$
$$x = 4(-5) - 2$$
$$x = -20 - 2$$
$$x = -22$$

The solution is $(-22, -5)$.

37. (1) $x + y = A$

(2) $x = A - y$

Substitute $A - y$ for x in Equation (1).

$$x + y = A$$
$$(A - y) + y = A$$
$$A = A$$

The solution is dependent.

Objective B Exercises

39. x = amount invested at 8%
y = amount invested at 6.5%

$x + y = 10{,}000$ represents the fact that the sum of the two investments is \$10,000.

$0.08x + 0.065y = 710$ represents the fact that the total interest earned by the two investments is \$710.

41. Strategy:

→Amount invested at 5%: x

→Amount invested at 7.5%: y

	Principal	Rate	Interest
Amount at 5%	x	0.05	$0.05x$
Amount at 7.5%	y	0.075	$0.075y$

→The total amount invested is \$3500:

$x + y = 3500$

The total interest earned is \$215:

$0.05x + 0.075y = 215$

Solution:

$$x + y = 3500$$
$$y = 3500 - x$$

$$0.05x + 0.075y = 215$$
$$0.05x + 0.075(3500 - x) = 215$$
$$0.05x + 262.50 - 0.075x = 215$$
$$-0.025x = -47.50$$
$$x = 1900$$

$y = 3500 - x = 3500 - 1900 = 1600$

The amounts invested should be \$1900 at 5% and \$1600 at 7.5%.

43. Strategy:

→Amount invested at 9%: x

→Amount invested at 6%: y

	Principal	Rate	Interest
Amount at 9%	x	0.09	$0.09x$
Amount at 6%	y	0.06	$0.06y$

→The total amount invested is \$6000:

$x+y=6000$

The interest earned on each account is the same:

$0.06x=0.09y$

Solution:

$x+y=6000$

$y=6000-x$

$0.09x=0.06y$

$0.09x=0.06(6000-x)$

$0.09x=360-0.06x$

$0.15x=360$

$x=2400$

$y=6000-x=6000-2400=3600$

The amounts invested were \$2400 at 9% and \$3600 at 6%.

45. Strategy:

→Amount invested at 8%: x

→Amount invested at 11%: y

	Principal	Rate	Interest
Amount at 8%	x	0.08	$0.08x$
Amount at11%	y	0.11	$0.11y$

→The total amount invested is \$6000:

$x+y=6000$

The interest earned on the 8% account is twice the amount of the interest earned on the 11% account:

$0.08x=2(0.11y)$

Solution:

$x+y=6000$

$y=6000-x$

$0.08x=0.22y$

$0.08x=0.22(6000-x)$

$0.08x=1320-0.22x$

$0.30x=1320$

$x=4400$

$y=6000-x=6000-4400=1600$

The amounts invested should b \$4400 at 8% and \$1600 at 11%.

47. Strategy:

→Amount invested at 6.5%: x

→Amount invested at 8.5%: y

	Principal	Rate	Interest
Amount at 6.5%	x	0.065	$0.065x$
Amount at 8.5%	y	0.085	$0.085y$

→The amount invested at 8.5% is twice the amount invested at 6.5%: $y=2x$

The total interest earned is \$4935:

$0.065x+0.085y=4935$

Solution:

$0.065x+0.085y=4935$

$0.065x+0.085(2x)=4935$

$0.065x+0.17x=4935$

$0.235x=4935$

$x=21,000$

The amount invested at 6.5% was \$21,000.

49. Strategy:

→Amount invested at 8%: x

→Amount invested at 7%: y

	Principal	Rate	Interest
Amount at 8%	x	0.08	$0.08x$
Amount at 7%	y	0.07	$0.07y$

→The total amount invested is 10($2000) = $20,000: $x+y=20{,}000$

The total annual interest earned is $1520: $0.08x+0.07y=1520$

Solution:

$$x+y=20{,}000$$
$$y=20{,}000-x$$

$$0.08x+0.07y=1520$$
$$0.08x+0.07(20{,}000-x)=1520$$
$$0.08x+1400-0.07x=1520$$
$$0.01x=120$$
$$x=12{,}000$$

$$y=20{,}000-x=20{,}000-12{,}000=8000$$

The amounts invested were $12,000 at 8% and $8000 at 7%.

51. Strategy:

→Amount invested at 7.5%: x

→Amount invested at 9%: y

	Principal	Rate	Interest
Amount at 7.5%	x	0.075	$0.075x$
Amount at 9%	y	0.09	$0.09y$

→The amount invested at 9% (trust deed) is one-half the amount invested at 7.5% (real estate trust): $y=\frac{1}{2}x$

The total annual interest earned is $900: $0.075x+0.09y=900$

Solution:

$$0.075x+0.09y=900$$
$$0.075x+0.09\left(\frac{1}{2}x\right)=900$$
$$0.075x+0.045x=900$$
$$0.12x=900$$
$$x=7500$$

$$y=\frac{1}{2}x=\frac{1}{2}(7500)=3750$$

The amount invested in the trust deed was $3750.

Critical Thinking 13.2

53. $10(0.8x - 0.1y = 0.3) \quad 8x - y = 3 \ (1)$

$10(0.5x - 0.2y = -0.5) \quad 5x - 2y = -5 \ (2)$

Solve Equation (1) for y.

$8x - y = 3$

$-y = -8x + 3$

$y = 8x - 3$

Substitute $8x - 3$ for y in equation (2).

$5x - 2y = -5$

$5x - 2(8x - 3) = -5$

$5x - 16x + 6 = -5$

$-11x = -11$

$x = 1$

Substitute 1 for x in Equation (1).

$8x - y = 3$

$8(1) - y = 3$

$8 - y = 3$

$-y = -5$

$y = 5$

The solution is (1, 5).

55. $10(-0.1x + 0.3y = 1.1) \quad -x + 3y = 11 \ (1)$

$10(0.4x - 0.1y = -2.2) \quad 4x - y = -22 \ (2)$

Solve Equation (1) for x.

$-x + 3y = 11$

$-x = -3y + 11$

$x = 3y - 11$

Substitute $3y - 11$ for x in equation (2).

$4x - y = -22$

$4(3y - 11) - y = -22$

$12y - 44 - y = -22$

$11y = 22$

$y = 2$

Substitute 2 for y in Equation (1).

$-x + 3y = 11$

$-x + 3(2) = 11$

$-x + 6 = 11$

$-x = 5$

$x = -5$

The solution is (–5, 2).

57. $100(1.25 - 0.01y = 1.5) \quad 125x - y = 150 \ (1)$

$100(0.24x - 0.02y = -1.52) \quad 24x - 2y = -152 \ (2)$

Solve Equation (1) for y.

$125x - y = 150$

$-y = -125x + 150$

$y = 125x - 150$

Substitute $125x - 150$ for y in equation (2).

$24x - 2y = -152$

$24x - 2(125x - 150) = -152$

$24x - 250x + 300 = -152$

$-226x = -452$

$x = 2$

Substitute 2 for x in Equation (1).

$125x - y = 150$

$125(2) - y = 150$

$250 - y = 150$

$-y = -100$

$y = 100$

The solution is (2, 100).

59. Strategy:

→Total amount invested: x

→Amount invested at 9%: $0.2x$

→Amount invested at 8%: $0.35x$

→Amount invested at 9.5%: $0.095(0.45x)$

	Principal	Rate	Interest
Amount at 9%	$0.2x$	0.09	$0.09(2x)$
Amount at 8%	$0.35x$	0.08	$0.08(0.35x)$
Amount at 9.5%	$0.45x$	0.95	$0.095(0.45x)$

→The interest totals \$5325.

Solution:

$0.09(0.2x) + 0.08(0.35x) + 0.095(0.45x) = 5325$

$0.018x + 0.028x + 0.04275x = 5325$

$0.08875x = 5325$

$x = 60000$

$0.2(60000) = 12000$

$0.35(60000) = 21000$

$0.45(60000) = 27000$

\$12,000 was invested at 9%. \$21,000 was invested at 8%. \$27,000 was invested at 9.5%.

Projects or Group Activities 13.2

61. (1) $2x-3y=7$
$-3y=-2x+7$
$y=\frac{2}{3}x-\frac{7}{3}$

(2) $kx-3y=4$
$-3y=-kx+4$
$y=\frac{k}{3}x-\frac{4}{3}$

For there to be no solution, the slopes must be the same and the y-intercepts different.

$\frac{2}{3}=\frac{k}{3}$

$3\left(\frac{2}{3}\right)=3\left(\frac{k}{3}\right)$

$2=k$

k must be 2.

63. (1) $x=4y+4$
$-4y=-x+4$
$y=\frac{1}{4}x-1$

(2) $kx-8y=4$
$-8y=-kx+4$
$y=\frac{k}{8}x-\frac{1}{2}$

For there to be no solution, the slopes must be the same and the y-intercepts different.

$\frac{1}{4}=\frac{k}{8}$

$8\left(\frac{1}{4}\right)=8\left(\frac{k}{8}\right)$

$2=k$

k must be 2.

65. Strategy:
To find the value in 5 years, let y_1 equal the value after the first year, y_2 equal the value after the second year, y_3 equal the value after the third year, y_4 equal the value after the fourth year, and y_5 equal the value after the fifth year.

→Find the value after the first year using the simple interest formula $y_1=P+Prt$ using $P=2500$, $r=0.075$, and $t=1$.

→Find the value of each successive year using the formula $y_2=P+Prt$ using $P=y_1$, $r=0.075$, and $t=1$.

Solution:

$y_1=P+Prt$
$y_1=2500+2500(0.075)1$
$y_1=2687.50$

$y_2=y_1+y_1(0.075)$
$y_2=2687.50+2687.50(0.075)$
$y_2=2889.06$

$y_3=y_2+y_2(0.075)$
$y_3=2889.06+2889.06(0.075)$
$y_3=3105.74$

$y_4=y_3+y_3(0.075)$
$y_4=3105.74+3105.74(0.075)$
$y_4=3338.67$

$y_5=y_4+y_4(0.075)$
$y_5=3338.67+3338.67(0.075)$
$y_5=3589.07$

The value after 5 years is \$3589.07.

Check Your Progress: Chapter 13

1.

The solution is (4, 1).

2.

Dependent. The solutions are the ordered-pair solutions of $y = -\frac{4}{3}x + 4$.

3.

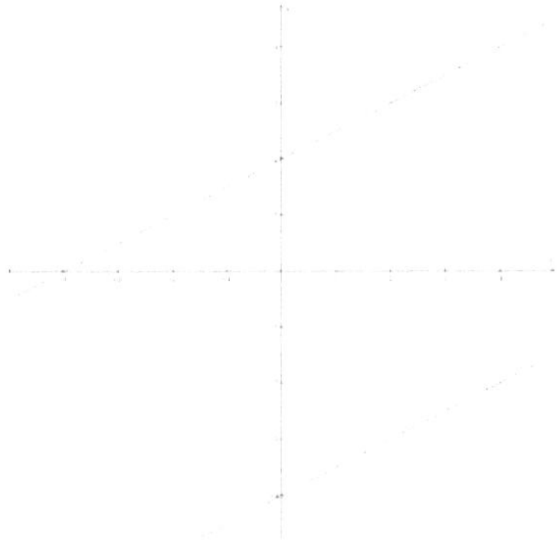

Inconsistent. There is no solution.

4.

$y = 2x - 7$	
-3	$2(2) - 7$
-3	-3

$-3 = -3$

$y = 3x - 9$	
-3	$3(2) - 9$
-3	-3

$-3 = -3$

Yes, (2, −3) is a solution of the system of equations.

5.
(1) $x - 3y = -14$
(2) $3x + 8y = 26$

Solve equation (1) for x.

$$x - 3y = -14$$
$$x = 3y - 14$$

Substitute $3y - 14$ for x in Equation (2).

$$3x + 8y = 26$$
$$3(3y - 14) + 8y = 26$$
$$9y - 52 + 8y = 26$$
$$17y = 78$$
$$y = 4$$

Substitute 4 for y in Equation (1).

$$x - 3y = -14$$
$$x - 3(4) = -14$$
$$x - 12 = -14$$
$$x = -2$$

The solution is (−2, 4).

6.
(1) $x - 3y = -12$
(2) $y = \frac{1}{3}x + 4$

Substitute $\frac{1}{3}x + 4$ for y in Equation (1).

$$x - 3y = -12$$
$$x - 3\left(\frac{1}{3}x + 4\right) = -12$$
$$x - x - 12 = 12$$
$$-12 = -12$$

Dependent. The solutions are the ordered-pair solutions of $y = \frac{1}{3}x + 4$.

7. Strategy:

→Amount invested at 6.5%: x

→Amount invested at 8.5%: y

	Principal	Rate	Interest
Amount at 6.5%	x	0.065	$0.065x$
Amount at 8.5%	y	0.085	$0.085y$

→The total amount invest is $30,000:

$x+y=30{,}000$

The total annual interest earned is $2190:

$0.065x+0.085y=2190$

Solution:

$x+y=30{,}000$

$y=30{,}000-x$

$$0.065x+0.085y=2190$$
$$0.065x+0.085(30{,}000-x)=2190$$
$$0.065x+2550-0.085x=2190$$
$$-0.02x=-360$$
$$x=18{,}000$$

$y=30{,}000-x=30{,}000-18{,}000=12{,}000$

The amounts invested were $18,000 at 6.5% and $12,000 at 8.5%.

Section 13.3

Concept Check

1. **a.** 3 **b.** –4

Objective A Exercises

3. (1) $x+y=4$

(2) $x-y=6$

Add the equations.

$2x=10$

$x=5$

Replace 5 for x in Equation (1).

$x+y=5$

$5+y=4$

$y=-1$

The solution is (5, −1).

5. (1) $x+y=4$

(2) $2x+y=5$

Eliminate y.

$-1(x+y)=-1\cdot 4$

$-x-y=-4$

$2x+y=5$

Add the equations.

$x=1$

Replace 1 for x in Equation (2).

$2x+y=5$

$2(1)+y=5$

$2+y=5$

$y=3$

The solution is (1, 3).

7. (1) $2x-y=1$

(2) $x+3y=4$

Eliminate y.

$3(2x-y)=3\cdot 1$

$6x-3y=3$

$x+3y=4$

Add the equations.

$7x=7$

$x=1$

Replace 1 for x in Equation (2).

$x+3y=4$

$x+3(1)=4$

$x+3=4$

$x=1$

The solution is (1, 1).

9. (1) $4x-5y=22$

(2) $x+2y=-1$

Eliminate x.

$$4x-5y=22$$
$$-4(x+2y)=-4(-1)$$
$$4x-5y=22$$
$$-4x-8y=4$$

Add the equations.

$$-13y=26$$
$$y=-2$$

Replace -2 for y in Equation (1).

$$4x-5y=22$$
$$4x-5(-2)=22$$
$$4x+10=22$$
$$4x=12$$
$$x=3$$

The solution is (3, −2).

11. (1) $2x-y=1$

(2) $4x-2y=2$

Eliminate y.

$$-2(2x-y)=-2\cdot 1$$
$$-4x+2y=-2$$
$$4x-2y=2$$

Add the equations.

$$0=0$$

The equations are dependent. The solutions are the ordered pairs that satisfy the equation $2x-y=1$.

13. (1) $4x+3y=15$

(2) $2x-5y=1$

Eliminate y.

$$5(4x+3y)=5\cdot 15$$
$$3(2x-5y)=3\cdot 1$$
$$20x+15y=75$$
$$6x-15y=3$$

Add the equations.

$$26x=78$$
$$x=3$$

Replace 3 for x in Equation (1).

$$4x+3y=15$$
$$4(3)+3y=15$$
$$12+3y=15$$
$$3y=3$$
$$y=1$$

The solution is (3, 1).

15. (1) $2x-3y=1$

(2) $4x-6y=2$

Eliminate x.

$$-2(2x-3y)=-2\cdot 1$$
$$-4x+6y=-2$$
$$4x-6y=2$$

Add the equations.

$$0=0$$

The system is dependent. The solutions are the ordered pairs that satisfy the equation $2x-3y=1$.

17. (1) $3x-6y=-1$

(2) $6x-4y=2$

Eliminate x.

$-2(3x-6y)=-2(-1)$

$-6x+12y=2$

$6x-4y=2$

Add the equations.

$8y=4$

$y=\frac{1}{2}$

Replace $\frac{1}{2}$ for y in Equation (1).

$3x-6y=-1$

$3x-6\left(\frac{1}{2}\right)=-1$

$3x-3=-1$

$3x=2$

$x=\frac{2}{3}$

The solution is $\left(\frac{2}{3},\frac{1}{2}\right)$.

19. (1) $5x+7y=10$

(2) $3x-14y=6$

Eliminate y.

$2(5x+7y)=2\cdot 10$

$10x+14y=20$

$3x-14y=6$

Add the equations.

$13x=26$

$x=2$

Replace 2 for x in Equation (2).

$3x-14y=6$

$3(2)-14y=6$

$6-14y=6$

$-14y=0$

$y=0$

The solution is (2, 0).

21. (1) $3x-2y=0$

(2) $6x+5y=0$

Eliminate x.

$-2(3x-2y)=-2(0)$

$-6x+4y=0$

$6x+5y=0$

Add the equations.

$9y=0$

$y=0$

Replace 0 for y in Equation (2).

$6x+5y=0$

$6x+5(0)=0$

$6x=0$

$x=0$

The solution is (0, 0).

23. (1) $2x-3y=16$

(2) $3x+4y=7$

Eliminate y.

$4(2x-3y)=4\cdot 16$

$3(3x+4y)=3\cdot 7$

$8x-12y=64$

$9x+12y=21$

Add the equations.

$17x=85$

$x=5$

Replace 5 for x in Equation (1).

$2x-3y=16$

$2(5)-3y=16$

$10-3y=16$

$-3y=16$

$y=-2$

The solution is (5, −2).

25. (1) $5x+3y=7$

(2) $2x+5y=1$

Eliminate x.

$$-2(5x+3y)=-2\cdot 7$$
$$5(2x+5y)=5\cdot 1$$
$$-10x-6y=-14$$
$$10x+25y=5$$

Add the equations.

$$19y=-9$$
$$y=-\frac{9}{19}$$

Replace $-\frac{9}{19}$ for y in Equation (2).

$$2x+5y=1$$
$$2x+5\left(-\frac{9}{19}\right)=1$$
$$2x-\frac{45}{19}=1$$
$$2x=\frac{19}{19}+\frac{45}{19}$$
$$2x=\frac{64}{19}$$
$$x=\frac{32}{19}$$

The solution is $\left(\frac{32}{19}, -\frac{9}{19}\right)$.

27. (1) $3x+4y=4$

(2) $5x+12y=5$

Eliminate y.

$$-3(3x+4y)=-3\cdot 4$$
$$-9x-12y=-12$$
$$5x+12y=5$$

Add the equations.

$$-4x=-7$$
$$x=\frac{7}{4}$$

Replace $\frac{7}{4}$ for x in Equation (1).

$$3x+4y=4$$
$$3\left(\frac{7}{4}\right)+4y=4$$
$$\frac{21}{4}+4y=4$$
$$4y=-\frac{5}{4}$$
$$y=-\frac{5}{16}$$

The solution is $\left(\frac{7}{4}, -\frac{5}{16}\right)$.

29. (1) $8x-3y=11$

(2) $6x-5y=11$

Eliminate y.

$$5(8x-3y)=5\cdot 11$$
$$-3(6x-5y)=-3\cdot 11$$
$$40x-15y=55$$
$$-18x+15y=-33$$

Add the equations.

$$22x=22$$
$$x=1$$

Replace 1 for x in Equation (1).

$$8x-3y=11$$
$$8(1)-3y=11$$
$$8-3y=11$$
$$-3y=3$$
$$y=-1$$

The solution is $(1, -1)$.

31. (1) $5x+15y=20$

(2) $2x+6y=12$

Eliminate x.

$2(5x+15y)=2\cdot 20$

$-5(2x+6y)=-5\cdot 12$

$10x+30y=40$

$-10x-30y=-60$

Add the equations.

$0=-20$

This is not a true statement. The system is inconsistent. There is no solution.

33. (1) $3x=2y+7$

(2) $5x-2y=13$

Write Equation (1) in the form $Ax+By=C$.

$3x=2y+7$

$3x-2y=7$

Eliminate y.

$-1(3x-2y)=-1\cdot 7$

$-3x+2y=-7$

$5x-2y=13$

Add the equations.

$2x=6$

$x=3$

Replace 3 for x in Equation (2).

$5x-2y=13$

$5(3)-2y=13$

$15-2y=13$

$-2y=-2$

$y=1$

The solution is (3, 1).

35. (1) $2x+9y=16$

(2) $5x=1-3y$

Write Equation (2) in the form $Ax+By=C$.

$5x=1-3y$

$5x+3y=1$

Eliminate y.

$2x+9y=16$

$-3(5x+3y)=-3\cdot 1$

$2x+9y=16$

$-15x-9y=-3$

Add the equations.

$-13x=13$

$x=-1$

Replace -1 for x in Equation (1).

$2x+9y=16$

$2(-1)+9y=16$

$-2+9y=16$

$9y=18$

$y=2$

The solution is (−1, 2).

37. (1) $2x+3y=7-2x$

(2) $7x+2y=9$

Write Equation (1) in the form $Ax+By=C$.

$2x+3y=7-2x$

$2x+5y=7$

Eliminate y.

$2(4x+3y)=2\cdot 7$

$-3(7x+2y)=-3\cdot 9$

$8x+6y=14$

$-21x-6y=-27$

Add the equations.

$-13x=-13$

$x=1$

Replace 1 for x in Equation (2).

$7x+2y=9$

$7(1)+2y=9$

$7+2y=9$

$2y=2$

$y=1$

The solution is (1, 1).

39. (1) $Ax + By = C$

(2) $2Ax + 2By = 2C$

Eliminate x.

$-2(Ax + By) = C$

$-2Ax - 2By = -2C$

$2Ax + 2By = 2C$

Add the equations.

$0 = 0$

The system is dependent.

41. (1) $Ax + By = C$

(2) $Bx + Ay = 2C$

Eliminate x.

$-B(Ax + By) = -B \cdot C$

$A(Bx + Ay) = A \cdot 2C$

$-ABx - B^2y = -BC$

$ABx + A^2y = 2AC$

Add the equations.

$A^2y - B^2y = -BC - 2AC$

$y(A^2 - B^2) = -BC - 2AC$

$y = \dfrac{-BC - 2AC}{A^2 - B^2}$

The system is independent.

Critical Thinking 13.3

43. (1) $Ax - 4y = 9$, $(-1, 3)$

$A(-1) - 4(-3) = 9$

$-A + 12 = 9$

$-A = -3$

$A = 3$

(2) $4x + By = -1$, $(-1, 3)$

$4(-1) + B(-3) = -1$

$-4 - 3B = -1$

$-3B = 3$

$B = -1$

Section 13.4

Concept Check

1. True

3. False

5. $100x + 50y$

Objective A Exercises

7. Traveling with the wind, a plane can fly further in the same amount of time than against the wind, so $m > n$.

9. Strategy:

→Rate of rowing in calm water: r

→Rate of current: c

	Rate	Time	Distance
With current	$r + c$	2	$2(r + c)$
Against current	$r - c$	2	$2(r - c)$

→The distance traveled with the current is 40 km. The distance traveled against the current is 16 km.

Solution:

$2(r + c) = 40$ $\quad$ $2(r - c) = 16$

$r + c = 20$ $\quad$ $r - c = 8$

$r + c = 20$

$r - c = 8$

$2r = 28$

$r = 14$

$r + c = 20$

$14 + c = 20$

$c = 6$

The rate rowing in calm water is 14 km/h. The rate of the current is 6 km/h.

11. Strategy:

→Rate of whale in calm water: r

→Rate of ocean current: c

	Rate	Time	Distance
With current	$r+c$	1.5	$1.5(r+c)$
Against current	$r-c$	2	$2(r-c)$

→The distance traveled with the current is 60 mi. The distance traveled against the current is 16 mi.

Solution:

$$1.5(r+c)=60 \qquad 2(r-c)=60$$
$$r+c=40 \qquad r-c=30$$

$$\begin{aligned} r+c&=40\\ r-c&=30\\ \hline 2r&=70\\ r&=35\\ r+c&=40\\ 35+c&=40\\ c&=5 \end{aligned}$$

The rate whale in calm water is 35 mph.

The rate of the current is 5 mph.

13. Strategy:

→Rate of Lear jet in calm water: r

→Rate of wind: w

	Rate	Time	Distance
With current	$r+w$	2	$2(r+w)$
Against current	$r-w$	2	$2(r-w)$

→The distance traveled with the wind is 1120 mi. The distance traveled against the wind is 980 mi.

Solution:

$$2(r+w)=1120 \qquad 2(r-w)=980$$
$$2r+2w=1120 \qquad 2r-2w=980$$

$$\begin{aligned} 2r+2w&=1120\\ 2r-2w&=980\\ \hline 4w&=2100\\ w&=525\\ 2r+2w&=1120\\ 2(525)+2w&=1120\\ 1050+2w&=1120\\ 2w&=70\\ w&=35 \end{aligned}$$

The rate of the Learjet is 525 mph.

The rate of the wind is 35 mph.

15. Strategy:

→Rate of helicopter in calm air: r

→Rate of wind: w

	Rate	Time	Distance
With wind	$r+w$	$\frac{5}{3}$	$\frac{5}{3}(r+w)$
Against wind	$r-w$	2.5	$2.5(r-w)$

→The distance traveled with the wind is 450 mi. The distance traveled against the wind is 450 mi.

Solution:

$$\frac{5}{3}(p+w)=450$$

$$\frac{3}{5}\cdot\frac{5}{3}(p+w)=\frac{3}{5}\cdot 450$$

$$p+w=270$$

$$2.5(r-w)=450$$
$$r-w=180$$

$$r+w=270$$
$$r-w=180$$
$$2r=450$$
$$r=225$$
$$r+w=120$$
$$225+w=270$$
$$w=45$$

The rate of the helicopter in calm air is 225 mph. The rate of the wind is 45 mph.

17. Strategy:

→Rate of boat in calm water: r

→Rate of current: c

	Rate	Time	Distance
With current	$r+c$	3.5	$3.5(r+c)$
Against current	$r-c$	3	$3(r-c)$

→The distance traveled with the current is 35 mi. The distance traveled against the current is 12 mi.

Solution:

$$3.5(r+c)=35 \qquad 3(r-c)=12$$
$$r+c=10 \qquad r-c=4$$

$$r+c=10$$
$$r-c=4$$
$$2r=14$$
$$r=7$$
$$r+c=10$$
$$7+c=10$$
$$c=3$$

The rate of the boat in calm water is 7 mph. The rate of the current is 3 mph.

Objective B Exercises

19. Less than

21. x is the cost of the adult ticket and y is the cost of a child ticket.
$4x+2y=320$ represents the fact that you spent \$320 on 4 adult tickets and 2 child tickets.
$2x+3y=240$ represents the fact that you r neighbor spent \$240 on 2 adult tickets and 3 child tickets.

23. Strategy:

→Cost per pound of the wheat flour: x

→Cost per pound of the rye flour: y

1st Purchase:

	Number	Cost	Total Cost
Wheat	12	x	$12x$
Rye	15	y	$15y$

2nd Purchase:

	Number	Cost	Total Cost
Wheat	15	x	$15x$
Rye	10	y	$10y$

→The first purchase cost \$18.30. The second purchase cost \$16.75.

Solution:

$$12x+15y=18.30$$
$$15x+10y=16.75$$

$$5(12x+15y)=5(18.30)$$
$$-4(15x+10y)=-4(16.75)$$

$$60x+75y=91.50$$
$$-60x-40y=-67.00$$

$$35y=24.50$$
$$y=0.70$$

$$12x+15y=18.30$$
$$12x+15(0.70)=18.30$$
$$12x+10.50=18.30$$
$$12x=7.80$$
$$x=0.65$$

The cost per pound of the wheat flour is \$.65. The cost per pound of the rye flour is \$.70.

25. Strategy:

→Cost of turkey sandwich: x

→Cost of French fries: y

1st Day:

	Number	Cost	Total Cost
Turkey	4	x	$4x$
French fries	7	y	$7y$

2nd Day:

	Number	Cost	Total Cost
Turkey	5	x	$5x$
French fries	5	y	$5y$

→The cost the first day was \$38.30. The cost the second day was \$40.75.

Solution:

$$4x+7y=38.30 \qquad 5(4x+7y)=5(38.30)$$
$$5x+5y=40.75 \qquad -4(5x+5y)=-4(40.75)$$

$$20x+35y=191.50$$
$$-20x-20y=-163.00$$

$$15y=28.50$$
$$y=1.90$$

$$4x+7y=38.30$$
$$4x+7(1.90)=38.30$$
$$4x+13.30=38.30$$
$$4x=25$$
$$x=6.25$$

The cost of the turkey sandwich is \$6.25. The cost for an order of French fries is \$1.90.

27. Strategy:

→Amount of 20% sugar solution: x

→Amount of 40% sugar solution: y

	Amount	Percent	Quantity
20% sugar	x	0.20	$0.20x$
40% sugar	y	0.40	$0.40y$
34% sugar	50	0.34	0.35(50)

→The sum of the quantities before mixing is equal to the quantity after mixing. The total amount of the two solutions is 50 oz.

Solution:

$$x + y = 50 \qquad -0.20(x+y) = -0.20(50)$$
$$0.20x + 0.40y = 17 \qquad 0.20x + 0.40y = 17$$
$$-0.20x - 0.20y = -10$$
$$0.20x + 0.40y = 17$$
$$0.20y = 7$$
$$y = 35$$

$$x + y = 50$$
$$x + 35 = 50$$
$$x = 15$$

The amount of 20% solution was 15 oz. The amount of 40% solution was 35 oz.

29. Strategy:

→To find the height, consider the two equations a system of equations. Solve the system for x by using substitution.

Solution:

$$W = 49 + 1.7x \qquad 49 + 1.7x = 53.1 + 1.36x$$
$$W = 53.1 + 1.36x \qquad 0.34x = 4.1$$
$$x \approx 12$$

The height is $60 + x = 60 + 12 = 72$.
A woman who is 72 in. tall would have the same ideal weight from both formulas.

31. **a.** Strategy:

→Cost of a Lincoln stamp: x

→Cost of a Jefferson stamp: y

→Cost of a Henry Clay stamp: z

1st Value:

	Stamps	Cost	Total cost
Lincoln	3	x	$3x$
Jefferson	5	y	$5y$

2nd Value:

	Stamps	Cost	Total cost
Lincoln	2	x	$2x$
Jefferson	3	y	$3y$

→The total cost of the first was \$3.20. The total cost of the second was \$2.10.

Solution:

$$3x + 5y = 3.20 \qquad -2(3x+5y) = -2(3.20)$$
$$2x + 3y = 2.10 \qquad 3(2x+3y) = 3(2.10)$$
$$-6x - 10y = -6.40$$
$$6x + 9y = 6.30$$
$$-y = -0.10$$
$$y = 0.10$$

$$3x + 5y = 3.20$$
$$3x + 5(0.10) = 3.20$$
$$3x + 0.50 = 3.20$$
$$3x = 2.70$$
$$x = 0.90$$

The Lincoln stamp cost \$.90 and the Jefferson stamp cost \$.10.

b. Strategy:

→The total cost of all three stamps was \$1.12.

Solution:

$$x + y + z = 1.12$$
$$0.90 + 0.10 + z = 1.12$$
$$1.00 + z = 1.12$$
$$z = 0.12$$

The Henry Clay stamp cost \$.12.

Critical Thinking 13.4

33. Strategy:
→Measure of smaller angle: x
→Measure of larger angle: y
→The measure of the larger angle is four times the measure of the smaller angle. The sum of the measures of the angles are 90°.

Solution:

$$y = 4x \qquad x + 4x = 90$$
$$x + y = 90 \qquad 5x = 90$$
$$x = 18$$
$$y = 4x = 4(18) = 72$$

The measures of the angles are 18° and 72°.

Projects and Group Activities 13.4

35. Strategy:

	Rate	Time	Distance
Hour hand	30	t	$30t = d$
Minute hand	360	t	$360t = d + 360$

→Solve the system of equations.

Solution:

$$30t = d$$
$$360t = d + 360$$

$$360t = 30t + 360$$
$$330t = 360$$
$$t = 1.09$$

1.09 hr = 1.09(60) = 65.5 min
The hands will align after 65.5 min.

37. (2.875, 1.1875)

39. (–19, –7)

Chapter Review Exercises

1.

$5x + 4y = -17$		$2x - y = 1$	
$5(-1) + 4(-3)$	-17	$2(-1) - (-3)$	1
$-5 - 12$	-17	$-2 + 3$	1
$-17 = -17$		$1 = 1$	

Yes, (–1, –3) is a solution of the system of equations.

2.

$-x + 9y = 2$		$6x - 4y = 12$	
$-(-2) + 9(0)$	2	$6(-2) - 0$	12
$2 + 0$	2	-12	12
$2 = 2$		$-12 \neq 12$	

No, (–2, 0) is not a solution of the system of equations.

3.

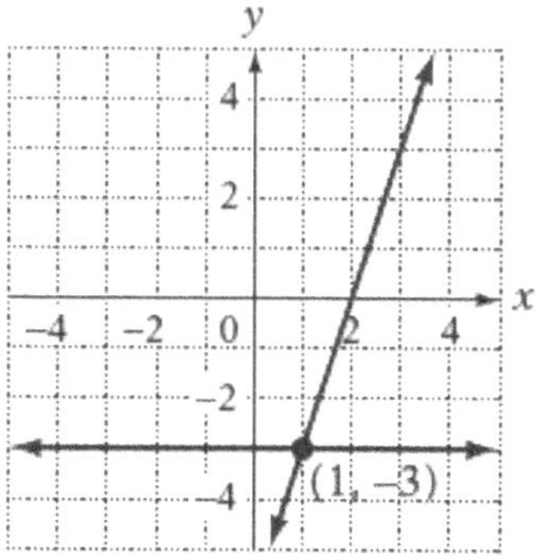

The solution is (1, –3).

4.

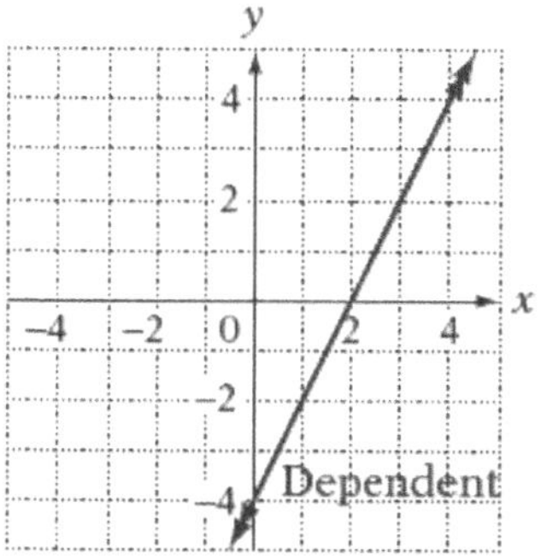

The solutions are the ordered-pair solution of $y = 2x - 4$.

5.

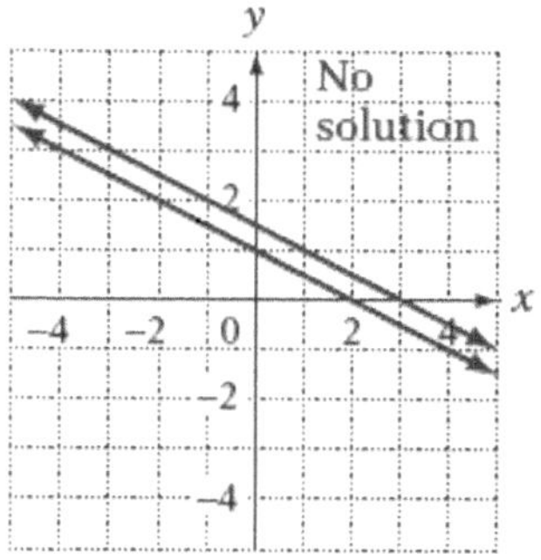

No solution

6. (1) $4x+7y=3$

(2) $x=y-2$

Substitute $y-2$ for x in Equation (1).

$$4x+7y=3$$
$$4(y-2)+7y=3$$
$$4y-8+7y=3$$
$$11y-8=3$$
$$11y=11$$
$$y=1$$

Substitute 1 for y in Equation (2).

$$x=y-2$$
$$x=1-2$$
$$x=-1$$

The solution is $(-1, 1)$.

7. (1) $6x-y=0$

(2) $7x-y=1$

Solve Equation (1) for y.

$$6x-y=0$$
$$-y=-6x$$
$$y=6x$$

Substiture $6x$ for y in Equation (2).

$$7x-y=1$$
$$7x-6x=1$$
$$x=1$$

Substitute 1 for x in Equation (1).

$$6x-y=0$$
$$6(1)-y=0$$
$$6-y=0$$
$$-y=-6$$
$$y=6$$

The solution is $(1, 6)$.

8. (1) $3x+8y=-1$

(2) $x-2y=-5$

Eliminate y by adding the equations.

$$3x+8y=-1 \qquad 3x+8y=-1$$
$$4x-8y=-5 \qquad 4(x-2y)=4(-5)$$

$$3x+8y=-1$$
$$\underline{4x-8y=-20}$$
$$7x=-21$$
$$x=-3$$

Substitute -3 for x in Equation (2).

$$x-2y=-5$$
$$-3-2y=-5$$
$$-2y=-2$$
$$y=1$$

The solution is $(-3, 1)$.

9. (1) $6x+4y=-3$

(2) $12x-10y=-15$

Eliminate x by adding the equations.

$$6x+4y=-3 \qquad -2(6x+4y)=-2(-3)$$
$$12x-10y=-15 \qquad 12x-10y=-15$$

$$-12x-8y=6$$
$$\underline{12x-10y=-15}$$
$$-18y=-9$$
$$y=\frac{1}{2}$$

Substitute $\frac{1}{2}$ for y in Equation (1).

$$6x+4y=-3$$
$$6x+4\left(\frac{1}{2}\right)=-3$$
$$6x+2=-3$$
$$6x=-5$$
$$x=-\frac{5}{6}$$

The solution is $\left(-\frac{5}{6}, \frac{1}{2}\right)$.

10. (1) $12x - 9y = 18$

(2) $y = \frac{4}{3}x - 3$

Substitute $\frac{4}{3}x - 3$ for y in Equation (1).

$$12x - 9y = 18$$
$$12x - 9\left(\frac{4}{3}x - 3\right) = 18$$
$$12x - 12x + 27 = 18$$
$$27 \neq 18$$

The system is inconsistent and has no solution.

11. (1) $8x - y = 2$

(2) $y = 5x + 1$

Substitute $5x + 1$ for y in Equation (1).

$$8x - y = 2$$
$$8x - (5x + 1) = 2$$
$$8x - 5x - 1 = 2$$
$$3x - 1 = 2$$
$$3x = 3$$
$$x = 1$$

Substitute 1 for x in Equation (2).

$$y = 5x + 1$$
$$y = 5(1) + 1$$
$$y = 5 + 1$$
$$y = 6$$

The solution is (1, 6).

12. (1) $4x - y = 9$

(2) $2x + 3y = -13$

Eliminate y by adding the equations.

$4x - y = 9 \quad 3(4x - y) = 3(9)$

$2x + 3y = 12 \quad 2x + 3y = -13$

$$12x - 3y = 27$$
$$2x + 3y = -13$$
$$14x = 14$$
$$x = 1$$

Substitute 1 for x in Equation (1).

$$4x - y = 9$$
$$4(1) - y = 9$$
$$4 - y = 9$$
$$-y = 5$$
$$y = -5$$

The solution is (1, −5).

13. (1) $5x + 7y = 21$

(2) $20x + 28y = 63$

Eliminate y by adding the equations.

$5x + 7y = 21 \quad -4(5x + 7y) = -4(21)$

$20x + 28y = 63 \quad 20x + 28y = 63$

$$-20x - 28y = -84$$
$$20x + 28y = 63$$
$$0 \neq -21$$

The system is inconsistent and has no solution.

14. (1) $4x + 3y = 12$

(2) $y = -\frac{4}{3}x + 4$

Substitute $-\frac{4}{3}x + 4$ for y in Equation (1).

$$4x + 3y = 12$$
$$4x + 3\left(-\frac{4}{3}x + 4\right) = 12$$
$$4x - 4x + 12 = 12$$
$$12 = 12$$

The system is of equations is dependent. The solutions are the ordered-pairs that satisfy the equation $y = -\frac{4}{3}x + 4$.

15. (1) $7x + 3y = -16$

(2) $x - 2y = 5$

Solve Equation (2) for x.

$$x - 2y = 5$$
$$x = 2y + 5$$

Substiture $2y + 5$ for x in Equation (1).

$$7x + 3y = -16$$
$$7(2y + 5) + 3y = -16$$
$$14y + 35 + 3y = -16$$
$$17y + 35 = -16$$
$$17y = -51$$
$$y = -3$$

Substitute −3 for y in Equation (2).

$$x - 2y = 5$$
$$x - 2(-3) = 5$$
$$x + 6 = 5$$
$$x = -1$$

The solution is (−1, −3).

16. (1) $3x+y=-2$

(2) $-9x-3y=6$

Eliminate x by adding the equations.

$$3x+y=-2 \qquad 3(3x+y)=3(-2)$$
$$-9x-3y=6 \qquad -9x-3y=6$$

$$9x+3y=-6$$
$$\underline{-9x-3y=6}$$
$$0=0$$

The system is dependent. The solutions are the ordered pairs that statisfy the equation $3x+y=-2$.

17. (1) $6x-18y=7$

(2) $9x+24y=2$

Eliminate x by adding the equations.

$$6x-18y=7 \qquad 3(6x-18y)=3(7)$$
$$9x+24y=2 \qquad -2(9x+24y)=-2(2)$$

$$18x-54y=21$$
$$\underline{-18x-48y=-4}$$
$$-102y=17$$
$$y=-\frac{1}{6}$$

Substitute $-\frac{1}{6}$ for y in Equation (1).

$$6x-18y=7$$
$$6x-18\left(-\frac{1}{6}\right)=7$$
$$6x+3=7$$
$$6x=4$$
$$x=\frac{2}{3}$$

The solution is $\left(\frac{2}{3}, -\frac{1}{6}\right)$.

18. Strategy:

→Rate of team in calm water: r

→Rate of current: c

	Rate	Time	Distance
With current	$r+c$	2	$2(r+c)$
Against current	$r-c$	3	$3(r-c)$

→The distance traveled with the current is 24 mi. The distance traveled against the current is 18 mi.

Solution:

$$2(r+c)=24 \qquad 3(r-c)=18$$
$$r+c=12 \qquad r-c=6$$

$$r+c=12$$
$$\underline{r-c=6}$$
$$2r=18$$
$$r=9$$
$$r+c=12$$
$$9+c=12$$
$$c=3$$

The rate of the team in calm water is 9 mph. The rate of the current is 3 mph.

19. Strategy:

→Number of shares at \$6: x

→Number of shares at \$25: y

	Number	Cost	Total cost
\$6 shares	x	6	$6x$
\$25 shares	y	25	$25y$

→The total number of shares is 1500. The total purchase price is \$12,800.

Solution:

$$x+y=1500 \qquad -6(x+y)=-6(1500)$$
$$6x+25y=12{,}800 \qquad 6x+25y=12800$$

$$-6x-6y=-9000$$
$$\underline{6x+25y=12{,}800}$$
$$19y=3800$$
$$y=200$$
$$x+y=1500$$
$$x+200=1500$$
$$x=1300$$

Number of \$6 shares purchased is 1300. Number of \$25 shares is 200.

20. Strategy:

→Rate of flight crew in calm air: p

→Rate of wind: w

	Rate	Time	Distance
With wind	$p + w$	3	$3(p + w)$
Against wind	$p - w$	4	$4(p - w)$

→The distance traveled with the wind is 420 km. The distance traveled against the wind is 440 km.

Solution:

$3(p+w)=420 \qquad 4(p-w)=440$

$p+w=140 \qquad p-w=110$

$p+w=140$

$\underline{p-w=110}$

$2p=250$

$p=125$

$p+w=140$

$125+w=140$

$w=15$

The rate of the flight crew in calm air is 125 km/h.

The rate of the wind is 15 km/h.

21. Strategy:

→Rate of plane in calm air: p

→Rate of wind: w

	Rate	Time	Distance
With wind	$p + w$	3	$3(r + w)$
Against wind	$p - w$	4	$4(r - w)$

→The distance traveled with the wind and against the wind is 360 mi.

Solution:

$3(p+w)=360 \qquad 4(p-w)=360$

$p+w=120 \qquad p-w=90$

$p+w=120$

$\underline{p-w=90}$

$2p=210$

$p=105$

$p+w=120$

$105+w=120$

$w=15$

The rate of the plane in calm air is 105 mph.

The rate of the wind is 15 mph.

22. Strategy:

→Regular rate: x

→Premium rate: y

1st Customer	Minutes	Rate	Total Cost
Prime	9	x	$9x$
Non-prime	2	y	$2y$

2nd Customer	Minutes	Rate	Total Cost
Prime	6	x	$6x$
Non-prime	3	y	$3y$

→The first customer's total cost was $14. The second customer's total cost was $13.50.

Solution:

$9x+2y=14 \qquad 2(9x+2y)=2(14)$

$6x+3y=13.50 \qquad -3(6x+3y)=-3(13.50)$

$18x+4y=28$

$\underline{-18x-9y=-40.50}$

$-5y=-12.50$

$y=2.50$

$9x+2y=14$

$9x+2(2.50)=14$

$9x+5=14$

$9x=9$

$x=1$

The regular cost is $1 per hour. The premium cost is $2.50 per hour.

23. Strategy:
→Amount invested at 8.5%: x
→Amount invested at 7%: y

	P	R	I
Investment at 8.5%	x	0.085	$0.085x$
Investment at 7%	y	0.07	$0.07y$

→The total amount invested is $12,000. The interest earned is $915.

Solution:

$$x + y = 12{,}000$$
$$y = 12{,}000 - x$$
$$0.085x + 0.07y = 915$$
$$0.085x + 0.07(12{,}000 - x) = 915$$
$$0.085 + 840 - 0.07x = 915$$
$$0.015x = 75$$
$$x = 5000$$
$$y = 12{,}000 - x = 12{,}000 - 5000 = 7000$$

The amounts invested are $7000 at 7% and $5000 at 8.5%.

24. Strategy:
→Number of bushels of lentils: x
→Number of bushels of corn: y
If 50 bushels of lentils were added, there would be twice as many bushels of lentils as of corn: $x + 50 = 2y$
If 150 bushels of corn were added, there would be the same number of corn as lentils: $y + 150 = x$

Solution:

$$x + 50 = 2y \qquad y + 150 + 50 = 2y$$
$$y + 150 = x \qquad 200 = y$$

$$y + 150 = x$$
$$200 + 150 = x$$
$$350 = x$$

There were originally 350 bushels of lentils and 200 bushels of corn.

25. Strategy:
→Amount invested at 5.4%: x
Amount invested at 6.6%: y

	P	R	I
Investment at 5.4%	x	0.054	$0.054x$
Investment at 6.6%	y	0.066	$0.066y$

→The total amount invested is $300,000.
$x + y = 300{,}000$
The interest earned in each account is the same. $0.054x = 0.066y$

Solution:

$$x + y = 300{,}000$$
$$y = 300{,}000 - x$$
$$0.054x = 0.066y$$
$$0.054x = 0.066(300{,}0000x)$$
$$0.054x = 19{,}800 - 0.066x$$
$$0.12x = 19{,}800$$
$$x = 165{,}000$$
$$y = 300{,}000 - x = 300{,}000 - 165{,}000 = 135{,}000$$

The amounts invested are $165,000 at 5.4% and $135,000 at 6.6%.

Chapter Test

1.

$2x + 5y = 11$		$x + 3y = 7$	
$2(-2) + 5(3)$	11	$-2 + 3(3)$	7
$-4 + 15$	11	$-2 + 9$	7
$11 = 11$		$7 = 7$	

Yes, (−2, 3) is a solution.

2.

$3x-2y=9$	
$3(1)-2(-3)$	9
$3+6$	9
$9=9$	

$4x+y=1$	
$4(1)-3$	1
$4-3$	1
$1=1$	

Yes, (−2, 3) is a solution.

3.

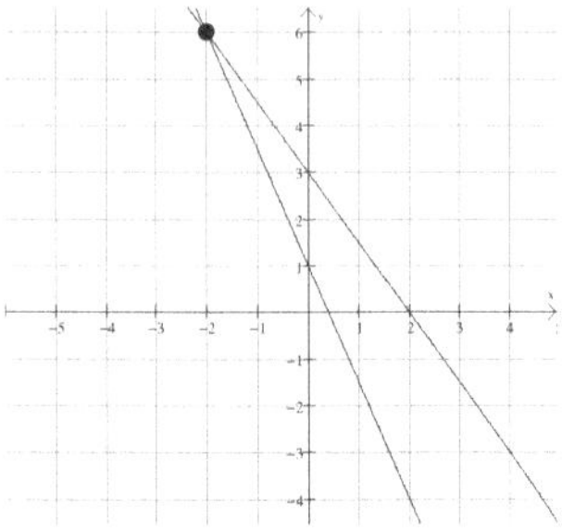

The solution is (−2, 6).

4. (1) $4x-y=11$

(2) $y=2x-5$

Substiture $2x-5$ for y in Equation (1).

$$4x-y=11$$
$$4x-(2x-5)=11$$
$$4x-2x+5=11$$
$$2x+5=11$$
$$2x=6$$
$$x=3$$

Substitute 3 for x in Equation (2).

$$y=2x-5$$
$$y=2(3)-5$$
$$y=6-5$$
$$y=1$$

The solution is (3, 1).

5. (1) $x=2y+3$

(2) $3x-2y=5$

Substiture $2y+3$ for x in Equation (2).

$$3x-2y=5$$
$$3(2y+3)-2y=5$$
$$6y+9-2y=5$$
$$4y+9=5$$
$$4y=-4$$
$$y=-1$$

Substitute −1 for y in Equation (1).

$$x=2y+3$$
$$x=2(-1)+3$$
$$x=-2+3$$
$$x=1$$

The solution is (1, −1).

6. (1) $3x+5y=1$

(2) $2x-y=5$

Solve Equation (2) for y.

$$2x-y=5$$
$$-y=-2x+5$$
$$y=2x-5$$

Substitute $2x-5$ for y in Equation (1).

$$3x+5y=1$$
$$3x+5(2x-5)=1$$
$$3x+10x-25=1$$
$$13x-25=1$$
$$13x=26$$
$$x=2$$

Substitute 2 for x in Equation (2).

$$2x-y=5$$
$$2(2)-y=5$$
$$4-y=5$$
$$-y=1$$
$$y=-1$$

The solution is (2, −1).

7. (1) $3x - 5y = 13$

(2) $x + 3y = 1$

Solve Equation (2) for x.

$$x + 3y = 1$$
$$x = -3y + 1$$

Substiture $-3y + 1$ for x in Equation (1).

$$3x - 5y = 13$$
$$3(-3y + 1) - 5y = 13$$
$$-9y + 3 - 5y = 13$$
$$-14y = 10$$
$$y = -\frac{5}{7}$$

Substitute $-\frac{5}{7}$ for y in Equation (2).

$$x + 3y = 1$$
$$x + 3\left(-\frac{5}{7}\right) = 1$$
$$x - \frac{15}{7} = 1$$
$$x = \frac{15}{7} + 1$$
$$x = \frac{15}{7} + \frac{7}{7}$$
$$x = \frac{22}{7}$$

The solution is $\left(\frac{22}{7}, -\frac{5}{7}\right)$.

8. (1) $2x - 4y = 1$

(2) $y = \frac{1}{2}x + 3$

Substiture $\frac{1}{2}x + 3$ for y in Equation (1).

$$2x - 4y = 1$$
$$2x - 4\left(\frac{1}{2}x + 3\right) = 1$$
$$2x - 2x - 12 = 1$$
$$-12 \neq 1$$

The system is inconsistent and has no solution.

9. (1) $4x + 3y = 11$

(2) $5x - 3y = 7$

Eliminate y by adding the equations.

$$4x + 3y = 11$$
$$\underline{5x - 3y = 7}$$
$$9x = 18$$
$$x = 2$$

Substitute 2 for x in Equation (1).

$$4x + 3y = 11$$
$$4(2) + 3y = 11$$
$$8 + 3y = 11$$
$$3y = 3$$
$$y = 1$$

The solution is (2, 1).

10. (1) $2x - 5y = 6$

(2) $4x + 3y = -1$

Eliminate x by adding the equations.

$$-2(2x - 5y) = -2(6) \qquad -4x + 10y = -12$$
$$4x + 3y = -1 \qquad \underline{4x + 3y = -1}$$
$$13y = -13$$
$$y = -1$$

Substitute -1 for y in Equation (1).

$$2x - 5y = 6$$
$$2x - 5(-1) = 6$$
$$2x + 5 = 6$$
$$2x = 1$$
$$x = \frac{1}{2}$$

The solution is $\left(\frac{1}{2}, -1\right)$.

11. (1) $x + 2y = 8$

(2) $3x + 6y = 24$

Eliminate x by adding the equations.

$$-3(x + 2y) = -3(8) \qquad -3x - 6y = -24$$
$$3x + 6y = 24 \qquad \underline{3x + 6y = 24}$$
$$0 = 0$$

The system is dependent. The solutiions are the ordered pairs that satisfy the equation $x + 2y = 8$.

12. (1) $7x+3y=11$
(2) $2x-5y=9$

Eliminate y by adding the equations.

$$5(7x+3y)=5(11) \qquad 35x+15y=55$$
$$3(2x-5y)=3(9) \qquad 6x-15y=27$$
$$41x=82$$
$$x=2$$

Substitute 2 for x in Equation (1).

$$7x+3y=11$$
$$7(2)+3y=11$$
$$14+3y=11$$
$$3y=-3$$
$$y=-1$$

The solution is (2, −1).

13. (1) $5x+6y=-7$
(2) $3x+4y=-5$

Eliminate x by adding the equations.

$$3(5x+6y)=3(-7) \qquad 15x+18y=-21$$
$$-5(3x+4y)=-5(-5) \qquad -15x-20y=25$$
$$-2y=4$$
$$y=-2$$

Substitute −2 for y in Equation (1).

$$5x+6y=-7$$
$$5x+6(-2)=-7$$
$$5x-12=-7$$
$$5x=5$$
$$x=1$$

The solution is (1, −2).

14. Strategy:
→Rate of plane in calm air: p
→Rate of wind: w

	Rate	Time	Distance
With wind	$p+w$	2	$2(p+w)$
Against wind	$p-w$	3	$3(p-w)$

→The distance traveled in both directions is 240 mi.

Solution:

$$2(p+w)=240 \qquad 3(p-w)=240$$
$$p+w=120 \qquad p-w=80$$

$$p+w=120$$
$$p-w=80$$
$$2p=200$$
$$p=100$$
$$p+w=120$$
$$100+w=120$$
$$w=20$$

The rate of the plane in calm air is 100 mph.
The rate of the wind is 20 mph.

15. Strategy:

→Reserved seat cost: x

→General–admission cost: y

1st Performance	Number	Cost	Total Cost
Reserved	50	x	$50x$
General	80	y	$80y$

2nd Performance	Number	Cost	Total Cost
Prime	60	x	$60x$
Non-prime	90	y	$90y$

→The first performance receipts were \$980. The second performance receipts were \$1140.

Solution:

$$50x+80y=980 \qquad 6(50x+80y)=6(980)$$
$$60x+90y=1140 \qquad -5(60x+90y)=-5(1140)$$

$$300x+480y=5880$$
$$-300x-450y=-5700$$
$$30y=180$$
$$y=6$$
$$50x+80y=980$$
$$50x+80(6)=980$$
$$50x+480=980$$
$$50x=500$$
$$x=10$$

The reserved-seat ticket cost is \$10. The general-admission ticket cost is \$6.

16. Strategy:

→Amount invested at 7.6%: x

→Amount invested at 6.4%: y

	P	R	I
Investment at 7.6%	x	0.076	$0.076x$
Investment at 6.4%	y	0.064	$0.064y$

→The total amount invested is \$28,000.

$x+y=28,000$

The interest earned in each account is the same. $0.076x=0.064y$

Solution:

$$x+y=28,000$$
$$y=28,000-x$$
$$0.076x=0.064y$$
$$0.076x=0.064(28,0000-x)$$
$$0.076x=1792-0.064x$$
$$0.14x=1792$$
$$x=15,200$$
$$y=28,000-x=280,000-12,800=15,200$$

The amounts invested are \$15,200 at 6.4% and \$12,800 at 7.6%.

Cumulative Review Exercises

1. $\dfrac{a^2-b^2}{2a}$

$$\frac{4^2-(-2)^2}{2(4)}=\frac{16-4}{8}=\frac{12}{8}=\frac{3}{2}$$

2.

$$-\frac{3}{4}x=\frac{9}{8}$$
$$-\frac{4}{3}\left(-\frac{3}{4}x\right)=-\frac{\overset{1}{\cancel{4}}}{\underset{1}{\cancel{3}}}\left(\frac{\overset{3}{\cancel{9}}}{\underset{2}{\cancel{8}}}\right)$$
$$x=-\frac{3}{2}$$

The solution is $-\dfrac{3}{2}$.

3.

$$f(x)=x^2+2x-1$$
$$f(2)=2^2+2(2)-1$$
$$f(2)=4+4-1$$
$$f(2)=7$$

4. $(2a^2-3a+1)(2-3a)$

$$\begin{array}{r} 2a^2-3a+1 \\ \underline{-3a+2} \\ 4a^2-6a+2 \\ \underline{-6a^3+9a^2-3a} \\ -6a^3+13a^2-9a+2 \end{array}$$

5. $\dfrac{(-2x^2y)^4}{-8x^3y^2}=\dfrac{16x^8y^4}{-8x^3y^2}=-2x^5y^2$

6.

$$\begin{array}{r} 2b-1 \\ 2b-3\overline{)4b^2-8b+4} \\ \underline{4b^2-6b} \\ -2b+4 \\ \underline{-2b+3} \\ 1 \end{array}$$

$(4b^2-8b+4)\div(2b-3)=2b-1+\dfrac{1}{2b-3}$

7. $\dfrac{8x^{-2}y^5}{-2xy^4}=\dfrac{-4y^{5-4}}{x^{2+1}}=-\dfrac{4y}{x^3}$

8. $4x^2y^4-64y^2$

The GCF is $4y^2$: $4y^2(x^2y^2-16)$

$4((xy)^2-4^2)=4y^2(xy+4)(xy-4)$

9.

$(x-5)(x+2)=-6$

$x^2-3x-10=-6$

$x^2-3x-4=0$

$(x-4)(x+1)=0$

$x-4=0 \quad x+1=0$

$x=4 \quad x=-1$

The solutions are 4 and −1.

10. $\dfrac{x^2-6x+8}{2x^3+6x^2}\div\dfrac{2x-8}{4x^3+12x^2}$

$=\dfrac{x^2-6x+8}{2x^3+6x^2}\cdot\dfrac{4x^3+12x^2}{2x-8}$

$=\dfrac{\overset{1}{\cancel{(x-4)}}(x-2)}{2x^2\underset{1}{\cancel{(x+3)}}}\cdot\dfrac{4x^2\overset{1}{\cancel{(x+3)}}}{2\underset{1}{\cancel{(x-4)}}}=x-2$

11. $\dfrac{x-1}{x+2}+\dfrac{2x+1}{x^2+x-2}$

$=\dfrac{x-1}{x+2}+\dfrac{2x+1}{(x+2)(x-1)}$

$=\dfrac{x-1}{x+2}\cdot\dfrac{x-1}{x-1}+\dfrac{2x+1}{(x+2)(x-1)}$

$=\dfrac{x^2-2x+1}{(x+2)(x-1)}+\dfrac{2x+1}{(x+2)(x-1)}$

$=\dfrac{x^2-2x+1+2x+1}{(x+2)(x-1)}$

$=\dfrac{x^2+2}{(x+2)(x-1)}$

12. $\dfrac{x+4-\dfrac{7}{x-2}}{x+8+\dfrac{21}{x-2}}=\dfrac{x+4-\dfrac{7}{x-2}}{x+8+\dfrac{21}{x-2}}\cdot\dfrac{x-2}{x-2}$

$=\dfrac{(x-2)(x+4)-7}{(x+8)(x-2)+21}=\dfrac{x^2+2x-8-7}{x^2+6x-16+21}$

$=\dfrac{x^2+2x-15}{x^2+6x+5}=\dfrac{\overset{1}{\cancel{(x+5)}}(x-3)}{\underset{1}{\cancel{(x+5)}}(x+1)}=\dfrac{x-3}{x+1}$

13.

$\dfrac{x}{2x-3}+2=\dfrac{-7}{2x-3}$

$(2x-3)\left(\dfrac{x}{2x-3}+2\right)=(2x-3)\left(\dfrac{-7}{2x-3}\right)$

$x+(2x-3)2=-7$

$x+4x-6=-7$

$5x-6=-7$

$5x=-1$

$x=-\dfrac{1}{5}$

The solution is $-\dfrac{1}{5}$.

14.

$A=P+Prt$

$A-P=Prt$

$\dfrac{A-P}{Pt}=r$

15. $2x-3y=12$

x-intercept	y-intercept
$2x-3(0)=12$	$2(0)-3y=12$
$2x=12$	$-3y=12$
$x=6$	$y=-4$
(6, 0)	(0, −4)

16. $m=\frac{y_2-y_1}{x_2-x_1}=\frac{4-(-3)}{-3-2}=\frac{7}{-5}=-\frac{7}{5}$

The slope of the line is $-\frac{7}{5}$.

17. $m=-\frac{3}{2}, (x_1,y_1)=(-2,3)$

$$y-y_1=m(x-x_1)$$
$$y-3=-\frac{3}{2}(x-(-2))$$
$$y-3=-\frac{3}{2}(x+2)$$
$$y-3=-\frac{3}{2}x-3$$
$$y=-\frac{3}{2}x$$

The equation of the line is $y=-\frac{3}{2}x$.

18.

$5x-3y=10$	
$5(2)-3(0)$	10
$10-0$	10

$10=10$

$4x+7y=8$	
$4(2)+7(0)$	8
$8+0$	8

$8=8$

Yes, (2, 0) is a solution.

19. (1) $3x-5y=-23$

(2) $x+2y=-4$

Solve Equation (2) for x.

$$x+2y=-4$$
$$x=-2y-4$$

Substiture $2y-4$ for x in Equation (1).

$$3x-5y=-23$$
$$3(-2y-4)-5y=-23$$
$$-6y-12-5y=-23$$
$$-11y=-11$$
$$y=1$$

Substitute 1 for y in Equation (2).

$$x+2y=-4$$
$$x+2(1)=-4$$
$$x+2=-4$$
$$x=-6$$

The solution is (−6, 1).

20. (1) $5x-3y=29$

(2) $4x+7y=-5$

Eliminate y by adding the equations.

$7(5x-3y)=7(29)$ $\quad$ $35x-21y=203$

$3(4x+7y)=3(-5)$ $\quad$ $12x+21y=-15$

$$47x=188$$
$$x=4$$

Substitute 4 for x in Equation (1).

$$5x-3y=29$$
$$5(4)-3y=29$$
$$20-3y=29$$
$$-3y=9$$
$$y=-3$$

The solution is (4, −3).

21. $2x-3y=6$

$$-3y=-2x+6$$
$$y=\frac{2}{3}x-2$$

x	y
0	−2
3	0
−3	−4

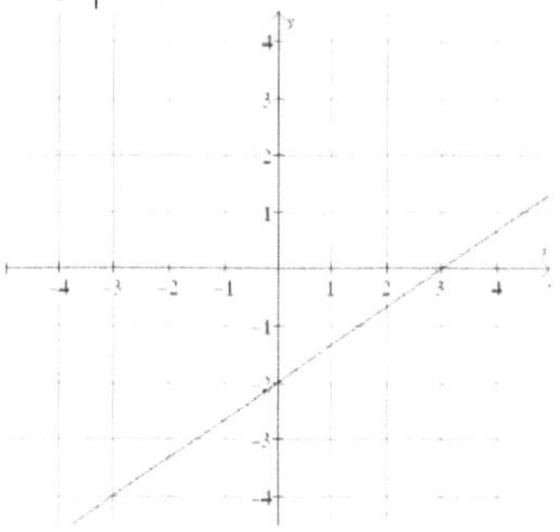

22.

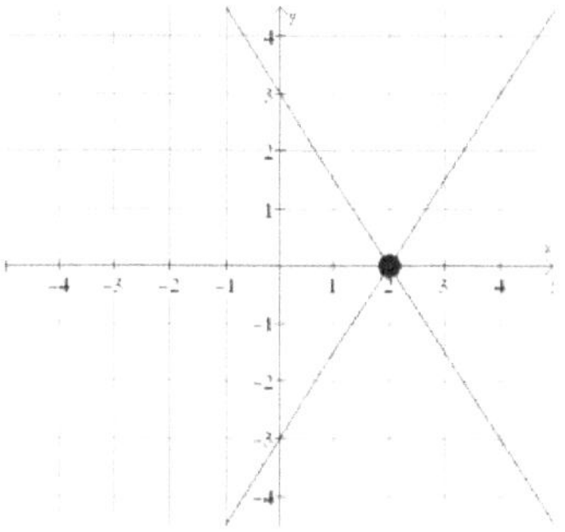

The solution of the system is (2, 0).

23. Strategy:

→Amount invested at 9.6%: x

→Amount invested at 7.2%: y

	P	R	I
Investment at 9.6%	x	0.096	$0.096x$
Investment at 7.2%	y	0.072	$0.072y$

→The total amount invested is $8750.

$x+y=8750$

The interest earned in each account is the same.

$0.096x=0.072y$

Solution:

$$x+y=8750$$
$$y=8750-x$$
$$0.096x=0.072y$$
$$0.096x=0.072(8750-x)$$
$$0.096x=630-0.072x$$
$$0.168x=630$$
$$x=3750$$
$$y=8750-x=8750-3750=5000$$

The amounts invested are $3750 at 9.6% and $5000 at 7.2%.

24. Strategy:

→Rate of passenger train: x

→Rate of freight train: $x-8$

	Rate	Time	Distance
Passenger train	x	3	$3x$
Freight train	$x-8$	3.5	$3.5(x-8)$

→The distances the trains traveled are the same.

Solution:

$$3x=3.5(x-8)$$
$$3x=3.5x-28$$
$$-0.5x=-28$$
$$x=56$$
$$x-8=56-8=48$$

The rate of the passenger train is 56 mph.
The rate of the freight train is 48 mph.

25. Strategy:

→Side of the original square: x

→Side of the new square: $x+4$

→Use the equation for the area of a square.

Solution:

$$s^2=A$$
$$(x+4)^2=144$$
$$x^2+8x+16=144$$
$$x^2+8x-128=0$$
$$(x+16)(x-8)=0$$
$$x+16=0 \qquad x-8=0$$
$$x=-16 \qquad x=8$$

Since the side of a square cannot be negative, −16 cannot be a solution. The side of the original square is 8 in.

26. Strategy:

→Rate of wind: x

	Time	= Rate	÷ Distance
With wind	$\frac{570}{160+x}$	$160+x$	570
Against wind	$\frac{390}{160-x}$	$160-x$	390

→The time with the wind and against the wind are equal.

Solution:

$$\frac{570}{160+x}=\frac{390}{160-x}$$
$$(160+x)(160-x)\frac{570}{160+x}$$
$$=(160+x)(160-x)\frac{390}{160-x}$$
$$(160-x)570=(160+x)390$$
$$91{,}200-570x=62{,}400+390x$$
$$28{,}800=960x$$
$$30=x$$

The rate of wind is 30 mph.

27. Strategy:

→Rate of motorboat in calm water: r

→Rate of current: c

	Rate	Time	Distance
With current	$r+c$	3	$3(r+c)$
Against current	$r-c$	4	$4(r-c)$

→The distance traveled with the current and against the current is 48 mi.

Solution:

$$3(r+c)=48 \qquad 4(r-c)=48$$
$$r+c=16 \qquad r-c=12$$

$$\begin{aligned} r+c&=16\\ \underline{r-c}&\underline{=12}\\ 2r&=28\\ r&=14 \end{aligned}$$

The rate the motorboat in calm water is 14 mph.

28. Use the basic percent equation.

$$\frac{156}{205}\approx 0.76=76\%$$

76% of the voting-age population was registered to vote.

Chapter 14: Inequalities

Prep Test

1. $-45 < -27$

2. $3x - 5(2x - 3) = 3x - 10x + 15$
$= -7x + 15$

3. The same number can be added to each side of an equation without changing the solution of the equation.

4. Each side of an equation can be multiplied by the same nonzero number without changing the solution of the equation.

5. Strategy:
To find the amount of fat in the hamburger, multiply the number of pounds by 0.15.

Solution:
$(0.15)3 = 0.45$
There are 0.45 lb of fat in 3 lb of this grade of hamburger.

6. $7 - 2(2x - 3) = 3x - 1$
$7 - 4x + 6 = 3x - 1$
$-4x + 13 = 3x - 1$
$-7x = -14$
$x = 2$
The solution is 2.

7. $y = \frac{2}{3}x - 3$

x	y
0	−3
3	−1
−3	−5

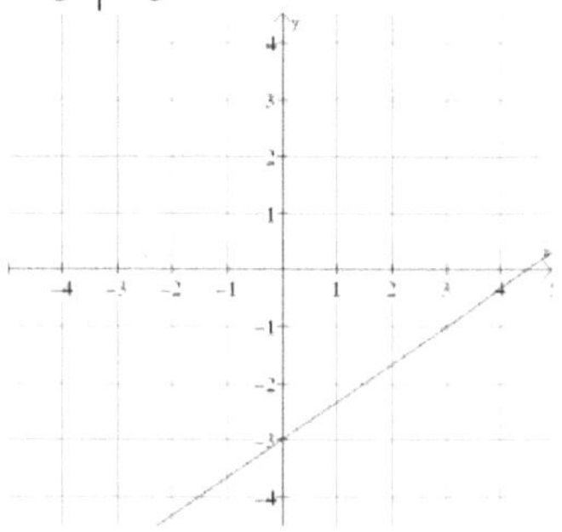

Section 14.1

Concept Check

1. roster; set-builder; interval

Objective A Exercises

3. $A = \{16, 17, 18, 19, 20, 21\}$

5. $A = \{9, 11, 13, 15, 17\}$

7. $A \cup B = \{3, 4, 5, 6\}$

9. $A \cup B = \{-10, -9, -8, 8, 9, 10\}$

11. $A \cup B = \{1, 3, 7, 9, 11, 13\}$

13. $A \cap B = \{4, 5\}$

15. $A \cap B = \varnothing$

17. $A \cap B = \{c, d, e\}$

Objective B Exercises

19. $\{x \mid x > -5, x \in \text{negative integer}\}$

21. $\{x \mid x > 30, x \in \text{integer}\}$

23. $\{x \mid x > 8, x \in \text{real numbers}\}$

25. $(1, 2)$

27. $(3, \infty)$

29. $[-4, 5)$

31. $(-\infty, 2]$

33. $[-3, 1]$

35. $\{x \mid -5 < x < -3\}$

37. $\{x \mid x \le -2\}$

39. $\{x \mid -3 \le x \le -2\}$

41. $\{x \mid x \le 6\}$

43. $[-5, 4]$

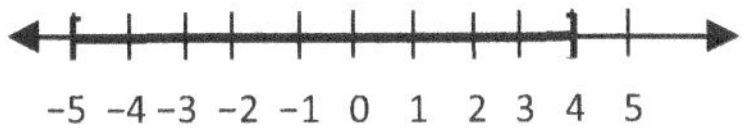

45. $\{x \mid x < 4\}$

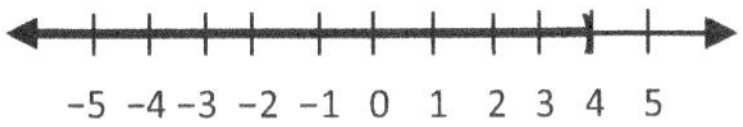

47. $\{x \mid x \le -4\}$

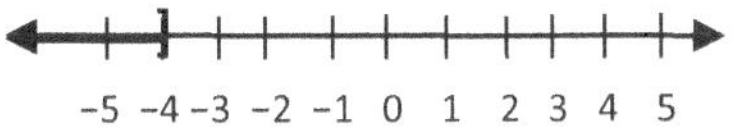

49. $(-\infty, 3]$

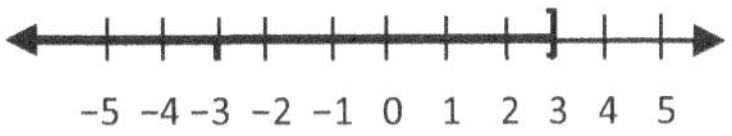

51. $[-1, 3)$

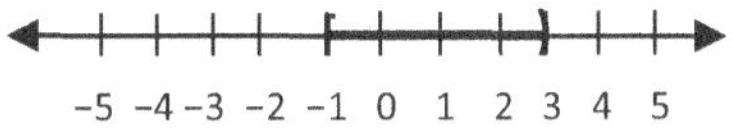

53. $\{x \mid -3 < x < 3\}$

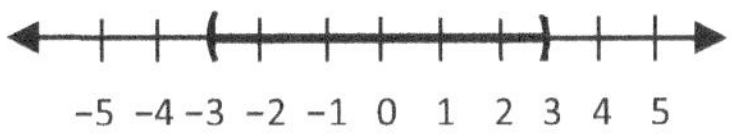

55. $\{x \mid 2 \le x \le 4\}$

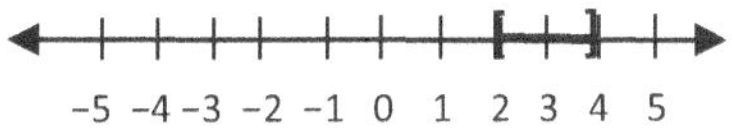

57. $\{x \mid -\infty < x < \infty\}$

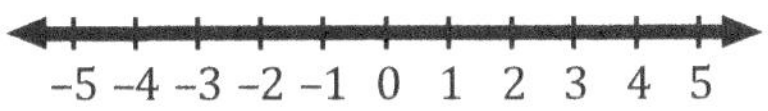

59. None

Critical Thinking 14.1

61. $m \ge 250$

63. True

Projects and Group Activities 14.1

65. Answer will vary. For example, $A = \{1, 2, 3, 4\}$ and $B = \{1, 2, 3, 4\}$.

Section 14.2

Concept Check

1. **a.** No

b. No

c. Yes

d. No

e. No

f. Yes

3. **(i)** $-4x \le 12$

$-4(0)$	12
0	12

$0 \le 12$ is true. 0 is a solution.

(i) $-4x \le 12$

$-4(3)$	12
-12	12

$-12 \le 12$ is true. 3 is a solution.

(iii) $-4x \le 12$

$-4(-3)$	12
12	12

$12 \le 12$ is true. -3 is a solution.

(iv) $-4x \le 12$

$-4(-4)$	12
16	12

$16 \le 12$ is false. -4 is not a solution.

5. does not include; >

Objective A Exercises

7.
$$x+1<3$$
$$x+1-1<3-1$$
$$x<2$$
$$\{x \mid x<2\}$$

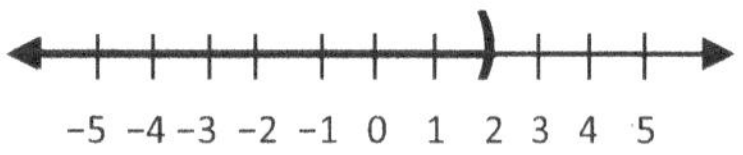

9.
$$x-5>-2$$
$$x-5+5>-2+5$$
$$x>3$$
$$\{x \mid x>3\}$$

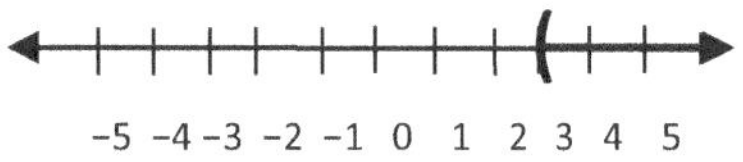

11.
$$7\le n+4$$
$$7+(-4)\le n+4+(-4)$$
$$3\le n$$
$$n\ge 3$$
$$\{n \mid n\ge 3\}$$

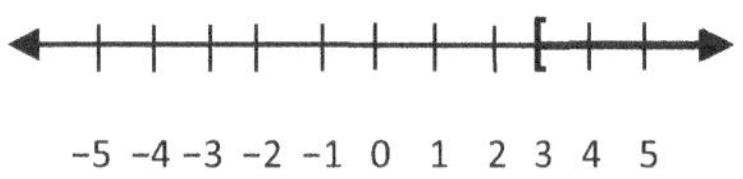

13.
$$x-6\le -10$$
$$x-6+6\le -10+6$$
$$x\le -4$$
$$\{x \mid x\le -4\}$$

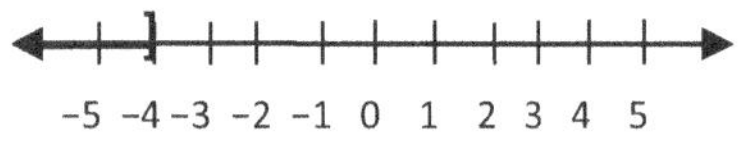

15.
$$y-3\ge -12$$
$$y-3+3\ge -12+3$$
$$y\ge -9$$
$$[-9,\infty)$$

17.
$$3x-5<2x+7$$
$$3x+(-2x)-5<2x+(-2x)+7$$
$$x-5<7$$
$$x-5+5<7+5$$
$$x<12$$
$$(-\infty, 12)$$

19.
$$8x-7\ge 7x-2$$
$$8x+(-7x)-7\ge 7x+(-7x)-2$$
$$x-7\ge -2$$
$$x-7+7\le -2+7$$
$$x\ge 5$$
$$[5, \infty)$$

21.
$$2x+4<x-7$$
$$2x+(-x)+4<x+(-x)-7$$
$$x+4<-7$$
$$x+4+(-4)<-7+(-4)$$
$$x<-11$$
$$(-\infty, -11)$$

23.
$$4x-8\le 2+3x$$
$$4x+(-3x)-8\le 2+3x+(-3x)$$
$$x-8\le 2$$
$$x-8+8\le 2+8$$
$$x\le 10$$
$$(-\infty, 10]$$

25.
$$6x+4\ge 5x-2$$
$$6x+(-5x)+4\ge 5x+(-5x)-2$$
$$x+4\ge -2$$
$$x+4+(-4)\ge -2+(-4)$$
$$x\ge -6$$
$$[-6, \infty)$$

27.
$$2x-12>x-10$$
$$2x+(-x)-12>x+(-x)-10$$
$$x-12>-10$$
$$x-12+12>-10+12$$
$$x>2$$
$$\{x \mid x>2\}$$

29.
$$d+\frac{1}{2}<\frac{1}{3}$$
$$d+\frac{1}{2}+\left(-\frac{1}{2}\right)<\frac{1}{3}+\left(-\frac{1}{2}\right)$$
$$d<\frac{2}{6}-\frac{3}{6}$$
$$d<-\frac{1}{6}$$
$$\left\{d \mid d<-\frac{1}{6}\right\}$$

31.

$$x+\frac{5}{8}\ge-\frac{2}{3}$$
$$x+\frac{5}{8}+\left(-\frac{5}{8}\right)\ge-\frac{2}{3}+\left(-\frac{5}{8}\right)$$
$$x\ge-\frac{16}{24}-\frac{15}{24}$$
$$x\ge-\frac{31}{24}$$
$$\left\{x\mid x\ge-\frac{31}{24}\right\}$$

33.

$$x-\frac{3}{8}<\frac{1}{4}$$
$$x-\frac{3}{8}+\frac{3}{8}<\frac{1}{4}+\frac{3}{8}$$
$$x<\frac{2}{8}+\frac{3}{8}$$
$$x<\frac{5}{8}$$
$$\left\{x\mid x<\frac{5}{8}\right\}$$

35.

$$2x-\frac{1}{2}<x+\frac{3}{4}$$
$$2x+(-x)-\frac{1}{2}<x+(-x)+\frac{3}{4}$$
$$x-\frac{1}{2}<\frac{3}{4}$$
$$x-\frac{1}{2}+\frac{1}{2}<\frac{3}{4}+\frac{1}{2}$$
$$x<\frac{3}{4}+\frac{2}{4}$$
$$x<\frac{5}{4}$$
$$\left\{x\mid x<\frac{5}{4}\right\}$$

37.

$$x-3.5<2.1$$
$$x-3.5+3.5<2.1+3.5$$
$$x<5.6$$
$$\{x\mid x<5.6\}$$

39.

$$x+n<a$$
$$x+n+(-n)<a+(-n)$$
$$x<a-n$$

Since $n > a$, $a - n$ will be negative so all values that are less $a - n$ will be negative.

41.

$$x+n<a$$
$$x+n+(-n)<a+(-n)$$
$$x<a-n$$

Since $n < a$, $a - n$ will be positive, so all values that are less $a - n$ will contain both positive and negative numbers.

Objective B Exercises

43.

$$3x<12$$
$$\frac{3x}{3}<\frac{12}{3}$$
$$x<4$$
$$\{x\mid x<4\}$$

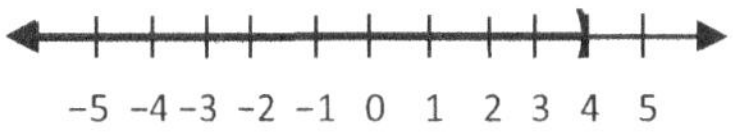

45.

$$15\le 5y$$
$$\frac{15}{5}\le\frac{5y}{5}$$
$$3\le y$$
$$y\ge 3$$
$$\{y\mid y\ge 3\}$$

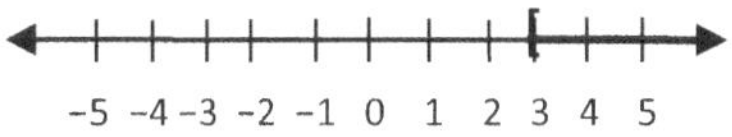

47.

$$16x\le 16$$
$$\frac{16x}{16}\le\frac{16}{16}$$
$$x\le 1$$
$$\{x\mid x\le 1\}$$

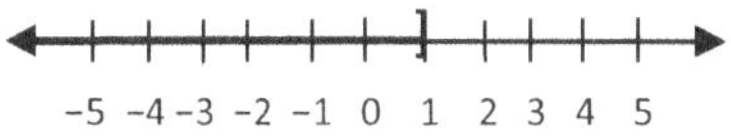

49.

$$-8x>8$$
$$\frac{-8x}{-8}<\frac{8}{-8}$$
$$x<-1$$
$$\{x\mid x<-1\}$$

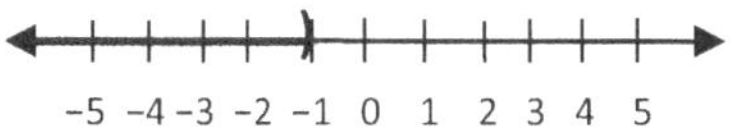

51.
$$-6b > 24$$
$$\frac{-6b}{-6} < \frac{24}{-6}$$
$$b < -4$$
$$\{b \mid b < -4\}$$

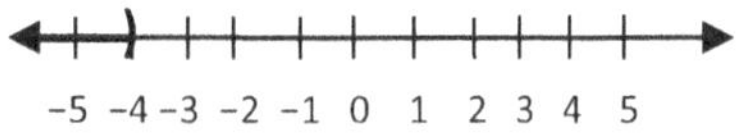

53.
$$-5y \geq 0$$
$$\frac{-5y}{-5} \leq \frac{0}{-5}$$
$$y \leq 0$$
$$(-\infty, 0]$$

55.
$$7x > 2$$
$$\frac{7x}{7} > \frac{2}{7}$$
$$x > \frac{2}{7}$$
$$\left(\frac{2}{7}, \infty\right)$$

57.
$$-x \geq 3$$
$$\frac{-x}{-1} \leq \frac{3}{-1}$$
$$x \leq -3$$
$$(-\infty, -3]$$

59.
$$2 > -y$$
$$\frac{2}{-1} < \frac{-y}{-1}$$
$$-2 < y$$
$$y > -2$$
$$(-2, \infty)$$

61.
$$\frac{5}{6}n < 15$$
$$\frac{6}{5} \cdot \frac{5}{6}n < \frac{6}{5} \cdot 15$$
$$n < 18$$
$$(-\infty, 18)$$

63.
$$10 \leq \frac{5}{8}x$$
$$\frac{8}{5} \cdot 10 \leq \frac{8}{5} \cdot \frac{5}{8}x$$
$$16 \leq x$$
$$x \geq 16$$
$$[16, \infty)$$

65.
$$-\frac{2}{11}b \geq -6$$
$$-\frac{11}{2}\left(-\frac{2}{11}b\right) \leq -\frac{11}{2} \cdot (-6)$$
$$b \leq 33$$
$$\{b \mid b \leq 33\}$$

67.
$$-\frac{3}{5}x < 0$$
$$-\frac{5}{3}\left(-\frac{3}{5}x\right) > -\frac{5}{3}(0)$$
$$x > 0$$
$$\{x \mid x > 0\}$$

69.
$$-\frac{3}{8}x \geq \frac{9}{14}$$
$$-\frac{8}{3}\left(-\frac{3}{8}x\right) \leq -\frac{8}{3}\left(\frac{9}{14}\right)$$
$$x \leq -\frac{12}{7}$$
$$\left\{x \mid x \leq -\frac{12}{7}\right\}$$

71.
$$-\frac{4}{5}x < -\frac{8}{15}$$
$$-\frac{5}{4}\left(-\frac{4}{5}x\right) > -\frac{5}{4}\left(-\frac{8}{15}\right)$$
$$x > \frac{2}{3}$$
$$\left\{x \mid x > \frac{2}{3}\right\}$$

73.
$$1.5x \leq 6.30$$
$$\frac{1.5x}{1.5} \leq \frac{6.30}{1.5}$$
$$x \leq 4.2$$
$$\{x \mid x \leq 4.2\}$$

75.
$$4.25m > -34$$
$$\frac{4.25m}{4.25} > \frac{-34}{4.25}$$
$$m > -8$$
$$\{m \mid m > -8\}$$

77. (ii)

79. (iii)

Objective C Exercises

81. Strategy:
→Monthly income: i
→Payment: \$2050
→Payment ≤ 38% of monthly income

Solution:

$$2050 \le 0.38i$$
$$\frac{2050}{0.38} \le \frac{0.38i}{0.38}$$
$$5395 \le i$$
$$i \ge 5395$$

The monthly income must be \$5395 or more.

83. Strategy:
→Amount of cholesterol to be reduced: x
The maximum amount of cholesterol is 200 units.

Solution:

$$275 - x \le 200$$
$$-x \le -75$$
$$\frac{-x}{-1} \ge \frac{-75}{-1}$$
$$x \ge 75$$

The patient must reduce his cholesterol level by at least 75 units to reach the recommended level.

85. Strategy:
→Score on the fifth test: x
The minimum grade of 80 points is required to earn a B grade.

Solution:

$$\frac{75+83+86+78+x}{5} \ge 80$$
$$\frac{322+x}{5} \ge 80$$
$$322 + x \ge 400$$
$$x \ge 78$$

The student must earn a grade of 78 or higher to receive a B grade.

87. Strategy:
→Height of the tower: x
→Distance from the property line: 68
The distance from the property line must be greater than 150% of the height of the tower.

Solution:

$$68 > 1.50h$$
$$45.3 > h$$
$$h < 45.3$$

The maximum height is 45 ft.

Critical Thinking 14.2

89. $x \le 4$ includes the element 4; $x < 4$ does not

91. $x \le -2$

Project and Group Activities 14.2

93. $|x| < 3$

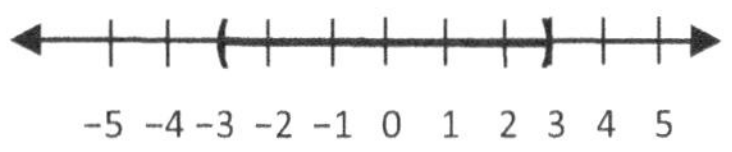

95. $|x| > 2$

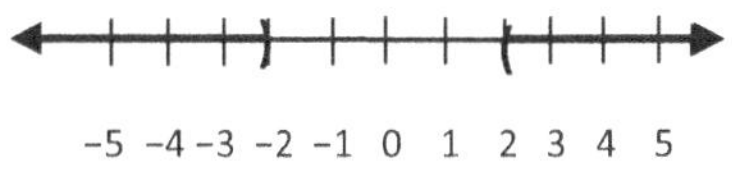

Check Your Progress: Chapter 14

1. $\{-11, -10\}$

2. $A \cup B = \{4, 8, 16, 20, 24\}$

3. $A \cap B = \{20, 40\}$

4. $A \cap B = \varnothing$

5. $[-1, 3]$

6. $(8, \infty)$

7. $\{x \mid x \ge -9\}$

8. $\{x \mid -4 < x \le 7\}$

9. $(-5, \infty)$

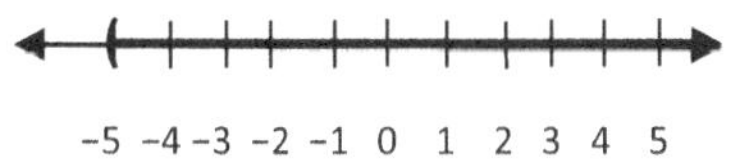

10. $\{x \mid -3 \le x < 5\}$

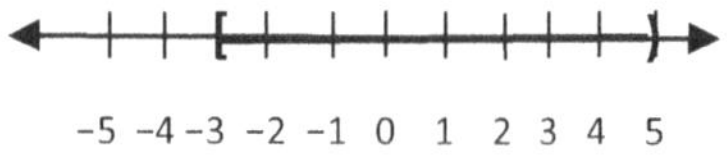

11. $x - 5 \ge -3$

$x \ge 2$

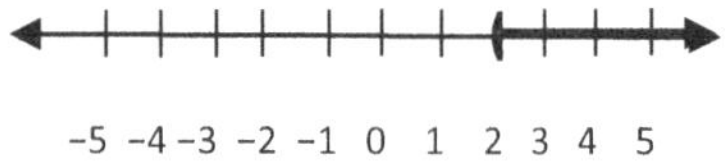

12. $4 < x + 7$

$-3 < x$

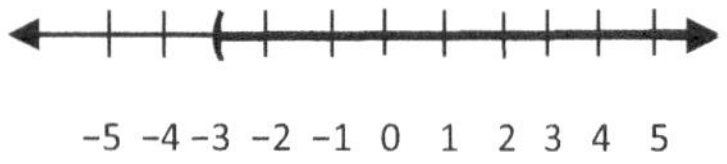

13.
$$\begin{aligned} 10x - 15 &\le 9x - 20 \\ 10x - 9x - 15 &\le 9x - 9x - 20 \\ x - 15 &\le -20 \\ x - 15 + 15 &\le -20 + 15 \\ x &\le -5 \end{aligned}$$

$\{x \mid x \le -5\}$

14.
$$\begin{aligned} -12 &\le -3x \\ \frac{-12}{-3} &\ge \frac{-3x}{-3} \\ 4 &\ge x \end{aligned}$$

$(-\infty, 4]$

15. Strategy:

→Score on the fifth test: x

The minimum total of 400 points is required to earn an A grade.

Solution:

$$\begin{aligned} x + 88 + 91 + 83 + 76 &\ge 400 \\ x + 338 &\ge 400 \\ x &\ge 62 \end{aligned}$$

The student must earn a grade of 62 or higher to receive an A grade.

Section 14.3

Concept Check

1. False

3. True

Objective A Exercises

5.
$$\begin{aligned} 4x - 8 &< 2x \\ 4x + (-2x) - 8 &< 2x + (-2x) \\ 2x - 8 &< 0 \\ 2x - 8 + 8 &< 0 + 8 \\ 2x &< 8 \\ \frac{2x}{2} &< \frac{8}{2} \\ x &< 4 \end{aligned}$$

$(-\infty, 4)$

7.
$$\begin{aligned} 2x - 8 &> 4x \\ 2x + (-4x) - 8 &> 4x + (-4x) \\ -2x - 8 &> 0 \\ -2x - 8 + 8 &> 0 + 8 \\ -2x &> 8 \\ \frac{-2x}{-2} &< \frac{8}{-2} \\ x &< -4 \end{aligned}$$

$(-\infty, -4)$

9.
$$\begin{aligned} 8 - 3x &\le 5x \\ 8 - 3x + (-5x) &\le 5x + (-5x) \\ 8 - 8x &\le 0 \\ 8 + (-8) - 8x &\le 0 + (-8) \\ -8x &\le -8 \\ \frac{-8x}{-8} &\ge \frac{-8}{-8} \\ x &\ge 1 \end{aligned}$$

$[1, \infty)$

11.
$$\begin{aligned} 3x + 2 &> 5x - 8 \\ 3x + (-5x) + 2 &> 5x + (-5x) - 8 \\ -2x + 2 &> -8 \\ -2x + 2 + (-2) &> -8 + (-2) \\ -2x &> -10 \\ \frac{-2x}{-2} &< \frac{-10}{-2} \\ x &< 5 \end{aligned}$$

$(-\infty, 5)$

13.
$$\begin{aligned}5x-2&<3x-2\\5x+(-3x)-2&<3x+(-3x)-2\\2x-2&<-2\\2x-2+2&<-2+2\\2x&<0\\\frac{2x}{2}&<\frac{0}{2}\\x&<0\end{aligned}$$
$(-\infty, 0)$

15.
$$\begin{aligned}0.1(180+x)&>x\\18+0.1x&>x\\18+0.1x+(-x)&>x+(-x)\\18-0.9x&>0\\18+(-18)-0.9x&>0+(-18)\\-0.9x&>-18\\\frac{-0.9x}{-0.9}&<\frac{-18}{-0.9}\\x&<20\end{aligned}$$
$\{x \mid x<20\}$

17.
$$\begin{aligned}2(2y-5)&\le 3(5-2y)\\4y-10&\le 15-6y\\4y+6y-10&\le 15-6y+6y\\10y-10&\le 15\\10y-10+10&\le 15+10\\10y&\le 25\\\frac{10y}{10}&\le\frac{25}{10}\\y&\le\frac{5}{2}\end{aligned}$$
$\left\{y \mid y<\frac{5}{2}\right\}$

19.
$$\begin{aligned}5(2-x)&>3(2x-5)\\10-5x&>6x-15\\10-5x+(-6x)&>6x+(-6x)-15\\10-11x&>-15\\10+(-10)-11x&>-15+(-10)\\-11x&>-25\\\frac{-11x}{-11}&<\frac{-25}{-11}\\x&<\frac{25}{11}\end{aligned}$$
$\left\{x \mid x<\frac{25}{11}\right\}$

21.
$$\begin{aligned}4-3(3-n)&\le 3(2-5n)\\4-9+3n&\le 6-15n\\-5+3n&\le 6-15n\\-5+3n+15n&\le 6-15n+15n\\-5+18n&\le 6\\-5+5+18n&\le 6+5\\18n&\le 11\\\frac{18n}{18}&\le\frac{11}{18}\\n&\le\frac{11}{18}\end{aligned}$$
$\left\{n \mid n\le\frac{11}{18}\right\}$

23.
$$\begin{aligned}&2x-3(x-4)\ge 4-2(x-7)\\&2x-3x+12\ge 4-2x+14\\&-x+12\ge 18-2x\\&-x+2x+12\ge 18-2x+2x\\&x+12\ge 18\\&x+12+(-12)\ge 18+(-12)\\&x\ge 6\end{aligned}$$
$\{x \mid x\ge 6\}$

25.
$$\begin{aligned}-7x-2&>-4x+1\\-7x+7x-2&>-4x+7x+1\\-2&>3x+1\\-2+(-1)&>3x+1+(-1)\\-3&>3x\\3x&<-3\end{aligned}$$
(iii) and (iv)

Objective B Exercises

27. Strategy:
→To find the dollar amount, write and solve an inequality using x to represent the total sales the agent made.
The agent's salary must be the greater of \$3200 or \$1000 plus an 11% commission on the selling price of each item.

Solution:
$$\begin{aligned}1000+11\%x&\le 3200\\1000+0.11x&\le 3200\\0.11x&\le 2200\\x&\le 20{,}000\end{aligned}$$
In one month the agent expects to make sales totaling at most \$20,000.

29. Strategy:
→Required number of computers: x
The number of computers multiplied by 200 must be greater than \$1500.

Solution:
$200x > 1500$
$x > 7.5$
The company must have at least 8 computers for the site license to be more economical.

31. Strategy:
→Number of ounces of artificial flavors: x
The number of ounces of artificial flavors that can be added to 32 oz of real orange juice so that the mixture is 80% real orange juice must be less than or equal to 32 oz.

Solution:
$0.80(x+32) \le 32$
$0.80x + 25.6 \le 32$
$0.80x \le 6.4$
$x \le 8$
The amount of artificial flavors that can be added is 8 oz or less.

Critical Thinking 14.3

33. Strategy:
To find the number that is a solution to $3x-4 \ge 5$ but not a solution to $3x-4>5$, find where $3x - 4$ is equal to 5.

Solution:
$3x-4=5$
$3x-4+4=5+4$
$3x=9$
$\frac{3x}{3}=\frac{9}{3}$
$x=3$
3 is a solution to $3x-4\ge 5$ but not a solution to $3x-4>5$.

35. Strategy:
The third side must be less than the sum of 10 and 18 (28). The sum of the third side and the smaller side (10) must be greater than the other side (18).

Solution:
$x+10>18$
$x+10-10>18-10$
$x>8$
The third side must be between 8 in and 28 in.

Section 14.4

Concept Check

1.

$y < -5x + 2$	
0	$-5(0)+2$
0	2

$0 < 2$ is true. (0, 0) is a solution.

3.

$y \le \frac{1}{4}x - 5$	
0	$\frac{1}{4}(0)-5$
0	-5

$0 \le -5$ is false. (0, 0) is not a solution.

Objective A Exercises

5. $y > -x + 4$
Graph $y = -x + 4$ as a dotted line. Shade the upper half-plane.

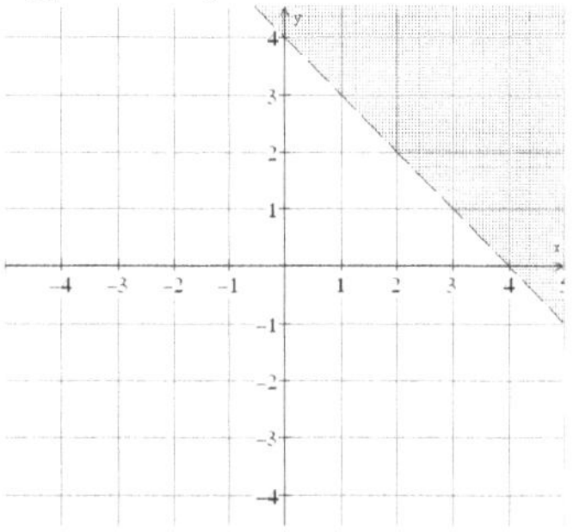

7. $y > 2x + 3$

Graph $y = 2x + 3$ as a dotted line. Shade the upper half-plane.

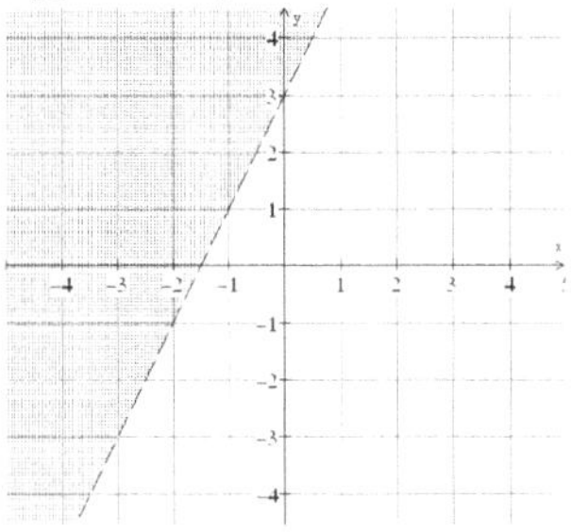

9.

$$2x + y \geq 4$$
$$2x - 2x + y \geq -2x + 4$$
$$y \geq -2x + 4$$

Graph $y = -2x + 4$ as a solid line. Shade the upper half-plane.

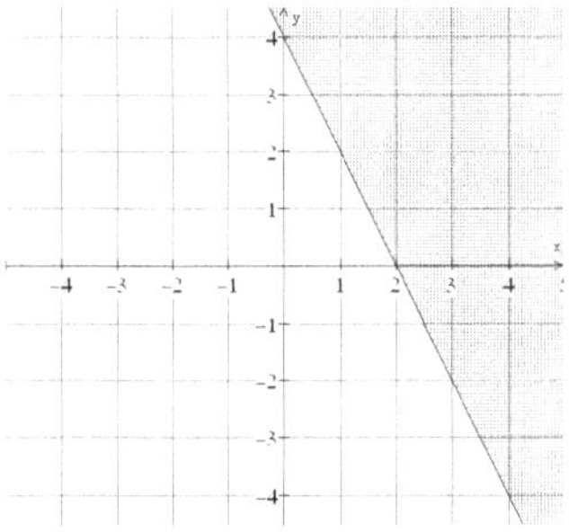

11. $y \leq -2$

Graph $y = -2$ as a solid line. Shade the lower half-plane.

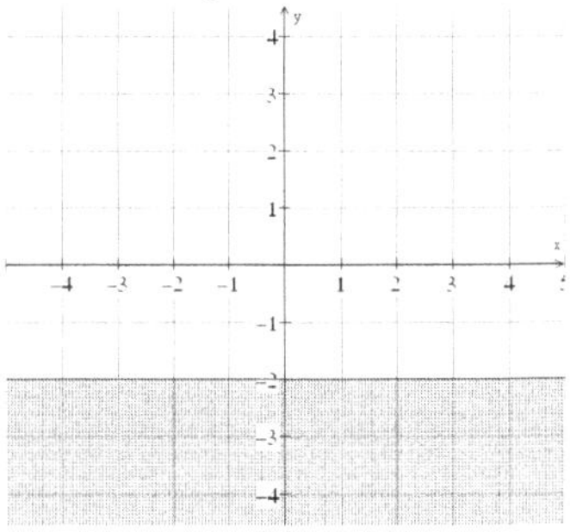

13.

$$3x - 2y < 8$$
$$3x - 3x - 2y < -3x + 8$$
$$\frac{-2y}{-2} > \frac{-3x + 8}{-2}$$
$$y > \frac{3}{2}x - 4$$

Graph $y = \frac{3}{2}x - 4$ as a dotted line. Shade the upper half-plane.

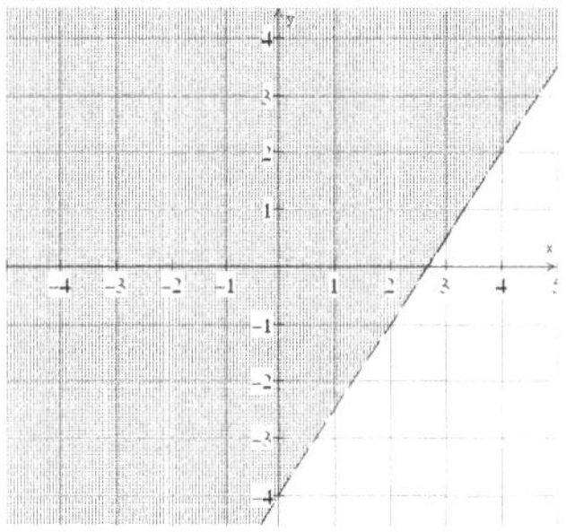

15.

$$-3x - 4y \geq 4$$
$$-3x + 3x - 4y \geq 3x + 4$$
$$-4y \geq 3x + 4$$
$$\frac{-4y}{-4} \leq \frac{3x + 4}{-4}$$
$$y \leq \frac{-3}{4}x - 1$$

Graph $y = -\frac{3}{4}x - 1$ as a solid line. Shade the lower half-plane.

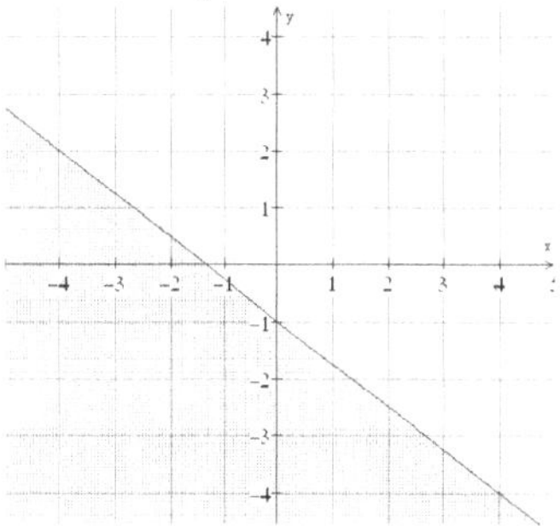

17.

$$
\begin{aligned}
6x + 5y &\le -10 \\
6x - 6x + 5y &\le -6x - 10 \\
5y &\le -6x - 10 \\
\frac{5y}{5} &\le \frac{-6x - 10}{5} \\
y &\le -\frac{6}{5}x - 2
\end{aligned}
$$

Graph $y = -\frac{6}{5}x - 2$ as a solid line. Shade the lower half-plane.

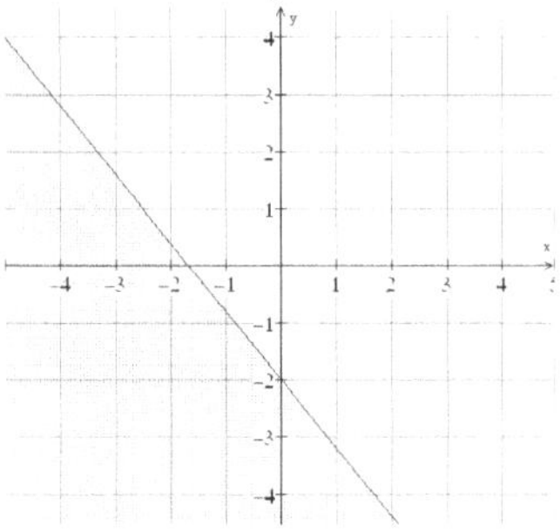

19.

$$
\begin{aligned}
-4x + 3y &< -12 \\
-4x + 4x + 3y &< 4x - 12 \\
3y &< 4x - 12 \\
\frac{3y}{3} &< \frac{4x - 12}{3} \\
y &< \frac{4}{3}x - 4
\end{aligned}
$$

Graph $y = \frac{4}{3}x - 4$ as a dotted line. Shade the lower half-plane.

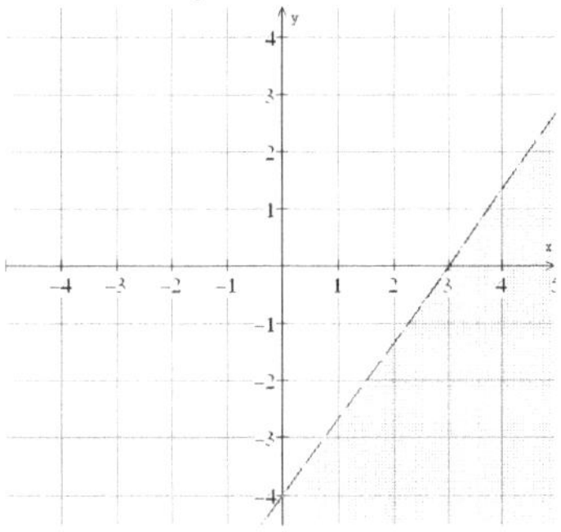

21.

$$
\begin{aligned}
-2x + 3y &\le 6 \\
-2x + 2x + 3y &\le 2x + 6 \\
3y &\le 2x + 6 \\
\frac{3y}{3} &\le \frac{2x + 6}{3} \\
y &\le \frac{2}{3}x + 2
\end{aligned}
$$

Graph $y = \frac{2}{3}x + 2$ as a solid line. Shade the lower half-plane.

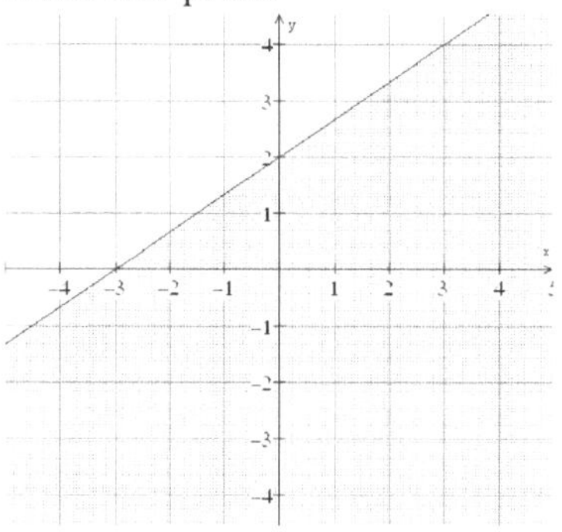

23.

$$
\begin{aligned}
Ax + By &> C \\
A(0) + B(0) &> C \\
0 &> C
\end{aligned}
$$

C must be negative.

Critical Thinking 14.4

25. The line is $y = 2$, dotted. The inequality is $y > 2$.

27. Slope = $-\frac{2}{3}$, y-intercept: 2

The line is $y = -\frac{2}{3}x + 2$, solid. The inequality is $y \le -\frac{2}{3}x + 2$.

Projects or Group Activities 14.4

29. Strategy:
To find if any points satisfy both inequalities, graph both inequalities on the same set of axis.
Graph $y < 2x - 3$ as a dotted line. Shade the lower half of the plane.
Graph $y > -\frac{1}{4}x + 1$ as a dotted line. Shade the upper half of the plane.
The intersection of the two shades provides the solution to both inequalities.

Solution:

Answers for the three points will vary. For example, (3, 1), (4, 1), and (4, 2).

Chapter Review Exercises: Chapter 14

1.
$$2x - 3 > x + 15$$
$$2x - x - 3 > x - x + 15$$
$$x - 3 > 15$$
$$x - 3 + 3 > 15 + 3$$
$$x > 18$$
$$\{x \mid x > 18\}$$

2. $A \cap B = \varnothing$

3. $\{x \mid x > -8,\ x \in \text{odd integers}\}$

4. $A \cup B = \{2, 4, 6, 8, 10\}$

5. $A = \{1, 3, 5, 7\}$

6.
$$12 - 4(x - 1) \le 5(x - 4)$$
$$12 - 4x + 4 \le 5x - 20$$
$$16 - 4x \le 5x - 20$$
$$16 - 4x - 5x \le 5x - 5x - 20$$
$$16 - 9x \le -20$$
$$16 - 16 - 9x \le -20 - 16$$
$$-9x \le -36$$
$$\frac{-9x}{-9} \ge \frac{-36}{-9}$$
$$x \ge 4$$
$[4, \infty)$

7. $\{x \mid x > 3\}$

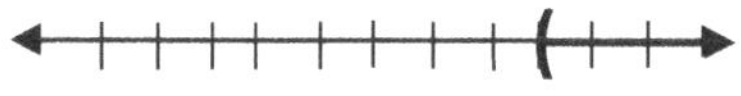

8.
$$3x + 4 \ge -8$$
$$3x \ge -12$$
$$\frac{3x}{3} \ge \frac{-12}{3}$$
$$x \ge -4$$
$$\{x \mid x \ge -4\}$$

9.
$$\begin{aligned} 3x+2y &\le 12 \\ 3x-3x+2y &\le -3x+12 \\ 2y &\le -3x+12 \\ \frac{2y}{2} &\le \frac{-3x+12}{2} \\ y &\le -\frac{3}{2}x+6 \end{aligned}$$

Graph $y=-\frac{3}{2}x+6$ as a solid line. Shade the lower half-plane.

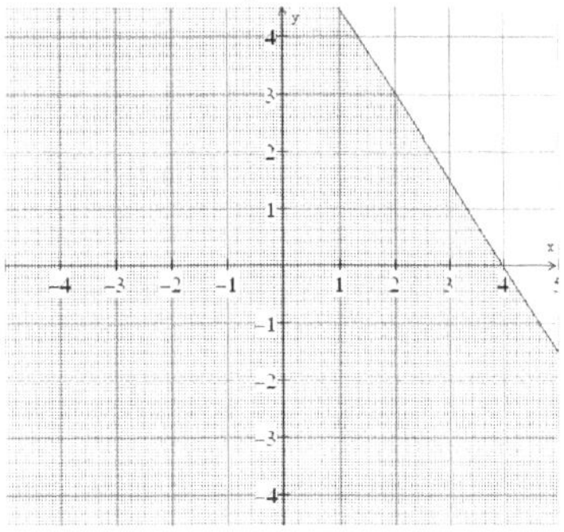

10.
$$\begin{aligned} 5x+2y &< 6 \\ 2y &< -5x+6 \\ \frac{2y}{2} &< \frac{-5x+6}{2} \\ y &< -\frac{5}{2}x+3 \end{aligned}$$

Graph $y=-\frac{5}{2}x+3$ as a dotted line. Shade the lower half-plane.

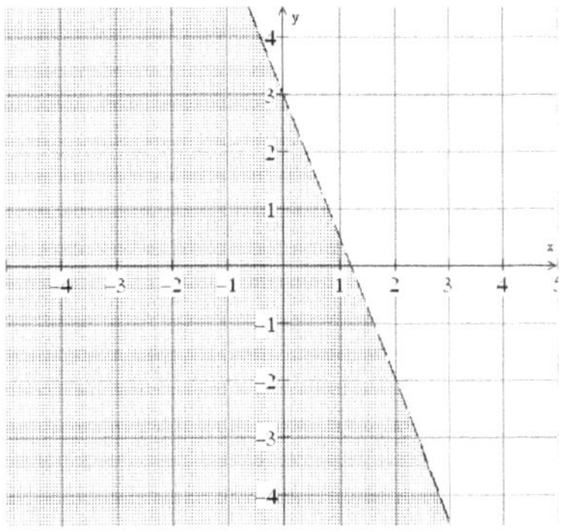

11. $(-4, \infty)$

12.
$$\begin{aligned} x-3 &> -1 \\ x &> 2 \end{aligned}$$
$\{x \mid x>2\}$

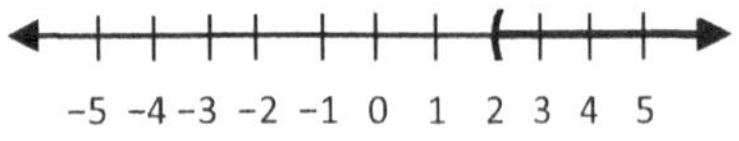

13. $A \cap B=\{1,5,9\}$

14. [1, 4]

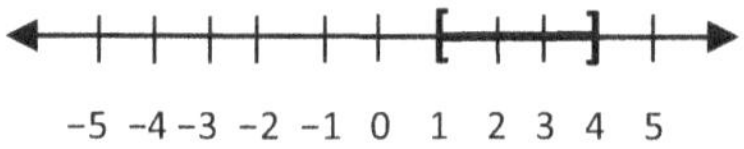

15. $\{x \mid -1<x\le 2\}$

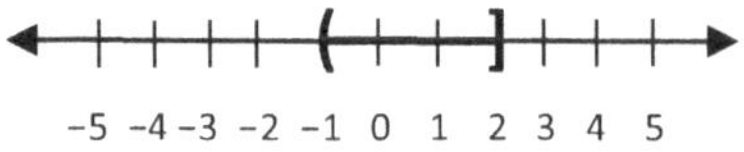

16.
$$\begin{aligned} -15x &\le 45 \\ \frac{-15x}{-15} &\ge \frac{45}{-15} \\ x &\ge -3 \end{aligned}$$
$\{x \mid x \ge -3\}$

17.
$$\begin{aligned} 6x-9 &< 4x+3(x+3) \\ 6x-9 &< 4x+3x+9 \\ 6x-9 &< 7x+9 \\ 6x-7x-9 &< 7x-7x+9 \\ -x-9 &< 9 \\ -x &< 18 \\ \frac{-x}{-1} &> \frac{18}{-1} \\ x &> -18 \end{aligned}$$
$(-18, \infty)$

18.
$$\begin{aligned} 5-4(x+9) &> 11(12x-9) \\ 5-4x-36 &> 132x-99 \\ -4x-31 &> 132x-99 \\ -136x &> -68 \\ \frac{-136x}{-136} &< \frac{-68}{-136} \\ x &< \frac{1}{2} \end{aligned}$$
$\left\{x \mid x<\frac{1}{2}\right\}$

19.
$$\begin{aligned} -\frac{3}{4}x &> \frac{2}{3} \\ -\frac{4}{3}\left(-\frac{3}{4}x\right) &< -\frac{4}{3}\cdot\frac{2}{3} \\ x &< -\frac{8}{9} \end{aligned}$$
$\left\{x \mid x<-\frac{8}{9}\right\}$

20.

$$7x - 2(x+3) \geq x + 10$$
$$7x - 2x - 6 \geq x + 10$$
$$5x - 6 \geq x + 10$$
$$4x \geq 16$$
$$\frac{4x}{4} \geq \frac{16}{4}$$
$$x \geq 4$$

$[4, \infty)$

21.

$$2x - 3y < 9$$
$$-3y < -2x + 9$$
$$\frac{-3y}{-3} > \frac{-2x+9}{-3}$$
$$y > \frac{2}{3}x - 3$$

Graph $y = \frac{2}{3}x - 3$ as a dotted line. Shade the upper half-plane.

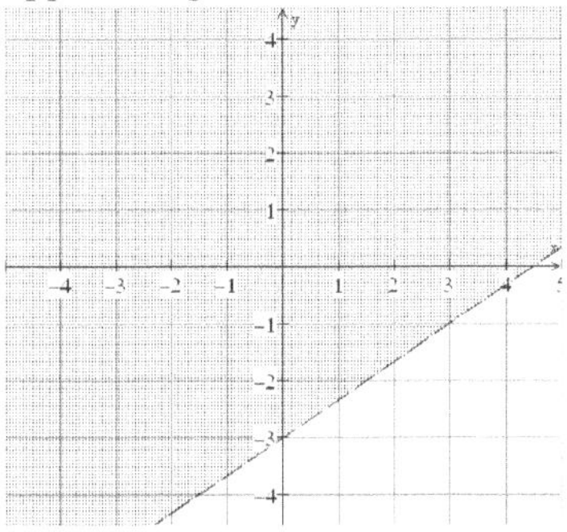

22. Strategy:

→Number of residents in nursing home: x
Cost of florist B must be less than cost of florist A.

Solution:

$$18 + 64x < 6 + 70x$$
$$18 + 64x - 64x < 6 + 70x - 64x$$
$$18 < 6 + 6x$$
$$18 - 6 < 6 - 6 + 6x$$
$$12 < 6x$$
$$2 < x$$

For florist B to be more economical, there must be 3 or more residents in the nursing home.

23. Strategy:
To find the minimum length:

→Replace the variables in the area formula by the given values and solve for x.

→Replace the variable in the expression $3x + 5$ with the value found for x.

Solution:

$$LW > A$$
$$(3x+5)12 > 276$$
$$36x + 60 > 276$$
$$36x > 216$$
$$x > 6$$
$$3x + 5 = \text{length}$$
$$3(6) + 5 = \text{length}$$
$$23 = \text{length}$$

The minimum length is 24 ft.

24. Strategy:
→The smallest integer that satisfies the inequality: x

Solution:

$$x - 6 > 25$$
$$x > 31$$

32 is the smallest integer that satisfies the inequality.

25. Strategy:
→Required score: x

Solution:

$$68 + 82 + 73 + 95 + x \geq 480$$
$$408 + x \geq 480$$
$$x \geq 72$$

72 is the lowest score that the student can make and still achieve a minimum of 480 points.

Chapter Test

1. (0, 5)

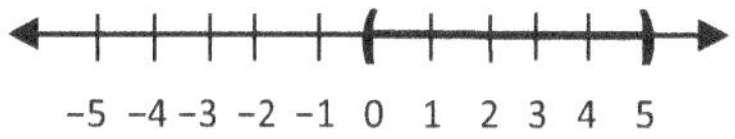

2. $\{x \mid x < 50, x \in \text{ positive integers}\}$

3. $A = \{4, 6, 8\}$

4.
$$\begin{aligned}3(2x-5)&\ge 8x-9\\6x-15&\ge 8x-9\\-2x&\ge 6\\\frac{-2x}{-2}&\le\frac{6}{-2}\\x&\le -3\end{aligned}$$
$(-\infty, -3]$

5.
$$\begin{aligned}x+\frac{1}{2}&>\frac{5}{8}\\x&>\frac{5}{8}-\frac{1}{2}\\x&>\frac{5}{8}-\frac{4}{8}\\x&>\frac{1}{8}\end{aligned}$$
$\left\{x \mid x>\frac{1}{8}\right\}$

6. $\{x \mid x>-2\}$

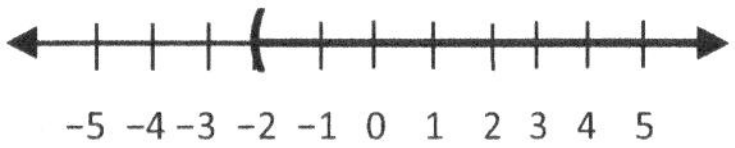

7.
$$\begin{aligned}5-3x&>8\\-3x&>3\\\frac{-3x}{-3}&<\frac{3}{-3}\\x&<-1\end{aligned}$$
$(-\infty, -1)$

8. $\{x \mid x>-23\}$

9.
$$\begin{aligned}3x+y&>4\\y&>-3x+4\end{aligned}$$
Graph the line $y=-3x+4$ as a dotted line.

Shade the upper-half plane.

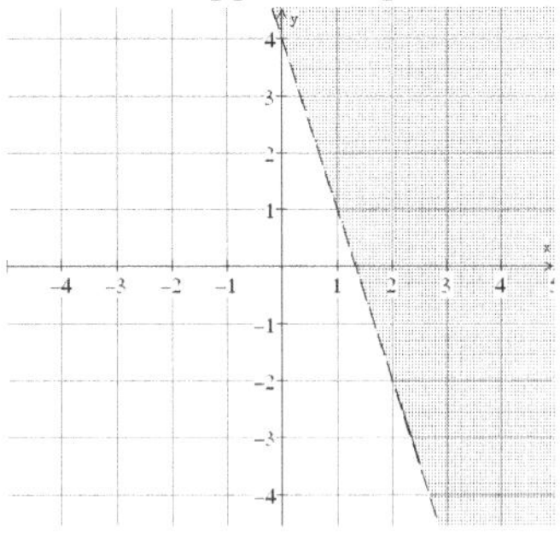

10.
$$\begin{aligned}4x-5y&\ge 15\\-5y&\ge -4x+15\\\frac{-5y}{-5}&\le\frac{-4x+15}{-5}\\y&\le\frac{4}{5}x-3\end{aligned}$$
Graph the line $y=\frac{4}{5}x-3$ as a solid line.

Shade the lower-half plane.

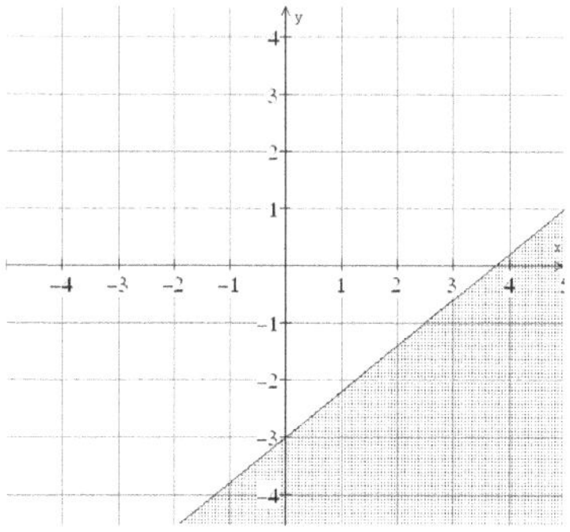

11. $A\cap B=\{12\}$

12.
$$\begin{aligned}4+x&<1\\x&<-3\end{aligned}$$
$\{x \mid x<-3\}$

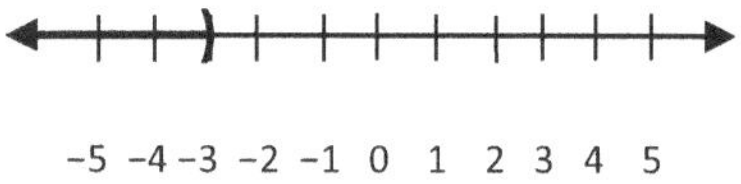

13.
$$\begin{aligned}-\frac{3}{8}x&\le 5\\-\frac{8}{3}\left(-\frac{3}{8}x\right)&\ge -\frac{8}{3}\cdot 5\\x&\ge -\frac{40}{3}\end{aligned}$$
$\left\{x \mid x\ge -\frac{40}{3}\right\}$

14.
$$\begin{aligned}6x-3(2-3x)&<4(2x-7)\\6x-6+9x&<8x-28\\15x-6&<8x-28\\7x&<-22\\\frac{7x}{7}&<\frac{-22}{7}\\x&<-\frac{22}{7}\end{aligned}$$
$\left(-\infty, -\frac{22}{7}\right)$

15.
$$\frac{2}{3}x \ge 2$$
$$\frac{3}{2}\cdot\frac{2}{3}x \ge \frac{3}{2}\cdot 2$$
$$x \ge 3$$
$[3, \infty)$

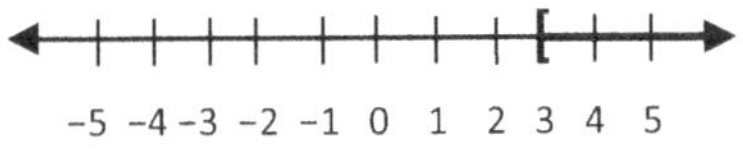

16.
$$2x-7 \le 6x+9$$
$$-4x \le 16$$
$$\frac{-4x}{-4} \ge \frac{16}{-4}$$
$$x \ge -4$$
$$\{x \mid x \ge -4\}$$

17. Strategy:
→Additional inches to grow: x
The child must be at least 48 in. tall.

Solution:
$$x+43 \ge 48$$
$$x \ge 5$$
The child must grow 5 or more in.

18. Strategy:
→To find the width, solve an inequality using the formula $LW < A$.

Solution:
$$15(2x-4) < 180$$
$$30x-60 < 180$$
$$30x < 240$$
$$x < 8$$
$$2x-4 = 16-4 = 12$$
$$2x-4 < 12$$
The width must be less than or equal to 11 ft.

19. Strategy:
→Required diameter: d
→Circumference: πd
The circumference must be between 0.1220 in. and 0.1240 in.

Solution:

$$0.1220 < \pi d \qquad \pi d < 0.1240$$
$$0.1220 < 3.14d \qquad 3.14d < 0.1240$$
$$\frac{0.1220}{3.14} < \frac{3.14d}{3.14} \qquad \frac{3.14d}{3.14} < \frac{0.1240}{3.14}$$
$$0.0389 < d \qquad d < 0.0395$$
The diameter must be between 0.0389 and 0.0395 in.

20. Stretegy:
→Total value of stock: x
The salary of \$5000 must be greater than or equal to \$1000 plus 2% of the total value of the stock.

Solution:
$$1000+0.02x \le 5000$$
$$0.02x \le 4000$$
$$x \le 200{,}000$$
The total value of the stock processed by the broker was \$200,000 or less.

Cumulative Review

1.
$$2[5a-3(2-5a)-8] = 2[5a-6+15a-8]$$
$$= 2[20a-14] = 40a-28$$

2.
$$\frac{5}{8}-4x = \frac{1}{8}$$
$$-4x = \frac{1}{8}-\frac{5}{8}$$
$$-4x = -\frac{1}{2}$$
$$-\frac{1}{4}(-4x) = -\frac{1}{4}\left(-\frac{1}{2}\right)$$
$$x = \frac{1}{8}$$
The solution is $\frac{1}{8}$.

3.
$$\begin{aligned}2x-3[x-2(x-3)]&=2\\2x-3[x-2x+6]&=2\\2x-3[-x+6]&=2\\2x+3x-18&=2\\5x-18&=2\\5x&=20\\x&=4\end{aligned}$$
The solution is 4.

4. $(-3a)(-2a^3b^2)^2=(-3a)(4a^6b^4)=-12a^7b^4$

5. $\dfrac{27a^3b^2}{(-3ab^2)^3}=\dfrac{27a^3b^2}{-27a^3b^6}=-\dfrac{1}{b^4}$

6.
$$\begin{array}{r}4x-2\\4x-1\overline{)16x^2-12x-2}\\\underline{16x^2-4x}\\-8x-2\\\underline{-8x+2}\\-4\end{array}$$
$(16x^2-12x-2)\div(4x-1)=4x-2-\dfrac{4}{4x-1}$

7.
$$\begin{aligned}f(x)&=x^2-4x-5\\f(-1)&=(-1)^2-4(-1)-5\\f(-1)&=1+4-5\\f(-1)&=0\end{aligned}$$

8. $27a^2x^2-3a^2$

The GCF is $3a^2$.
$$\begin{aligned}3a^2(9x^2-1)&=3a^2((3x^2)-1^2)\\&=3a^2(3x+1)(3x-1)\end{aligned}$$

9.
$$\frac{x^2-2x}{x^2-2x-8}\div\frac{x^3-5x^2+6x}{x^2-7x+12}$$
$$=\frac{x(x-2)}{(x-4)(x+2)}\cdot\frac{(x-4)(x-3)}{x(x^2-5x+6)}$$
$$=\frac{x\overset{1}{\cancel{(x-2)}}}{\underset{1}{\cancel{(x-4)}}(x+2)}\cdot\frac{\overset{1}{\cancel{(x-4)}}\,\overset{1}{\cancel{(x-3)}}}{x\underset{1}{\cancel{(x-3)}}\,\underset{1}{\cancel{(x-2)}}}$$
$$=\frac{1}{x+2}$$

10. $\dfrac{4a}{2a-3}-\dfrac{2a}{a+3}$

LCM is $(2a-3)(a+3)$.
$$\frac{4a}{2a-3}\cdot\frac{a+3}{a+3}-\frac{2a}{a+3}\cdot\frac{2a-3}{2a-3}$$
$$=\frac{4a^2+12a}{(2a-3)(a+3)}+\frac{-4a^2+6a}{(2a-3)(a+3)}$$
$$=\frac{18a}{(2a-3)(a+3)}$$

11.
$$\begin{aligned}\frac{5y}{6}-\frac{5}{9}&=\frac{y}{3}-\frac{5}{6}\\18\left(\frac{5y}{6}-\frac{5}{9}\right)&=18\left(\frac{y}{3}-\frac{5}{6}\right)\\15y-10&=6y-15\\9y&=-5\\y&=-\frac{5}{9}\end{aligned}$$
The solution is $-\dfrac{5}{9}$.

12.
$$\begin{aligned}R&=\frac{C-S}{t}\\t\cdot R&=t\cdot\frac{C-S}{t}\\tR&=C-S\\tR+S&=C\\C&=tR+S\end{aligned}$$

13. $m=\dfrac{y_2-y_1}{x_2-x_1}=\dfrac{4-(-3)}{-1-2}=\dfrac{7}{-3}=-\dfrac{7}{3}$

The slope is $-\dfrac{7}{3}$.

14. $m=-\dfrac{3}{2},(x_1,y_2)=(1,-3)$
$$\begin{aligned}y-y_1&=m(x-x_1)\\y-(-3)&=-\frac{3}{2}(x-1)\\y+3&=-\frac{3}{2}x+\frac{3}{2}\\y&=-\frac{3}{2}x+\frac{3}{2}-3\\y&=-\frac{3}{2}x+\frac{3}{2}-\frac{6}{2}\\y&=-\frac{3}{2}x-\frac{3}{2}\end{aligned}$$

15. (1) $x = 3y + 1$

(2) $2x + 5y = 13$

Substitute $3y + 1$ for x in equation (2).

$$\begin{aligned} 2x + 5y &= 13 \\ 2(3y+1) + 5y &= 13 \\ 6y + 2 + 5y &= 13 \\ 11y + 2 &= 13 \\ 11y &= 11 \\ y &= 1 \end{aligned}$$

Substitute 1 for y in equation (1).

$$\begin{aligned} x &= 3y + 1 \\ x &= 3(1) + 1 \\ x &= 3 + 1 \\ x &= 4 \end{aligned}$$

The solution is (4, 1).

16. (1) $9x - 2y = 17$

(2) $5x + 3y = -7$

Eliminate y by adding the equations.

$3(9x - 2y) = 3(17)$ $\quad 27x - 6y = 51$

$2(5x + 3y) = 2(-7)$ $\quad 10x + 6y = -14$

$$37x = 37$$

$$x = 1$$

Substitute 1 for x in equation (1).

$$\begin{aligned} 9x - 2y &= 17 \\ 9(1) - 2y &= 17 \\ -2y &= 8 \\ y &= -4 \end{aligned}$$

The solution is (1, –4).

17. $A \cup B = \{-10,\ -2,\ 0,\ 1,\ 2\}$

18. $\{x \mid x < 48,\ x \in \text{ real numbers}\}$

19. $(-\infty, 4)$

20.
$$\begin{aligned} \frac{3}{8}x &> -\frac{3}{4} \\ \frac{8}{3}\cdot\frac{3}{8}x &> \frac{8}{3}\cdot\left(-\frac{3}{4}\right) \\ x &> -2 \end{aligned}$$

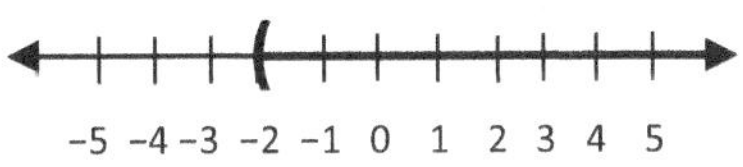

21.
$$\begin{aligned} -\frac{4}{5}x &> 12 \\ -\frac{5}{4}\left(-\frac{4}{5}x\right) &< -\frac{5}{4}\cdot 12 \\ x &< -15 \end{aligned}$$

22.
$$\begin{aligned} 15 - 3(5x - 7) &< 2(7 - 2x) \\ 15 - 15x + 21 &< 14 - 4x \\ -15x + 36 &< 14 - 4x \\ -11x &< -22 \\ \frac{-11x}{-11} &> \frac{-22}{-11} \\ x &> 2 \end{aligned}$$

23. $y = 2x - 1$

x	y
0	$2(0) - 1 = -1$
1	$2(1) - 1 = 1$
–1	$2(-1) - 1 = -3$

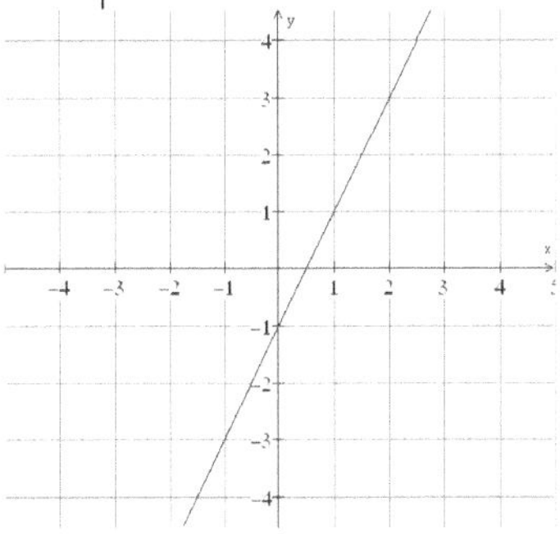

24.
$$\begin{aligned} 6x - 3y &\ge 6 \\ -3y &\ge -6x + 6 \\ \frac{-3y}{-3} &\le \frac{-6x + 6}{-3} \\ y &\le 2x - 2 \end{aligned}$$

Graph the line $y = 2x - 2$ as a solid line.

Shade the lower-half plane.

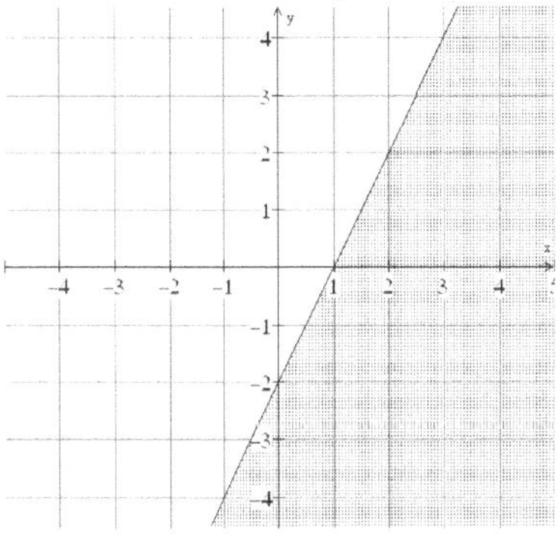

25. Strategy:
Unknown number: x

Solution:

$$\frac{3}{5}x < -15$$

$$\frac{5}{3}\cdot\frac{3}{5}x < \frac{5}{3}(-15)$$

$$x < -25$$

$$\{x \mid x \le -26,\ x \in \text{ integers}\}$$

26. Strategy:
→Number of miles: x
Company A's cost must be less than company B's cost.

Solution:

$$6 + 0.25x < 15 + 0.10x$$

$$0.15x < 9$$

$$x < 60$$

For 6 days:

$$x < 6\cdot 60$$

$$x < 360$$

For 6 days' rental, the maximum number of miles is 359 mi.

27. Strategy:
→Write and solve a proportion using x to represent the number of fish in the lake.

Solution:

$$\frac{3}{150} = \frac{100}{x}$$

$$150x\cdot\frac{3}{150} = 150x\cdot\frac{100}{x}$$

$$3x = 15000$$

$$x = 5000$$

There are an estimated 5000 fish in the lake.

28. Strategy:
→First angle: $x + 30$
→Second angle: x
→Third angle: $2x + 10$
Sum of the angle measures must equal 180°.

Solution:

$$(x+30) + x + (2x+10) = 180$$

$$4x + 40 = 180$$

$$4x = 140$$

$$x = 35$$

$x + 30 = 35 + 30 = 65$
$2x + 10 = 2(35) + 10 = 70 + 10 = 80$
The angle measures are 65°, 35°, and 80°.

Chapter 15: Radical Expressions

Prep Test

1. $-|-14| = -(14) = -14$

2. $3x^2y - 4xy^2 - 5x^2y = -2x^2y - 4xy^2$

3. $1.5h = 21$
$\frac{1.5h}{1.5} = \frac{21}{1.5}$
$h = 14$
The solution is 14.

4. $3x - 2 = 5 - 2x$
$5x = 7$
$x = \frac{7}{5}$
The solution is $\frac{7}{5}$.

5. $x^3 \cdot x^3 = x^{3+3} = x^6$

6. $(x+y)^2 = x^2 + 2xy + y^2$

7. $(2x-3)^2 = (2x)^2 - 2(2x)(3) + (3)^2$
$= 4x^2 - 12x + 9$

8. $(2-3v)(2+3v) = 4 - 9v^2$

9. $(a-5)(a+5) = a^2 - 25$

10. $\frac{2x^4y^3}{18x^2y} = \frac{x^{4-2}y^{3-1}}{9} = \frac{x^2y^2}{9}$

Section 15.1

Concept Check

1. radical sign; radicand

3. A variable exponential function is a perfect square if the numerical coefficient is a perfect square and the exponent on each variable is an even number.

Objective A Exercises

5. $\sqrt{16} = 4$

7. $\sqrt{49} = 7$

9. $\sqrt{32} = \sqrt{16 \cdot 2} = \sqrt{16} \cdot \sqrt{2} = 4\sqrt{2}$

11. $\sqrt{8} = \sqrt{4 \cdot 2} = \sqrt{4} \cdot \sqrt{2} = 2\sqrt{2}$

13. $-6\sqrt{18} = -6\sqrt{9 \cdot 2} = -6\sqrt{9} \cdot \sqrt{2}$
$= -6 \cdot 3\sqrt{2} = -18\sqrt{2}$

15. $5\sqrt{40} = 5\sqrt{4 \cdot 10} = 5\sqrt{4} \cdot \sqrt{10}$
$= 5 \cdot 2\sqrt{10} = 10\sqrt{10}$

17. $\sqrt{15} = \sqrt{3 \cdot 5} = \sqrt{15}$

19. $\sqrt{29}$

21. $-9\sqrt{72} = -9\sqrt{4 \cdot 9 \cdot 2} = -9\sqrt{4} \cdot \sqrt{9} \cdot \sqrt{2}$
$= -9 \cdot 2 \cdot 3\sqrt{2} = -54\sqrt{2}$

23. $\sqrt{45} = \sqrt{9 \cdot 5} = \sqrt{9} \cdot \sqrt{5} = 3\sqrt{5}$

25. $\sqrt{0} = 0$

27. $6\sqrt{128} = 6\sqrt{64 \cdot 2} = 6\sqrt{64} \cdot \sqrt{2}$
$= 6 \cdot 8\sqrt{2} = 48\sqrt{2}$

29. $-\sqrt{115}$
$-(11)^2 = -121 \qquad -(10)^2 = -100$
$-\sqrt{115}$ is between -11 and -10.

31. $\sqrt{\sqrt{64}} = \sqrt{8}$
$2^2 = 4 \qquad 3^2 = 9$
$\sqrt{\sqrt{64}}$ is between 2 and 3.

33. $\sqrt{240} \approx 15.492$

35. $\sqrt{288} \approx 16.971$

37. $\sqrt{350} \approx 18.708$

Objective B Exercises

39. $\sqrt{x^{14}} = x^7$

41. $\sqrt{y^{15}} = \sqrt{y^{14} \cdot y} = \sqrt{y^{14}} \cdot \sqrt{y} = y^7\sqrt{y}$

43. $\sqrt{a^{20}} = a^{10}$

45. $\sqrt{x^4y^4} = x^2y^2$

47. $\sqrt{4x^4} = 2x^2$

49. $\sqrt{24x^2} = \sqrt{4x^2 \cdot 6} = \sqrt{4x^2}\sqrt{6} = 2x\sqrt{6}$

51. $\sqrt{60x^5} = \sqrt{4x^4 \cdot 3 \cdot 5x}$
$= \sqrt{4x^4}\sqrt{3 \cdot 5x} = 2x^2\sqrt{15x}$

53. $\sqrt{49a^4b^8} = 7a^2b^4$

55. $\sqrt{18x^5y^7} = \sqrt{9x^4y^6 \cdot 2xy}$
$= \sqrt{9x^4y^6}\sqrt{2xy} = 3x^2y^3\sqrt{2xy}$

57. $\sqrt{40x^{11}y^7} = \sqrt{4x^{10}y^6 \cdot 2 \cdot 5xy}$
$= \sqrt{4x^{10}y^6}\sqrt{2 \cdot 5xy} = 2x^5y^3\sqrt{10xy}$

59. $\sqrt{80a^9b^{10}} = \sqrt{16a^8b^{10} \cdot 5a}$
$= \sqrt{16a^8b^{10}}\sqrt{5a} = 4a^4b^5\sqrt{5a}$

61. $-2\sqrt{16a^2b^3} = -2\sqrt{16a^2b^2 \cdot b} = -2\sqrt{16a^2b^2}\sqrt{b}$
$= -2 \cdot 4ab\sqrt{b} = -8ab\sqrt{b}$

63. $x\sqrt{x^4y^2} = x \cdot x^2y = x^3y$

65. $-4\sqrt{20a^4b^7} = -4\sqrt{4a^4b^6 \cdot 5b} = -4\sqrt{4a^4b^6}\sqrt{5b}$
$= -4 \cdot 2a^2b^3\sqrt{5b} = -8a^2b^3\sqrt{5b}$

67. $3x\sqrt{12x^2y^7} = 3x\sqrt{4x^2y^6 \cdot 3y} = 3x\sqrt{4x^2y^6}\sqrt{3y}$
$= 3x \cdot 2xy^3\sqrt{3y} = 6x^2y^3\sqrt{3y}$

69. $2x^2\sqrt{8x^2y^3} = 2x^2\sqrt{4x^2y^2 \cdot 2y}$
$= 2x^2\sqrt{4x^2y^2}\sqrt{2y}$
$= 2x^2 \cdot 2xy\sqrt{2y} = 4x^3y\sqrt{2y}$

71. $\sqrt{25(a+4)^2} = 5(a+4) = 5a+20$

73. $\sqrt{4(x+2)^4} = 2(x+2)^2$
$= 2(x^2+4x+4)$
$= 2x^2+8x+8$

75. $\sqrt{x^2+4x+4} = \sqrt{(x+2)^2} = x+2$

77. $\sqrt{y^2+2y+1} = \sqrt{(y+1)^2} = y+1$

79. $\sqrt{100a^6} = 10a^3$ Rational

81. $\sqrt{\sqrt{25a^6}} = \sqrt{5a^3} = \sqrt{a^2 \cdot 5a}$
$= \sqrt{a^2}\sqrt{4a} = a\sqrt{5a}$

Irrational

Critical Thinking 15.1

83. Strategy:
→To find the average number of credit cards, replace S in the formula with 1 for a first-year student, 2 for a sophomore, 3 for a junior and 4 for a senior.

a. Solution:
$N = 2.3\sqrt{S}$
$N = 2.3\sqrt{1}$
$N = 2.3$
The first-year students have an average of 2.3 credit cards.

b. Solution:
$N = 2.3\sqrt{S}$
$N = 2.3\sqrt{2}$
$N \approx 3.3$
The sophomores have an average of 3.3 credit cards.

c. Solution:
$N = 2.3\sqrt{S}$
$N = 2.3\sqrt{3}$
$N \approx 4.0$
The juniors have an average of 4.0 credit cards.

d. Solution:
$N = 2.3\sqrt{S}$
$N = 2.3\sqrt{4}$
$N = 4.6$
The seniors have an average of 4.6 credit cards.

Projects and Group Activities 15.1

85. **a.** Strategy:
Evaluate the expression $S=\sqrt{30fl}$ when $f = 1.2$ and $l = 60$.

Solution:

$$
\begin{aligned}
S &= \sqrt{30fl}\\
&= \sqrt{30\cdot 1.2\cdot 60}\\
&= \sqrt{2160}\\
&= \sqrt{144\cdot 15}\\
&= 12\sqrt{15}
\end{aligned}
$$

The speed of the car was $12\sqrt{15}$ mph.

b. $12\sqrt{15}\approx 46$ mph

Section 15.2

Concept Check

1. 2, 20, and 50 are not perfect squares.

3. Yes

5. Yes

Objective A Exercises

7. $2\sqrt{2}+\sqrt{2}=(2+1)\sqrt{2}=3\sqrt{2}$

9. $-3\sqrt{7}+2\sqrt{7}=(-3+2)\sqrt{7}=-\sqrt{7}$

11. $-3\sqrt{11}-8\sqrt{11}=(-3-8)\sqrt{11}=-11\sqrt{11}$

13. $2\sqrt{x}+8\sqrt{x}=(2+8)\sqrt{x}=10\sqrt{x}$

15. $8\sqrt{y}-10\sqrt{y}=(8-10)\sqrt{y}=-2\sqrt{y}$

17. $-2\sqrt{3b}-9\sqrt{3b}=(-2-9)\sqrt{3b}=-11\sqrt{3b}$

19. $3x\sqrt{2}-x\sqrt{2}=(3x-x)\sqrt{2}=2x\sqrt{2}$

21. $2a\sqrt{3a}-5a\sqrt{3a}=(2a-5a)\sqrt{3a}=-3a\sqrt{3a}$

23. $3\sqrt{xy}-8\sqrt{xy}=(3-8)\sqrt{xy}=-5\sqrt{xy}$

25.
$$
\begin{aligned}
\sqrt{45}+\sqrt{125} &= \sqrt{9\cdot 5}+\sqrt{25\cdot 5}\\
&= \sqrt{9}\sqrt{5}+\sqrt{25}\sqrt{5}\\
&= 3\sqrt{5}+5\sqrt{5}\\
&= (3+5)\sqrt{5}=8\sqrt{5}
\end{aligned}
$$

27.
$$
\begin{aligned}
2\sqrt{2}+3\sqrt{8} &= 2\sqrt{2}+3\sqrt{4\cdot 2}\\
&= 2\sqrt{2}+3\sqrt{4}\sqrt{2}\\
&= 2\sqrt{2}+3\cdot 2\sqrt{2}\\
&= (2+6)\sqrt{2}\\
&= 8\sqrt{2}
\end{aligned}
$$

29.
$$
\begin{aligned}
5\sqrt{18}-2\sqrt{75} &= 5\sqrt{9\cdot 2}-2\sqrt{25\cdot 3}\\
&= 5\sqrt{9}\sqrt{2}-2\sqrt{25}\sqrt{3}\\
&= 5\cdot 3\sqrt{2}-2\cdot 5\sqrt{3}\\
&= 15\sqrt{2}-10\sqrt{3}
\end{aligned}
$$

31.
$$
\begin{aligned}
5\sqrt{4x}-3\sqrt{9x} &= 5\cdot 2\sqrt{x}-3\cdot 3\sqrt{x}\\
&= 10\sqrt{x}-9\sqrt{x}\\
&= (10-9)\sqrt{x}=\sqrt{x}
\end{aligned}
$$

33.
$$
\begin{aligned}
3\sqrt{3x^2}-5\sqrt{27x^2} &= 3\sqrt{x^2 3}-5\sqrt{9x^2\cdot 3}\\
&= 3x\sqrt{3}-5\cdot 3x\sqrt{3}\\
&= 3x\sqrt{3}-15x\sqrt{3}\\
&= (3x-15x)\sqrt{3}=-12x\sqrt{3}
\end{aligned}
$$

35.
$$
\begin{aligned}
2x\sqrt{xy^2}-3y\sqrt{x^2y} &= 2x\sqrt{y^2}\sqrt{x}-3y\sqrt{x^2}\sqrt{y}\\
&= 2xy\sqrt{x}-3xy\sqrt{y}
\end{aligned}
$$

37.
$$
\begin{aligned}
3x\sqrt{12x}-5\sqrt{27x^2} &= 3x\sqrt{4\cdot 3x}-5\sqrt{9x^2\cdot 3x}\\
&= 3x\sqrt{4}\sqrt{3x}-5\sqrt{9x^2}\sqrt{3x}\\
&= 3x\cdot 2\sqrt{3x}-5\cdot 3x\sqrt{3x}\\
&= 6x\sqrt{3x}-15x\sqrt{3x}\\
&= (6x-15x)\sqrt{3x}\\
&= -9x\sqrt{3x}
\end{aligned}
$$

39.
$$
\begin{aligned}
&4y\sqrt{8y^3}-7\sqrt{18y^5}\\
&= 4y\sqrt{4y^2\cdot 2y}-7\sqrt{9y^4\cdot 2y}\\
&= 4y\sqrt{4y^2}\sqrt{2y}-7\sqrt{9y^2}\sqrt{2y}\\
&= 4y\cdot 2y\sqrt{2y}-7\cdot 3y^2\sqrt{2y}\\
&= 8y^2\sqrt{2y}-21y^2\sqrt{2y}\\
&= (8y^2-21y^2)\sqrt{2y}\\
&= -13y^2\sqrt{2y}
\end{aligned}
$$

41. $b^2\sqrt{a^5b}+3a^2\sqrt{ab^5}$
$=b^2\sqrt{a^4\cdot ab}+3a^2\sqrt{b^4\cdot ab}$
$=b^2\sqrt{a^4}\sqrt{ab}+3a^2\sqrt{b^4}\sqrt{ab}$
$=b^2a^2\sqrt{ab}+3a^2b^2\sqrt{ab}$
$=\left(a^2b^2+3a^2b^2\right)\sqrt{ab}=4a^2b^2\sqrt{ab}$

43. $4\sqrt{2}-5\sqrt{2}+8\sqrt{2}=(4-5+8)\sqrt{2}=7\sqrt{2}$

45. $5\sqrt{x}-8\sqrt{x}+9\sqrt{x}=(5-8+9)\sqrt{x}=6\sqrt{x}$

47. $8\sqrt{2}-3\sqrt{y}+8\sqrt{2}=(8-8)\sqrt{2}-3\sqrt{y}$
$=0\sqrt{2}-3\sqrt{y}=-3\sqrt{y}$

49. $8\sqrt{8}-4\sqrt{32}-9\sqrt{50}$
$=8\sqrt{4\cdot 2}-4\sqrt{16\cdot 2}-9\sqrt{25\cdot 2}$
$=8\sqrt{4}\sqrt{2}-4\sqrt{16}\sqrt{2}-9\sqrt{25}\sqrt{2}$
$=8\cdot 2\sqrt{2}-4\cdot 4\sqrt{2}-9\cdot 5\sqrt{2}$
$=16\sqrt{2}-16\sqrt{2}-45\sqrt{2}$
$=(16-16-45)\sqrt{2}=-45\sqrt{2}$

51. $-2\sqrt{3}+5\sqrt{27}-4\sqrt{45}$
$=-2\sqrt{3}+5\sqrt{9\cdot 3}-4\sqrt{9\cdot 5}$
$=-2\sqrt{3}+5\sqrt{9}\sqrt{3}-4\sqrt{9}\sqrt{5}$
$=-2\sqrt{3}+5\cdot 3\sqrt{3}-4\cdot 3\sqrt{5}$
$=-2\sqrt{3}+15\sqrt{3}-12\sqrt{5}$
$=(-2+15)\sqrt{3}-12\sqrt{5}=13\sqrt{3}-12\sqrt{5}$

53. $4\sqrt{75}+3\sqrt{48}-\sqrt{99}$
$=4\sqrt{25\cdot 3}+3\sqrt{16\cdot 3}-\sqrt{9\cdot 11}$
$=4\sqrt{25}\sqrt{3}+3\sqrt{16}\sqrt{3}-\sqrt{9}\sqrt{11}$
$=4\cdot 5\sqrt{3}+3\cdot 4\sqrt{3}-3\sqrt{11}$
$=20\sqrt{3}+12\sqrt{3}-3\sqrt{11}$
$=(20+12)\sqrt{3}-3\sqrt{11}$
$=32\sqrt{3}-3\sqrt{11}$

55. $\sqrt{25x}-\sqrt{9x}+\sqrt{16x}=5\sqrt{x}-3\sqrt{x}+4\sqrt{x}$
$=(5-3+4)\sqrt{x}=6\sqrt{x}$

57. $3\sqrt{3x}+\sqrt{27x}-8\sqrt{75x}$
$=3\sqrt{3x}+\sqrt{9\cdot 3x}-8\sqrt{25\cdot 3x}$
$=3\sqrt{3x}+\sqrt{9}\sqrt{3x}-8\sqrt{25}\sqrt{3x}$
$=3\sqrt{3x}+3\sqrt{3x}-8\cdot 5\sqrt{3x}$
$=3\sqrt{3x}+3\sqrt{3x}-40\sqrt{3x}$
$=(3+3-40)\sqrt{3x}=-34\sqrt{3x}$

59. $2a\sqrt{75b}-a\sqrt{20b}+4a\sqrt{45b}$
$=2a\sqrt{25\cdot 3b}-a\sqrt{4\cdot 5b}+4a\sqrt{9\cdot 5b}$
$=2a\sqrt{25}\sqrt{3b}-a\sqrt{4}\sqrt{5b}+4a\sqrt{9}\sqrt{5b}$
$=2a\cdot 5\sqrt{3b}-a\cdot 2\sqrt{5b}+4a\cdot 3\sqrt{5b}$
$=10a\sqrt{3b}-2a\sqrt{5b}+12a\sqrt{5b}$
$=10a\sqrt{3b}+(-2a+12a)\sqrt{5b}$
$=10a\sqrt{3b}+10a\sqrt{5b}$

61. $x\sqrt{3y^2}-2y\sqrt{12x^2}+xy\sqrt{3}$
$=x\sqrt{y^2}\sqrt{3}-2y\sqrt{4x^2}\sqrt{3}+xy\sqrt{3}$
$=xy\sqrt{3}-2y\cdot 2x\sqrt{3}+xy\sqrt{3}$
$=xy\sqrt{3}-4xy\sqrt{3}+xy\sqrt{3}$
$=(xy-4xy+xy)\sqrt{3}=-2xy\sqrt{3}$

63. $\sqrt{2ab}+\sqrt{2ab}=(1+1)\sqrt{2ab}=2\sqrt{2ab}$
(iv) $\sqrt{8ab}=\sqrt{4\cdot 2ab}=\sqrt{4}\sqrt{2ab}=2\sqrt{2ab}$

Critical Thinking 15.2

65. $\frac{a}{3}\sqrt{54ab^3}+\frac{b}{4}\sqrt{96a^3b}$
$=\frac{a}{3}\sqrt{9b^2\cdot 6ab}+\frac{b}{4}\sqrt{16a^2\cdot 6ab}$
$=\frac{a}{3}\sqrt{9b^2}\sqrt{6ab}+\frac{b}{4}\sqrt{16a^2}\sqrt{6ab}$
$=\frac{a}{3}\cdot 3b\sqrt{6ab}+\frac{b}{4}\cdot 4a\sqrt{6ab}$
$=ab\sqrt{6ab}+ab\sqrt{6ab}=2ab\sqrt{6ab}$

67. Strategy:
To find the perimeter, add all four sides together.

Solution:
$P=3\sqrt{2}+\sqrt{2}+3\sqrt{2}+\sqrt{2}$
$P=8\sqrt{2}$
The perimeter is $8\sqrt{2}$ cm.

Projects and Group Activities 15.2

69. To add two monomials like $5x + 3x$, the terms must be like. The variables must be the same and raised to the same power. Then we add the numerical coefficients. $(5 + 3)$ and affix the sum to the variable. So, $5x + 3x = 8x$. To add two radical expressions, the terms must have the same variables under the radical. Then add the numerical coefficients. So, $3\sqrt{x}+5\sqrt{x}=8\sqrt{x}$.

Section 15.3

Concept Check

1. $3-\sqrt{5}$

3. $\sqrt{2a}+8$

5. $(5-\sqrt{y})(5+\sqrt{y})=25+5\sqrt{y}-5\sqrt{y}-\sqrt{y^2}$
$=25-y$

7. $\dfrac{\sqrt{x}}{\sqrt{x}}$

9. $\dfrac{\sqrt{3}}{3}$ is in simplest form because the radicand does not contain a perfect-square factor greater than 1, and there is no radical expression in the denominator. $\dfrac{1}{\sqrt{3}}$ is not in simplest form because there is a radial expression in the denominator.

Objective A Exercises

11. $\sqrt{5}\cdot\sqrt{5}=\sqrt{25}=5$

13. $\sqrt{3}\cdot\sqrt{12}=\sqrt{36}=6$

15. $\sqrt{x}\cdot\sqrt{x}=\sqrt{x^2}=x$

17. $\sqrt{xy^3}\cdot\sqrt{x^5y}=\sqrt{x^6y^4}=x^3y^2$

19. $\sqrt{3a^2b^5}\cdot\sqrt{6ab^7}=\sqrt{18a^3b^{12}}=\sqrt{9a^2b^{12}\cdot 2a}$
$=\sqrt{9a^2b^{12}}\sqrt{2a}=3ab^6\sqrt{2a}$

21. $\sqrt{6a^3b^2}\cdot\sqrt{24a^5b}=\sqrt{144a^8b^3}$
$=\sqrt{144a^8b^2\cdot b}$
$=\sqrt{144a^8b^2}\sqrt{b}$
$=12a^4b\sqrt{b}$

23. $\sqrt{2ac}\cdot\sqrt{5ab}\cdot\sqrt{10cb}=\sqrt{100a^2b^2c^2}=10abc$

25. $\sqrt{2}(\sqrt{2}-\sqrt{3})=\sqrt{4}-\sqrt{6}=2-\sqrt{6}$

27. $\sqrt{x}(\sqrt{x}-\sqrt{y})=\sqrt{x^2}-\sqrt{xy}=x-\sqrt{xy}$

29. $\sqrt{5}(\sqrt{10}-\sqrt{x})=\sqrt{50}-\sqrt{5x}$
$=\sqrt{25\cdot 2}-\sqrt{5x}$
$=\sqrt{25}\sqrt{2}-\sqrt{5x}$
$=5\sqrt{2}-\sqrt{5x}$

31. $\sqrt{3a}(\sqrt{3a}-\sqrt{3b})=\sqrt{9a^2}-\sqrt{9ab}$
$=3a-3\sqrt{ab}$

33. $(\sqrt{x}-3)^2=\sqrt{x^2}-2\cdot 3\sqrt{x}+3^2$
$=x-6\sqrt{x}+9$

35. $(\sqrt{5}+3)(2\sqrt{5}-4)=2\sqrt{25}-4\sqrt{5}+6\sqrt{5}-12$
$=2\cdot 5+2\sqrt{5}-12$
$=10-4\sqrt{5}-12=-2+2\sqrt{5}$

37. $(4+\sqrt{8})(3+\sqrt{2})=12+4\sqrt{2}+3\sqrt{8}+\sqrt{16}$
$=12+4\sqrt{2}+3\sqrt{4\cdot 2}+4$
$=16+4\sqrt{2}+3\cdot 2\sqrt{2}$
$=16+4\sqrt{2}+6\sqrt{2}$
$=16+10\sqrt{2}$

39. $(2\sqrt{x}+4)(3\sqrt{x}-1)=6\sqrt{x^2}-2\sqrt{x}+12\sqrt{x}-4$
$=6x+10\sqrt{x}-4$

41. $(3\sqrt{x}-2y)(5\sqrt{x}-4y)$
$=15\sqrt{x^2}-12y\sqrt{x}-10y\sqrt{x}+8y^2$
$=15x-22y\sqrt{x}+8y^2$

43. $(3-\sqrt{5})(3+\sqrt{5})=9-\sqrt{25}=9-5=4$

45. $(3\sqrt{x}-4)(3\sqrt{x}+4)=9\sqrt{x^2}-16=9x-16$

47. $(\sqrt{a}-1)(\sqrt{a}+1)=\sqrt{a^2}-1=a-1$

is less than a.

Objective B Exercises

49. $\dfrac{\sqrt{32}}{\sqrt{2}}=\sqrt{\dfrac{32}{2}}=\sqrt{16}=4$

51. $\dfrac{\sqrt{98}}{\sqrt{2}}=\sqrt{\dfrac{98}{2}}=\sqrt{49}=7$

53. $\dfrac{\sqrt{27a}}{\sqrt{3a}}=\sqrt{\dfrac{27a}{3a}}=\sqrt{9}=3$

55. $\dfrac{\sqrt{15x^3y}}{\sqrt{3xy}}=\sqrt{\dfrac{15x^3y}{3xy}}=\sqrt{5x^2}=x\sqrt{5}$

57. $\dfrac{\sqrt{2a^5b^4}}{\sqrt{98ab^4}}=\sqrt{\dfrac{2a^5b^4}{98ab^4}}=\sqrt{\dfrac{a^4}{49}}=\dfrac{\sqrt{a^4}}{\sqrt{49}}=\dfrac{a^2}{7}$

59. $\dfrac{\sqrt{9xy^2}}{\sqrt{27x}}=\sqrt{\dfrac{9xy^2}{27x}}=\sqrt{\dfrac{y^2}{3}}=\dfrac{y}{\sqrt{3}}$

$=\dfrac{y}{\sqrt{3}}\cdot\dfrac{\sqrt{3}}{\sqrt{3}}=\dfrac{y\sqrt{3}}{\sqrt{9}}=\dfrac{y\sqrt{3}}{3}$

61. $\dfrac{\sqrt{16x^3y^2}}{\sqrt{8x^3y}}=\sqrt{\dfrac{16x^3y^2}{8x^3y}}=\sqrt{2y}$

63. $\dfrac{1}{\sqrt{2}-3}=\dfrac{1}{\sqrt{2}-3}\cdot\dfrac{\sqrt{2}+3}{\sqrt{2}+3}$

$=\dfrac{\sqrt{2}+3}{2-9}$

$=\dfrac{\sqrt{2}+3}{-7}=-\dfrac{\sqrt{2}+3}{7}$

65. $\dfrac{3}{5+\sqrt{5}}=\dfrac{3}{5+\sqrt{5}}\cdot\dfrac{5-\sqrt{5}}{5-\sqrt{5}}$

$=\dfrac{15-3\sqrt{5}}{25-5}=\dfrac{15-3\sqrt{5}}{20}$

67. $\dfrac{7}{\sqrt{2}-7}=\dfrac{7}{\sqrt{2}-7}\cdot\dfrac{\sqrt{2}+7}{\sqrt{2}+7}$

$=\dfrac{7\sqrt{2}+49}{2-49}$

$=\dfrac{7\sqrt{2}+49}{-47}=-\dfrac{7\sqrt{2}+49}{47}$

69. $\dfrac{-\sqrt{15}}{3-\sqrt{12}}=\dfrac{-\sqrt{15}}{3-\sqrt{12}}\cdot\dfrac{3+\sqrt{12}}{3+\sqrt{12}}$

$=\dfrac{-3\sqrt{15}-\sqrt{180}}{9-12}$

$=\dfrac{-3\sqrt{15}-\sqrt{9\cdot4\cdot5}}{-3}$

$=\dfrac{-3\sqrt{15}-6\sqrt{5}}{-3}=\sqrt{15}+2\sqrt{5}$

71. $\dfrac{\sqrt{xy}}{\sqrt{x}-\sqrt{y}}=\dfrac{\sqrt{xy}}{\sqrt{x}-\sqrt{y}}\cdot\dfrac{\sqrt{x}+\sqrt{y}}{\sqrt{x}+\sqrt{y}}$

$=\dfrac{\sqrt{x^2y}+\sqrt{xy^2}}{x-y}$

$=\dfrac{x\sqrt{y}+y\sqrt{x}}{x-y}$

73. $\dfrac{3-\sqrt{6}}{5-2\sqrt{6}}=\dfrac{3-\sqrt{6}}{5-2\sqrt{6}}\cdot\dfrac{5+2\sqrt{6}}{5+2\sqrt{6}}$

$=\dfrac{15+\sqrt{6}-2\sqrt{36}}{25-4(6)}$

$=\dfrac{15+\sqrt{6}-12}{25-24}=3+\sqrt{6}$

75. $\dfrac{\sqrt{2}+2\sqrt{6}}{2\sqrt{2}-3\sqrt{6}}=\dfrac{\sqrt{2}+2\sqrt{6}}{2\sqrt{2}-3\sqrt{6}}\cdot\dfrac{2\sqrt{2}+3\sqrt{6}}{2\sqrt{2}+3\sqrt{6}}$

$=\dfrac{2\sqrt{4}+3\sqrt{12}+4\sqrt{12}+6\sqrt{36}}{4\sqrt{4}-9\sqrt{36}}$

$=\dfrac{4+7\sqrt{4\cdot3}+36}{8-54}$

$=\dfrac{40+14\sqrt{3}}{-46}=-\dfrac{20+7\sqrt{3}}{23}$

77. $\dfrac{3+\sqrt{x}}{2-\sqrt{x}}=\dfrac{3+\sqrt{x}}{2-\sqrt{x}}\cdot\dfrac{2+\sqrt{x}}{2+\sqrt{x}}$

$=\dfrac{6+5\sqrt{x}+x}{4-x}$

79. $\dfrac{a}{\sqrt{a}}=\dfrac{a}{\sqrt{a}}\cdot\dfrac{\sqrt{a}}{\sqrt{a}}=\dfrac{a\sqrt{a}}{a}=\sqrt{a}$ is equal to $\sqrt{a}$.

Critical Thinking 15.3

81. Strategy:
To find the area of a rectangle, use the formula $A = lw$, where $l = 8+\sqrt{5}$ and $w = 8-\sqrt{5}$.

Solution:
$A = lw$
$A = (8+\sqrt{5})(8-\sqrt{5})$
$A = 64 - 8\sqrt{5} + 8\sqrt{5} - \sqrt{25}$
$A = 64 - 5$
$A = 59$
The area is 59 m^2.

83. $\sqrt{\frac{5}{8}}\sqrt{\frac{5}{8}} = \sqrt{\left(\frac{5}{8}\right)^2} = \frac{5}{8}$

85. $\sqrt{1\frac{9}{16}} = \sqrt{\frac{25}{16}} = \frac{5}{4}$

87.

$x^2 - 2x - 5 = 0$	
$(1+\sqrt{6})^2 - 2(1+\sqrt{6}) - 5$	0
$(1+\sqrt{6})(1+\sqrt{6}) - 2 - 2\sqrt{6} - 5$	0
$1+\sqrt{6}+\sqrt{6}+6-2-2\sqrt{6}-5$	0
0	0

$x^2 - 2x - 5 = 0$	
$(1-\sqrt{6})^2 - 2(1-\sqrt{6}) - 5$	0
$(1-\sqrt{6})(1-\sqrt{6}) - 2 + 2\sqrt{6} - 5$	0
$1-\sqrt{6}-\sqrt{6}+6-2+2\sqrt{6}-5$	0
0	0

Yes, both are solutions.

Check Your Progress: Chapter 15

1. $\sqrt{300} = \sqrt{100\cdot 3} = \sqrt{100}\sqrt{3} = 10\sqrt{3}$

2. $5\sqrt{180} = 5\sqrt{36\cdot 5} = 5\sqrt{36}\sqrt{5}$
$= 5\cdot 6\sqrt{5} = 30\sqrt{5}$

3. $\sqrt{64x^{10}} - 8x^5$

4. $\sqrt{18x^5y^4} = \sqrt{9x^4y^4\cdot 2x} = \sqrt{9x^4y^4}\sqrt{2x}$
$= 3x^2y^2\sqrt{2x}$

5. $5\sqrt{5} - 8\sqrt{5} = -3\sqrt{5}$

6. $5\sqrt{a^3b} + a\sqrt{4ab} - 3\sqrt{49a^3b}$
$= 5\sqrt{a^2\cdot ab} + a\sqrt{4\cdot ab} - 3\sqrt{49a^2\cdot ab}$
$= 5\sqrt{a^2}\sqrt{ab} + a\sqrt{4}\sqrt{ab} - 3\sqrt{49a^2}\sqrt{ab}$
$= 5a\sqrt{ab} + 2a\sqrt{ab} - 3\cdot 7a\sqrt{ab}$
$= 5a\sqrt{ab} + 2a\sqrt{ab} - 21a\sqrt{ab}$
$= -14a\sqrt{ab}$

7. $\sqrt{2a^3b^5}\sqrt{32ab^5} = \sqrt{64a^4b^{10}} = 8a^2b^5$

8. $\sqrt{8}(\sqrt{2} - \sqrt{5}) = \sqrt{16} - \sqrt{40} = 4 - \sqrt{4\cdot 10}$
$= 4 - \sqrt{4}\sqrt{10} = 4 - 2\sqrt{10}$

9. $(\sqrt{a} - 3)(3\sqrt{a} + 2) = 3\sqrt{a^2} + 2\sqrt{a} - 9\sqrt{a} - 6$
$= 3a - 7\sqrt{a} - 6$

10. $(\sqrt{2y} + 5)(\sqrt{2y} - 5)$
$= \sqrt{4y^2} - 5\sqrt{2y} + 5\sqrt{2y} - 25$
$= 2y - 25$

11. $\frac{\sqrt{50}}{\sqrt{2}} = \sqrt{\frac{50}{2}} = \sqrt{25} = 5$

12. $\frac{6}{\sqrt{12x}} = \frac{6}{\sqrt{12x}}\cdot\frac{\sqrt{12x}}{\sqrt{12x}} = \frac{6\sqrt{12x}}{12x} = \frac{\sqrt{12x}}{2x}$
$= \frac{\sqrt{4\cdot 3x}}{2x} = \frac{2\sqrt{3x}}{2x} = \frac{\sqrt{3x}}{x}$

13. $\frac{7}{7-\sqrt{2}} = \frac{7}{7-\sqrt{2}}\cdot\frac{7+\sqrt{2}}{7+\sqrt{2}}$
$= \frac{49+7\sqrt{2}}{49-2} = \frac{49+7\sqrt{2}}{47}$

14. $\frac{5+\sqrt{2}}{3-\sqrt{2}} = \frac{5+\sqrt{2}}{3-\sqrt{2}}\cdot\frac{3+\sqrt{2}}{3+\sqrt{2}} = \frac{15+5\sqrt{2}+3\sqrt{2}+2}{9-2}$
$= \frac{17+8\sqrt{2}}{7}$

Section 15.4

Concept Check

1. No

3. Yes

5. Sometimes true

7. Always true

9. 90°; legs

Objective A Exercises

11. $\sqrt{x}=5$
$(\sqrt{x})^2=5^2$
$x=25$
The solution is 25.

Check:
$\sqrt{x}=5$
$\sqrt{25} \mid 5$
$5=5$

13. $\sqrt{a}=12$
$(\sqrt{a})^2=12^2$
$a=144$
The solution is 144.

Check:
$\sqrt{a}=12$
$\sqrt{144} \mid 12$
$12=12$

15. $\sqrt{5x}=5$
$(\sqrt{5x})^2=(5)^2$
$5x=25$
$x=5$
The solution is 5.

Check:
$\sqrt{5x}=5$
$\sqrt{5\cdot 5} \mid 5$
$\sqrt{25}$
5
$5=5$

17. $\sqrt{4x}=8$
$(\sqrt{4x})^2=(8)^2$
$4x=64$
$x=16$
The solution is 16.

Check:
$\sqrt{4x}=8$
$\sqrt{4\cdot 16} \mid 8$
$\sqrt{64} \mid 8$
$8 \mid 8$
$8=8$

19. $\sqrt{2x}-4=0$
$\sqrt{2x}=4$
$(\sqrt{2x})^2=(4)^2$
$2x=16$
$x=8$
The solution is 8.

Check:
$\sqrt{2x}-4=0$
$\sqrt{2\cdot 8}-4 \mid 0$
$\sqrt{16}-4 \mid 0$
$4-4 \mid 0$
$0 \mid 0$
$0=0$

21. $\sqrt{4x}+5=2$
$\sqrt{4x}=-3$
$(\sqrt{4x})^2=(-3)^2$
$4x=9$
$x=\frac{9}{4}$

Check:
$\sqrt{4x}+5=2$
$\sqrt{4\cdot\frac{9}{4}}+5 \mid 2$
$\sqrt{9}+5 \mid 2$
$3+5 \mid 2$
$8 \mid 2$
$8\neq 2$

The equation has no solution.

23. $\sqrt{3x-2}=4$
$(\sqrt{3x-2})^2=4^2$
$3x-2=16$
$3x=18$
$x=6$
The solution is 6.

Check:
$\sqrt{3x-2}=4$
$\sqrt{3\cdot 6-2} \mid 4$
$\sqrt{18-2} \mid 4$
$\sqrt{16} \mid 4$
$4=4$

25. $\sqrt{2x+1}=7$
$(\sqrt{2x+1})^2=7^2$
$2x+1=49$
$2x=48$
$x=24$
The solution is 24.

Check:
$\sqrt{2x+1}=7$
$\sqrt{2\cdot 24+1} \mid 7$
$\sqrt{48+1} \mid 7$
$\sqrt{49} \mid 7$
$7=7$

27. $\sqrt{5x+2}=0$
$(\sqrt{5x+2})^2=0^2$
$5x+2=0$
$5x=-2$
$x=-\frac{2}{5}$
The solution is $-\frac{2}{5}$.

Check:
$\sqrt{5x+2}=0$
$\sqrt{5\left(-\frac{2}{5}\right)+2} \mid 0$
$\sqrt{-2+2} \mid 0$
$\sqrt{0} \mid 0$
$0=0$

29. $\sqrt{3x}-6=-4$
$\sqrt{3x}=2$
$(\sqrt{3x})^2=2^2$
$3x=4$
$x=\frac{4}{3}$
The solution is $\frac{4}{3}$.

Check:
$\sqrt{3x}-6=-4$
$\sqrt{3\left(\frac{4}{3}\right)}-6 \mid -4$
$\sqrt{4}-6 \mid -4$
$-4 \mid -4$
$-4=-4$

31.

$$\begin{aligned} 0&=2-\sqrt{3-x}\\ \sqrt{3-x}&=2\\ \left(\sqrt{3-x}\right)^2&=2^2\\ 3-x&=4\\ -x&=1\\ x&=-1 \end{aligned}$$

Check:

$0=2-\sqrt{3-x}$	
0	$2-\sqrt{3-(-1)}$
0	$2-\sqrt{3+1}$
0	$2-\sqrt{4}$
0	$2-2$

$0=0$

The solution is –1.

33.

$$\begin{aligned} 0&=\sqrt{3x-9}-6\\ 6&=\sqrt{3x-9}\\ (6)^2&=\left(\sqrt{3x-9}\right)^2\\ 36&=3x-9\\ 45&=3x\\ 15&=x \end{aligned}$$

Check:

$0=\sqrt{3x-9}-6$	
0	$\sqrt{3(15)-9}-6$
0	$\sqrt{45-9}-6$
0	$\sqrt{36}-6$
0	$6-6$
0	0

$0=0$

The solution is 15.

35.

$$\begin{aligned} \sqrt{5x-1}&=\sqrt{3x+9}\\ \left(\sqrt{5x-1}\right)^2&=\left(\sqrt{3x+9}\right)^2\\ 5x-1&=3x+9\\ 2x-1&=9\\ 2x&=10\\ x&=5 \end{aligned}$$

Check:

$\sqrt{5x-1}=\sqrt{3x+9}$	
$\sqrt{5(5)-1}$	$\sqrt{3(5)-}$
$\sqrt{25-1}$	$\sqrt{15+9}$
$\sqrt{24}$	$\sqrt{24}$

$\sqrt{24}=\sqrt{24}$

The solution is 5.

37.

$$\begin{aligned} \sqrt{5x-3}&=\sqrt{4x-2}\\ \left(\sqrt{5x-3}^2\right)&=\left(\sqrt{4x-2}\right)^2\\ 5x-3&=4x-2\\ x-3&=-2\\ x&=1 \end{aligned}$$

Check:

$\sqrt{5x-3}=\sqrt{4x-2}$	
$\sqrt{5(1)-3}$	$\sqrt{4(1)-2}$
$\sqrt{5-3}$	$\sqrt{4-2}$
$\sqrt{2}$	$\sqrt{2}$

$\sqrt{2}=\sqrt{2}$

The solution is 1.

39.

$$\begin{aligned} \sqrt{x^2-5x+6}&=\sqrt{x^2-8x+9}\\ \left(\sqrt{x^2-5x+6}\right)^2&=\left(\sqrt{x^2-8x+9}\right)^2\\ x^2-5x+6&=x^2-8x+9\\ -5x+6&=-8x+9\\ 3x+6&=9\\ 3x&=3\\ x&=1 \end{aligned}$$

Check:

$\sqrt{x^2-5x+6}=\sqrt{x^2-8x+9}$	
$\sqrt{(1)^2-5(1)+6}$	$\sqrt{(1)^2-8(1)+9}$
$\sqrt{1-5+6}$	$\sqrt{1-8+9}$
$\sqrt{2}$	$\sqrt{2}$

$\sqrt{2}=\sqrt{2}$

The solution is 1.

41. $\sqrt{x} = \sqrt{x+3} - 1$

$$\left(\sqrt{x}\right)^2 = \left(\sqrt{x+3} - 1\right)^2$$
$$x = x + 3 - 2\sqrt{x+3} + 1$$
$$x = x + 4 - 2\sqrt{x+3}$$
$$-4 = -2\sqrt{x+3}$$
$$2 = \sqrt{x+3}$$
$$(2)^2 = \left(\sqrt{x+3}\right)^2$$
$$4 = x + 3$$
$$1 = x$$

Check:

$\sqrt{x}$	$= \sqrt{x+3} - 1$
$\sqrt{1}$	$\sqrt{1+3} - 1$
1	$\sqrt{4} - 1$
1	$2 - 1$
1	1

$1 = 1$

The solution is 1.

43. $\sqrt{2x+5} = 5 - \sqrt{2x}$

$$\left(\sqrt{2x+5}\right)^2 = \left(5 - \sqrt{2x}\right)^2$$
$$2x + 5 = 25 - 10\sqrt{2x} + 2x$$
$$-20 = -10\sqrt{2x}$$
$$2 = \sqrt{2x}$$
$$(2)^2 = \left(\sqrt{2x}\right)^2$$
$$4 = 2x$$
$$2 = x$$

Check:

$\sqrt{2x+5}$	$= 5 - \sqrt{2x}$
$\sqrt{2(2)+5}$	$5 - \sqrt{2(2}$
$\sqrt{4+5}$	$5 - \sqrt{4}$
$\sqrt{9}$	$5 - 2$
3	3

$3 = 3$

The solution is 2.

45. $\sqrt{3x} - \sqrt{3x+7} = 1$

$$\left(\sqrt{3x} - \sqrt{3x+7}\right)^2 = (1)^2$$
$$3x - 2\sqrt{3x(3x+7)} + 3x + 7 = 1$$
$$-2\sqrt{3x(3x+7)} + 6x + 7 = 1$$
$$-2\sqrt{3x(3x+7)} = -6x - 6$$
$$\sqrt{3x(3x+7)} = 3x + 3$$
$$\left(\sqrt{3x(3x+7)}\right)^2 = (3x+3)^2$$
$$9x^2 + 21x = 9x^2 + 18x + 9$$
$$3x = 9$$
$$x = 3$$

Check:

$\sqrt{3x} - \sqrt{3x+7}$	$= 1$
$\sqrt{3(3)} - \sqrt{3(3)+7}$	1
$\sqrt{9} - \sqrt{9+7}$	1
$3 - \sqrt{16}$	1
$3 - 4$	1
-1	1

$-1 \neq 1$

There is no solution to the equation.

47. (iii) can have no solution.
The left side of the equation is a positive value and the right side is negative.

Objective B Exercises

49. Strategy:
Let x = the number.
Square root of the sum of a number and 5: $\sqrt{x+5}$

Solution:

$$2 + \sqrt{x+5} = 6$$
$$\sqrt{x+5} = 4$$
$$\left(\sqrt{x+5}\right)^2 = (4)^2$$
$$x + 5 = 16$$
$$x = 11$$

Check:

$2 + \sqrt{x+5}$	$= 6$
$2 + \sqrt{11+5}$	6
$2 + \sqrt{16}$	6
$2 + 4$	6
6	6

$6 = 6$

The number is 11.

51. Strategy:
Solve the formula
$\text{Hypotenuse} = \sqrt{(\text{leg})^2 + (\text{leg})^2}$ using 5 for one leg and 9 for the other leg.

Solution:

$$\begin{aligned}\text{Hypotenuse} &= \sqrt{(\text{leg})^2 + (\text{leg})^2}\\ &= \sqrt{(5)^2 + (9)^2}\\ &= \sqrt{25+81}\\ &= \sqrt{106}\\ &\approx 10.30\end{aligned}$$

The hypotenuse is 10.30 cm.

53. Strategy:
Solve the formula
$\text{Leg} = \sqrt{(\text{hypotenuse})^2 - (\text{leg})^2}$ using 12 for the hypotenuse and 7 for the leg.

Solution:

$$\begin{aligned}\text{Leg} &= \sqrt{(\text{hypotenuse})^2 - (\text{leg})^2}\\ &= \sqrt{(12)^2 - (7)^2}\\ &= \sqrt{144-49}\\ &= \sqrt{95}\\ &\approx 9.75\end{aligned}$$

The leg is 9.75 ft.

55. Strategy:
Solve the formula
$\text{Hypotenuse} = \sqrt{(\text{leg})^2 + (\text{leg})^2}$ using 5 for one leg and 11 for the other leg.

Solution:

$$\begin{aligned}\text{Hypotenuse} &= \sqrt{(\text{leg})^2 + (\text{leg})^2}\\ &= \sqrt{(5)^2 + (11)^2}\\ &= \sqrt{25+121}\\ &= \sqrt{146}\\ &\approx 12.1\end{aligned}$$

The diagonal is 12.1 mi.

57. Strategy:
To find the depth of the water, solve the equation $C = \sqrt{32H}$, where C is 20.

Solution:

$$\begin{aligned}C &= \sqrt{32H}\\ 20 &= \sqrt{32H}\\ 20^2 &= \left(\sqrt{32H}\right)^2\\ 400 &= 32H\\ \frac{400}{32} &= H\\ 12.5 &= H\end{aligned}$$

The depth of the water is 12.5 ft.

59. Strategy:
To find the length of the diagonal between home plate and second base, use the equation $c = \sqrt{a^2 + b^2}$ and compare $\frac{1}{2}c$ with 46.

Solution:

$$\begin{aligned}c &= \sqrt{a^2+b^2}\\ c &= \sqrt{60^2+60^2}\\ c &= \sqrt{3600+3600}\\ c &= \sqrt{7200}\\ c &\approx 84.85\end{aligned}$$

$$\frac{1}{2}c \approx 42.43$$

$42.43 > 40.5$

The pitcher's mound is more than halfway between home plate and second base.

61. Strategy:
To find distance from the dock, use the equation $c = \sqrt{a^2 + b^2}$ and compare the distance to 5 mi.

Solution:

$$\begin{aligned}c &= \sqrt{a^2+b^2}\\ c &= \sqrt{2.5^2+4^2}\\ c &= \sqrt{6.25+16}\\ c &= \sqrt{22.25}\\ c &\approx 4.72\end{aligned}$$

$4.72 < 5$

Yes, Marta will be able to use the walkie-talkie.

63. Strategy:
To find the height of the screen, use the equation $a=\sqrt{c^2-b^2}$, where $b = 20.5$ and $c = 41$.

Solution:

$$a=\sqrt{c^2-b^2}$$
$$a=\sqrt{41^2-20.5^2}$$
$$a=\sqrt{1681-420.25}$$
$$a=\sqrt{1260.75}$$
$$a\approx 35.5$$

The height of the screen is 35.5 in.

65. Strategy:
To find the length of the pendulum, replace T in the equation with 1.5 and solve for L. Use 3.14 for π.

Solution:

$$T=2\pi\sqrt{\frac{L}{32}}$$
$$1.5=2\pi\sqrt{\frac{L}{32}}$$
$$\frac{1.5}{2\pi}=\sqrt{\frac{L}{32}}$$
$$\left(\frac{1.5}{2\pi}\right)^2=\left(\sqrt{\frac{L}{32}}\right)^2$$
$$\frac{1.5^2}{(2\pi)^2}=\frac{L}{32}$$
$$\frac{32(1.5^2)}{(2\pi)^2}=L$$
$$1.83\approx L$$

The length of the pendulum is 1.83 ft.

67. Strategy:
To find the height of the bridge, solve the formula $T=\sqrt{\frac{d}{16}}$ for d when T is 2.

Solution:

$$T=\sqrt{\frac{d}{16}}$$
$$2=\sqrt{\frac{d}{16}}$$
$$(2)^2=\left(\sqrt{\frac{d}{16}}\right)^2$$
$$4=\frac{d}{16}$$
$$16(4)=16\left(\frac{d}{16}\right)$$
$$64=d$$

The height of the bridge is 64 ft.

Critical Thinking 15.4

69.
$$\sqrt{\frac{5y+2}{3}}=3$$
$$\left(\sqrt{\frac{5y+2}{3}}\right)^2=(3)^2$$
$$\frac{5y+2}{3}=9$$
$$3\left(\frac{5y+2}{3}\right)=3(9)$$
$$5y+2=27$$
$$5y=25$$
$$y=5$$

Check:

$\sqrt{\frac{5y+2}{3}}=3$	
$\sqrt{\frac{5(5)+2}{3}}$	3
$\sqrt{\frac{25+2}{3}}$	
$\sqrt{\frac{27}{3}}$	
$\sqrt{9}$	
3	

$3=3$

The solution is 5.

71. $\sqrt{9x^2+49}+1=3x+2$

$$\sqrt{9x^2+49}=3x+1$$
$$\left(\sqrt{9x^2+49}\right)^2=(3x+1)^2$$
$$9x^2+49=9x^2+6x+1$$
$$48=6x$$
$$8=x$$

Check:

$\sqrt{9x^2+49}+1=3x+2$	
$\sqrt{9(8)^2+49}+1$	$3(8)+2$
$\sqrt{8(64)+49}+1$	$24+2$
$\sqrt{576+49}+1$	26
$\sqrt{625}+1$	
$25+1$	
26	

$$26=26$$

The solution is 8.

73. $a=4\sqrt{2},\ c=5\sqrt{2}$

$$b=\sqrt{c^2-a^2}$$
$$c=\sqrt{\left(5\sqrt{2}\right)^2-\left(4\sqrt{2}\right)^2}=\sqrt{25(2)-16(2)}$$
$$=\sqrt{50-32}=\sqrt{18}=\sqrt{9\cdot 2}=3\sqrt{2}$$

a. Perimeter $=a+b+c=4\sqrt{2}+5\sqrt{2}+3\sqrt{2}$
$$=12\sqrt{2}\text{ cm}$$

b. Area $=\frac{1}{2}bh$
$$=\frac{1}{2}\left(4\sqrt{2}\right)\left(3\sqrt{2}\right)=\frac{1}{2}\left(12\sqrt{4}\right)$$
$$=\frac{1}{2}(24)=12\text{ cm}^2$$

75. No. The Pythagorean Theorem is true only for right triangles.

Projects or Group Activities 15.4

77. Strategy:
To find the length of the diagonal *AB*, find length of the bottom leg of the triangle and the length of the height of the triangle and use the formula $c=\sqrt{a^2+b^2}$.

Solution:

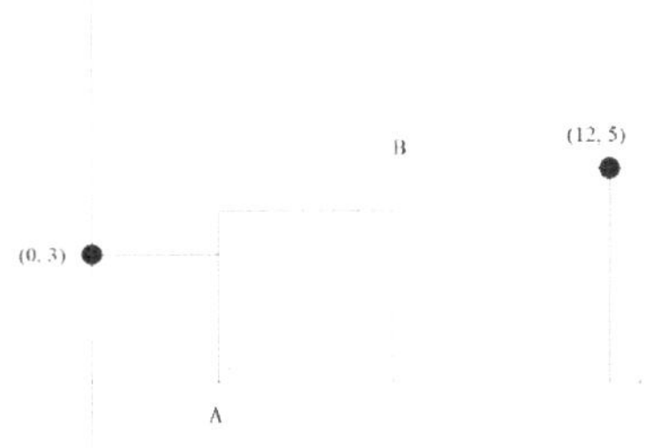

Since the smallest square is 3 units high, each side of the smallest square must be 3 units. So the distance from (0, 0) to *A* must be 3 units. Since the largest square is 5 units high, each side of the larger square must be 5 units. That gives us the following lengths.

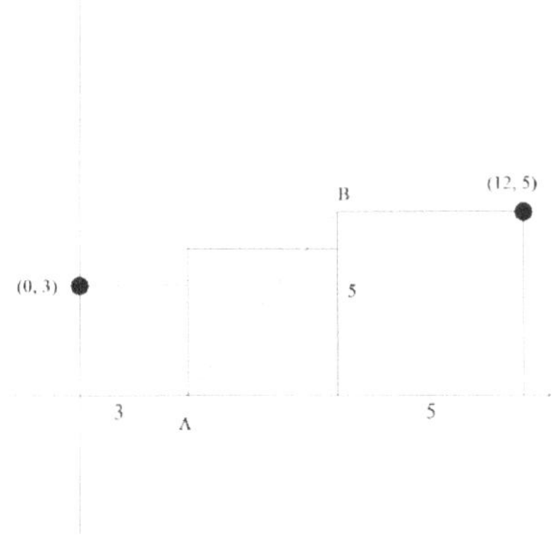

The length of all three squares is 12 units, so the length of the middle square is 12 – 3 – 5 = 4.

$$c=\sqrt{a^2+b^2}$$
$$c=\sqrt{4^2+5^2}$$
$$c=\sqrt{16+25}$$
$$c=\sqrt{41}$$
$$c\approx 6.4$$

The length of *AB* is 6.4 units.

Chapter Review Exercises

1. $\sqrt{3}\left(\sqrt{12}-\sqrt{3}\right)=\sqrt{36}-\sqrt{9}=6-3=3$

2. $$\begin{aligned}3\sqrt{18a^5b}&=3\sqrt{9a^4\cdot 2ab}\\&=3\sqrt{9a^4}\sqrt{2ab}=3\cdot 3a^2\sqrt{2ab}\\&=9a^2\sqrt{2ab}\end{aligned}$$

3. $2\sqrt{36}=2\cdot 6=12$

4. $$\begin{aligned}\sqrt{6a}\left(\sqrt{3a}+\sqrt{2a}\right)&=\sqrt{18a^2}+\sqrt{12a^2}\\&=\sqrt{9a^2\cdot 2}+\sqrt{4a^2\cdot 3}\\&=\sqrt{9a^2}\sqrt{2}+\sqrt{4a^2}\sqrt{3}\\&=3a\sqrt{2}+2a\sqrt{3}\end{aligned}$$

5. $\dfrac{12}{\sqrt{6}}=\dfrac{12}{\sqrt{6}}\cdot\dfrac{\sqrt{6}}{\sqrt{6}}=\dfrac{12\sqrt{6}}{36}=\dfrac{12\sqrt{6}}{6}=2\sqrt{6}$

6. $$\begin{aligned}2\sqrt{8}-3\sqrt{32}&=2\sqrt{4\cdot 2}-3\sqrt{16\cdot 2}\\&=2\sqrt{4}\sqrt{2}-3\sqrt{16}\sqrt{2}\\&=2\cdot 2\sqrt{2}-3\cdot 4\sqrt{2}\\&=4\sqrt{2}-12\sqrt{2}=-8\sqrt{2}\end{aligned}$$

7. $$\begin{aligned}\left(3-\sqrt{7}\right)\left(3+\sqrt{7}\right)&=3^2-\left(\sqrt{7}\right)^2\\&=9-7=2\end{aligned}$$

8. $$\begin{aligned}\sqrt{x+3}-\sqrt{x}&=1\\\sqrt{x+3}&=\sqrt{x}+1\\\left(\sqrt{x+3}\right)^2&=\left(\sqrt{x}+1\right)^2\\x+3&=x+2\sqrt{x}+1\\2&=2\sqrt{x}\\1&=\sqrt{x}\\1^2&=\left(\sqrt{x}\right)^2\\1&=x\end{aligned}$$

Check:

$\sqrt{x+3}-\sqrt{x}=1$	
$\sqrt{1+3}-\sqrt{1}$	1
$\sqrt{4}-\sqrt{1}$	1
$2-1$	1

$1=1$

The solution is 1.

9. $$\begin{aligned}\frac{2x}{\sqrt{3}-\sqrt{5}}&=\frac{2x}{\sqrt{3}-\sqrt{5}}\cdot\frac{\sqrt{3}+\sqrt{5}}{\sqrt{3}+\sqrt{5}}\\&=\frac{2x\left(\sqrt{3}+\sqrt{5}\right)}{3-5}\\&=\frac{2x\left(\sqrt{3}+\sqrt{5}\right)}{-2}\\&=-x\left(\sqrt{3}+\sqrt{5}\right)=-x\sqrt{3}-x\sqrt{5}\end{aligned}$$

10. $$\begin{aligned}-3\sqrt{120}&=-3\sqrt{4\cdot 30}=-3\sqrt{4}\sqrt{30}\\&=-3\cdot 2\sqrt{30}=-6\sqrt{30}\end{aligned}$$

11. $$\begin{aligned}\sqrt{5x}&=10\\\left(\sqrt{5x}\right)^2&=10^2\\5x&=100\\x&=20\end{aligned}$$

The solution is 20.

Check:

$\sqrt{5x}=10$	
$\sqrt{5\cdot 20}$	10
$\sqrt{100}$	10

$10=10$

12. $$\begin{aligned}5\sqrt{48}&=5\sqrt{16\cdot 3}=5\sqrt{16}\sqrt{3}\\&=5\cdot 4\sqrt{3}=20\sqrt{3}\end{aligned}$$

13. $\dfrac{\sqrt{98x^7y^9}}{\sqrt{2x^3y}}=\sqrt{\dfrac{98x^7y^9}{2x^3y}}=\sqrt{49x^4y^8}=7x^2y^4$

14. $$\begin{aligned}3-\sqrt{7x}&=5\\-\sqrt{7x}&=2\\\sqrt{7x}&=-2\end{aligned}$$

The equation has no solution.

15. $$\begin{aligned}&6a\sqrt{80b}-\sqrt{180a^2b}+5a\sqrt{b}\\&=6a\sqrt{16\cdot 5b}-\sqrt{36a^2\cdot 5b}+5a\sqrt{b}\\&=6a\sqrt{16}\sqrt{5b}-\sqrt{36a^2}\sqrt{5b}+5a\sqrt{b}\\&=6a\cdot 4\sqrt{5b}-6a\sqrt{5b}+5a\sqrt{b}\\&=24a\sqrt{5b}-6a\sqrt{5b}+5a\sqrt{b}\\&=18a\sqrt{5b}+5a\sqrt{b}\end{aligned}$$

16. $$\begin{aligned}4\sqrt{250}&=4\sqrt{25\cdot 10}=4\sqrt{25}\sqrt{10}\\&=4\cdot 5\sqrt{10}=20\sqrt{10}\end{aligned}$$

17. $$\begin{aligned}&2x\sqrt{60x^3y^3}+3x^2y\sqrt{15xy}\\&=2x\sqrt{4x^2y^2\cdot 15xy}+3x^2y\sqrt{15xy}\\&=2x\sqrt{4x^2y^2}\sqrt{15xy}+3x^2y\sqrt{15xy}\\&=2x\cdot 2xy\sqrt{15xy}+3x^2y\sqrt{15xy}\\&=4x^2y\sqrt{15xy}+3x^2y\sqrt{15xy}\\&=7x^2y\sqrt{15xy}\end{aligned}$$

18. $\left(4\sqrt{y}-\sqrt{5}\right)\left(2\sqrt{y}+3\sqrt{5}\right)$
$=8y+12\sqrt{5y}-2\sqrt{5y}-3(5)$
$=8y+10\sqrt{5y}-15$

19. $3\sqrt{12x}+5\sqrt{48x}=3\sqrt{4\cdot 3x}+5\sqrt{16\cdot 3x}$
$=3\sqrt{4}\sqrt{3x}+5\sqrt{16}\sqrt{3x}$
$=3\cdot 2\sqrt{3x}+5\cdot 4\sqrt{3x}$
$=6\sqrt{3x}+20\sqrt{3x}$
$=26\sqrt{3x}$

20. $\sqrt{2x-3}+4=0$
$\sqrt{2x-3}=-4$
The equation has no solution.

21. $\dfrac{8}{\sqrt{x}-3}=\dfrac{8}{\sqrt{x}-3}\cdot\dfrac{\sqrt{x}+3}{\sqrt{x}+3}$
$=\dfrac{8\sqrt{x}+24}{x-9}$

22. $4y\sqrt{243x^{17}y^9}=4y\sqrt{81x^{16}y^8\cdot 3xy}$
$=4y\sqrt{81x^{16}y^8}\sqrt{3xy}$
$=4y\cdot 9x^8y^4\sqrt{3xy}$
$=36x^8y^5\sqrt{3xy}$

23. $y\sqrt{24y^6}=y\sqrt{4y^6\cdot 6}=y\sqrt{4y^6}\sqrt{6}=2y^4\sqrt{6}$

24. $2x+4=\sqrt{x^2+3}$
$(2x+4)^2=\left(\sqrt{x^2+3}\right)^2$
$4x^2+16x+16=x^2+3$
$3x^2+16x+13=0$
$(x+1)(3x+13)=0$
$x+1=0 \qquad 3x+13=0$
$x=-1 \qquad x=-\dfrac{13}{3}$

Check:

$2x+4=\sqrt{x^2+3}$	
$2(-1)+4$	$\sqrt{(-1)^2+3}$
$-2+4$	$\sqrt{4}$

$2=2$

Check:

$2x+4=\sqrt{x^2+3}$	
$2\left(-\dfrac{13}{3}\right)+4$	$\sqrt{\left(-\dfrac{13}{3}\right)^2+3}$
$-\dfrac{26}{3}+\dfrac{12}{3}$	$\sqrt{\dfrac{169}{9}+\dfrac{27}{9}}$
$-\dfrac{14}{3}$	$\sqrt{\dfrac{196}{3}}$
$-\dfrac{14}{3}$	$\dfrac{14}{3}$

$-\dfrac{14}{3}\neq\dfrac{14}{3}$

The solution is –1.

25. $2x^2\sqrt{18x^2y^5}+6y\sqrt{2x^6y^3}$
$-9xy^2\sqrt{8x^4y}$
$=2x^2\sqrt{9x^2y^4\cdot 2y}+6y\sqrt{x^6y^2\cdot 2y}$
$-9xy^2\sqrt{4x^4\cdot 2y}$
$=2x^2\sqrt{9x^2y^4}\sqrt{2y}+6y\sqrt{x^6y^2}\sqrt{2y}$
$-9xy^2\sqrt{4x^4}\sqrt{2y}$
$=2x^2\cdot 3xy^2\sqrt{2y}+6y\cdot x^3y\sqrt{2y}$
$-9xy^2\cdot 2x^2\sqrt{2y}$
$=6x^3y^2\sqrt{2y}+6x^3y^2\sqrt{2y}$
$-18x^3y^2\sqrt{2y}$
$=-6x^3y^2\sqrt{2y}$

26. $\frac{16}{\sqrt{a}}=\frac{16}{\sqrt{a}}\cdot\frac{\sqrt{a}}{\sqrt{a}}=\frac{16\sqrt{a}}{a}$

27. Strategy:
To find the distance, d, across the pond, use the equation $c=\sqrt{a^2+b^2}$ where $c=d$, $a=25$, and $b=35$.

Solution:

$c=\sqrt{a^2+b^2}$

$d=\sqrt{25^2+35^2}$

$d=\sqrt{625+1225}$

$d=\sqrt{1850}$

$d\approx 43$

The distance across the pond is approximately 43 ft.

28. Strategy:
To find the weight, use the formula

$d=4000\sqrt{\frac{W_0}{W_d}}-4000$ where

$d=4000$ and $W_d=36$.

Solution:

$d=4000\sqrt{\frac{W_0}{W_d}}-4000$

$4000=4000\sqrt{\frac{W_0}{36}}-4000$

$8000=4000\sqrt{\frac{W_0}{36}}$

$2=\sqrt{\frac{W_0}{36}}$

$2^2=\left(\sqrt{\frac{W_0}{36}}\right)^2$

$4=\frac{W_0}{36}$

$144=W_0$

The explorer weighs 144 lb on the surface of Earth.

29. Strategy:
To find the depth, use the formula $V=3\sqrt{d}$ where $V=30$.

Solution:

$V=3\sqrt{d}$

$30=3\sqrt{d}$

$10=\sqrt{d}$

$10^2=\left(\sqrt{d}\right)^2$

$100=d$

The depth of the water is 100 ft.

30. Strategy:
To find the radius, use the formula $v=4\sqrt{r}$ where $v=25$.

Solution:

$v=4\sqrt{4}$

$20=4\sqrt{r}$

$5=\sqrt{r}$

$5^2=\left(\sqrt{r}\right)^2$

$25=r$

The radius of the sharpest corner is 25 ft.

Chapter Test

1. $\sqrt{121x^8y^2}=11x^4y$

2. $\sqrt{3x^2y}\sqrt{6xy^2}\sqrt{2x}=\sqrt{36x^4y^3}$
$=\sqrt{36x^4y^2}\sqrt{y}$
$=6x^2y\sqrt{y}$

3. $5\sqrt{8}-3\sqrt{50}=5\sqrt{4}\sqrt{2}-3\sqrt{25}\sqrt{2}$
$=5\cdot 2\sqrt{2}-3\cdot 5\sqrt{2}$
$=10\sqrt{2}-15\sqrt{2}=-5\sqrt{2}$

4. $\sqrt{45}=\sqrt{9}\sqrt{5}=3\sqrt{5}$

5. $\frac{\sqrt{162}}{\sqrt{2}}=\sqrt{\frac{162}{2}}=\sqrt{81}=9$

6. $\sqrt{9x}+3=18$

$\sqrt{9x}=15$

$\left(\sqrt{9x}\right)^2=15^2$

$9x=225$

$x=25$

The solution is 25.

Check:

$\sqrt{9x}+3=18$	
$\sqrt{9\cdot 25}+3$	18
$\sqrt{225}+3$	18
$15+3$	18

$18=18$

7. $\sqrt{32a^5b^{11}}=\sqrt{16a^4b^{10}}\sqrt{2ab}=4a^2b^5\sqrt{2ab}$

8. $\dfrac{\sqrt{98a^6b^4}}{\sqrt{2a^3b^2}}=\sqrt{\dfrac{98a^6b^4}{2a^3b^2}}=\sqrt{49a^3b^2}$

$=\sqrt{49a^2b^2}\sqrt{a}=7ab\sqrt{a}$

9. $\dfrac{2}{\sqrt{3}-1}=\dfrac{2}{\sqrt{3}-1}\cdot\dfrac{\sqrt{3}+1}{\sqrt{3}+1}$

$=\dfrac{2\left(\sqrt{3}+1\right)}{3-1}=\dfrac{2\left(\sqrt{3}+1\right)}{2}=\sqrt{3}+1$

10. $\sqrt{8x^3y}\sqrt{10xy^4}=\sqrt{80x^4y^5}=\sqrt{16x^4y^4}\sqrt{5y}$

$=4x^2y^2\sqrt{5y}$

11. $\sqrt{x-5}+\sqrt{x}=5$

$\sqrt{x-5}=5-\sqrt{x}$

$\left(\sqrt{x-5}\right)^2=\left(5-\sqrt{x}\right)^2$

$x-5=25-10\sqrt{x}+x$

$-30=-10\sqrt{x}$

$3=\sqrt{x}$

$3^2=\left(\sqrt{x}\right)^2$

$9=x$

Check:

$\sqrt{x-5}+\sqrt{x}=5$	
$\sqrt{9-5}+\sqrt{9}$	5
$\sqrt{4}+3$	5
$2+3$	5

$5=5$

The solution is 9.

12. $3\sqrt{8y}-2\sqrt{72x}+5\sqrt{18y}$

$=3\sqrt{4}\sqrt{2y}-2\sqrt{36}\sqrt{2x}+5\sqrt{9}\sqrt{2y}$

$=3\cdot 2\sqrt{2y}-2\cdot 6\sqrt{2x}+5\cdot 3\sqrt{2y}$

$=6\sqrt{2y}-12\sqrt{2x}+15\sqrt{2y}$

$=21\sqrt{2y}-12\sqrt{2x}$

13. $\sqrt{72x^7y^2}=\sqrt{36x^6y^2}\sqrt{2x}=6x^3y\sqrt{2x}$

14. $\left(\sqrt{y}-3\right)\left(\sqrt{y}+5\right)=y+2\sqrt{y}-15$

15. $2x\sqrt{3xy^3}-2y\sqrt{12x^3y}-3xy\sqrt{xy}$

$=2x\sqrt{y^2}\sqrt{3xy}-2y\sqrt{4x^2}\sqrt{3xy}-3xy\sqrt{xy}$

$=2xy\sqrt{3xy}-4xy\sqrt{3xy}-3xy\sqrt{xy}$

$=-2xy\sqrt{3xy}-3xy\sqrt{xy}$

16. $\dfrac{2-\sqrt{5}}{6+\sqrt{5}}=\dfrac{2-\sqrt{5}}{6+\sqrt{5}}\cdot\dfrac{6-\sqrt{5}}{6-\sqrt{5}}=\dfrac{12-8\sqrt{5}+5}{36-5}$

$=\dfrac{17-8\sqrt{5}}{31}$

17. $\sqrt{a}\left(\sqrt{a}-\sqrt{b}\right)=a-\sqrt{ab}$

18. $\sqrt{75}=\sqrt{25}\sqrt{3}=5\sqrt{3}$

19. Strategy:
To find the length, L, of the pendulum, use the equation $T=2\pi\sqrt{\dfrac{L}{32}}$, where $T=3$, and use $\pi=3.14$.

Solution:

$T=2\pi\sqrt{\dfrac{L}{32}}$

$3=2\pi\sqrt{\dfrac{L}{32}}$

$\dfrac{3}{2\pi}=\sqrt{\dfrac{L}{32}}$

$\left(\dfrac{3}{2\pi}\right)^2=\sqrt{\dfrac{L}{32}}$

$\left(\dfrac{3}{2\pi}\right)^2=\left(\sqrt{\dfrac{L}{32}}\right)^2$

$\dfrac{9}{4\pi^2}=\dfrac{L}{32}$

$32\cdot\dfrac{9}{4\pi^2}=L$

$\dfrac{72}{\pi^2}=L$

$7.30\approx L$

The length of the pendulum is 7.30 ft.

20. Strategy:
To find the distance, x, from the base of the pole, use the equation $a=\sqrt{c^2-b^2}$, where $a = x$, $c = 8$, and $b = 4$.

Solution:

$$a=\sqrt{c^2-b^2}$$
$$x=\sqrt{8^2-4^2}$$
$$x=\sqrt{64-16}$$
$$x=\sqrt{48}$$
$$x\approx 6.93$$

The rope should be secured about 7 ft from the base of the pole.

Cumulative Review Exercises

1.
$$\left(\frac{2}{3}\right)^2\cdot\left(\frac{3}{4}-\frac{3}{2}\right)+\left(\frac{1}{2}\right)^2=\frac{4}{9}\left(\frac{3}{4}-\frac{6}{4}\right)+\frac{1}{4}$$
$$=\frac{4}{9}\left(-\frac{3}{4}\right)+\frac{1}{4}$$
$$=-\frac{1}{3}+\frac{1}{4}$$
$$=-\frac{4}{12}+\frac{3}{12}=-\frac{1}{12}$$

2.
$$-3[x-2(3-2x)-5x]+2x$$
$$=-3[x-6+4x-5x]+2x$$
$$=-3[-6]+2x$$
$$=2x+18$$

3.
$$2x-4[3x-2(1-3x)]=2(3-4x)$$
$$2x-4[3x-2+6x]=6-8x$$
$$2x-4[9x-2]=6-8x$$
$$2x-36x+8=6-8x$$
$$-34x+8=6-8x$$
$$-26x=-2$$
$$x=\frac{1}{13}$$

The solution is $\frac{1}{13}$.

4.
$$(-3x^2y)(-2x^3y^4)=6x^5y^5$$

5.
$$\frac{12b^4-6b^2+2}{-6b^2}=\frac{12b^4}{-6b^2}-\frac{6b^2}{-6b^2}+\frac{2}{-6b^2}$$
$$=-2b^2+1-\frac{1}{3b^2}$$

6.
$$f(x)=\frac{2x}{x-3}$$
$$f(-3)=\frac{2(-3)}{-3-3}=\frac{-6}{-6}=1$$

7.
$$2b^3-16a^2+30a$$
The GCF is $2a$:
$$2a(a^2-8a+15)$$
$$2a(a-5)(a-3)$$

8.
$$\frac{3x^3-6x^2}{4x^2+4x}\cdot\frac{3x-9}{9x^3-45x^2+54x}$$
$$=\frac{3x^2(x-2)}{4x(x+1)}\cdot\frac{3(x-3)}{9x(x^2-5x+6)}$$
$$=\frac{3x^2\overset{1}{\cancel{(x-2)}}}{4x(x+1)}\cdot\frac{3\overset{1}{\cancel{(x-3)}}}{9x\underset{1}{\cancel{(x-3)}}\underset{1}{\cancel{(x-2)}}}$$
$$=\frac{1}{4(x+1)}$$

9.
$$\frac{x+2}{x-4}-\frac{6}{(x-4)(x-3)}$$
$$\text{LCM}=(x-4)(x-3)$$
$$=\frac{x+2}{x-4}\cdot\frac{(x-3)}{(x-3)}-\frac{6}{(x-4)(x-3)}$$
$$=\frac{x^2-x-6}{(x-4)(x-3)}+\frac{-6}{(x-4)(x-3)}$$
$$=\frac{x^2-x-12}{(x-4)(x-3)}$$
$$=\frac{\overset{1}{\cancel{(x-4)}}(x+3)}{\underset{1}{\cancel{(x-4)}}(x-3)}=\frac{x+3}{x-3}$$

10.
$$\frac{x}{2x-5}-2=\frac{3x}{2x-5}$$
$$(2x-5)\left(\frac{x}{2x-5}-2\right)=(2x-5)\left(\frac{3x}{2x-5}\right)$$
$$x-2(2x-5)=3x$$
$$x-4x+10=3x$$
$$-3x+10=3x$$
$$10=6x$$
$$\frac{5}{3}=x$$

The solution is $\frac{5}{3}$.

11. $m = \frac{1}{2}, \quad (x_1, y_1) = (-2, -3)$

$$y - y_1 = m(x - x_1)$$
$$y - (-3) = \frac{1}{2}(x - (-2))$$
$$y + 3 = \frac{1}{2}(x + 2)$$
$$y + 3 = \frac{1}{2}x + 1$$
$$y = \frac{1}{2}x - 2$$

The equation is $y = \frac{1}{2}x - 2$.

12. (1) $4x - 3y = 1$

(2) $2x + y = 3$

Solve Equation (2) for y.

$$2x + y = 3$$
$$y = 3 - 2x$$

Substitute $3 - 2x$, for y in Equation (1).

$$4x - 3y = 1$$
$$4x - 3(3 - 2x) = 1$$
$$4x - 9 + 6x = 1$$
$$10x - 9 = 1$$
$$10x = 10$$
$$x = 1$$

Substitute 1 for x in Equation (2).

$$2x + y = 3$$
$$2(1) + y = 3$$
$$y = 1$$

The solution is (1, 1).

13. (1) $5x + 4y = 7$

(2) $3x - 2y = 13$

Eliminate y.

$$5x + 4y = 7 \qquad 5x + 4y = 7$$
$$2(3x - 2y) = 2(13) \qquad 6x - 4y = 26$$
$$11x = 33$$
$$x = 3$$

Substitute 3 for x in Equation (1).

$$5x + 4y = 7$$
$$5(3) + 4y = 7$$
$$15 + 4y = 7$$
$$4y = -8$$
$$y = -2$$

The solution is (3, –2).

14. $3(x - 7) \geq 5x - 12$

$$3x - 21 \geq 5x - 12$$
$$-2x \geq 9$$
$$\frac{-2x}{-2} \leq \frac{9}{-2}$$
$$x \leq -\frac{9}{2}$$

15. $\sqrt{108} = \sqrt{36}\sqrt{3} = 6\sqrt{3}$

16. $3\sqrt{32} - 2\sqrt{128}$

$$= 3\sqrt{16}\sqrt{2} - 2\sqrt{64}\sqrt{2}$$
$$= 3 \cdot 4\sqrt{2} - 2 \cdot 8\sqrt{2}$$
$$= 12\sqrt{2} - 16\sqrt{2} = -4\sqrt{2}$$

17. $2a\sqrt{2ab^3} + b\sqrt{8a^3b} - 5ab\sqrt{ab}$

$$= 2a\sqrt{b^2}\sqrt{2ab} + b\sqrt{4a^2}\sqrt{2ab} - 5ab\sqrt{ab}$$
$$= 2ab\sqrt{2ab} + 2ab\sqrt{2ab} - 5ab\sqrt{ab}$$
$$= 4ab\sqrt{2ab} - 5ab\sqrt{ab}$$

18. $\sqrt{2a^9b}\sqrt{98ab^3}\sqrt{2a} = \sqrt{392a^{11}b^4}$

$$= \sqrt{49 \cdot 4a^{10}b^4}\sqrt{2a}$$
$$= 14a^5b^2\sqrt{2a}$$

19. $\sqrt{3}\left(\sqrt{6} - \sqrt{x^2}\right) = \sqrt{18} - \sqrt{3x^2}$

$$= \sqrt{9}\sqrt{2} - \sqrt{x^2}\sqrt{3}$$
$$= 3\sqrt{2} - x\sqrt{3}$$

20. $\frac{\sqrt{320}}{\sqrt{5}} = \sqrt{\frac{320}{5}} = \sqrt{64} = 8$

21. $\frac{3}{2 - \sqrt{5}} = \frac{3}{2 - \sqrt{5}} \cdot \frac{2 + \sqrt{5}}{2 + \sqrt{5}} = \frac{3\left(2 + \sqrt{5}\right)}{4 - 5}$

$$= -3\left(2 + \sqrt{5}\right) = -6 - 3\sqrt{5}$$

22. $\sqrt{3x - 2} - 4 = 0$

$$\sqrt{3x - 2} = 4$$
$$\left(\sqrt{3x - 2}\right)^2 = 4^2$$
$$3x - 2 = 16$$
$$3x = 18$$
$$x = 6$$

Check:

$\sqrt{3x-2} - 4 = 0$	
$\sqrt{3(6) - 2} - 4$	0
$\sqrt{16} - 4$	0
$4 - 4$	0

$$0 = 0$$

The solution is 6.

23. Strategy:
To find the cost, C, use the equation $S = C + rC$, where $S = 59.40$, and $r = 20\%$.

Solution:

$$S = C + rC$$
$$59.40 = C + 0.20C$$
$$59.40 = 1.20C$$
$$\frac{59.40}{1.20} = C$$
$$49.50 = C$$

The cost of the book is \$49.50.

24. Strategy:
Rate of faster cyclist: r
Rate of slower cyclist: $r - 4$

	Rate	Time	Distance
Faster cyclist	r	2	$2r$
Slower cyclist	$r - 4$	2	$2(r - 4)$

The distance apart is 52 mi.

Solution:

$$2r + 2(r - 4) = 52$$
$$2r + 2r - 8 = 52$$
$$4r = 60$$
$$r = 15$$

The faster cyclist is going 15 mph.

25. Strategy:
First number: x
Second number: $21 - x$

The product of the two numbers	is	one hundred four

Solution:

$$x(21 - x) = 104$$
$$21x - x^2 = 104$$
$$0 = x^2 - 21x + 104$$
$$0 = (x - 8)(x - 13)$$

$x - 8 = 0 \qquad x - 13 = 0$

$x = 8 \qquad x = 13$

The numbers are 8 and 13.

26. Strategy:
Time for the large pipe: t
Time for the small pipe: $2t$

	Rate	Time	Part
Small pipe	$\frac{1}{2t}$	16	$\frac{16}{2t}$
Large pipe	$\frac{1}{t}$	16	$\frac{16}{t}$

The sum of the parts of the task completed by each pipe must equal 1.

Solution:

$$\frac{16}{2t} + \frac{16}{t} = 1$$
$$2t\left(\frac{16}{2t} + \frac{16}{t}\right) = 2t(1)$$
$$16 + 32 = 2t$$
$$48 = 2t$$
$$24 = t$$

It would take the small pipe working alone 48 h.

27.

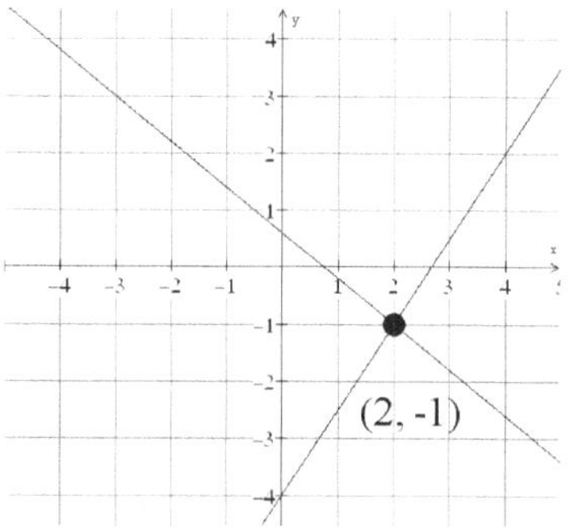

The solution is $(2, -1)$.

28. $3x + y \le 2$

$y \le -3x + 2$

Graph $y = -3x + 2$ using a solid line.
Shade below the line.

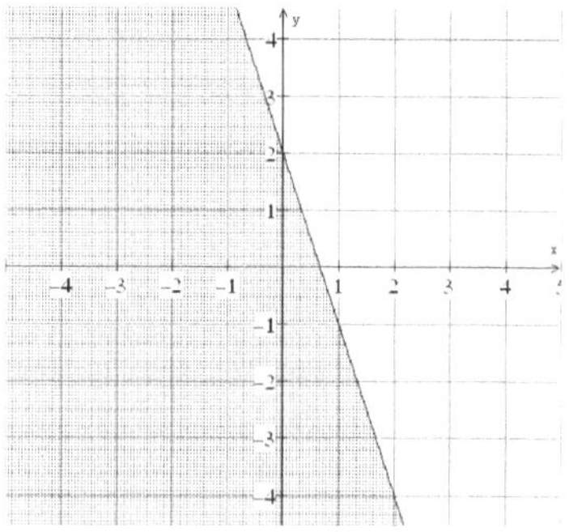

29. Strategy:
First integer: n
Second integer: $n + 1$

The square root of the sum of two consecutive integers	is equal to	9

Solution:

$$\sqrt{n+n+1} = 9$$
$$\sqrt{2n+1} = 9$$
$$\left(\sqrt{2n+1}\right)^2 = 9^2$$
$$2n+1 = 81$$
$$2n = 80$$
$$n = 40$$
$$n+1 = 41$$

The smaller integer is 40.

30. Strategy:
To find the height, d, of the building, use the equation $T = \sqrt{\frac{d}{16}}$, where $T = 5$.

Solution:

$$T = \sqrt{\frac{d}{16}}$$
$$5 = \sqrt{\frac{d}{16}}$$
$$5^2 = \left(\sqrt{\frac{d}{16}}\right)^2$$
$$25 = \frac{d}{16}$$
$$400 = d$$

The height of the building is 400 ft.

Chapter 16: Quadratic Equations

Prep Test

1. $b^2-4ac;\ a=2,\ b=-3,\ c=-4$

$$\begin{aligned} b^2-4ac &= (-3)^2-4(2)(-4) \\ &= 9+32 \\ &= 41 \end{aligned}$$

2. $$\begin{aligned} 5x+4 &= 3 \\ 5x &= -1 \\ x &= -\frac{1}{5} \end{aligned}$$

The solution is $-\frac{1}{5}$.

3. $x^2+x-12=(x+4)(x-3)$

4. $4x^2-12x+9=(2x-3)^2$

5. Yes, since $x^2-10x+25=(x-5)^2$, $x^2-10x+25$ is a perfect square trinomial.

6. $$\begin{aligned} \frac{5}{x-2} &= \frac{15}{x} \\ (x-2)(x)\left(\frac{5}{x-2}\right) &= (x-2)(x)\left(\frac{15}{x}\right) \\ 5x &= 15x-30 \\ 30 &= 10x \\ 3 &= x \end{aligned}$$

The solution is 3.

7. y-intercept = (0, b) = (0, 3)

$$m=-2=\frac{-2}{1}=\frac{\text{change in } y}{\text{change in } x}$$

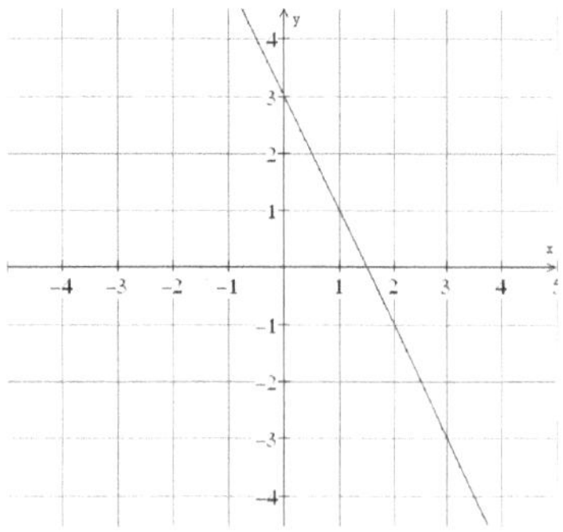

8. $\sqrt{28}=\sqrt{4\cdot 7}=\sqrt{4}\sqrt{7}=2\sqrt{7}$

9. $\sqrt{a^2}=|a|$

10. Strategy:
→Time of the first half of the hike: t

	Rate	Time	Distance
1st half of hike	4.5	t	$4.5t$
2nd half of hike	3	$2-t$	$3(2-t)$

→The distance hiked at the different rates is the same.
→After finding the time, calculate the distance.

Solution:

$$\begin{aligned} 4.5t &= 3(2-t) \\ 4.5t &= 6-3t \\ 7.5t &= 6 \\ t &= 0.8 \text{ h} \\ 4.5t &= 4.5(0.8)=3.6 \end{aligned}$$

The hiking trial is 3.6 mi long.

Section 16.1

Concept Check

1. $3x+4;\ x-7$

3. $x^2-3x-8=0$

5. $$\begin{aligned} x+5 &= x(x-3) \\ x+5 &= x^2-3x \\ 0 &= x^2-4x-5 \end{aligned}$$

Objective A Exercises

7. $(2x+5)(3x-1)=0$

$2x+5=0 \qquad 3x-1=0$

$2x=-5 \qquad 3x=1$

$x=-\frac{5}{2} \qquad x=\frac{1}{3}$

The solutions are $-\frac{5}{2}$ and $\frac{1}{3}$.

9. $x^2+2x-15=0$

$(x-3)(x+5)=0$

$x-3=0 \qquad x+5=0$

$x=3 \qquad x=-5$

The solutions are –5 and 3.

11. $z^2-4z+3=0$

$(z-3)(z-1)=0$

$z-3=0 \quad z-1=0$

$z=3 \quad z=1$

The solutions are 1 and 3.

13. $p^2+3p+2=0$

$(p+1)(p+2)=0$

$p+1=0 \quad p+2=0$

$p=-1 \quad p=-2$

The solutions are –2 and –1.

15. $x^2-6x+9=0$

$(x-3)(x-3)=0$

$x-3=0 \quad x-3=0$

$x=3 \quad x=3$

3 is a double root of the equation. The solution is 3.

17. $12y^2+8y=0$

$4y(3y+2)=0$

$4y=0 \quad 3y+2=0$

$y=0 \quad 3y=-2$

$y=-\frac{2}{3}$

The solutions are $-\frac{2}{3}$ and 0.

19. $r^2-10=3r$

$r^2-3r-10=0$

$(r-5)(r+2)=0$

$r-5=0 \quad r+2=0$

$r=5 \quad r=-2$

The solutions are –2 and 5.

21. $3v^2-5v+2=0$

$(3v-2)(v-1)=0$

$3v-2=0 \quad v-1=0$

$3v=2 \quad v=1$

$v=\frac{2}{3}$

The solutions are $\frac{2}{3}$ and 1.

23. $3s^2+8s=3$

$3s^2+8s-3=0$

$(3s-1)(s+3)=0$

$3s-1=0 \quad s+3=0$

$3s=1 \quad s=-3$

$s=\frac{1}{3}$

The solutions are –3 and $\frac{1}{3}$.

25. $\frac{3}{4}z^2-z=-\frac{1}{3}$

$\frac{3}{4}z^2-z+\frac{1}{3}=0$

$12\left(\frac{3}{4}z^2-z+\frac{1}{3}\right)=12(0)$

$9z^2-12z+4=0$

$(3z-2)(3z-2)=0$

$3z-2=0 \quad 3z-2=0$

$3z=2 \quad 3z=2$

$z=\frac{2}{3} \quad z=\frac{2}{3}$

$\frac{2}{3}$ is a double root of the equation. The solution is $\frac{2}{3}$.

27. $4t^2=4t+3$

$4t^2-4t-3=0$

$(2t+1)(2t-3)=0$

$2t+1=0 \quad 2t-3=0$

$2t=-1 \quad 2t=3$

$t=-\frac{1}{2} \quad t=\frac{3}{2}$

The solutions are $-\frac{1}{2}$ and $\frac{3}{2}$.

29. $4v^2-4v+1=0$

$(2v-1)(2v-1)=0$

$2v-1=0 \quad 2v-1=0$

$2v=1 \quad 2v=1$

$v=\frac{1}{2} \quad v=\frac{1}{2}$

$\frac{1}{2}$ is a double root of the equation. The solution is $\frac{1}{2}$.

31.
$$x^2 - 9 = 0$$
$$(x-3)(x+3) = 0$$
$$x-3=0 \qquad x+3=0$$
$$x=3 \qquad x=-3$$
The solutions are –3 and 3.

33.
$$4y^2 - 1 = 0$$
$$(2y-1)(2y+1) = 0$$
$$2y-1=0 \qquad 2y+1=0$$
$$2y=1 \qquad 2y=-1$$
$$y=\frac{1}{2} \qquad y=-\frac{1}{2}$$
The solutions are $-\frac{1}{2}$ and $\frac{1}{2}$.

35.
$$x+15 = x(x-1)$$
$$x+15 = x^2 - x$$
$$0 = x^2 - 2x - 15$$
$$0 = (x+3)(x-5)$$
$$x+3=0 \qquad x-5=0$$
$$x=-3 \qquad x=5$$
The solutions are –3 and 5.

37.
$$r^2 - r - 2 = (2r-1)(r-3)$$
$$r^2 - r - 2 = 2r^2 - 7r + 3$$
$$0 = r^2 - 6r + 5$$
$$0 = (r-1)(r-5)$$
$$r-1=0 \qquad r-5=0$$
$$r=1 \qquad r=5$$
The solutions are 1 and 5.

39. **(i)** No
$$x^2 - a^2 = 0$$
$$x^2 = a^2$$
$$\sqrt{x^2} = \sqrt{a^2}$$
$$x = \pm a$$

(ii) No
$$x^2 + 2ax + a^2 = 0$$
$$(x+a)(x+a) = 0$$
$$x+a=0 \qquad x+a=0$$
$$x=-a \qquad x=-a$$
Since a is positive, $-a$ is a negative root.

(iii) Yes
$$x^2 - 2ax + a^2 = 0$$
$$(x-a)(x-a) = 0$$
$$x-a=0 \qquad x-a=0$$
$$x=a \qquad x=a$$
a is a positive double root.

Objective B Exercises

41.
$$y^2 = 49$$
$$\sqrt{y^2} = \pm\sqrt{49}$$
$$y = \pm 7$$
The solutions are –7 and 7.

43.
$$z^2 - 64 = 0$$
$$z^2 = 64$$
$$\sqrt{z^2} = \pm\sqrt{64}$$
$$z = \pm 8$$
The solutions are –8 and 8.

45.
$$9w^2 - 64 = 0$$
$$9w^2 = 64$$
$$w^2 = \frac{64}{9}$$
$$\sqrt{w^2} = \pm\sqrt{\frac{64}{9}}$$
$$w = \pm\frac{8}{3}$$
The solutions are $-\frac{8}{3}$ and $\frac{8}{3}$.

47.

$$4z^2 = 25$$
$$z^2 = \frac{25}{4}$$
$$\sqrt{z^2} = \pm\sqrt{\frac{25}{4}}$$
$$z = \pm\frac{5}{2}$$

The solutions are $-\frac{5}{2}$ and $\frac{5}{2}$.

49.

$$25x^2 - 64 = 0$$
$$25x^2 = 64$$
$$x^2 = \frac{64}{25}$$
$$\sqrt{x^2} = \pm\sqrt{\frac{64}{25}}$$
$$x = \pm\frac{8}{5}$$

The solutions are $-\frac{8}{5}$ and $\frac{8}{5}$.

51.

$$z^2 + 49 = 0$$
$$z^2 = -49$$
$$\sqrt{z^2} = \pm\sqrt{-49}$$

$\sqrt{-49}$ is not a real number. The equation has no real number solution.

53.

$$v^2 - 48 = 0$$
$$v^2 = 48$$
$$\sqrt{v^2} = \pm\sqrt{48}$$
$$v = \pm\sqrt{16}\sqrt{3}$$
$$v = \pm 4\sqrt{3}$$

The solutions are $-4\sqrt{3}$ and $4\sqrt{3}$.

55.

$$(y+2)^2 = 49$$
$$\sqrt{(y+2)^2} = \pm\sqrt{49}$$
$$y + 2 = \pm 7$$
$$y + 2 = 7 \qquad y + 2 = -7$$
$$y = 5 \qquad y = -9$$

The solutions are –9 and 5.

57.

$$4(z-3)^2 = 100$$
$$(z-3)^2 = 25$$
$$\sqrt{(z-3)^2} = \pm\sqrt{25}$$
$$z - 3 = \pm 5$$
$$z - 3 = 5 \qquad z - 3 = -5$$
$$z = 8 \qquad z = -2$$

The solutions are –2 and 8.

59.

$$4(y+3)^2 - 81 = 0$$
$$4(y+3)^2 = 81$$
$$(y+3)^2 = \frac{81}{4}$$
$$\sqrt{(y+3)^2} = \pm\sqrt{\frac{81}{4}}$$
$$y + 3 = \pm\frac{9}{2}$$
$$y + 3 = \frac{9}{2} \qquad y + 3 = -\frac{9}{2}$$
$$y = \frac{9}{2} - 3 \qquad y = -\frac{9}{2} - 3$$
$$y = \frac{9}{3} - \frac{6}{2} \qquad y = -\frac{9}{2} - \frac{6}{2}$$
$$y = \frac{3}{2} \qquad y = -\frac{15}{2}$$

The solutions are $-\frac{15}{2}$ and $\frac{3}{2}$.

61.

$$81(y-2)^2 - 64 = 0$$
$$81(y-2)^2 = 64$$
$$(y-2)^2 = \frac{64}{81}$$
$$\sqrt{(y-2)^2} = \pm\sqrt{\frac{64}{81}}$$
$$y - 2 = \pm\frac{8}{9}$$
$$y - 2 = \frac{8}{9} \qquad y - 2 = -\frac{8}{9}$$
$$y = \frac{8}{9} + 2 \qquad y = -\frac{8}{9} + 2$$
$$y = \frac{8}{9} + \frac{18}{9} \qquad y = -\frac{8}{9} + \frac{18}{9}$$
$$y = \frac{26}{9} \qquad y = \frac{10}{9}$$

The solutions are $\frac{10}{9}$ and $\frac{26}{9}$.

63.

$$(y+5)^2-50=0$$
$$(y+5)^2=50$$
$$\sqrt{(y+5)^2}=\pm\sqrt{50}$$
$$y+5=\pm\sqrt{25}\sqrt{2}$$
$$y+5=\pm5\sqrt{2}$$
$$y+5=5\sqrt{2} \qquad y+5=-5\sqrt{2}$$
$$y=-5+5\sqrt{2} \qquad y=-5-5\sqrt{2}$$

The solutions are $-5-5\sqrt{2}$ and $-5+5\sqrt{2}$.

65.

$$2\left(z-\frac{1}{2}\right)^2=12$$
$$\left(z-\frac{1}{2}\right)^2=6$$
$$\sqrt{\left(z-\frac{1}{2}\right)^2}=\pm\sqrt{6}$$
$$z-\frac{1}{2}=\pm\sqrt{6}$$
$$z-\frac{1}{2}=\sqrt{6} \qquad z-\frac{1}{2}=-\sqrt{6}$$
$$z=\frac{1}{2}+\sqrt{6} \qquad z=\frac{1}{2}-\sqrt{6}$$

The solutions are $\frac{1}{2}-\sqrt{6}$ and $\frac{1}{2}+\sqrt{6}$.

67.

$$(x+a)^2=0$$
$$\sqrt{(x+a)^2}=\pm\sqrt{0}$$
$$x+a=0$$
$$x=-a$$

There is one real number solution.

69.

$$(x+a)^2=b$$
$$\sqrt{(x+a)^2}=\pm\sqrt{b}$$
$$x+a=\pm\sqrt{b}$$
$$x+a=\sqrt{b} \qquad x+a=-\sqrt{b}$$
$$x=-a+\sqrt{b} \qquad x=-a-\sqrt{b}$$

There are two real number solutions.

Critical Thinking 16.1

71.

$$n(n-2)=15$$
$$n^2-2n=15$$
$$n^2-2n-15=0$$
$$(n-5)(n+3)=0$$
$$n-5=0 \qquad n+3=0$$
$$n=5 \qquad n=-3$$

$$2n^2-7n-4 \qquad 2n^2-7n-4$$
$$2(5)^2-7(5)-4 \qquad 2(-3)^2-7(-3)-4$$
$$2(25)-35-4 \qquad 2(9)+21-4$$
$$50-35-4=11 \qquad 18+21-4=35$$

Projects and Group Activities 16.1

73. Strategy:
To find the rate, r, use the equation $P=A(1+r)^2$, where $P = 1782.15$ and $A = 1500$.

Solution:

$$P=A(1+r)^2$$
$$1782.15=1500(1+r)^2$$
$$1.1881=(1+r)^2$$
$$\pm\sqrt{1.1881}=\sqrt{(1+r)^2}$$
$$\pm1.09=1+r$$
$$1.09=1+r \qquad -1.09=1+r$$
$$0.09=r \qquad -2.09=r$$

–2.09 is not possible.
The annual percentage rate is 9%.

Section 16.2

Concept Check

1. perfect-square trinomial

3. $x^2 - 6x$

$$\left(-\frac{6}{2}\right)^2 = (-3)^2 = 9$$

$$x^2 - 6x + 9 = (x-3)^2$$

5. $x^2 - 5x$

$$\left(-\frac{5}{2}\right)^2 = \frac{25}{4}$$

$$x^2 - 5x + \frac{25}{4} = \left(x - \frac{5}{2}\right)^2$$

Objective A Exercises

7.
$$x^2 + 2x - 3 = 0$$
$$x^2 + 2x = 3$$

Complete the square. $\left(\frac{2}{2}\right)^2 = (1)^2 = 1$

$$x^2 + 2x + 1 = 3 + 1$$
$$(x+1)^2 = 4$$
$$\sqrt{(x+1)^2} = \pm\sqrt{4}$$
$$x + 1 = \pm 2$$
$$x + 1 = 2 \qquad x + 1 = -2$$
$$x = 1 \qquad x = -3$$

The solutions are –3 and 1.

9.
$$z^2 - 6z - 16 = 0$$
$$z^2 - 6z = 16$$

Complete the square. $\left(-\frac{6}{2}\right)^2 = (-3)^2 = 9$

$$z^2 - 6z + 9 = 16 + 9$$
$$(z-3)^2 = 25$$
$$\sqrt{(z-3)^2} = \pm\sqrt{25}$$
$$z - 3 = \pm 5$$
$$z - 3 = 5 \qquad z - 3 = -5$$
$$z = 8 \qquad z = -2$$

The solutions are –2 and 8.

11.
$$x^2 = 4x - 4$$
$$x^2 - 4x = -4$$

Complete the square. $\left(-\frac{4}{2}\right)^2 = (-2)^2 = 4$

$$x^2 - 4x + 4 = -4 + 4$$
$$(x-2)^2 = 0$$
$$\sqrt{(x-2)^2} = \sqrt{0}$$
$$x - 2 = 0$$
$$x = 2$$

The solution is 2.

13.
$$v^2 - 6v + 13 = 0$$
$$v^2 - 6v = 13$$

Complete the square. $\left(-\frac{6}{2}\right)^2 = (-3)^2 = 9$

$$v^2 - 6v + 9 = -13 + 9$$
$$(v-3)^2 = -4$$
$$\sqrt{(v-3)^2} = \sqrt{-4}$$

$\sqrt{-4}$ is not a real number. The equation has no real number solution.

15.
$$y^2 + 5y + 4 = 0$$
$$y^2 + 5y = -4$$

Complete the square. $\left(\frac{5}{2}\right)^2 = \frac{25}{4}$

$$y^2 + 5y + \frac{25}{4} = -4 + \frac{25}{4}$$
$$\left(y + \frac{5}{2}\right)^2 = \frac{9}{4}$$
$$\sqrt{\left(y + \frac{5}{2}\right)^2} = \pm\sqrt{\frac{9}{4}}$$
$$y + \frac{5}{2} = \pm\frac{3}{2}$$
$$y + \frac{5}{2} = \frac{3}{2} \qquad y + \frac{5}{2} = -\frac{3}{2}$$
$$y = -1 \qquad y = -4$$

The solutions are –4 and –1.

17. $w^2+7w=8$

Complete the square. $\left(\frac{7}{2}\right)^2=\frac{49}{4}$

$$w^2+7w+\frac{49}{4}=8+\frac{49}{4}$$
$$\left(w+\frac{7}{2}\right)^2=\frac{81}{4}$$
$$\sqrt{\left(w+\frac{7}{2}\right)^2}=\pm\sqrt{\frac{81}{4}}$$
$$w+\frac{7}{2}=\pm\frac{9}{2}$$
$$w+\frac{7}{2}=\frac{9}{2} \qquad w+\frac{7}{2}=-\frac{9}{2}$$
$$w=1 \qquad w=-8$$

The solutions are –8 and 1.

19. $v^2+4v+1=0$

$v^2+4v=-1$

Complete the square. $\left(\frac{4}{2}\right)^2=(2)^2=4$

$$v^2+4v+4=-1+4$$
$$(v+2)^2=3$$
$$\sqrt{(v+2)^2}=\pm\sqrt{3}$$
$$v+2=\pm\sqrt{3}$$
$$v+2=\sqrt{3} \qquad v+2=-\sqrt{3}$$
$$v=-2+\sqrt{3} \qquad v=-2-\sqrt{3}$$

The solutions are $-2-\sqrt{3}$ and $-2+\sqrt{3}$.

21. $x^2+6x=5$

Complete the square. $\left(\frac{6}{2}\right)^2=(3)^2=9$

$$x^2+6x+9=5+9$$
$$(x+3)^2=14$$
$$\sqrt{(x+3)^2}=\pm\sqrt{14}$$
$$x+3=\pm\sqrt{14}$$
$$x+3=\sqrt{14} \qquad x+3=-\sqrt{14}$$
$$x=-3+\sqrt{14} \qquad x=-3-\sqrt{14}$$

The solutions are $-3-\sqrt{14}$ and $-3+\sqrt{14}$.

23. $\frac{z^2}{2}=z+\frac{1}{2}$

$$2\left(\frac{z^2}{2}\right)=2\left(z+\frac{1}{2}\right)$$
$$z^2=2z+1$$
$$z^2-2z=1$$

Complete the square. $\left(\frac{2}{2}\right)^2=(1)^2=1$

$$z^2-2z+1=1+1$$
$$(z-1)^2=2$$
$$\sqrt{(z-1)^2}=\pm\sqrt{2}$$
$$z-1=\pm\sqrt{2}$$
$$z-1=\sqrt{2} \qquad z-1=-\sqrt{2}$$
$$z=1+\sqrt{2} \qquad z=1-\sqrt{2}$$

The solutions are $1-\sqrt{2}$ and $1+\sqrt{2}$.

25. $p^2+3p=1$

Complete the square. $\left(\frac{3}{2}\right)^2=\frac{9}{4}$

$$p^2+3p+\frac{9}{4}=1+\frac{9}{4}$$
$$\left(p+\frac{3}{2}\right)^2=\frac{13}{4}$$
$$\sqrt{\left(p+\frac{3}{2}\right)^2}=\pm\sqrt{\frac{13}{4}}$$
$$p+\frac{3}{2}=\pm\frac{\sqrt{13}}{2}$$
$$p+\frac{3}{2}=\frac{\sqrt{13}}{2} \qquad p+\frac{3}{2}=-\frac{\sqrt{13}}{2}$$
$$p=-\frac{3}{2}+\frac{\sqrt{13}}{2} \qquad p=-\frac{3}{2}-\frac{\sqrt{13}}{2}$$
$$p=\frac{-3+\sqrt{13}}{2} \qquad p=\frac{-3-\sqrt{13}}{2}$$

The solutions are $\frac{-3-\sqrt{13}}{2}$ and $\frac{-3+\sqrt{13}}{2}$.

27. $t^2 - 3t = -2$

Complete the square. $\left(\frac{3}{2}\right)^2 = \frac{9}{4}$

$$t^2 - 3t + \frac{9}{4} = -2 + \frac{9}{4}$$
$$\left(t - \frac{3}{2}\right)^2 = \frac{1}{4}$$
$$\sqrt{\left(t - \frac{3}{2}\right)^2} = \pm\sqrt{\frac{1}{4}}$$
$$t - \frac{3}{2} = \pm\frac{1}{2}$$
$$t - \frac{3}{2} = \frac{1}{2} \qquad t - \frac{3}{2} = -\frac{1}{2}$$
$$t = 2 \qquad t = 1$$

The solutions are 2 and 1.

29. $v^2 + v - 3 = 0$

$v^2 + v = 3$

Complete the square. $\left(\frac{1}{2}\right)^2 = \frac{1}{4}$

$$v^2 + v + \frac{1}{4} = 3 + \frac{1}{4}$$
$$\left(v + \frac{1}{2}\right)^2 = \frac{13}{4}$$
$$\sqrt{\left(v + \frac{1}{2}\right)^2} = \pm\sqrt{\frac{13}{4}}$$
$$v + \frac{1}{2} = \pm\frac{\sqrt{13}}{2}$$
$$v + \frac{1}{2} = \frac{\sqrt{13}}{2} \qquad v + \frac{1}{2} = -\frac{\sqrt{13}}{2}$$
$$v = -\frac{1}{2} + \frac{\sqrt{13}}{2} \qquad v = -\frac{1}{2} - \frac{\sqrt{13}}{2}$$
$$v = \frac{-1 + \sqrt{13}}{2} \qquad v = \frac{-1 - \sqrt{13}}{2}$$

The solutions are $\frac{-1 - \sqrt{13}}{2}$ and $\frac{-1 + \sqrt{13}}{2}$.

31. $y^2 = 7 - 10y$

$y^2 + 10y = 7$

Complete the square. $\left(\frac{10}{2}\right)^2 = (5)^2 = 25$

$$y^2 + 10y + 25 = 7 + 25$$
$$(y + 5)^2 = 32$$
$$\sqrt{(y + 5)^2} = \pm\sqrt{32}$$
$$y + 5 = \pm\sqrt{16}\sqrt{2}$$
$$y + 5 = \pm 4\sqrt{2}$$
$$y + 5 = 4\sqrt{2} \qquad y + 5 = -4\sqrt{2}$$
$$y = -5 + 4\sqrt{2} \qquad y = -5 - \sqrt{4}\sqrt{2}$$

The solutions are $-5 - 4\sqrt{2}$ and $-5 + 4\sqrt{2}$.

33. $r^2 - 3r = 5$

Complete the square. $\left(-\frac{3}{2}\right)^2 = \frac{9}{4}$

$$r^2 - 3r + \frac{9}{4} = 5 + \frac{9}{4}$$
$$\left(r - \frac{3}{2}\right)^2 = \frac{29}{4}$$
$$\sqrt{\left(r - \frac{3}{2}\right)^2} = \sqrt{\frac{29}{4}}$$
$$r - \frac{3}{2} = \pm\frac{\sqrt{29}}{2}$$
$$r - \frac{3}{2} = \frac{\sqrt{29}}{2} \qquad r - \frac{3}{2} = -\frac{\sqrt{29}}{2}$$
$$r = \frac{3}{2} + \frac{\sqrt{29}}{2} \qquad r = \frac{3}{2} - \frac{\sqrt{29}}{2}$$
$$r = \frac{3 + \sqrt{29}}{2} \qquad r = \frac{3 - \sqrt{29}}{2}$$

The solutions are $\frac{3 - \sqrt{29}}{2}$ and $\frac{3 + \sqrt{29}}{2}$.

35. $t^2 - t = 4$

Complete the square. $\left(-\frac{1}{2}\right)^2 = \frac{1}{4}$

$$t^2 - t + \frac{1}{4} = 4 + \frac{1}{4}$$
$$\left(t - \frac{1}{2}\right)^2 = \frac{17}{4}$$
$$\sqrt{\left(t - \frac{1}{2}\right)^2} = \sqrt{\frac{17}{4}}$$
$$t - \frac{1}{2} = \pm\frac{\sqrt{17}}{2}$$
$$t - \frac{1}{2} = \frac{\sqrt{17}}{2} \qquad t - \frac{1}{2} = -\frac{\sqrt{17}}{2}$$
$$t = \frac{1}{2} + \frac{\sqrt{17}}{2} \qquad t = \frac{1}{2} - \frac{\sqrt{17}}{2}$$
$$t = \frac{1+\sqrt{17}}{2} \qquad t = \frac{1-\sqrt{17}}{2}$$

The solutions are $\frac{1-\sqrt{17}}{2}$ and $\frac{1+\sqrt{17}}{2}$.

37. $x^2 - 3x + 5 = 0$

$$x^2 - 3x = -5$$

Complete the square. $\left(-\frac{3}{2}\right)^2 = \frac{9}{4}$

$$x^2 - 3x + \frac{9}{4} = -5 + \frac{9}{4}$$
$$\left(x - \frac{3}{2}\right)^2 = -\frac{11}{4}$$
$$\sqrt{\left(x - \frac{3}{2}\right)^2} = \sqrt{-\frac{11}{4}}$$

$\sqrt{-\frac{11}{4}}$ is not a real number. The equation has no real number solution.

39. $2t^2 - 3t + 1 = 0$

$$2t^2 - 3t = -1$$
$$\frac{1}{2}\left(2t^2 - 3t\right) = \frac{1}{2}(1)$$
$$t^2 - \frac{3}{2}t = \frac{1}{2}$$

Complete the square.

$$\left[\frac{1}{2}\left(-\frac{3}{2}\right)\right]^2 = \left[-\frac{3}{4}\right]^2 = \frac{9}{16}$$
$$t^2 - \frac{3}{2}t + \frac{9}{16} = -\frac{1}{2} + \frac{9}{16}$$
$$\left(t - \frac{3}{4}\right)^2 = \frac{1}{16}$$
$$\sqrt{\left(t - \frac{3}{4}\right)^2} = \sqrt{\frac{1}{16}}$$
$$t - \frac{3}{4} = \frac{1}{4}$$
$$t - \frac{3}{4} = \frac{1}{4} \qquad t - \frac{3}{4} = -\frac{1}{4}$$
$$t = 1 \qquad t = \frac{1}{2}$$

The solutions are $\frac{1}{2}$ and 1.

41. $2r^2 + 5r = 3$

$$\frac{1}{2}\left(2r^2 + 5r\right) = \frac{1}{2}(3)$$
$$r^2 + \frac{5}{2}r = \frac{3}{2}$$

Complete the square.

$$\left[\frac{1}{2}\left(\frac{5}{2}\right)\right]^2 = \left[\frac{5}{4}\right]^2 = \frac{25}{16}$$
$$r^2 + \frac{5}{2}r + \frac{25}{16} = \frac{3}{2} + \frac{25}{16}$$
$$\left(r + \frac{5}{4}\right)^2 = \frac{49}{16}$$
$$\sqrt{\left(r + \frac{5}{4}\right)^2} = \sqrt{\frac{49}{16}}$$
$$r + \frac{5}{4} = \pm\frac{7}{4}$$
$$r + \frac{5}{4} = \frac{7}{4} \qquad r + \frac{5}{4} = -\frac{7}{4}$$
$$r = \frac{1}{2} \qquad r = -3$$

The solutions are –3 and $\frac{1}{2}$.

43.

$$2s^2 = 7s - 1$$
$$2s^2 - 7s = -1$$
$$\frac{1}{2}\left(2x^2 - 7s\right) = \frac{1}{2}(-1)$$
$$y^2 - \frac{7}{2}s = -\frac{1}{2}$$

Complete the square.

$$\left[\frac{1}{2}\left(-\frac{7}{2}\right)\right]^2 = \left[-\frac{7}{4}\right]^2 = \frac{49}{16}$$
$$s^2 - \frac{7}{2}s + \frac{49}{16} = -\frac{1}{2} + \frac{49}{16}$$
$$\left(s - \frac{7}{4}\right)^2 = \frac{41}{16}$$
$$\sqrt{\left(s - \frac{7}{4}\right)^2} = \sqrt{\frac{41}{16}}$$
$$s - \frac{7}{4} = \pm\frac{\sqrt{41}}{4}$$
$$s = \frac{7 \pm \sqrt{41}}{4}$$

The solutions are $\frac{7-\sqrt{41}}{4}$ and $\frac{7+\sqrt{41}}{4}$.

45.

$$6s^2 + s = 3$$
$$\frac{1}{6}\left(6s^2 + s\right) = \frac{1}{6}(3)$$
$$s^2 + \frac{1}{6}s = \frac{1}{2}$$

Complete the square.

$$\left[\frac{1}{2}\left(\frac{1}{6}\right)\right]^2 = \left[\frac{1}{12}\right]^2 = \frac{1}{144}$$
$$s^2 + \frac{1}{6}s + \frac{1}{144} = \frac{1}{2} + \frac{1}{144}$$
$$\left(s + \frac{1}{12}\right)^2 = \frac{73}{144}$$
$$\sqrt{\left(s + \frac{1}{12}\right)^2} = \sqrt{\frac{73}{144}}$$
$$s + \frac{1}{12} = \pm\frac{\sqrt{73}}{12}$$
$$s = \frac{-1 \pm \sqrt{73}}{12}$$

The solutions are $\frac{-1-\sqrt{73}}{12}$ and $\frac{-1+\sqrt{73}}{12}$.

47.

$$6p^2 = 5p + 4$$
$$6p^2 - 5p = 4$$
$$\frac{1}{6}\left(6p^2 - 5p\right) = \frac{1}{6}(4)$$
$$p^2 - \frac{5}{6}p = \frac{2}{3}$$

Complete the square.

$$\left[\frac{1}{2}\left(-\frac{5}{6}\right)\right]^2 = \left[-\frac{5}{12}\right]^2 = \frac{25}{144}$$
$$p^2 - \frac{5}{6}p + \frac{25}{144} = \frac{2}{3} + \frac{25}{144}$$
$$\left(p - \frac{5}{12}\right)^2 = \frac{121}{144}$$
$$\sqrt{\left(p - \frac{5}{12}\right)^2} = \sqrt{\frac{121}{144}}$$
$$p - \frac{5}{12} = \pm\frac{11}{12}$$
$$p - \frac{5}{12} = \frac{11}{12} \qquad p - \frac{5}{12} = -\frac{11}{12}$$
$$p = \frac{4}{3} \qquad p = -\frac{1}{2}$$

The solutions are $-\frac{1}{2}$ and $\frac{4}{3}$.

49. $2+\sqrt{7}$ will be a positive number.

$\sqrt{7} > \sqrt{4}$

$\sqrt{7} > 2$ so $2-\sqrt{7}$

will be a negative number.

There will be one negative answer and one positive answer.

51. $w^2+5w=2$

Complete the square. $\left(\frac{5}{2}\right)^2=\frac{25}{4}$

$$w^2+5y+\frac{25}{4}=2+\frac{25}{4}$$
$$\left(w+\frac{5}{2}\right)^2=\frac{33}{4}$$
$$\sqrt{\left(w+\frac{5}{2}\right)^2}=\sqrt{\frac{33}{4}}$$
$$w+\frac{5}{2}=\pm\frac{\sqrt{33}}{2}$$

$$w+\frac{5}{2}=\frac{\sqrt{33}}{2} \qquad w+\frac{5}{2}=-\frac{\sqrt{33}}{2}$$
$$w=-\frac{5}{2}+\frac{\sqrt{33}}{2} \qquad w=-\frac{5}{2}-\frac{\sqrt{33}}{2}$$
$$w=\frac{-5+\sqrt{33}}{2} \qquad w=\frac{-5-\sqrt{33}}{2}$$
$$w\approx 0.372 \qquad w\approx -5.372$$

The solutions are approximately –5.372 and 0.372.

53. $2x^2+3x=11$

$$\frac{1}{2}(2x^2+3x)=\frac{1}{2}(11)$$
$$x^2+\frac{3}{2}x=\frac{11}{2}$$

Complete the square.

$$\left[\frac{1}{2}\left(\frac{3}{2}\right)\right]^2=\left[\frac{3}{4}\right]^2=\frac{9}{16}$$
$$x^2+\frac{3}{2}z+\frac{9}{16}=\frac{11}{2}+\frac{9}{16}$$
$$\left(x+\frac{3}{4}\right)^2=\frac{97}{16}$$
$$\sqrt{\left(x+\frac{3}{4}\right)^2}=\sqrt{\frac{97}{16}}$$
$$x+\frac{3}{4}=\pm\frac{\sqrt{97}}{4}$$

$$x+\frac{3}{4}=\frac{\sqrt{97}}{4} \qquad x+\frac{3}{4}=-\frac{\sqrt{97}}{4}$$
$$x=-\frac{3}{4}+\frac{\sqrt{97}}{4} \qquad x=-\frac{3}{4}-\frac{\sqrt{97}}{4}$$
$$x=\frac{-3+\sqrt{97}}{4} \qquad x=\frac{-3-\sqrt{97}}{4}$$
$$x\approx 1.712 \qquad x\approx -3.212$$

The solutions are approximately –3.212 and 1.712.

Critical Thinking 16.2

55. $x^2+8x-14=0$

$$x^2+8x=14$$

Complete the square. $\left(\frac{8}{2}\right)^2=(4)^2=16$

$$x^2+8x+16=14+16$$
$$(x+4)^2=30$$
$$\sqrt{(x+4)^2}=\pm\sqrt{30}$$
$$x+4=\pm\sqrt{30}$$
$$x=-4\pm\sqrt{30}$$

The solutions are $-4+\sqrt{30}$ and $-4-\sqrt{30}$.

57. $\frac{x+1}{2}+\frac{3}{x-1}=4$

$$2(x-1)\left[\frac{x+1}{2}+\frac{3}{x-1}\right]=2(x-1)[4]$$
$$x^2-1+6=8x-8$$
$$x^2-8x=-13$$

Complete the square.

$$\left(-\frac{8}{2}\right)^2=(-4)^2=16$$
$$x^2-8x+16=-13+16$$
$$(x-4)^2=3$$
$$\sqrt{(x-4)^2}=\sqrt{3}$$
$$x-4=\pm\sqrt{3}$$

$$x-4=\sqrt{3} \qquad x-4=-\sqrt{3}$$
$$x=4+\sqrt{3} \qquad x=4-\sqrt{3}$$

The solutions are approximately $4-\sqrt{3}$ and $4+\sqrt{3}$.

Projects and Group Activities 16.2

59. The square of any real number is positive or zero. Therefore, the expression $(x-2)^2$ is greater than or equal to zero; it cannot be equal to a negative number. The equation $(x-2)^2=-4$ does not have a real number solution.

61. Strategy:

Use the formula $h = -16t^2 + 76t + 5$, where $h = 0$.

Solution:

$$h = -16t^2 + 76t + 5$$
$$0 = -16t^2 + 76t + 5$$
$$16t^2 - 76t = 5$$
$$\frac{1}{16}\left(16t^2 - 76t\right) = \frac{1}{16}(5)$$
$$t^2 - \frac{19}{4}t = \frac{5}{16}$$

Complete the square.

$$\left[\frac{1}{2}\left(\frac{19}{4}\right)\right]^2 = \left[\frac{19}{8}\right]^2 = \frac{361}{64}$$
$$t^2 - \frac{19}{4}t + \frac{361}{64} = \frac{5}{16} + \frac{361}{64}$$
$$\left(t - \frac{19}{8}\right)^2 = \frac{381}{64}$$
$$\sqrt{\left(t - \frac{19}{8}\right)^2} = \sqrt{\frac{381}{64}}$$
$$t - \frac{19}{8} = \frac{\sqrt{381}}{8}$$

$$t - \frac{19}{8} = \frac{\sqrt{381}}{8} \qquad t - \frac{19}{8} = -\frac{\sqrt{381}}{8}$$
$$t = \frac{19}{8} + \frac{\sqrt{381}}{8} \qquad t = \frac{19}{8} - \frac{\sqrt{381}}{8}$$
$$t = \frac{19 + \sqrt{381}}{8} \qquad t = \frac{19 - \sqrt{381}}{8}$$
$$t \approx 4.81 \qquad t \approx -0.065$$

-0.065 cannot be a solution.

The ball hits the ground approximately 4.81 s after it is hit.

Section 16.3

Concept Check

1. The solutions are $\frac{1+\sqrt{13}}{2}$ and $\frac{1-\sqrt{13}}{2}$.

3. The quadratic formula can be used to solve any quadratic equation in the form $ax^2 + bx + c = 0$ where $a \neq 0$.

Objective A Exercises

5. $x^2 - 4x - 5 = 0$

$a = 1,\ b = -4,\ c = -5$

$$x = \frac{-(-4) \pm \sqrt{(-4)^2 - 4(1)(-5)}}{2(1)} = \frac{4 \pm \sqrt{16+20}}{2}$$
$$= \frac{4 \pm \sqrt{36}}{2} = \frac{4 \pm 6}{2}$$
$$x = \frac{4+6}{2} \qquad x = \frac{4-6}{2}$$
$$x = \frac{10}{2} = 5 \qquad x = \frac{-2}{2} = -1$$

The solutions are -1 and 5.

7. $y^2 = 2y + 3$

$y^2 - 2y - 3 = 0$

$a = 1,\ b = -2,\ c = -3$

$$y = \frac{-(-2) \pm \sqrt{(-2)^2 - 4(1)(-3)}}{2(1)} = \frac{2 \pm \sqrt{4+12}}{2}$$

$$= \frac{2 \pm \sqrt{16}}{2} = \frac{2 \pm 4}{2}$$

$$y = \frac{2+4}{2} \qquad y = \frac{2-4}{2}$$

$$y = \frac{6}{2} = 3 \qquad y = \frac{-2}{2} = -1$$

The solutions are –1 and 3.

9. $2y^2 - y - 1 = 0$

$a = 2,\ b = -1,\ c = -1$

$$y = \frac{-(-1) \pm \sqrt{(-1)^2 - 4(2)(-1)}}{2(2)} = \frac{1 \pm \sqrt{1+8}}{2}$$

$$= \frac{1 \pm \sqrt{9}}{2} = \frac{1 \pm 3}{2}$$

$$y = \frac{1+3}{4} \qquad y = \frac{1-3}{4}$$

$$y = \frac{4}{4} = 1 \qquad y = \frac{-2}{4} = -\frac{1}{2}$$

The solutions are $-\frac{1}{2}$ and 1.

11. $w^2 + 3w + 5 = 0$

$a = 1,\ b = 3,\ c = 5$

$$w = \frac{-(3) \pm \sqrt{(3)^2 - 4(1)(5)}}{2(1)} = \frac{-3 \pm \sqrt{9-20}}{2}$$

$$= \frac{-3 \pm \sqrt{-11}}{2}$$

$\sqrt{-11}$ is not a real number. The quadratic equation has no real number solution.

13. $4y^2 + 4y = 15$

$4y^2 + 4y - 15 = 0$

$a = 4,\ b = 4,\ c = -15$

$$y = \frac{-(4) \pm \sqrt{(4)^2 - 4(4)(-15)}}{2(4)}$$

$$= \frac{-4 \pm \sqrt{16+240}}{8}$$

$$= \frac{-4 \pm \sqrt{256}}{8} = \frac{-4 \pm 16}{8}$$

$$y = \frac{-4+16}{8} \qquad y = \frac{-4-16}{8}$$

$$y = \frac{12}{8} = \frac{3}{2} \qquad y = \frac{-20}{8} = -\frac{5}{2}$$

The solutions are $-\frac{5}{2}$ and $\frac{3}{2}$.

15. $2x^2 + x + 1 = 0$

$a = 2,\ b = 1,\ c = 1$

$$x = \frac{-(1) \pm \sqrt{(1)^2 - 4(2)(1)}}{2(2)}$$

$$= \frac{-1 \pm \sqrt{1-8}}{4} = \frac{-1 \pm \sqrt{-7}}{4}$$

$\sqrt{-7}$ is not a real number. The quadratic equation has no real number solution.

17. $\frac{1}{2}t^2 - t = \frac{5}{2}$

$$2\left(\frac{1}{2}t^2 - t\right) = \frac{1}{2}\left(\frac{5}{2}\right)$$

$t^2 - 2t = 5$

$t^2 - 2t - 5 = 0$

$a = 1,\ b = -2,\ c = -5$

$$t = \frac{-(-2) \pm \sqrt{(-2)^2 - 4(1)(-5)}}{2(1)}$$

$$= \frac{2 \pm \sqrt{4+20}}{2} = \frac{2 \pm \sqrt{24}}{2} = \frac{2 \pm \sqrt{4}\sqrt{6}}{2}$$

$$= \frac{2 \pm 2\sqrt{6}}{2} = 1 \pm \sqrt{6}$$

The solutions are $1 - \sqrt{6}$ and $1 + \sqrt{6}$.

19. $\frac{1}{3}t^2 + 2t - \frac{1}{3} = 0$

$3\left(\frac{1}{3}t^2 + 2t - \frac{1}{3}\right) = 3(0)$

$t^2 + 6t - 1 = 0$

$a = 1,\ b = 6,\ c = -1$

$$t = \frac{-(6) \pm \sqrt{(6)^2 - 4(1)(-1)}}{2(1)}$$

$$= \frac{-6 \pm \sqrt{36+4}}{2} = \frac{-6 \pm \sqrt{40}}{2} = \frac{-6 \pm \sqrt{4}\sqrt{10}}{2}$$

$$= \frac{-6 \pm 2\sqrt{10}}{2} = -3 \pm \sqrt{10}$$

The solutions are $-3 - \sqrt{10}$ and $-3 + \sqrt{10}$.

21. $w^2 = 4w + 9$

$w^2 - 4w - 9 = 0$

$a = 1,\ b = -4,\ c = -9$

$$w = \frac{-(-4) \pm \sqrt{(-4)^2 - 4(1)(-9)}}{2(1)}$$

$$= \frac{4 \pm \sqrt{16+36}}{2} = \frac{4 \pm \sqrt{52}}{2} = \frac{4 \pm \sqrt{4}\sqrt{13}}{2}$$

$$= \frac{4 \pm 2\sqrt{13}}{2} = 2 \pm \sqrt{13}$$

The solutions are $2 - \sqrt{13}$ and $2 + \sqrt{13}$.

23. $9y^2 + 6y - 1 = 0$

$a = 9,\ b = 6,\ c = -1$

$$y = \frac{-(6) \pm \sqrt{(6)^2 - 4(9)(-1)}}{2(9)}$$

$$= \frac{-6 \pm \sqrt{36+36}}{18} = \frac{-6 \pm \sqrt{72}}{18} = \frac{-6 \pm \sqrt{36}\sqrt{2}}{18}$$

$$= \frac{-6 \pm 6\sqrt{2}}{18} = \frac{-1 \pm \sqrt{2}}{3}$$

The solutions are $\frac{-1 - \sqrt{2}}{3}$ and $\frac{-1 + \sqrt{2}}{3}$.

25. $4p^2 + 4p + 1 = 0$

$a = 4,\ b = 4,\ c = 1$

$$p = \frac{-(4) \pm \sqrt{(4)^2 - 4(4)(1)}}{2(4)}$$

$$= \frac{-4 \pm \sqrt{16-16}}{8} = \frac{-4 \pm \sqrt{0}}{8} = \frac{-4}{8} = -\frac{1}{2}$$

$-\frac{1}{2}$ is a double root of the equation.

The solution is $-\frac{1}{2}$.

27. $\frac{x^2}{2} = x - \frac{5}{4}$

$4\left(\frac{x^2}{2}\right) = 4\left(x - \frac{5}{4}\right)$

$2x^2 = 4x - 5$

$2x^2 - 4x + 5 = 0$

$a = 2,\ b = -4,\ c = 5$

$$x = \frac{-(-4) \pm \sqrt{(-4)^2 - 4(2)(5)}}{2(2)}$$

$$= \frac{4 \pm \sqrt{16-40}}{4} = \frac{4 \pm \sqrt{-24}}{4}$$

$\sqrt{-24}$ is not a real number. The quadratic equation has no real number solution.

29. $4p^2 + 16p = -11$

$4p^2 + 16p + 11 = 0$

$a = 4,\ b = 16,\ c = 11$

$$p = \frac{-(16) \pm \sqrt{(16)^2 - 4(4)(11)}}{2(4)}$$

$$= \frac{-16 \pm \sqrt{256-176}}{8} = \frac{-16 \pm \sqrt{80}}{8}$$

$$= \frac{-16 \pm \sqrt{16}\sqrt{5}}{8} = \frac{-16 \pm 4\sqrt{5}}{8} = \frac{-4 \pm \sqrt{5}}{2}$$

The solutions are $\frac{-4 - \sqrt{5}}{2}$ and $\frac{-4 + \sqrt{5}}{2}$.

31. $4x^2 = 4x + 11$

$4x^2 - 4x - 11 = 0$

$a = 4,\ b = -4,\ c = -11$

$$y = \frac{-(-4) \pm \sqrt{(-4)^2 - 4(4)(-11)}}{2(4)}$$
$$= \frac{4 \pm \sqrt{16 + 176}}{8} = \frac{4 \pm \sqrt{192}}{8}$$
$$= \frac{4 \pm \sqrt{64}\sqrt{3}}{8} = \frac{4 \pm 8\sqrt{3}}{8} = \frac{1 \pm 2\sqrt{3}}{2}$$

The solutions are $\frac{1 - 2\sqrt{3}}{2}$ and $\frac{1 + 2\sqrt{3}}{2}$.

33. $9v^2 = -30v - 23$

$9v^2 + 30v + 23 = 0$

$a = 9,\ b = 30,\ c = 23$

$$v = \frac{-(30) \pm \sqrt{(30)^2 - 4(9)(23)}}{2(9)}$$
$$= \frac{-30 \pm \sqrt{900 - 828}}{18} = \frac{-30 \pm \sqrt{72}}{18}$$
$$= \frac{-30 \pm \sqrt{36}\sqrt{2}}{18} = \frac{-30 \pm 6\sqrt{2}}{18}$$
$$= \frac{-5 \pm \sqrt{2}}{3}$$

The solutions are $\frac{-5 - \sqrt{2}}{3}$ and $\frac{-5 + \sqrt{2}}{3}$.

35. True

37. $x^2 - 2x - 21 = 0$

$a = 1,\ b = -2,\ c = -21$

$$x = \frac{-(-2) \pm \sqrt{(-2)^2 - 4(1)(-21)}}{2(1)}$$
$$= \frac{2 \pm \sqrt{4 + 84}}{2} = \frac{2 \pm \sqrt{88}}{2} = \frac{2 \pm \sqrt{4}\sqrt{22}}{2}$$
$$= \frac{2 \pm 2\sqrt{22}}{2} = 1 \pm \sqrt{22}$$

$x = 1 + \sqrt{22}$ ≈ 5.690 $\quad$ $x = 1 - \sqrt{22}$ ≈ -3.690

The solutions are approximately –3.690 and 5.690 .

39. $x^2 - 6x - 13 = 0$

$a = 1,\ b = -6,\ c = -13$

$$x = \frac{-(-6) \pm \sqrt{(-6)^2 - 4(1)(-13)}}{2(1)}$$
$$= \frac{6 \pm \sqrt{36 + 52}}{2} = \frac{6 \pm \sqrt{88}}{2} = \frac{6 \pm \sqrt{4}\sqrt{22}}{2}$$
$$= \frac{6 \pm 2\sqrt{22}}{2} = 3 \pm \sqrt{22}$$

$x = 3 + \sqrt{22}$ ≈ 7.690 $\quad$ $x = 3 - \sqrt{22}$ ≈ -1.690

The solutions are approximately –1.690 and 7.690.

41. $2p^2 - 7p - 10 = 0$

$a = 2,\ b = -7,\ c = -10$

$$p = \frac{-(-7) \pm \sqrt{(-7)^2 - 4(2)(-10)}}{2(2)}$$
$$= \frac{7 \pm \sqrt{49 + 80}}{4} = \frac{7 \pm \sqrt{129}}{4}$$

$x = \frac{7 + \sqrt{129}}{4}$ ≈ 4.589 $\quad$ $p = \frac{7 - \sqrt{129}}{4}$ ≈ -1.089

The solutions are approximately –1.089 and 4.589.

43. $4z^2 + 8z - 1 = 0$

$a = 4,\ b = 8,\ c = -1$

$$z = \frac{-(8) \pm \sqrt{(8)^2 - 4(4)(-1)}}{2(4)}$$
$$= \frac{-8 \pm \sqrt{64 + 16}}{8} = \frac{-8 \pm \sqrt{80}}{8}$$
$$= \frac{-8 \pm \sqrt{16}\sqrt{5}}{8} = \frac{-8 \pm 4\sqrt{5}}{8} = \frac{-2 \pm \sqrt{5}}{2}$$

$z = \frac{-2 + \sqrt{5}}{2}$ ≈ 0.118 $\quad$ $z = \frac{-2 - \sqrt{5}}{2}$ ≈ -2.118

The solutions are approximately –2.118 and 0.118 .

45. $5v^2 - v - 5 = 0$

$a = 5,\ b = -1,\ c = -5$

$$v = \frac{-(-1) \pm \sqrt{(-1)^2 - 4(5)(-5)}}{2(5)}$$

$$= \frac{1 \pm \sqrt{1+100}}{10} = \frac{1 \pm \sqrt{101}}{10}$$

$$v = \frac{1+\sqrt{101}}{10} \approx 1.105 \qquad v = \frac{1-\sqrt{101}}{10} \approx -0.905$$

The solutions are approximately –0.905 and 1.105 .

Critical Thinking 16.3

47. Strategy:

To find the difference between the larger root and the smaller root, solve the quadratic equation and subtract the smaller root from the larger.

Solution:

$x^2 - 6x = 14$

$x^2 - 6x - 14 = 0$

$a = 1,\ b = -6,\ c = -14$

$$x = \frac{-(-6) \pm \sqrt{(-6)^2 - 4(1)(-14)}}{2(1)}$$

$$= \frac{6 \pm \sqrt{36+56}}{2}$$

$$= \frac{6 \pm \sqrt{92}}{2}$$

$$= \frac{6 \pm \sqrt{4}\sqrt{23}}{2}$$

$$= \frac{6 \pm 2\sqrt{23}}{2}$$

$$= 3 \pm \sqrt{23}$$

$$\left(3+\sqrt{23}\right) - \left(3-\sqrt{23}\right)$$

$$= 3 + \sqrt{23} - 3 + \sqrt{23}$$

$$= 2\sqrt{23}$$

The difference between the larger root and the smaller root is $2\sqrt{23}$.

49.
$$\frac{x+2}{3} - \frac{4}{x-2} = 2$$

$$3(x-2)\left(\frac{x+2}{3}\right) - 3(x-2)\left(\frac{4}{x-2}\right) = 3(x-2)2$$

$$x^2 - 4 - 12 = 6x - 12$$

$$x^2 - 6x - 4 = 0$$

$a = 1,\ b = -6,\ c = -4$

$$x = \frac{-(-6) \pm \sqrt{(-6)^2 - 4(1)(-4)}}{2(1)}$$

$$= \frac{6 \pm \sqrt{36+16}}{2}$$

$$= \frac{6 \pm \sqrt{52}}{2}$$

$$= \frac{6 \pm \sqrt{4}\sqrt{13}}{2}$$

$$= \frac{6 \pm 2\sqrt{13}}{2}$$

$$= 3 \pm \sqrt{13}$$

The solutions are $3+\sqrt{13}$ and $3-\sqrt{13}$.

51. The equation $0x^2 + 3x + 4 = 0$ cannot be solved by using the quadratic formula because $a = 0$. The formula is divided by $2a$ and $2(0) = 0$ and division by zero is not defined. The $0x^2 + 3x + 4 = 0$ is not a quadratic equation.

Projects or Group Activities 16.3

53. Strategy:

→Compare the opposite of b to the sum of the proposed solution.

→Compare c to the product of the proposed solution.

Solution:

Opposite of b: 4

Sum of the solutions: $-3 + 7 = 4$

Product of the solution: $-3(7) = -21 = c$

Yes, the solutions are –3 and 7.

55. Strategy:

→Compare the opposite of b to the sum of the proposed solution.

→Compare c to the product of the proposed solution.

Solution:

Opposite of b: 4

Sum of the solutions:

$(2+\sqrt{3})+(2-\sqrt{3})=2+2=4$

Product of the solution:

$(2+\sqrt{3})(2-\sqrt{3})=4-3=1=c$

Yes, the solutions are $2+\sqrt{3}$ and $2-\sqrt{3}$.

57. Include these ideas:

Factoring:

Write the equation in the form $ax^2+bx+c=0$. Factor the left side. Set the factors equal to zero. Solve each equation.

Advantage: Easiest if factorable over the integers.

Disadvantage: Can't be used for all quadratics.

Complete the Square:

Write the equation in the form $x^2+\frac{b}{a}x=-\frac{c}{a}$. Add $\left(\frac{b}{2a}\right)^2$ to both sides of the equation. Solve the equation.

Advantage: Can be used for any quadratic.

Disadvantage: Sometimes can be difficult to use and time consuming.

Quadratic Formula:

Write the equation in form $ax^2+bx+c=0$. Determine a, b, and c. Use the quadratic formula to solve for x.

Advantage: Can be used for any quadratic.

Disadvantage: Need to memorize the formula; simplifying is sometimes difficult.

Check Your Progress: Chapter 16

1. $(x-9)(x+11)=0$

$x-9=0 \qquad x+11=0$

$x=9 \qquad x=-11$

The solutions are 9 and –11.

2. $6r^2=12-r$

$6r^2+r-12=0$

$(2r+3)(3r-4)=0$

$2r+3=0 \qquad 3r-4=0$

$2r=-3 \qquad 3r=4$

$r=-\frac{3}{2} \qquad r=\frac{4}{3}$

The solutions are $-\frac{3}{2}$ and $\frac{4}{3}$.

3. $2(x+1)^2=50$

$(x+1)^2=25$

$\sqrt{(x+1)^2}=\pm\sqrt{25}$

$x+1=\pm 5$

$x+1=5 \qquad x+1=-5$

$x=4 \qquad x=-6$

The solutions are 4 and –6.

4. $y^2-2y-5=0$

$y^2-2y=5$

Complete the square: $\left(\frac{-2}{2}\right)^2=(-1)^2=1$

$y^2-2y+1=5+1$

$(y-1)^2=6$

$\sqrt{(y-1)^2}=\pm\sqrt{6}$

$y-1=\pm\sqrt{6}$

$y=1\pm\sqrt{6}$

The solutions are $1-\sqrt{6}$ and $1+\sqrt{6}$.

5. $t^2 - 2t = 6$

$t^2 - 2t - 6 = 0$

$a = 1, b = -2, c = -6$

$$t = \frac{-(-2) \pm \sqrt{(-2)^2 - 4(1)(-6)}}{2(1)}$$

$$= \frac{2 \pm \sqrt{4+24}}{2}$$

$$= \frac{2 \pm \sqrt{28}}{2}$$

$$= \frac{2 \pm \sqrt{4}\sqrt{7}}{2}$$

$$= \frac{2 \pm 2\sqrt{7}}{2}$$

$$= 1 \pm \sqrt{7}$$

The solutions are $1-\sqrt{7}$ and $1+\sqrt{7}$.

6. $(2x-7)(3x+4) = 0$

$2x - 7 = 0 \qquad 3x + 4 = 0$

$2x = 7 \qquad 3x = -4$

$x = \frac{7}{2} \qquad x = -\frac{4}{3}$

The solutions are $-\frac{4}{3}$ and $\frac{7}{2}$.

7. $(x-4)^2 = 20$

$\sqrt{(x-4)^2} = \pm\sqrt{20}$

$x - 4 = \pm\sqrt{4}\sqrt{5}$

$x - 4 = \pm 2\sqrt{5}$

$x = 4 \pm 2\sqrt{5}$

The solutions are $4-2\sqrt{5}$ and $4+2\sqrt{5}$.

8. $x^2 + 6x + 4 = 0$

$x^2 + 6x = -4$

Complete the square: $\left(\frac{6}{2}\right)^2 = (3)^2 = 9$

$x^2 + 6x + 9 = -4 + 9$

$(x+3)^2 = 5$

$\sqrt{(x+3)^2} = \pm\sqrt{5}$

$x + 3 = \pm\sqrt{5}$

$x = 3 \pm \sqrt{5}$

The solutions are $-3-\sqrt{5}$ and $-3+\sqrt{5}$.

9. $4t^2 - 12t - 15 = 0$

$a = 4, b = -12, c = -15$

$$t = \frac{-(-12) \pm \sqrt{(-12)^2 - 4(4)(-15)}}{2(4)}$$

$$= \frac{12 \pm \sqrt{144+240}}{8}$$

$$= \frac{12 \pm \sqrt{384}}{8}$$

$$= \frac{12 \pm \sqrt{64}\sqrt{6}}{8}$$

$$= \frac{12 \pm 8\sqrt{6}}{8}$$

$$= \frac{3 \pm 2\sqrt{6}}{2}$$

The solutions are $\frac{3-2\sqrt{6}}{2}$ and $\frac{3+2\sqrt{6}}{2}$.

10. $y^2 - 12 = 0$

$y^2 = 12$

$\sqrt{y^2} = \pm\sqrt{12}$

$y = \pm\sqrt{4}\sqrt{3}$

$y = \pm 2\sqrt{3}$

The solutions are $-2\sqrt{3}$ and $2\sqrt{3}$.

11. $5y^2 + 11y = 12$

$5y^2 + 11y - 12 = 0$

$(5y-4)(y+3) = 0$

$5y - 4 = 0 \qquad y + 3 = 0$

$5y = 4 \qquad y = -3$

$y = \frac{4}{5}$

The solutions are -3 and $\frac{4}{5}$.

12. $2y^2 - 4y = 1$

$y^2 - 2y = \frac{1}{2}$

Complete the square: $\left(\frac{-2}{2}\right)^2 = (-1)^2 = 1$

$$y^2 - 2y + 1 = \frac{1}{2} + 1$$
$$(y-1)^2 = \frac{3}{2}$$
$$\sqrt{(y-1)^2} = \pm\sqrt{\frac{3}{2}}$$
$$y - 1 = \pm\frac{\sqrt{3}}{\sqrt{2}} \cdot \frac{\sqrt{2}}{\sqrt{2}}$$
$$y - 1 = \pm\frac{\sqrt{6}}{2}$$
$$y = 1 \pm \frac{\sqrt{6}}{2}$$
$$y = \frac{2}{2} \pm \frac{\sqrt{6}}{2}$$
$$y = \frac{2 \pm \sqrt{6}}{2}$$

The solutions are $\frac{2-\sqrt{6}}{2}$ and $\frac{2+\sqrt{6}}{2}$.

13.
$$\frac{x^2}{2} = 3x - \frac{7}{2}$$
$$2\left(\frac{x^2}{2}\right) = 2\left(3x - \frac{7}{2}\right)$$
$$x^2 = 6x - 7$$
$$x^2 - 6x + 7 = 0$$

$a = 1, b = -6, c = 7$

$$x = \frac{-(-6) \pm \sqrt{(-6)^2 - 4(1)(7)}}{2(1)}$$
$$= \frac{6 \pm \sqrt{36 - 28}}{2}$$
$$= \frac{6 \pm \sqrt{8}}{2}$$
$$= \frac{6 \pm \sqrt{4}\sqrt{2}}{2}$$
$$= \frac{6 \pm 2\sqrt{2}}{2}$$
$$= 3 \pm \sqrt{2}$$

The solutions are $3 + \sqrt{2}$ and $3 - \sqrt{2}$.

14. $(x+2)^2 + 49 = 0$

$$(x+2)^2 = -49$$
$$\sqrt{(x+2)^2} = \pm\sqrt{-49}$$

$\sqrt{-49}$ is not a real number. The quadratic equation has no real number solutions.

15. $4x^2 + 6x - 1 = 0$

$a = 4, b = 6, c = -1$

$$x = \frac{-(6) \pm \sqrt{(6)^2 - 4(4)(-1)}}{2(4)}$$
$$= \frac{-6 \pm \sqrt{36 + 16}}{8}$$
$$= \frac{-6 \pm \sqrt{52}}{8}$$
$$= \frac{-6 \pm \sqrt{4}\sqrt{13}}{8}$$
$$= \frac{-6 \pm 2\sqrt{13}}{8}$$
$$= \frac{-3 \pm \sqrt{13}}{4}$$

$x = \frac{-3 - \sqrt{13}}{4}$ $\quad$ $x = \frac{-3 + \sqrt{13}}{4}$

$x \approx -1.651$ $\quad$ $x \approx 0.151$

The solutions are approximately −1.651 and 0.151.

Section 16.4

Concept Check

1. parabola

3. Let $y = 0$ and solve for x.

5. Because the coefficient of x^2 is negative, the graph opens down.

7. Because the coefficient of x^2 is positive, the graph opens up.

Objective A Exercises

9. $f(x) = x^2 - 2x + 1$

$f(3) = 3^2 - 2(3) + 1 = 9 - 6 + 1 = 4$

11. $f(x) = 4 - x^2$

$f(-3) = 4 - (-3)^2 = 4 - 9 = -5$

13. $f(x) = -x^2 + 5x - 6$

$$f(-4) = -(-4)^2 + 5(-4) - 6$$
$$= -16 - 20 - 6 = -42$$

15. $y = x^2$

x	y
–2	4
–1	1
0	0
1	1
2	4

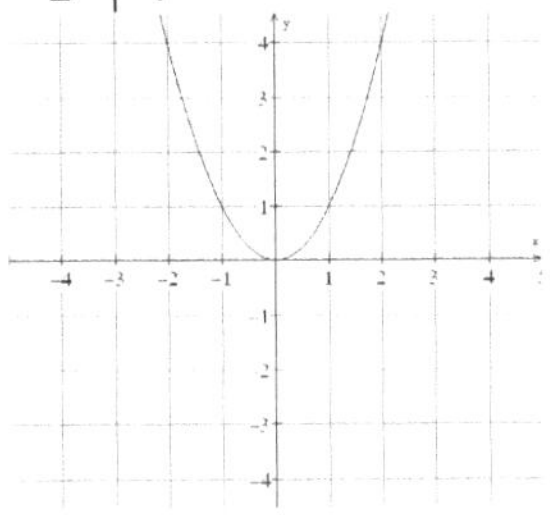

17. $y = -x^2 + 1$

x	y
–2	–3
–1	0
0	1
1	0
2	–3

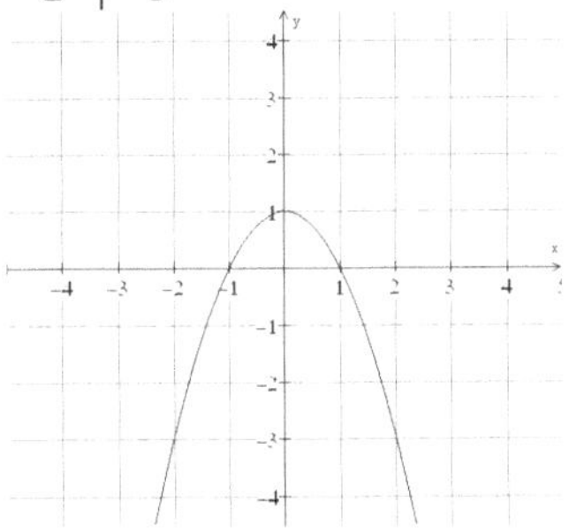

19. $f(x) = 2x^2$

x	y
–2	8
–1	2
0	0
1	2
2	8

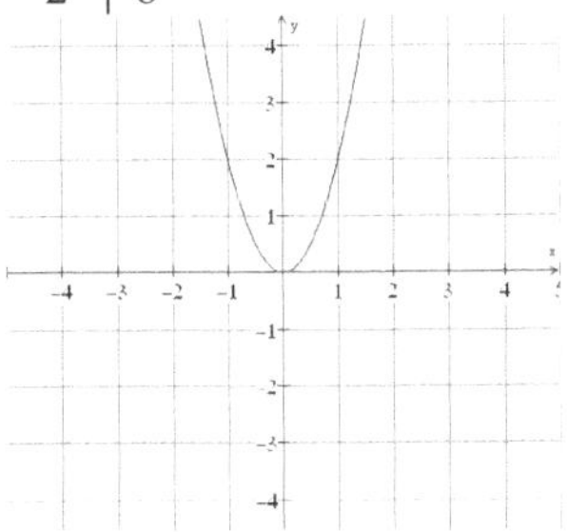

21. $y = -\frac{1}{2}x^2 + 1$

x	y
–2	–1
–1	$\frac{1}{2}$
0	1
1	$\frac{1}{2}$
2	–1

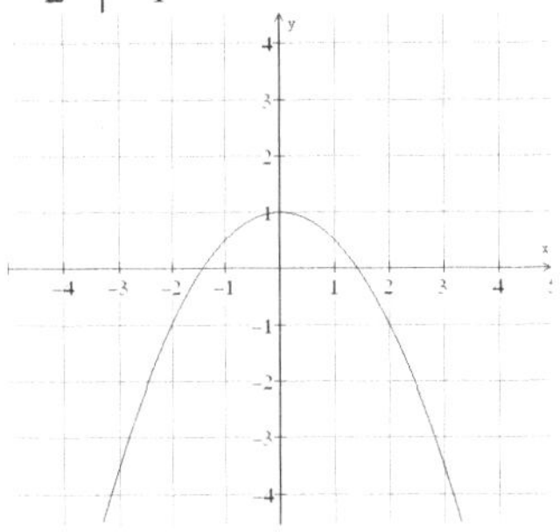

23. $y = x^2 - 4x$

x	y
–1	5
0	0
1	–3
2	–4
3	–3

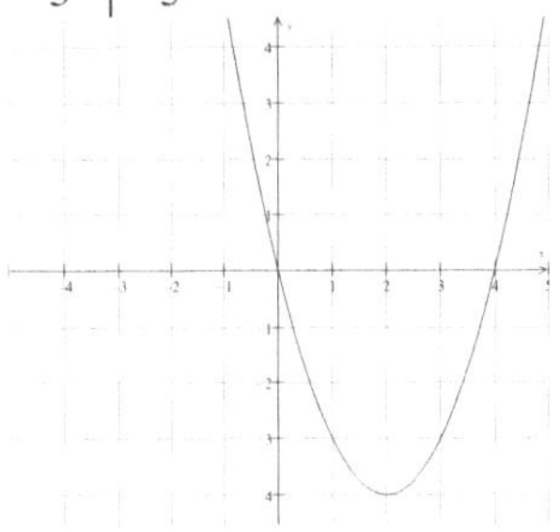

25. $y = x^2 - 2x + 3$

x	y
–2	11
–1	6
0	3
1	2
2	3

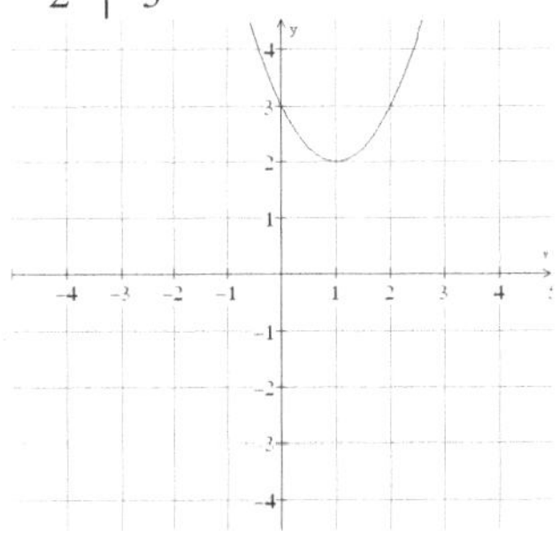

27. $y = -x^2 + 2x + 3$

x	y
−2	−5
−1	0
0	3
1	4
2	3

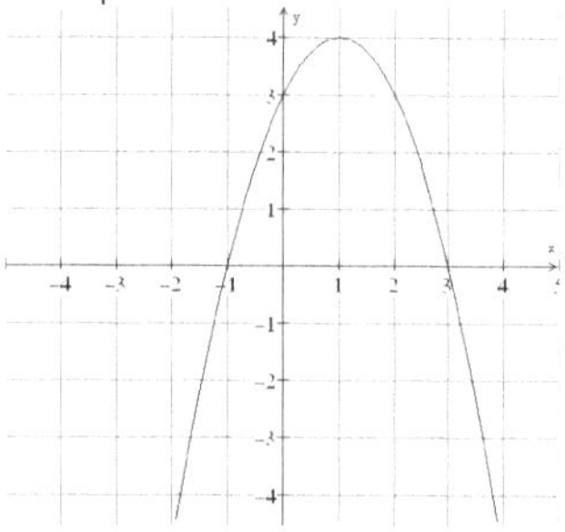

29. $y = -x^2 + 4x - 4$

x	y
0	−4
1	−1
2	0
3	−1
4	−4

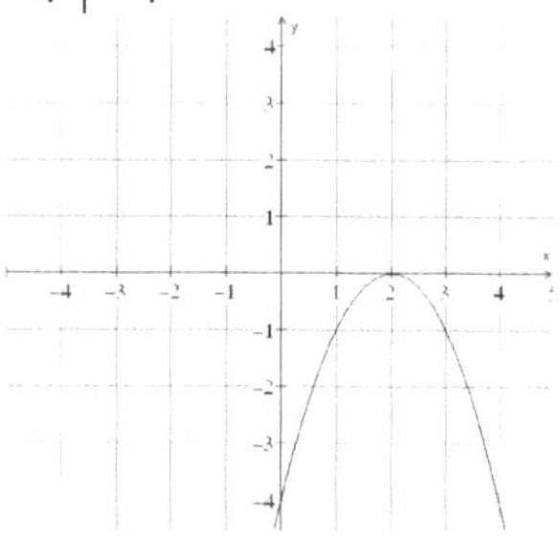

31. x-intercepts: $y = 0$

$x^2 + 5x - 6 = 0$

$(x+6)(x-1) = 0$

$x + 6 = 0 \qquad x - 1 = 0$

$x = -6 \qquad x = 1$

(−6, 0) and (1, 0)

y-intercept: $x = 0$

$y = 0^2 + 5(0) - 6 = -6$

(0, −6)

33. x-intercepts: $y = 0$

$x^2 + 12x + 36 = 0$

$(x+6)(x+6) = 0$

$x + 6 = 0 \qquad x + 6 = 0$

$x = -6 \qquad x = -6$

(−6, 0)

y-intercept: $x = 0$

$y = 0^2 + 12(0) + 36 = 36$

(0, 36)

35. x-intercepts: $f(x) = 0$

$x^2 + 4x - 2 = 0$

$$x = \frac{-(4) \pm \sqrt{(4)^2 - 4(1)(-2)}}{2(1)}$$

$$= \frac{-4 \pm \sqrt{16+8}}{2} = \frac{-4 \pm \sqrt{24}}{2}$$

$$= \frac{-4 \pm \sqrt{4}\sqrt{6}}{2} = \frac{-4 \pm 2\sqrt{6}}{2} = -2 \pm \sqrt{6}$$

$\left(-2-\sqrt{6},\ 0\right)$ and $\left(-2+\sqrt{6},\ 0\right)$

y-intercept: $x = 0$

$y = 0^2 + 4(0) - 2 = -2$

(0, −2)

37. x-intercepts: $y = 0$

$x^2 - x + 1 = 0$

$$x = \frac{-(-1) \pm \sqrt{(-1)^2 - 4(1)(1)}}{2(1)}$$

$$= \frac{1 \pm \sqrt{1-4}}{2} = \frac{1 \pm \sqrt{-3}}{2}$$

There are no real solutions, so there are no x-intercepts,

y-intercept: $x = 0$

$y = 0^2 - (0) + 1 = 1$

(0, 1)

39. x-intercepts: $f(x) = 0$

$2x^2 - 13x + 15 = 0$

$(2x - 3)(x - 5) = 0$

$2x - 3 = 0 \quad x - 5 = 0$

$2x = 3 \quad x = 5$

$x = \frac{3}{2}$

$\left(\frac{3}{2}, 0\right)$ and $(5, 0)$

y-intercept: $x = 0$

$y = 2(0)^2 - 13(0) + 15 = 15$

$(0, 15)$

41. x-intercepts: $y = 0$

$2 - 3x - 3x^2 = 0$

$0 = 3x^2 + 3x - 2$

$$x = \frac{-(3) \pm \sqrt{(3)^2 - 4(3)(-2)}}{2(3)}$$

$$= \frac{-3 \pm \sqrt{9 + 24}}{6} = \frac{-3 \pm \sqrt{33}}{6}$$

$\left(\frac{-3 - \sqrt{33}}{6}, 0\right)$ and $\left(\frac{-3 + \sqrt{33}}{6}, 0\right)$

y-intercept: $x = 0$

$y = 2 - 3(0) - 3(0)^2 = 2$

$(0, 2)$

43. negative

Critical Thinking 16.4

45. linear

47. neither

49. quadratic

51. **a.** (–2, 0), (2, 0)
b. (0, –4)
c. It is positive.
d. 0
e. –3

Projects or Group Activities 16.4

53. Answers will vary. For example:

$y = -x^2 + 6x - 8$

55. $y = 2x^2 - 3x - 2$

$12 = 2x^2 - 3x - 2$

$0 = 2x^2 - 3x - 14$

$0 = (2x - 7)(x + 2)$

$2x - 7 = 0 \quad x + 2 = 0$

$2x = 7 \quad x = -2$

$x = \frac{7}{2}$

The point $\left(\frac{7}{2}, 12\right)$ would not be in Quadrant II, so the solution is –2.

57. –1, 5

59. 0.5936307095, –0.3158529318

61. –1.766291404, 1.059184622

Section 16.5

Concept Check

1. $2W + 3$

3. $r - 30$

Objective A Exercises

5. Strategy:
This is a geometry problem.
→Length of the triangle base: x
Height of the triangle: $2x+2$
→Use the formula $A=\frac{1}{2}bh$.

Solution:
$$A=\frac{1}{2}bh$$
$$20=\frac{1}{2}x(2x+2)$$
$$20=x^2+x$$
$$0=x^2+x-20$$
$$0=(x+5)(x-4)$$
$$x+5=0 \qquad x-4=0$$
$$x=-5 \qquad x=4$$
The solution −5 is not possible.
$2x+2=2(4)+2=8+2=10$

The length is 4 m.
The height is 10 m.

7. Strategy:
This is a geometry problem.
→Width of the rectangle: x
Length of the rectangle: $x+2$
→Use the formula $A=LW$.

Solution:
$$A=LW$$
$$24=x(x+2)$$
$$24=x^2+2x$$
$$0=x^2+2x-24$$
$$0=(x+6)(x-4)$$
$$x+6=0 \qquad x-4=0$$
$$x=-6 \qquad x=4$$
The solution −6 is not possible.
$x+2=4+2=6$
The length is 6 ft.
The width is 4 ft.

9. Strategy:
This is a work problem.
→Time for second drain to empty tank: t
Time for first drain to empty tank: $t+16$

	Rate	Time	Part
2nd Drain	$\frac{1}{t}$	6	$\frac{6}{t}$
1st Drain	$\frac{1}{t+16}$	6	$\frac{6}{t+16}$

→The sum of the parts of the task completed by each drain must equal 1.

Solution:
$$\frac{6}{t}+\frac{6}{t+16}=1$$
$$t(t+16)\left(\frac{6}{t}+\frac{6}{t+16}\right)=t(t+1)(1)$$
$$6t+96+6t=t^2+16t$$
$$0=t^2+4t-96$$
$$0=(t+12)(t-8)$$
$$t+12=0 \qquad t-8=0$$
$$t=-12 \qquad t=8$$
The solution −12 is not possible.
$t+16=8+16=24$
The first drain can empty the tank in 24 min. The second drain can empty the tank in 8 min.

11. Strategy:
This is a geometry problem.
→Width of the rectangle: x
Length of the rectangle: $2x$
→Use the formula $A=LW$.

Solution:
$$A=LW$$
$$5000=x(2x)$$
$$5000=2x^2$$
$$2500=x^2$$
$$\sqrt{2500}=\sqrt{x^2}$$
$$\pm 50=x$$
The solution −50 is not possible.
$2x=2(50)=100$
The length is 100 ft.
The width is 50 ft.

13. Strategy:
This is a work problem.
→Time for using the second engine: t
Time for using the first engine: $t + 6$

	Rate	Time	Part
2nd Engine	$\frac{1}{t}$	4	$\frac{4}{t}$
1st Engine	$\frac{1}{t+6}$	4	$\frac{4}{t+6}$

→The sum of the parts of the task completed
by each engine must equal 1.

Solution:

$$\frac{4}{t}+\frac{4}{t+6}=1$$

$$t(t+6)\left(\frac{4}{t}+\frac{4}{t+6}\right)=t(t+6)(1)$$

$$4t+24+4t=t^2+6t$$

$$0=t^2-2t-24$$

$$0=(t+4)(t-6)$$

$t+4=0 \qquad t-6=0$

$t=-4 \qquad t=6$

The solution –4 is not possible.

$t+6=6+6=12$

Using the first engine would take 12 h.
Using the second engine would take 6 h.

15. Strategy:
This is a distance-rate problem.
→Rate of the plane in calm air: r

	Distance	Rate	Time
With wind	375	$r + 25$	$\frac{375}{r+25}$
Against wind	375	$r - 25$	$\frac{375}{r-25}$

→The time spent traveling against the wind is 2 h more than the time spent traveling with the wind.

Solution:

$$\frac{375}{r-25}=2+\frac{275}{r+25}$$

$$(r-25)(r+25)\left(\frac{375}{r-25}\right)$$

$$=(r-25)(r+25)\left(2+\frac{375}{r+25}\right)$$

$$375r+9375=2r^2-1250+375r-9375$$

$$2r^2-20,000=0$$

$$r^2-10,000=0$$

$$r^2=10,000$$

$$\sqrt{r^2}=\sqrt{10,000}$$

$$r=\pm 100$$

The solution –100 is not possible.
The rate of the plane is calm air is 100 mph.

17. Strategy:
This is a distance-rate problem.
→Rate before decrease: r

	Distance	Rate	Time
Before decrease	150	r	$\frac{150}{r}$
After decrease	35	$r-15$	$\frac{35}{r-15}$

→The total time spent traveling was 4 *h*.

Solution:

$$\frac{150}{r}+\frac{35}{r-15}=4$$
$$(r)(r-15)\left(\frac{150}{r}+\frac{35}{r-15}\right)=(r)(r-15)(4)$$
$$150r-2250+35r=4r^2-60r$$
$$0=4r^2-245r+2250$$
$$0=(r-50)(4r-45)$$
$$r-50=0$$
$$r=50$$
$$4r-45=0$$
$$4r=45$$
$$r=\frac{45}{4}$$
$$r=11.25$$

The solution 11.25 is not possible because if you decrease 15 mph, you would get a negative.
The rate of the motorcycle for the first 150 mi is 50 mph.

19. Strategy:
Use the formula $h=48t-16t^2$, where h = 32.

Solution:

$$h=48t-16t^2$$
$$32=48t-16t^2$$
$$16t^2-48t+32=0$$
$$16(t^2-3t+2)=0$$
$$16(t-2)(t-1)=0$$
$$t-2=0 \qquad t-1=0$$
$$t=2 \qquad t=1$$

The arrow will be 32 ft high at 1 s and at 2 s.

21. Strategy:
Use the function $A(t)=0.005\pi t^2$, where $A(t)=10$.

Solution:

$$A(t)=0.005\pi t^2$$
$$10=0.005\pi t^2$$
$$\frac{10}{0.005\pi}=t^2$$
$$\sqrt{\frac{10}{0.005\pi}}=\sqrt{t^2}$$
$$\pm\sqrt{\frac{10}{0.005\pi}}=t$$
$$t\approx 25.231$$

The weed will cover an area of 10 m^2 in 25 days.

23. Strategy:
Use the formula $y=0.03x^2+0.36x+34.6$, where y = 50.

Solution:

$$y=0.03x^2+0.36x+34.6$$
$$50=0.03x^2+0.36x+34.6$$
$$0=0.03x^2+0.36x-15.4$$
$$x=\frac{-(0.36)\pm\sqrt{(0.36)^2-4(0.03)(-15.4)}}{2(0.03)}$$
$$x=\frac{-0.36\pm\sqrt{1.9776}}{0.06}$$
$$x=\frac{-0.36+\sqrt{1.9776}}{0.06} \qquad x=\frac{-0.36-\sqrt{1.9776}}{0.06}$$
$$x\approx 17 \qquad x\approx -29$$

The solution −29 is not possible.
2000 + 17 = 2017
The population of people age 65 or older will be 50 million in the year 2017.

25. Strategy:
Use the formula $h=-16t^2+24t+4$, where $h=2$.

Solution:

$$h=-16t^2+24t+4$$
$$2=-16t^2+24t+4$$
$$16t^2-24t-2=0$$
$$x=\frac{-(-24)\pm\sqrt{(-24)^2-4(16)(-2)}}{2(16)}$$
$$x=\frac{24\pm\sqrt{576+128}}{32}$$
$$x=\frac{24\pm\sqrt{704}}{32}$$
$$x=\frac{24\pm\sqrt{64}\sqrt{11}}{32}$$
$$x=\frac{24\pm8\sqrt{11}}{32}$$
$$x=\frac{3\pm\sqrt{11}}{4}$$
$$x=\frac{3-\sqrt{11}}{4} \qquad x=\frac{3+\sqrt{11}}{4}$$
$$x\approx-0.8 \qquad x\approx1.58$$

The solution –0.8 is not possible.

The batter will hit the ball after 1.58 s.

27. Strategy:
Use the formula $s=-16t^2+88t+1$, where $s=0$.

Solution:

$$s=-16t^2+88t+1$$
$$0=-16t^2+88t+1$$
$$16t^2-88t-1=0$$
$$t=\frac{-(-88)\pm\sqrt{(-88)^2-4(16)(-1)}}{2(16)}$$
$$t=\frac{88\pm\sqrt{7808}}{32}$$
$$t=\frac{88+\sqrt{7808}}{32} \qquad t=\frac{88-\sqrt{7808}}{32}$$
$$t\approx5.5 \qquad t\approx-0.01$$

The hang time of the football is approximately 5.5 s.

29. False

Critical Thinking 16.5

31. Strategy:
This is a geometry problem.
→Length of the other leg: x
Length of one leg: $2x-1$
→Use the Pythagorean Theorem.

Solution:

$$x^2+(2x-1)^2=(\sqrt{13})^2$$
$$x^2+4x^2-4x+1=13$$
$$5x^2-4x-12=0$$
$$(5x+6)(x-2)=0$$
$$5x+6=0 \qquad x-2=0$$
$$x=-\frac{5}{6} \qquad x=2$$

The solution $-\frac{5}{6}$ is not possible.

$2x-1=2(2)-1=4-1=3$

The lengths of the legs are 2 cm and 3 cm.

33. Strategy:
Use the formula for the volume of a cone $V=\frac{1}{3}\pi r^2h$ and solve for r when h is 12 and $V=800$.

Solution:

$$V=\frac{1}{3}\pi r^2h$$
$$800=\frac{1}{3}\cdot3.14\cdot r^2\cdot12$$
$$800=12.56r^2$$
$$\frac{800}{12.56}=r^2$$
$$\pm\sqrt{\frac{800}{12.56}}=r$$
$$r=-\sqrt{\frac{800}{12.56}} \qquad r=\sqrt{\frac{800}{12.56}}$$
$$r\approx-7.98 \qquad r\approx7.98$$

The radius cannot be –7.98.
The radius is 7.98 cm.

35. Strategy:
This is a geometry problem.
→Length of the cardboard: $x-4$
Width of the cardboard: $x-4$
Height of the box: 2
→Use the formula $V = LWH$.

Solution:

$$V = LWH$$
$$512 = (x-4)(x-4)2$$
$$512 = 2(x-4)^2$$
$$256 = (x-4)^2$$
$$\sqrt{256} = \sqrt{(x-4)^2}$$
$$\pm 16 = x-4$$
$$x-4 = 16 \qquad x-4 = -16$$
$$x = 20 \qquad x = -12$$

The solution −12 is not possible.
The dimensions of the cardboard are 20 in. by 20 in.

Projects and Group Activities 16.5

37. **a.** $P = 2L + 2W$
$12 = 2L + 2W$

b. $A = LW$

c.
$$12 = 2L + 2W$$
$$12 - 2W = 2L$$
$$6 - W = L$$

$$A = LW$$
$$A = (6-W)W$$
$$A = 6W - W^2$$

d. W could be 1, 2, or 3.
$W = 1, L = 6 - 1 = 5$
$A = LW$
$A = 5 \cdot 1$
$A = 5$

$W = 2, L = 6 - 2 = 4$
$A = LW$
$A = 4 \cdot 2$
$A = 8$

$W = 3, L = 6 - 3 = 3$
$A = LW$
$A = 3 \cdot 3$
$A = 9$

Length 3 ft and width 3 ft gives the largest area, 9 ft^2.

Chapter Review Exercises

1.
$$6x^2 + 13x - 28 = 0$$
$$(2x+7)(3x-4) = 0$$
$$2x+7 = 0 \qquad 3x-4 = 0$$
$$x = -\frac{7}{2} \qquad x = \frac{4}{3}$$

The solutions are $-\frac{7}{2}$ and $\frac{4}{3}$.

2. $49x^2 = 25$

$$\frac{49x^2}{49} = \frac{25}{49}$$

$$x^2 = \frac{25}{49}$$

$$\sqrt{x^2} = \sqrt{\frac{25}{49}}$$

$$x = \pm\frac{5}{7}$$

$$x = \frac{5}{7} \qquad x = -\frac{5}{7}$$

The solutions are $-\frac{5}{7}$ and $\frac{5}{7}$.

3. $x^2 + 2x - 24 = 0$

$$x^2 + 2x = 24$$

Complete the square. $\left(\frac{2}{2}\right)^2 = (1)^2 = 1$

$$x^2 + 2x + 1 = 24 + 1$$

$$(x+1)^2 = 25$$

$$\sqrt{(x+1)^2} = \pm\sqrt{25}$$

$$x + 1 = \pm 5$$

$$x+1 = 5 \qquad x+1 = -5$$

$$x = 4 \qquad x = -6$$

The solutions are –6 and 4.

4. $x^2 + 5x - 6 = 0$

$a = 1,\ b = 5,\ c = -6$

$$x = \frac{-(5) \pm \sqrt{(5)^2 - 4(1)(-6)}}{2(1)}$$

$$x = \frac{-5 \pm \sqrt{25+24}}{2}$$

$$x = \frac{-5 \pm \sqrt{49}}{2}$$

$$x = \frac{-5 \pm 7}{2}$$

$$x = \frac{-5+7}{2} \qquad x = \frac{-5-7}{2}$$

$$= \frac{2}{2} = 1 \qquad = \frac{-12}{2} = -6$$

The solutions are –6 and 1.

5. $2x^2 + 5x = 12$

$$x^2 + \frac{5}{2}x = 6$$

Complete the square. $\left(\frac{5}{2}\cdot\frac{1}{2}\right)^2 = \left(\frac{5}{4}\right)^2 = \frac{25}{16}$

$$x^2 + \frac{5}{2}x + \frac{25}{16} = 6 + \frac{25}{16}$$

$$\left(x + \frac{5}{4}\right)^2 = \frac{96}{16} + \frac{25}{16}$$

$$\left(x + \frac{5}{4}\right)^2 = \frac{121}{16}$$

$$\sqrt{\left(x + \frac{5}{4}\right)^2} = \pm\sqrt{\frac{121}{16}}$$

$$x + \frac{5}{4} = \pm\frac{11}{4}$$

$$x + \frac{5}{4} = \frac{11}{4} \qquad x + \frac{5}{4} = -\frac{11}{4}$$

$$x = \frac{6}{4} = \frac{3}{2} \qquad x = -\frac{16}{4} = -4$$

The solutions are –4 and $\frac{3}{2}$.

6. $12x^2 + 10 = 29x$

$$12x^2 - 29x + 10 = 0$$

$$(12x-5)(x-2) = 0$$

$$12x - 5 = 0 \qquad x - 2 = 0$$

$$x = \frac{5}{12} \qquad x = 2$$

The solutions are $\frac{5}{12}$ and 2.

7. $(x+2)^2 - 24 = 0$

$$(x+2)^2 = 24$$

$$\sqrt{(x+2)^2} = \sqrt{24}$$

$$x + 2 = \pm\sqrt{4}\sqrt{6}$$

$$x + 2 = \pm 2\sqrt{6}$$

$$x = -2 \pm 2\sqrt{6}$$

The solutions are $-2 - 2\sqrt{6}$ and $-2 + 2\sqrt{6}$.

8. $2x^2+3=5x$

$2x^2-5x+3=0$

$a=2,\ b=-5,\ c=3$

$$x=\frac{-(-5)\pm\sqrt{(-5)^2-4(2)(3)}}{2(2)}$$

$$x=\frac{5\pm\sqrt{25-24}}{4}$$

$$x=\frac{5\pm\sqrt{1}}{4}$$

$$x=\frac{5\pm1}{4}$$

$$x=\frac{5+1}{4}=\frac{6}{4}=\frac{3}{2} \qquad x=\frac{5-1}{4}=\frac{4}{4}=1$$

The solutions are 1 and $\frac{3}{2}$.

9. $6x(x+1)=x-1$

$6x^2+6x=x-1$

$6x^2+5x+1=0$

$(3x+1)(2x+1)=0$

$3x+1=0 \qquad 2x+1=0$

$x=-\frac{1}{3} \qquad x=-\frac{1}{2}$

The solutions are $-\frac{1}{2}$ and $-\frac{1}{3}$.

10. $4y^2+9=0$

$4y^2=-9$

$y^2=\frac{-9}{4}$

$\sqrt{y^2}=\sqrt{-\frac{9}{4}}$

$\sqrt{-\frac{9}{4}}$ is not a real number. The equation has no real number solutions.

11. $x^2-4x+1=0$

$x^2-4x=-1$

Complete the square. $\left(\frac{-4}{2}\right)^2=(-2)^2=4$

$x^2-4x+4=-1+4$

$(x-2)^2=3$

$\sqrt{(x-2)^2}=\sqrt{3}$

$x-2=\pm\sqrt{3}$

$x=2\pm\sqrt{3}$

The solutions are $2-\sqrt{3}$ and $2+\sqrt{3}$.

12. $x^2-3x-5=0$

$a=1,\ b=-3,\ c=-5$

$$x=\frac{-(-3)\pm\sqrt{(-3)^2-4(1)(-5)}}{2(1)}$$

$$x=\frac{3\pm\sqrt{9+20}}{2}$$

$$x=\frac{3\pm\sqrt{29}}{2}$$

The solutions are $\frac{3-\sqrt{29}}{2}$ and $\frac{3+\sqrt{29}}{2}$.

13. $x^2+6x+12=0$

$x^2+6x=-12$

Complete the square. $\left(\frac{6}{2}\right)^2=(3)^2=9$

$x^2+6x+9=-12+9$

$(x+3)^2=-3$

$\sqrt{(x+3)^2}=\sqrt{-3}$

$\sqrt{-3}$ is not a real number. The equation has no real number solutions.

14. $(x+9)^2=x+11$

$x^2+18x+81=x+11$

$x^2+17x+70=0$

$(x+7)(x+10)=0$

$x+7=0 \qquad x+10=0$

$x=-7 \qquad x=-10$

The solutions are –10 and –7.

15. $\left(x-\frac{1}{2}\right)^2=\frac{9}{4}$

$\sqrt{\left(x-\frac{1}{2}\right)^2}=\sqrt{\frac{9}{4}}$

$x-\frac{1}{2}=\pm\frac{3}{2}$

$x-\frac{1}{2}=\frac{3}{2}$ $\qquad$ $x-\frac{1}{2}=-\frac{3}{2}$

$x=\frac{4}{2}=2$ $\qquad$ $x=-\frac{2}{2}=-1$

The solutions are –1 and 2.

16. $4x^2+16x=7$

$x^2+4x=\frac{7}{4}$

Complete the square. $\left(\frac{4}{2}\right)^2=(2)^2=4$

$x^2+4x+4=\frac{7}{4}+4$

$(x+2)^2=\frac{7}{4}+\frac{16}{4}$

$(x+2)^2=\frac{23}{4}$

$\sqrt{(x+2)^2}=\sqrt{\frac{23}{4}}$

$x+2=\pm\frac{\sqrt{23}}{2}$

$x=-2\pm\frac{\sqrt{23}}{2}$

$x=\frac{-4}{2}\pm\frac{\sqrt{23}}{2}$

$x=\frac{-4\pm\sqrt{23}}{2}$

The solutions are $\frac{-4-\sqrt{23}}{2}$ and $\frac{-4+\sqrt{23}}{2}$.

17. $x^2-4x+8=0$

$a=1,\ b=-4,\ c=8$

$x=\frac{-(-4)\pm\sqrt{(-4)^2-4(1)(8)}}{2(1)}$

$x=\frac{4\pm\sqrt{16-32}}{2}$

$x=\frac{4\pm\sqrt{-16}}{2}$

$\sqrt{-16}$ is not a real number. The equation has no real number solutions.

18. $2x^2+5x+2=0$

$a=2,\ b=5,\ c=2$

$x=\frac{-(5)\pm\sqrt{(5)^2-4(2)(2)}}{2(2)}$

$x=\frac{-5\pm\sqrt{25-16}}{4}$

$x=\frac{-5\pm\sqrt{9}}{4}$

$x=\frac{-5\pm3}{4}$

$x=\frac{-5+3}{4}$ $\qquad$ $x=\frac{-5-3}{4}$

$x=\frac{-2}{4}=-\frac{1}{2}$ $\qquad$ $x=\frac{-8}{4}=-2$

The solutions are –2 and $-\frac{1}{2}$.

19. $y=-3x^2$

x	y
–1	–3
0	0
1	–3

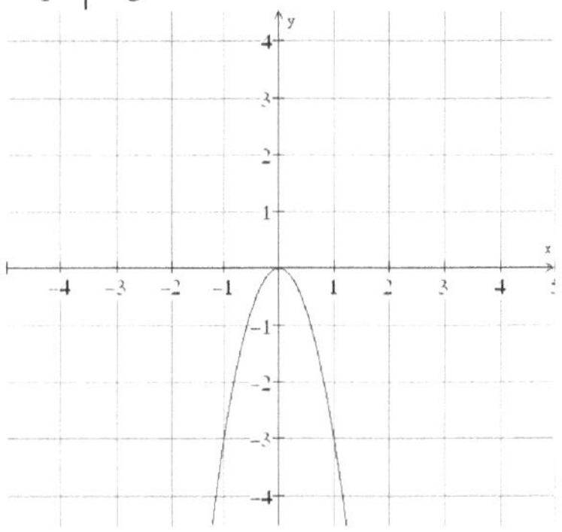

20. $y = -\frac{1}{4}x^2$

x	y
–4	–4
–2	–1
0	0
2	–1
4	–4

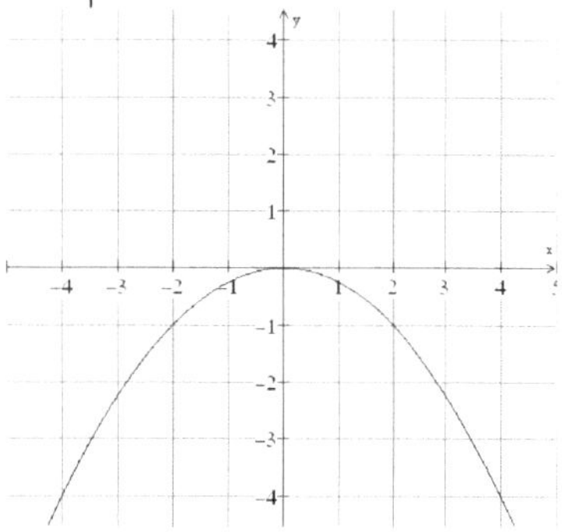

21. $y = 2x^2 + 1$

x	y
–1	3
0	1
1	3

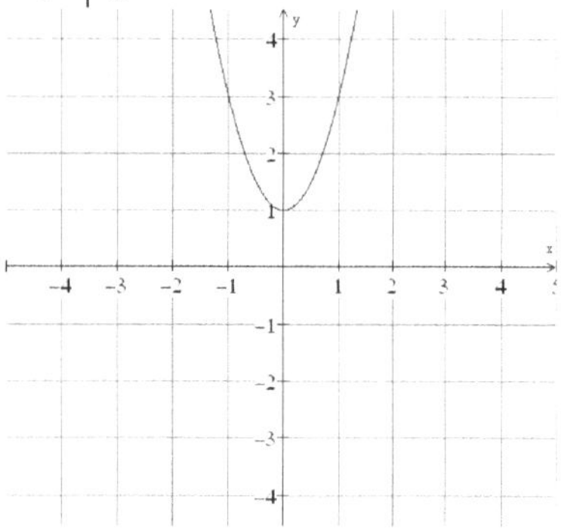

22. $y = x^2 - 4x + 3$

x	y
0	3
1	0
2	–1
3	0
4	3

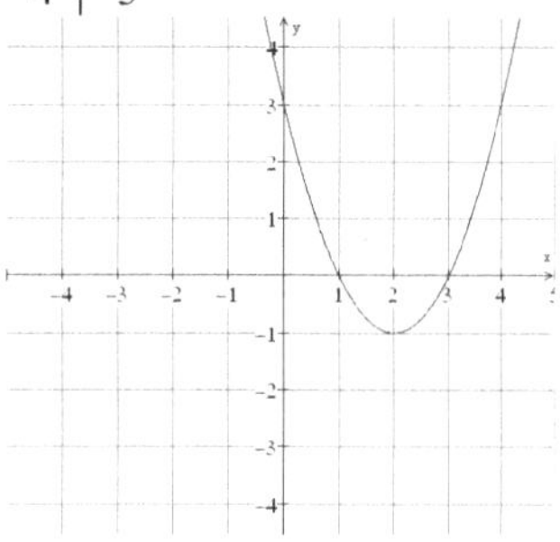

23. $y = -x^2 + 4x - 5$

x	y
0	–5
1	–2
2	–1
3	–2

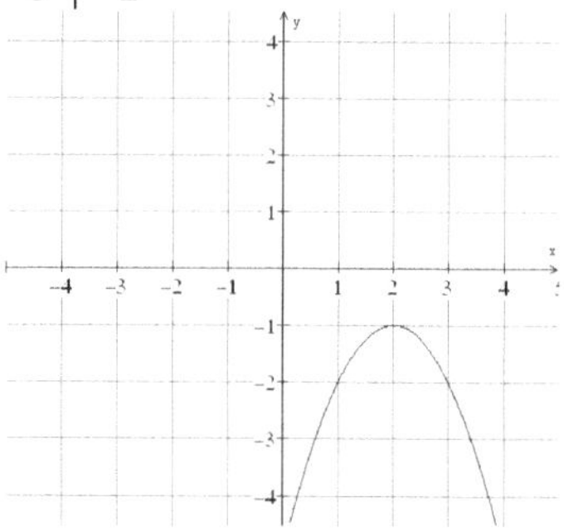

24. x-intercepts: $y = 0$

$x^2 - 2x - 15 = 0$

$(x+3)(x-5) = 0$

$x + 3 = 0 \qquad x - 5 = 0$

$x = -3 \qquad x = 5$

(–3, 0) and (5, 0)

y-intercept: $x = 0$

$y = 0^2 - 2(0) - 15 = -15$

(0, –15)

25. Strategy:
This is a distance-rate problem.
→Rate of the hawk in calm air: r

	Distance	Rate	Time
Against the wind	70	$r-5$	$\frac{70}{r-5}$
With the wind	40	$r+5$	$\frac{40}{r+5}$

→The time spent traveling against the wind is $\frac{1}{2}$ h more than the time spent traveling with the wind.

Solution:

$$\frac{70}{r-5}=\frac{1}{2}+\frac{40}{r+5}$$
$$2(r-5)(r+5)\left(\frac{70}{r-5}\right)=2(r-5)(r+5)\left(\frac{1}{2}+\frac{40}{r+5}\right)$$
$$70(2)(r+5)=(r-5)(r+5)+40(2)(r-5)$$
$$140r+700=r^2-25+80r-400$$
$$0=r^2-60r-1125$$
$$0=(r-75)(r+15)$$
$$r-75=0 \qquad r+15=0$$
$$r=75 \qquad r=-15$$

The solution –15 is not possible.
The rate of the hawk is calm air is 75 mph.

Chapter Test

1.
$$x^2-5x-6=0$$
$$(x-6)(x+1)=0$$
$$x-6=0 \qquad x+1=0$$
$$x=6 \qquad x=-1$$

The solutions are –1 and 6.

2.
$$3x^2+7x=20$$
$$3x^2+7x-20=0$$
$$(3x-5)(x+4)=0$$
$$3x-5=0 \qquad x+4=0$$
$$x=\frac{5}{3} \qquad x=-4$$

The solutions are –4 and $\frac{5}{3}$.

3.
$$2x^2+x=0$$
$$x(2x+1)=0$$
$$x=0 \qquad 2x+1=0$$
$$x=-\frac{1}{2}$$

The solutions are $-\frac{1}{2}$ and 0.

4.
$$4x^2-9=0$$
$$4x^2=9$$
$$\frac{4x^2}{4}=\frac{9}{4}$$
$$x^2=\frac{9}{4}$$
$$\sqrt{x^2}=\sqrt{\frac{9}{4}}$$
$$x=\pm\frac{3}{2}$$
$$x=\frac{3}{2} \qquad x=-\frac{3}{2}$$

The solutions are $-\frac{3}{2}$ and $\frac{3}{2}$.

5.
$$2(x-5)^2-50=0$$
$$2(x-5)^2=50$$
$$(x-5)^2=25$$
$$\sqrt{(x-5)^2}=\sqrt{25}$$
$$x-5=\pm5$$
$$x=5\pm5$$
$$x=5+5 \qquad x=5-5$$
$$x=10 \qquad x=0$$

The solutions are 0 and 10.

6.
$$3(x+4)^2-60=0$$
$$3(x+4)^2=60$$
$$(x+4)^2=20$$
$$\sqrt{(x+4)^2}=\sqrt{20}$$
$$x+4=\pm\sqrt{4}\sqrt{5}$$
$$x+4=\pm2\sqrt{5}$$
$$x=-4\pm2\sqrt{5}$$

The solutions are $-4-2\sqrt{5}$ and $-4+2\sqrt{5}$.

7. $x^2+4x-16=0$

$x^2+4x=16$

Complete the square. $\left(\frac{4}{2}\right)^2=(2)^2=4$

$x^2+4x+4=16+4$

$(x+2)^2=20$

$\sqrt{(x+2)^2}=\sqrt{20}$

$x+2=\pm\sqrt{4}\sqrt{5}$

$x+2=\pm2\sqrt{5}$

$x=-2\pm2\sqrt{5}$

The solutions are $-2-2\sqrt{5}$ and $-2+2\sqrt{5}$.

8. $x^2+3x=8$

Complete the square. $\left(\frac{3}{2}\right)^2=\frac{9}{4}$

$x^2+3x+\frac{9}{4}=8+\frac{9}{4}$

$x^2+3x+\frac{9}{4}=\frac{32}{4}+\frac{9}{4}$

$\left(x+\frac{3}{2}\right)^2=\frac{41}{4}$

$\sqrt{\left(x+\frac{3}{2}\right)^2}=\pm\sqrt{\frac{41}{4}}$

$x+\frac{3}{2}=\pm\frac{\sqrt{41}}{2}$

$x=-\frac{3}{2}\pm\frac{\sqrt{41}}{2}$

$x=\frac{-3\pm\sqrt{41}}{2}$

The solutions are $\frac{-3-\sqrt{41}}{2}$ and $\frac{-3+\sqrt{41}}{2}$.

9. $2x^2-6x+1=0$

$2x^2-6x=-1$

$x^2-3x=-\frac{1}{2}$

Complete the square. $\left(\frac{-3}{2}\right)^2=\frac{9}{4}$

$x^2-3x+\frac{9}{4}=-\frac{1}{2}+\frac{9}{4}$

$x^2-3x+\frac{9}{4}=-\frac{2}{4}+\frac{9}{4}$

$\left(x-\frac{3}{2}\right)^2=\frac{7}{4}$

$\sqrt{\left(x-\frac{3}{2}\right)^2}=\sqrt{\frac{7}{4}}$

$x-\frac{3}{2}=\pm\frac{\sqrt{7}}{2}$

$x=\frac{3}{2}\pm\frac{\sqrt{7}}{2}$

$x=\frac{3\pm\sqrt{7}}{2}$

The solutions are $\frac{3-\sqrt{7}}{2}$ and $\frac{3+\sqrt{7}}{2}$.

10. $2x^2+8x=3$

$x^2+4x=\frac{3}{2}$

Complete the square. $\left(\frac{4}{2}\right)^2=(2)^2=4$

$x^2+4x+4=\frac{3}{2}+4$

$x^2+4x+4=\frac{3}{2}+\frac{8}{2}$

$(x+2)^2=\frac{11}{2}$

$(x+2)^2=\frac{22}{4}$

$\sqrt{(x+2)^2}=\sqrt{\frac{22}{4}}$

$x+2=\pm\frac{\sqrt{22}}{2}$

$x=-2\pm\frac{\sqrt{22}}{2}$

$x=\frac{-4}{2}\pm\frac{\sqrt{22}}{2}$

$x=\frac{-4\pm\sqrt{22}}{2}$

The solutions are $\frac{-4-\sqrt{22}}{2}$ and $\frac{-4+\sqrt{22}}{2}$.

11. $x^2+4x+2=0$

$a=1,\ b=4,\ c=2$

$x=\frac{-(4)\pm\sqrt{(4)^2-4(1)(2)}}{2(1)}$

$x=\frac{-4\pm\sqrt{16-8}}{2}$

$x=\frac{-4\pm\sqrt{8}}{2}$

$x=\frac{-4\pm\sqrt{4}\sqrt{2}}{2}$

$x=\frac{-4\pm2\sqrt{2}}{2}$

$x=-2\pm\sqrt{2}$

The solutions are $-2-\sqrt{2}$ and $-2+\sqrt{2}$.

12. $x^2-3x=6$

$x^2-3x-6=0$

$a=1,\ b=-3,\ c=-6$

$x=\frac{-(-3)\pm\sqrt{(-3)^2-4(1)(-6)}}{2(1)}$

$x=\frac{3\pm\sqrt{9+24}}{2}$

$x=\frac{3\pm\sqrt{33}}{2}$

The solutions are $\frac{3-\sqrt{33}}{2}$ and $\frac{3+\sqrt{33}}{2}$.

13. $2x^2-5x-3=0$

$a=2,\ b=-5,\ c=-3$

$x=\frac{-(-5)\pm\sqrt{(-5)^2-4(2)(-3)}}{2(2)}$

$x=\frac{5\pm\sqrt{25+24}}{4}$

$x=\frac{5\pm\sqrt{49}}{4}$

$x=\frac{5\pm7}{4}$

$x=\frac{5+7}{4}\qquad x=\frac{5-7}{4}$

$x=\frac{12}{4}=3\qquad x=\frac{-2}{4}=-\frac{1}{2}$

The solutions are $-\frac{1}{2}$ and 3.

14. $3x^2-x=1$

$3x^2-x-1=0$

$a=3,\ b=-1,\ c=-1$

$x=\frac{-(-1)\pm\sqrt{(-1)^2-4(3)(-1)}}{2(3)}$

$x=\frac{1\pm\sqrt{1+12}}{6}$

$x=\frac{1\pm\sqrt{13}}{6}$

The solutions are $\frac{1-\sqrt{13}}{6}$ and $\frac{1+\sqrt{13}}{6}$.

15. $4x^2+6x-1=0$

$a=4,\ b=6,\ c=-1$

$$x=\frac{-(6)\pm\sqrt{(6)^2-4(4)(-1)}}{2(4)}$$

$$x=\frac{-6\pm\sqrt{36+16}}{8}$$

$$x=\frac{-6\pm\sqrt{52}}{8}$$

$$x=\frac{-6\pm\sqrt{4}\sqrt{13}}{8}$$

$$x=\frac{-6\pm 2\sqrt{13}}{8}$$

$$x=\frac{-3\pm\sqrt{13}}{4}$$

$$x=\frac{-3+\sqrt{13}}{4}\qquad x=\frac{-3-\sqrt{13}}{4}$$

$$x\approx 0.151\qquad x\approx -1.651$$

The solutions are –1.651 and 0.151.

16. $3x^2+2x-3=0$

$a=3,\ b=2,\ c=-3$

$$x=\frac{-(2)\pm\sqrt{(2)^2-4(3)(-3)}}{2(3)}$$

$$x=\frac{-2\pm\sqrt{4+36}}{6}$$

$$x=\frac{-2\pm\sqrt{40}}{6}$$

$$x=\frac{-2\pm\sqrt{4}\sqrt{10}}{6}$$

$$x=\frac{-2\pm 2\sqrt{10}}{6}$$

$$x=\frac{-1\pm\sqrt{10}}{3}$$

$$x=\frac{-1\pm\sqrt{10}}{3}\qquad x=\frac{-1-\sqrt{10}}{3}$$

$$x\approx 0.7208\qquad x\approx -1.3874$$

The solutions are –1.3874 and 0.7208.

17. $y=x^2+2x-4$

x	y
–4	4
–2	–4
–1	–5
0	–4
2	4

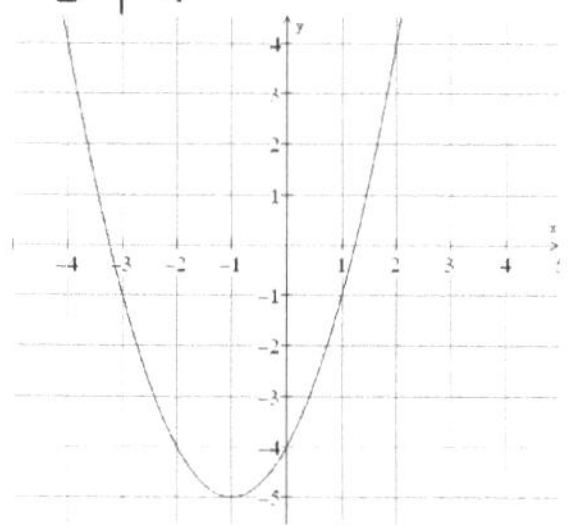

18. x-intercepts: $f(x)=0$

$$x^2+x-12=0$$

$$(x+4)(x-3)=0$$

$$x+4=0\qquad x-3=0$$

$$x=-4\qquad x=3$$

(–4, 0) and (3, 0)

y-intercept: $x=0$

$$f(0)=0^2+(0)-12=-12$$

(0, –12)

19. Strategy:

This is a geometry problem.

→Width: x

Length: $2x-2$

→Use the formula for the area of a rectangle.

Solution:

$$LW=A$$

$$x(2x-2)=40$$

$$2x^2-2x-40=0$$

$$x^2-x-20=0$$

$$(x-5)(x+4)=0$$

$$x-5=0\qquad x+4=0$$

$$x=5\qquad x=-4$$

The solution –4 is not possible.

$$2x-2=2(5)-2=10-2=8$$

The length is 8 ft and the width is 5 ft.

20. Strategy:
This is a distance-rate problem.
→Rate of the boat in calm water: r

	Distance	Rate	Time
Against current	60	$r-1$	$\frac{60}{r-1}$
With curent	60	$r+1$	$\frac{60}{r+1}$

→The time spent traveling against the current is 1 h more than the time spent traveling with the current.

Solution:

$$\frac{60}{r-1}=1+\frac{60}{r+1}$$
$$(r-1)(r+1)\left(\frac{60}{r-1}\right)=(r-1)(r+1)\left(1+\frac{60}{r+1}\right)$$
$$60r+60=r^2-1+60r-60$$
$$0=r^2-121$$
$$121=r^2$$
$$\sqrt{121}=\sqrt{r^2}$$
$$\pm 11=r$$

The solution −11 is not possible.
The rate of the boat is calm water is 11 mph.

Cumulative Review Exercises

1. $2x-3[2x-4(3-2x)+2]-3$
$=2x-3[2x-12+8x+2]-3$
$=2x-3[10x-10]-3$
$=2x-30x+30-3$
$=-28x+27$

2.
$$-\frac{3}{5}x=-\frac{9}{10}$$
$$-\frac{5}{3}\left(-\frac{3}{5}x\right)=-\frac{5}{3}\left(-\frac{9}{10}\right)$$
$$x=\frac{3}{2}$$

The solution is $\frac{3}{2}$.

3.
$$2x-3(4x-5)=-3x-6$$
$$2x-12x+15=-3x-6$$
$$-10x+15=-3x-6$$
$$-7x=-21$$
$$\frac{-7x}{-7}=\frac{-21}{-7}$$
$$x=3$$

The solution is 3.

4.
$$(2a^2b)^2(-3a^4b^2)=(4a^4b^2)(-3a^4b^2)$$
$$=-12a^8b^4$$

5.
$$\begin{array}{r} x+2 \\ x-2\overline{)x^2+0x-8} \\ \underline{x^2-2x} \\ 2x-8 \\ \underline{2x-4} \\ -4 \end{array}$$

$$(x^2-8)\div(x-2)=x+2-\frac{4}{x-2}$$

6.
$$3x^3+2x^2-8x$$
$$=x(3x^2+2x-8)$$
$$=x(3x-4)(x+2)$$

7.
$$\frac{3x^2-6x}{4x-6}\div\frac{2x^2+x-6}{6x^3-24x}$$
$$=\frac{3x^2-6x}{4x-6}\cdot\frac{6x^3-24x}{2x^2+x-6}$$
$$=\frac{3x(x-2)}{2(2x-3)}\cdot\frac{6x(x^2-4)}{(2x-3)(x+2)}$$
$$=\frac{3x(x-2)}{2(2x-3)}\cdot\frac{6x(x-2)\overset{1}{\cancel{(x+2)}}}{(2x-3)\underset{2}{\cancel{(x+2)}}}$$
$$=\frac{9x^2(x-2)^2}{(2x-3)^2}$$

8. $\dfrac{x}{2(x-1)}-\dfrac{1}{(x-1)(x+1)}$

$\text{LCM}=2(x-1)(x+1)$

$\dfrac{x}{2(x-1)}\cdot\dfrac{(x+1)}{(x+1)}-\dfrac{1}{(x-1)(x+1)}\cdot\dfrac{2}{2}$

$=\dfrac{x^2+x}{2(x-1)(x+1)}-\dfrac{2}{2(x-1)(x+1)}$

$=\dfrac{x^2+x-2}{2(x-1)(x+1)}$

$=\dfrac{(x+2)\cancel{(x-1)}}{2\cancel{(x-1)}(x+1)}=\dfrac{x+2}{2(x+1)}$

9. $\dfrac{1-\frac{7}{x}+\frac{12}{x^2}}{2-\frac{1}{x}-\frac{15}{x^2}}=\dfrac{1-\frac{7}{x}+\frac{12}{x^2}}{2-\frac{1}{x}-\frac{15}{x^2}}\cdot\dfrac{x^2}{x^2}$

$=\dfrac{x^2-7x+12}{2x^2-x-15}=\dfrac{\cancel{(x-3)}(x-4)}{(2x+5)\cancel{(x-3)}}$

$=\dfrac{x-4}{2x+5}$

10. $4x-3y=12$

x-intercept	y-intercept
$4x-3(0)=12$	$4(0)-3y=12$
$4x=12$	$-3y=12$
$x=3$	$y=-4$
(3, 0)	(0, –4)

11. $m=-\dfrac{4}{3},\ (x_1,\ y_1)=(-3,\ 2)$

$y-y_1=m(x-x_1)$

$y-2=-\dfrac{4}{3}(x-(-3))$

$y-2=-\dfrac{4}{3}(x+3)$

$y-2=-\dfrac{4}{3}x-4$

$y=-\dfrac{4}{3}x-2$

The equation of the line is $y=-\dfrac{4}{3}x-2$.

12. (1) $3x-y=5$

(2) $y=2x-3$

Substitute $2x-3$ for y in Equation (1).

$3x-y=5$

$3x-(2x-3)=5$

$3x-2x+3=5$

$x+3=5$

$x=2$

Substitute 2 for x in Equation (2).

$y=2x-3$

$y=2(2)-3$

$y=4-3$

$y=1$

The solution is (2, 1).

13. (1) $3x+2y=2$

(2) $5x-2y=14$

Eliminate y by adding the equations.

$3x+2y=2$

$\underline{5x-2y=14}$

$8x=16$

$x=2$

Substitute 2 for x in Equation (1).

$3x+2y=2$

$3(2)+2y=2$

$6+2y=2$

$2y=-4$

$y=-2$

The solution is (2, –2).

14. $2x-3(2-3x)>2x-5$

$2x-6+9x>2x-5$

$11x-6>2x-5$

$9x>1$

$x>\dfrac{1}{9}$

15. $(\sqrt{a}-\sqrt{2})(\sqrt{a}+\sqrt{2})=\sqrt{a^2}-\sqrt{4}=a-2$

16. $\dfrac{\sqrt{108a^7b^3}}{\sqrt{3a^4b}}=\sqrt{\dfrac{108a^7b^3}{3a^4b}}=\sqrt{36a^3b^2}$

$=\sqrt{36a^2b^2}\sqrt{a}=6ab\sqrt{a}$

17. $\frac{\sqrt{3}}{5+2\sqrt{3}}=\frac{\sqrt{3}}{5+2\sqrt{3}}\cdot\frac{5-2\sqrt{3}}{5-2\sqrt{3}}$

$=\frac{5\sqrt{3}-2\sqrt{9}}{25-4\sqrt{9}}=\frac{5\sqrt{3}-2\cdot 3}{25-4(3)}$

$=\frac{5\sqrt{3}-6}{25-12}=\frac{-6+5\sqrt{3}}{13}$

18. $3=8-\sqrt{5x}$

$\sqrt{5x}=5$

$\left(\sqrt{5x}\right)^2=5^2$

$5x=25$

$x=5$

The solution is 5.

Check:

$3=8-\sqrt{5x}$	
3	$8-\sqrt{5(5)}$
3	$8-\sqrt{25}$
3	$8-5$

$3=3$

19. $6x^2-17x=-5$

$6x^2-17x+5=0$

$(2x-5)(3x-1)=0$

$2x-5=0 \qquad 3x-1=0$

$2x=5 \qquad 3x=1$

$x=\frac{5}{2} \qquad x=\frac{1}{3}$

The solutions are $\frac{1}{3}$ and $\frac{5}{2}$.

20. $2(x-5)^2=36$

$(x-5)^2=18$

$\sqrt{(x-5)^2}=\sqrt{18}$

$x-5=\pm\sqrt{9}\sqrt{2}$

$x-5=\pm 3\sqrt{2}$

$x=5\pm 3\sqrt{2}$

The solutions are $5-3\sqrt{2}$ and $5+3\sqrt{2}$.

21. $3x^2+7x=-3$

$x^2+\frac{7}{3}x=-1$

Complete the square. $\left(\frac{7}{3}\cdot\frac{1}{2}\right)^2=\left(\frac{7}{6}\right)^2=\frac{49}{36}$

$x^2+\frac{7}{3}x+\frac{49}{36}=-1+\frac{49}{36}$

$x^2+\frac{7}{3}x+\frac{49}{36}=-\frac{36}{36}+\frac{49}{36}$

$\left(x+\frac{7}{6}\right)^2=\frac{13}{36}$

$\sqrt{\left(x+\frac{7}{6}\right)^2}=\sqrt{\frac{13}{36}}$

$x+\frac{7}{6}=\pm\frac{\sqrt{13}}{6}$

$x=-\frac{7}{6}\pm\frac{\sqrt{13}}{6}$

$x=\frac{-7\pm\sqrt{13}}{6}$

The equation solutions are $\frac{-7-\sqrt{13}}{6}$ and $\frac{-7+\sqrt{13}}{6}$.

22. $2x^2-3x-2=0$

$a=2,\ b=-3,\ c=-2$

$x=\frac{-(-3)\pm\sqrt{(-3)^2-4(2)(-2)}}{2(2)}$

$x=\frac{3\pm\sqrt{9+16}}{4}$

$x=\frac{3\pm\sqrt{25}}{4}$

$x=\frac{3\pm 5}{4}$

$x=\frac{3+5}{4} \qquad x=\frac{3-5}{4}$

$x=\frac{8}{4}=2 \qquad x=\frac{-2}{4}=-\frac{1}{2}$

The solutions are $-\frac{1}{2}$ and 2.

23. Strategy:

→Cost of mixture: x

	Amount	Cost	Value
Cashews	20	4.90	20(4.90)
Peanuts	50	2.10	50(2.10
Mixture	70	x	$70x$

→The sum of the values before mixing equals the value after mixing.

Solution:

$$20(4.90)+50(2.10)=70x$$
$$98+105=70x$$
$$203=70x$$
$$2.90=x$$

The cost of the mixture is $2.90 per pound.

24. Strategy:

Write and solve a proportion using x to represent the additional shares.

Solution:

$$\frac{100}{215}=\frac{x+100}{752.50}$$
$$(215)(752.50)\left(\frac{100}{215}\right)=(215)(752.50)\left(\frac{x+100}{752.50}\right)$$
$$75250=215x+21500$$
$$53{,}750=215x$$
$$250=x$$

250 additional shares are required.

25. Strategy:

→Rate of the plane in still air: p

Rate of the wind: w

	Rate	Time	Distance
With the wind	$p+w$	3	$3(p+w)$
Against the wind	$p-w$	4.5	$4.5(p-w)$

→The distance traveled with the wind is equal to the distance traveled against the wind.

Solution:

$$3(p+w)=720 \qquad \frac{1}{3}\cdot 3(p+w)=\frac{1}{3}(720)$$
$$4.5(p-w)=720 \qquad \frac{1}{4.5}\cdot 4.5(p-w)=\frac{1}{4.5}(720)$$

Add the equations and eliminate w.

$$p+w=240$$
$$\underline{p-w=160}$$
$$2p=400$$
$$p=200$$

Substitute 200 for p.

$$p+w=240$$
$$200+w=240$$
$$w=40$$

The rate of the plane in still air is 200 mph. The rate of the wind is 40 mph.

26. Strategy:

→Score on last test: x

→The sum of all the scores must be a minimum of 400 points.

Solution:

$$70+91+85+77+x\ge 400$$
$$323+x\ge 400$$
$$x\ge 77$$

The score on the last test must be 77 or better.

27. Strategy:
→First odd integer: x
Second odd integer: $x+2$
Third odd integer: $x+4$
→The sum of the squares of the three consecutive odd integers is 83.

Solution:

$$x^2+(x+2)^2+(x+4)^2=83$$
$$x^2+x^2+4x+4+x^2+8x+16=83$$
$$3x^2+12x+20=83$$
$$3x^2+12x-63=0$$
$$3(x^2+4x-21)=0$$
$$3(x+7)(x-3)=0$$

$x+7=0$ $\quad$ $x-3=0$
$x=-7$ $\quad$ $x=3$
$x+2=-7+2=-5$ $\quad$ $x+2=3+2=5$
$x+4=-7+4=-3$ $\quad$ $x+4=3+4=7$
The middle integer can be –5 or 5.

28. Strategy:
→Constant rate: r
Reduced rate: $r-3$

	Distance	Rate	Time
First 7 miles	7	r	$\frac{7}{r}$
Second 8 miles	8	$r-3$	$\frac{8}{r-3}$

→The total time for the trip was 3 h.

Solution:

$$\frac{7}{r}+\frac{8}{r-3}=3$$
$$r(r-3)\left(\frac{7}{r}+\frac{8}{r-3}\right)=r(r-3)3$$
$$7r-21+8r=3r^2-9r$$
$$0=3r^2-24r+21$$
$$0=3(r^2-8r+7)$$
$$0=3(r-7)(r-1)$$

$r-7=0$ $\quad$ $r-1=0$
$r=7$ $\quad$ $r=1$
$r-3=7-3=4$ $\quad$ $r-3=1-3=-2$
The solution –2 is not possible.
The rate for the last 8 mi is 4 mph.

29. $2x-3y>6$

$$-3y>-2x+6$$
$$\frac{-3y}{-3}<\frac{-2x+6}{-3}$$
$$y<\frac{2}{3}x-2$$

Graph the line $y<\frac{2}{3}x-2$ as a dashed line and shade below the line.

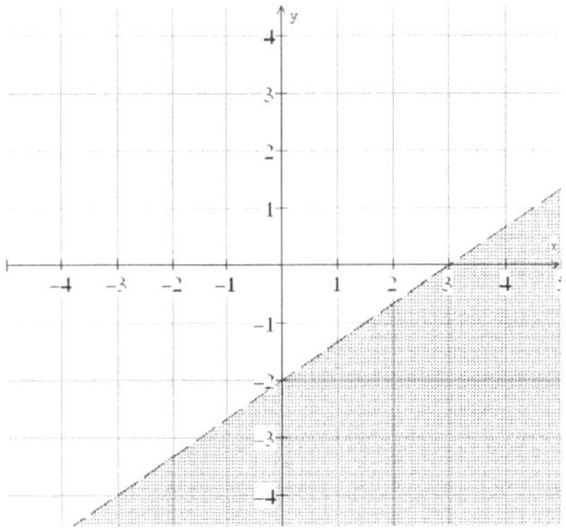

30. $y=x^2-2x-3$

x	y
–1	0
1	–4
3	0

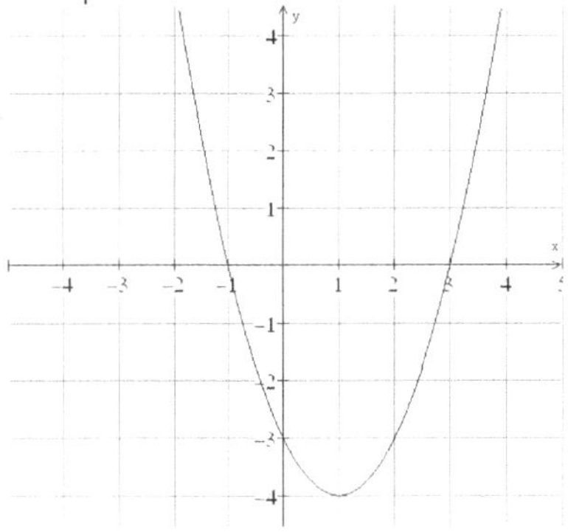

Final Exam

1. $-|-3| = -(3) = -3$

2. $-15-(-12)-3$
$= -15+12+(-3)$
$= -3+(-3)$
$= -6$

3. $-\frac{4}{5}-\left(-\frac{3}{10}\right) = -\frac{4}{5}+\frac{3}{10}$
$\frac{-8}{10}+\frac{3}{10} = \frac{-8+3}{10} = \frac{-5}{10} = -\frac{1}{2}$

4. $-7-\frac{12-15}{2-(-1)}\cdot(-4)$
$= -7-\frac{-3}{3}\cdot(-4)$
$= -7-(-1)(-4)$
$= -7-4 = -11$

5. $\frac{a^2-3b}{2a-2b^2}$
$\frac{3^2-3(-2)}{2(3)-2(-2)^2}$
$= \frac{9+6}{6-2(4)}$
$= \frac{15}{6-8} = -\frac{15}{2}$

6. $6x-(-4y)-(-3x)+2y$
$= 6x+4y+3x+2y$
$= 9x+6y$

7. $(-15z)\left(-\frac{2}{5}\right) = \left(\frac{-15z}{1}\right)\left(-\frac{2}{5}\right) = 6z$

8. $-2[5-3(2x-7)-2x]$
$= -2[5-6x+21-2x]$
$= -2[-8x+26]$
$= 16x-52$

9.
$20 = -\frac{2}{5}x$
$-\frac{5}{2}(20) = -\frac{5}{2}\left(-\frac{2}{5}x\right)$
$-50 = x$
The solution is –50.

10. $4-2(3x+1) = 3(2-x)+5$
$4-6x-2 = 6-3x+5$
$-6x+2 = -3x+11$
$-3x = 9$
$x = -3$
The solution is –3.

11. $\frac{1}{8} = 0.125 = 12.5\%$

12. $P \times B = A$
$19\% \times 80 = A$
$0.19 \times 80 = A$
$15.2 = A$

13. $(2x^2-5x+1)-(5x^2-2x-7)$
$2x^2-5x+1$
$\underline{-5x^2+2x+7}$
$-3x^2-3x+8$

14. $(-3xy^3)^4 = (-3)^4x^{1\cdot 4}y^{3\cdot 4} = 81x^4y^{12}$

15. $(3x^2-x-2)(2x+3)$
$3x^2-x-2$
$\underline{2x+3}$
$9x^2-3x-6$
$\underline{6x^3-2x^2-4x}$
$6x^3+7x^2-7x-6$

16. $\frac{(-2x^2y^3)^3}{(-4xy^4)^2} = \frac{-8x^6y^9}{16x^2y^8} = -\frac{x^4y}{2}$

17. $\dfrac{12x^2y-16x^3y^2-20y^2}{4xy^2}$

$=\dfrac{12x^2y}{4xy^2}-\dfrac{16x^3y^2}{4xy^2}-\dfrac{20y^2}{4xy^2}$

$=\dfrac{3x}{y}-4x^2-\dfrac{5}{x}$

18.

$$\begin{array}{r}
5x-12 \\
x+2\overline{)5x^2-2x-1} \\
\underline{5x^2+10x} \\
-12x-1 \\
\underline{-12x-24} \\
23
\end{array}$$

$\left(5x^2-2x-1\right)\div(x+2)=5x-12+\dfrac{23}{x+2}$

19. $\left(4x^{-2}y\right)^2\left(2xy^{-2}\right)^{-2}$

$=\left(16x^{-4}y^2\right)\left(2^{-2}x^{-2}y^4\right)$

$=\left(\dfrac{16y^2}{x^4}\right)\left(\dfrac{y^4}{4x^2}\right)=\dfrac{4y^6}{x^6}$

20. $f(t)=\dfrac{t}{t+1}$

$t(3)=\dfrac{3}{3+1}=\dfrac{3}{4}$

21. $x^2-5x-6=(x-6)(x+1)$

22. $6x^2-5x-6=(3x+2)(2x-3)$

23. $8x^3-28x^2+12x=4x\left(2x^2-7x+3\right)$

$=4x(2x-1)(x-3)$

24. $25x^2-16=(5x)^2-(4)^2$

$=(5x+4)(5x-4)$

25. $2a(4-x)-6(x-4)=2a(4-x)+6(4-x)$

$=(4-x)(2a+6)$

$=2(a+3)(4-x)$

26. $75y-12x^2y=3y\left(25-4x^2\right)$

$=3y\left[(5)^2-(2x)^2\right]$

$=3y(5+2x)(5-2x)$

27. $2x^2=7x-3$

$2x^2-7x+3=0$

$(2x-1)(x-3)=0$

$2x-1=0 \qquad x-3=0$

$2x=1 \qquad x=3$

$x=\dfrac{1}{2}$

The solutions are $\dfrac{1}{2}$ and 3.

28. $\dfrac{2x^2-3x+1}{4x^2-2x}\cdot\dfrac{4x^2+4x}{x^2-2x+1}$

$=\dfrac{\overset{1}{\cancel{(2x-1)}}\,\overset{1}{\cancel{(x-1)}}}{2x\,\underset{1}{\cancel{(2x-1)}}}\cdot\dfrac{4x(x+1)}{\underset{1}{\cancel{(x-1)}}\,(x-1)}$

$=\dfrac{2(x+1)}{x-1}$

29. $\dfrac{5}{x+3}-\dfrac{3x}{2x-5}$

LCD $=(x+3)(2x-5)$

$\dfrac{5}{x+3}\cdot\dfrac{(2x-5)}{(2x-5)}-\dfrac{3x}{2x-5}\cdot\dfrac{(x+3)}{(x+3)}$

$=\dfrac{10x-25-3x^2-9x}{(x+3)(2x-5)}=\dfrac{-3x^2+x-25}{(x+3)(2x-5)}$

30. $x-\dfrac{1}{1-\frac{1}{x}}=x-\dfrac{1}{1-\frac{1}{x}}\cdot\dfrac{x}{x}$

$=x-\dfrac{x}{x-1}=\dfrac{x}{1}\cdot\dfrac{x-1}{x-1}-\dfrac{x}{x-1}$

$=\dfrac{x^2-x}{x-1}-\dfrac{x}{x-1}=\dfrac{x^2-2x}{x-1}$

31. $\dfrac{5x}{3x-5}-3=\dfrac{7}{3x-5}$

$(3x-5)\left(\dfrac{5x}{3x-5}-3\right)=(3x-5)\left(\dfrac{7}{3x-5}\right)$

$5x-9x+15=7$

$-4x=-8$

$x=2$

The solution is 2.

32.

$$\begin{aligned} a &= 3a - 2b \\ -2a &= -2b \\ a &= b \end{aligned}$$

33.

$$\begin{aligned} m &= \frac{y_2 - y_1}{x_2 - x_1} \\ &= \frac{-1-(-3)}{2-(-1)} = \frac{-1+3}{2+1} = \frac{2}{3} \end{aligned}$$

34.

$$m = -\frac{2}{3},\ (x_1, y_1) = (3, -4)$$

$$\begin{aligned} y - y_1 &= m(x - x_1) \\ y-(-4) &= -\frac{2}{3}(x-3) \\ y+4 &= -\frac{2}{3}x+2 \\ y &= -\frac{2}{3}x-2 \end{aligned}$$

The equation of the line is $y = -\frac{2}{3}x - 2$.

35.

(1) $y = 4x - 7$

(2) $y = 2x + 5$

Substitute $4x - 7$ for y in Equation (2).

$$\begin{aligned} 4x-7 &= 2x+5 \\ 2x &= 12 \\ x &= 6 \end{aligned}$$

Substitute 6 for x in Equation (1).

$$\begin{aligned} y &= 4x-7 \\ y &= 4(6)-7 \\ y &= 24-7 \\ y &= 17 \end{aligned}$$

The solution is (6, 17).

36.

(1) $4x - 3y = 11$

(2) $2x + 5y = -1$

Eliminate x by adding the equations.

$$\begin{aligned} 4x-3y &= 11 \qquad & 4x-3y &= 11 \\ -2(2x+5y) &= -2(-1) \qquad & -4x-10y &= 2 \\ & & -13y &= 13 \\ & & y &= -1 \end{aligned}$$

Substitute -1 for y in Equation (2).

$$\begin{aligned} 2x+5y &= -1 \\ 2x+5(-1) &= -1 \\ 2x-5 &= -1 \\ 2x &= 4 \\ x &= 2 \end{aligned}$$

The solution is (2, −1).

37.

$$\begin{aligned} 4-x &\ge 7 \\ -x &\ge 3 \\ \frac{-x}{-1} &\le \frac{3}{-1} \\ x &\le -3 \end{aligned}$$

38.

$$\begin{aligned} 2-2(y-1) &\le 2y-6 \\ 2-2y+2 &\le 2y-6 \\ -2y+4 &\le 2y-6 \\ -4y &\le -10 \\ \frac{-4y}{-4} &\ge \frac{-10}{-4} \\ y &\ge \frac{5}{2} \end{aligned}$$

39.

$$\sqrt{49x^6} = 7x^3$$

40.

$$\begin{aligned} 2\sqrt{27a}+8\sqrt{48a} &= 2\sqrt{9\cdot 3a}+8\sqrt{16\cdot 3a} \\ &= 2\sqrt{9}\sqrt{3a}+8\sqrt{16}\sqrt{3a} \\ &= 2\cdot 3\sqrt{3a}+8\cdot 4\sqrt{3a} \\ &= 6\sqrt{3a}+32\sqrt{3a} \\ &= 38\sqrt{3a} \end{aligned}$$

41.

$$\begin{aligned} \frac{\sqrt{3}}{\sqrt{5}-2} &= \frac{\sqrt{3}}{\sqrt{5}-2}\cdot\frac{\sqrt{5}+2}{\sqrt{5}+2} \\ &= \frac{\sqrt{15}+2\sqrt{3}}{5-4} = \frac{\sqrt{15}+23}{1} \\ &= \sqrt{15}+2\sqrt{3} \end{aligned}$$

42.

$$\begin{aligned} \sqrt{2x-3}+4 &= 5 \\ \sqrt{2x-3} &= 1 \\ \left(\sqrt{2x-3}\right)^2 &= 1^2 \\ 2x-3 &= 1 \\ 2x &= 4 \\ x &= 2 \end{aligned}$$

The solution is 2.

Check:

$$\begin{array}{r|l} \multicolumn{2}{c}{\sqrt{2x-3}+4=5} \\ \hline \sqrt{2(2)-3}+4 & 5 \\ \sqrt{4-3}+4 & 5 \\ \sqrt{1}+4 & 5 \\ 1+4 & 5 \\ \multicolumn{2}{c}{5=5} \end{array}$$

43.
$$3x^2 - x = 4$$
$$3x^2 - x - 4 = 0$$
$$(3x-4)(x+1) = 0$$
$$3x - 4 = 0 \qquad x + 1 = 0$$
$$3x = 4 \qquad x = -1$$
$$x = \frac{4}{3}$$
The solutions are $\frac{4}{3}$ and -1.

44. $4x^2 - 2x - 1 = 0$

$a = 4,\ b = -2,\ c = -1$
$$x = \frac{-(-2) \pm \sqrt{(-2)^2 - 4(4)(-1)}}{2(4)}$$
$$x = \frac{2 \pm \sqrt{4+16}}{8}$$
$$x = \frac{2 \pm \sqrt{20}}{8}$$
$$x = \frac{2 \pm \sqrt{4}\sqrt{5}}{8}$$
$$x = \frac{2 \pm 2\sqrt{5}}{8}$$
$$x = \frac{1 \pm \sqrt{5}}{4}$$
The solutions are $\frac{1-\sqrt{5}}{4}$ and $\frac{1+\sqrt{5}}{4}$.

45. Strategy:
Unknown number: x

Solution:
$2x + 3(x-2) = 2x + 3x - 6 = 5x - 6$

46.
$$P \cdot B = A$$
$$80\% \cdot x = 4800$$
$$0.80x = 4800$$
$$x = 6000$$
The original value is $6000.

47. Strategy:
To find the mean snowfall:
→Determine the sum of the snowfall amounts.
→Divide the sum by 12.

Solution:
The sum of the snowfall amounts is
5.8+5.2+4.3+1.2+0.8+0.3+2.4+4.3
= 24.3
$\bar{x} = \frac{24.3}{12} = 2.025$
The mean snowfall is 2.025 in.

Strategy:
To find the median snowfall:
→Arrange the numbers from smallest to largest.
→The median is the middle term.

Solution:

0	0	0	0	0.3	0.8
1.2	2.4	4.3	4.3	5.2	5.8

median $= \frac{0.8+1.2}{2} = 1$
The median snowfall is 1 in.

Strategy:
To find the modal response, write down the category that received the most responses.

Solution:
Because 0 was recorded most frequently, the modal snowfall is 0 in.

48. Strategy:
→Amount of money to be invested at 11%: x

	Principal	Rate	Interest
At 8%	3000	0.08	0.08(3000)
At 11%	x	0.11	$0.11x$
Total	$x + 3000$	0.10	$0.10(x + 3000)$

→The total interest earned is 10% of the total investment.

Solution:
$$0.08(3000) + 0.11x = 0.10(x + 3000)$$
$$240 + 0.11x = 0.10x + 300$$
$$0.01x - 60$$
$$x = 6000$$
$6000 must be invested at 11%.

49. Strategy:

→Cost per pound of mixture: x

	Amount	Cost	Value
Peanuts	4	2.50	4(2.50)
Walnuts	2	7	2(7)
Mixture	6	x	$6x$

→The sum of the values before mixing equals the value after mixing.

Solution:

$$2.50(4)+7(2)=6x$$
$$10+14=6x$$
$$24=6x$$
$$4=x$$

The cost of the mixture is \$4.00 per pound.

50. Strategy:

To calculate the probability:

→Count the number of possible outcomes.

→Count the number of favorable outcomes.

→Use the probability formula.

Solution

There are 9 possible outcomes.

There are 6 favorable outcomes of not drawing a queen.

$$\text{Probability} = \frac{6}{9} = \frac{2}{3}$$

The probability of not choosing a queen is $\frac{2}{3}$.

51. Strategy:

→Rate on the first part of the trip: x

	Rate	Time	Distance
1st part	x	1	$1x$
2nd part	$2x$	1.5	$1.5(2x)$

→The total distance is 860 km.

Solution:

$$1x+1.5x(2x)=860$$
$$1x+3x=860$$
$$4x=860$$
$$x=215$$

The distance for the first part of the trip was 215 km.

52. Strategy:

→First angle: x

Second angle: $x + 10$

Third angle: $x + 20$

→The sum of the angles of a triangle is 180°.

Solution:

$$x+x+10+x+20=180$$
$$3x+30=180$$
$$3x=150$$
$$x=50$$
$$x+10=50+10=60$$
$$x+20=50+20=70$$

The angles are 50°, 60°, and 70°.

53. Strategy:

→First consecutive integer: x

Second consecutive integer: $x + 1$

Third consecutive integer: $x + 2$

→The sum of the squares of the three consecutive integers is 50.

Solution:

$$x^2+(x+1)^2+(x+2)^2=50$$
$$x^2+x^2+2x+1+x^2+4x+4=50$$
$$3x^2+6x-45=0$$
$$3(x^2+2x-15)=0$$
$$3(x+5)(x-3)=0$$

$x+5=0$ $\qquad$ $x-3=0$

$x=-5$ $\qquad$ $x=3$

$x+1=-5+1=-4$ $\qquad$ $x+1=3+1=4$

$x+2=-5+2=-3$ $\qquad$ $x+2=3+2=5$

The middle integer can be 4 or –4.

54. Strategy:
This is a work problem.
→Time for chef and apprentice working together: t

	Rate	Time	Part
Chef	$\frac{1}{60}$		$\frac{t}{60}$
Apprentice	$\frac{1}{90}$	t	$\frac{t}{90}$

→The sum of the parts of the task completed must equal 1.

Solution:

$$\frac{t}{60}+\frac{t}{90}=1$$
$$60\cdot 90\left(\frac{t}{60}+\frac{t}{90}\right)=60\cdot 90(1)$$
$$90t+60t=5400$$
$$150t=5400$$
$$t=36$$

$$\frac{36}{60}=0.6\text{ h}$$

Working together it would take them 36 min or 0.6 h.

55. Strategy:
→Width: x
Length: $x + 5$
→Use the equation for the area of a rectangle.

Solution:

$$A=LW$$
$$50=x(x+5)$$
$$50=x^2+5x$$
$$0=x^2+5x-50$$
$$0=(x+10)(x-5)$$
$$x+10=0 \qquad x-5=0$$
$$x=-10 \qquad x=5$$
$$x+5=5+5=10$$

The solution −10 is not possible. The width is 5 m. The length is 10 m.

56. Strategy:
Write and solve a proportion using x to represent the number of ounces of dye required.

Solution:

$$\frac{2}{15}=\frac{x}{120}$$
$$15\cdot 120\left(\frac{2}{15}\right)=15\cdot 120\left(\frac{x}{120}\right)$$
$$240=15x$$
$$16=x$$

16 oz of dye are required.

57. Strategy:
→Rate of the boat in calm water: x
Rate of the current: c

	Rate	Time	Distance
With current	$x+c$	2.5	$2.5(x+c)$
Against current	$x-c$	5	$5(x-c)$

→The distance traveled with the current is equal to the distance traveled against the current.

Solution:

$$2.5(x+c)=50 \qquad \frac{1}{2.5}(x+c)=\frac{1}{2.5}(50)$$
$$5(x-c)=50 \qquad \frac{1}{5}(x-c)=\frac{1}{5}(50)$$

Add the equations and eliminate c.

$$x+c=20$$
$$x-c=10$$
$$2x=30$$
$$x=15$$

Substitute 15 for x.

$$x+c=20$$
$$15+c=20$$
$$c=5$$

The rate of the boat in calm water is 15 mph.
The rate of the current is 5 mph.

58. Strategy:

→Rate of the wind: r

	Distance	Rate	Time
With wind	500	$225+r$	$\dfrac{500}{225+r}$
Against wind	500	$225-r$	$\dfrac{500}{225-r}$

→The time spent traveling against the wind is $\frac{1}{2}$ h more than the time spent traveling with the wind.

Solution:

$$\frac{500}{225+r}+\frac{1}{2}=\frac{500}{225-r}$$

$$2(225+r)(225-r)\left(\frac{500}{225+r}+\frac{1}{2}\right)$$
$$=2(225+r)(225-r)\left(\frac{500}{225-r}\right)$$
$$225{,}000-1000r+50{,}625-r^2$$
$$=225{,}000+1000r$$
$$0=r^2+2000r-50{,}625$$
$$0=(r+2025)(r-25)$$

$r+2025=0$ $r-25=0$

$r=-2025$ $r=25$

The solution –2025 is not possible.
The rate of the wind is 25 mph.

59. $\{x \quad 4\}$

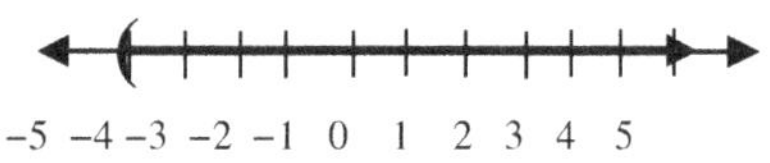

60. y-intercept = $(0, b) = (0, 4)$

$$m=-\frac{3}{2}=\frac{-3}{2}=\frac{\text{change in } y}{\text{change in } x}$$

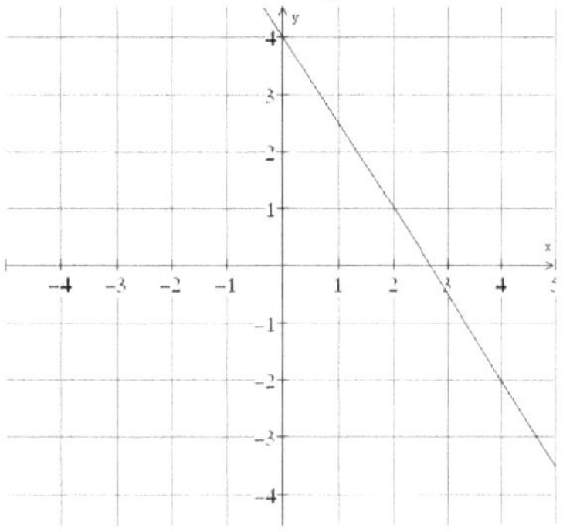

61. $x-4y=4$

x-intercept = (4, 0) y-intercept = (0, –1)

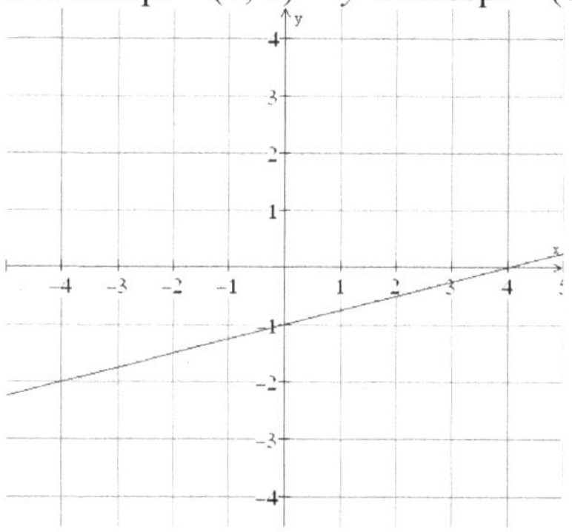

62. y-intercept = $(0, b) = (0, -3)$

$$m=-\frac{1}{2}=\frac{-1}{2}=\frac{\text{change in } y}{\text{change in } x}$$

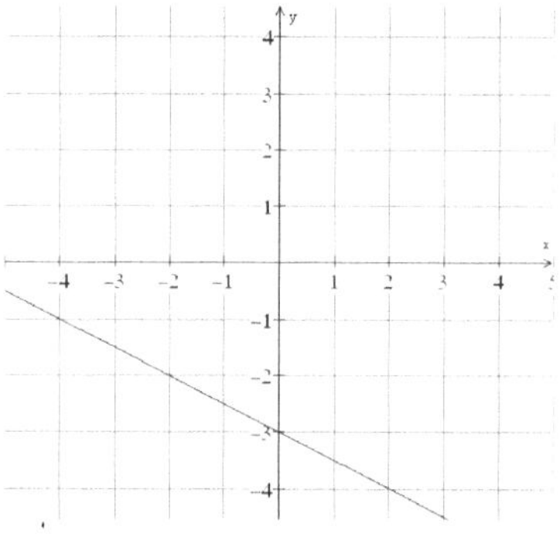

63. $5x-2y<10$

$$-2y<-5x+10$$
$$\frac{-2y}{-2}>\frac{-5x+10}{-2}$$
$$y>\frac{5}{2}x-5$$

Graph $y=\frac{5}{2}x-5$ as a dotted line. Shade above the line.

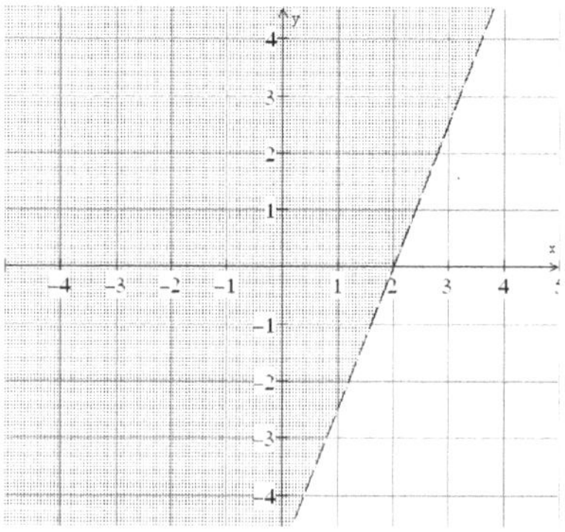

64.

x	y
0	3
1	0
2	−1
3	0
4	3

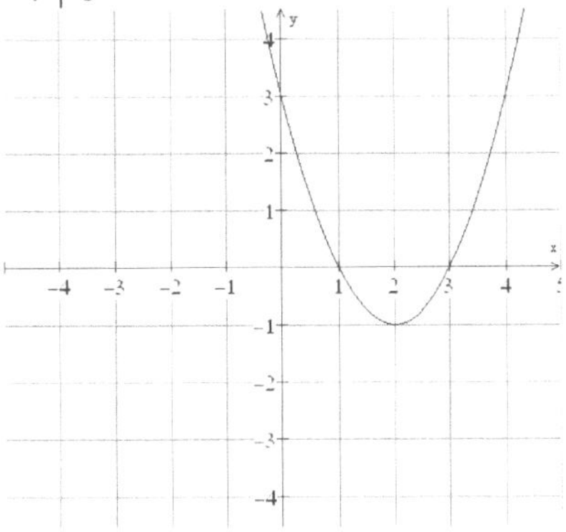

Chapter T: Transitioning to Algebra

Prep Test

1. $$\begin{array}{r} \overset{1}{2}8 \\ +43 \\ \hline 71 \end{array}$$

2. $x - y$
   $3005 - 38$
   $= 2618$

3. $3xy$
   $3(5)(9)$
   $=135$

4. $$\begin{array}{r} 16 \\ 6\overline{)96} \\ \underline{-6} \\ 36 \\ \underline{-36} \\ 0 \end{array}$$

5. $$\begin{array}{r|l} \multicolumn{2}{c}{78} \\ \hline 2 & 39 \\ 3 & 13 \\ 13 & 1 \end{array}$$
   $78 = 2 \cdot 3 \cdot 13$

6. $24 = \mathbf{2} \cdot \mathbf{2} \cdot \mathbf{2} \cdot \mathbf{3}$
   $36 = 2 \cdot 2 \cdot \mathbf{3} \cdot 3$
   $\text{LCM} = 2 \cdot 2 \cdot 2 \cdot 3 \cdot 3 = 72$

7. $24 = \mathbf{2} \cdot \mathbf{2} \cdot 2 \cdot \mathbf{3}$
   $36 = \mathbf{2} \cdot \mathbf{2} \cdot 3 \cdot \mathbf{3}$
   $\text{GCD} = 2 \cdot 2 \cdot 3 = 12$

8. $-37 < -28$

9. $|-27| = 27$

10. $-(-4) = 4$

Section T.1

Objective A Exercises

1. $\frac{2}{7} + \frac{1}{7} = \frac{2+1}{7} = \frac{3}{7}$

3. $\frac{1}{2} + \frac{1}{2} = \frac{1+1}{2} = \frac{2}{2} = 1$

5. $\frac{3}{8} + \frac{7}{8} + \frac{1}{8} = \frac{3+7+1}{8} = \frac{11}{8} = 1\frac{3}{8}$

7. $\frac{1}{2} + \frac{2}{3} = \frac{3}{6} + \frac{4}{6} = \frac{7}{6} = 1\frac{1}{6}$

9. $\frac{3}{14} + \frac{5}{7} = \frac{3}{14} + \frac{10}{14} = \frac{13}{14}$

11. $\frac{8}{15} + \frac{7}{20} = \frac{32}{60} + \frac{21}{60} = \frac{53}{60}$

13. $\frac{3}{20} + \frac{7}{30} = \frac{9}{60} + \frac{14}{60} = \frac{23}{60}$

15. $\frac{1}{3} + \frac{5}{6} + \frac{7}{9} = \frac{6}{18} + \frac{15}{18} + \frac{14}{18} = \frac{35}{18} = 1\frac{17}{18}$

17. $\frac{2}{3} + \frac{1}{5} + \frac{7}{12} = \frac{40}{60} + \frac{12}{60} + \frac{35}{60} = \frac{87}{60} = 1\frac{27}{60} = 1\frac{9}{20}$

19. $\frac{11}{12} - \frac{7}{12} = \frac{11-7}{12} = \frac{4}{12} = \frac{1}{3}$

21. $\frac{3}{4} - \frac{1}{8} = \frac{6}{8} - \frac{1}{8} = \frac{5}{8}$

23. $\frac{5}{9} - \frac{4}{15} = \frac{25}{45} - \frac{12}{45} = \frac{13}{45}$

25. $\frac{9}{20} - \frac{7}{20} = \frac{2}{20} = \frac{1}{10}$

27. $\frac{11}{24} - \frac{5}{24} = \frac{6}{24} = \frac{1}{4}$

29. $\frac{5}{7} - \frac{3}{14} = \frac{10}{14} - \frac{3}{14} = \frac{7}{14} = \frac{1}{2}$

31. $\frac{8}{15} - \frac{7}{20} = \frac{32}{60} - \frac{21}{60} = \frac{11}{60}$

33. $\frac{5}{6} - \frac{3}{4} = \frac{10}{12} - \frac{9}{12} = \frac{1}{12}$

35. $\frac{7}{8} - \frac{2}{3} = \frac{21}{24} - \frac{16}{24} = \frac{5}{24}$

Objective B Exercises

37. $\frac{2}{3} \cdot \frac{7}{8} = \frac{7}{12}$

39. $\frac{15}{16} \cdot \frac{7}{15} = \frac{7}{16}$

41. $\frac{2}{5} \cdot \frac{5}{6} = \frac{1}{3}$

43. $\frac{3}{5}\cdot\frac{10}{11}=\frac{6}{11}$

45. $\frac{8}{9}\cdot\frac{27}{4}=6$

47. $\frac{3}{8}\cdot\frac{5}{12}=\frac{5}{32}$

49. $\frac{7}{8}\cdot\frac{3}{14}=\frac{3}{16}$

51. $\frac{5}{6}\cdot\frac{4}{15}=\frac{2}{9}$

53. $\frac{2}{3}\cdot\frac{5}{4}\cdot\frac{1}{9}=\frac{5}{54}$

55. $\frac{2}{3}\cdot 6=\frac{2}{3}\cdot\frac{6}{1}=4$

57. $\frac{3}{7}\div\frac{3}{2}=\frac{3}{7}\cdot\frac{2}{3}=\frac{2}{7}$

59. $0\div\frac{1}{2}=0\cdot\frac{2}{1}=0$

61. $\frac{2}{15}\div\frac{3}{5}=\frac{2}{15}\cdot\frac{5}{3}=\frac{2}{9}$

63. $\frac{2}{5}\div\frac{4}{7}=\frac{2}{5}\cdot\frac{7}{4}=\frac{7}{10}$

65. $\frac{1}{2}\div\frac{1}{4}=\frac{1}{2}\cdot\frac{4}{1}=2$

67. $\frac{4}{15}\div\frac{2}{5}=\frac{4}{15}\cdot\frac{5}{2}=\frac{2}{3}$

69. $4\div\frac{2}{3}=\frac{4}{1}\cdot\frac{3}{2}=6$

71. $\frac{3}{2}\div 3=\frac{3}{2}\cdot\frac{1}{3}=\frac{1}{2}$

Section T.2

Objective A Exercises

1. $4+(-9)=-5$

3. $(-5)+(-12)=-17$

5. $-5+8=3$

7. $-14+(-6)=-20$

9. $-6+6=0$

11. $64+(-43)=21$

13. $8-15=8+(-15)=-7$

15. $-8-3=-8+(-3)=-11$

17. $7-(-1)=7+1=8$

19. $-9-(-9)=-9+9=0$

21. $-10-15=-10+(-15)=-25$

23. $(-11)-(-2)=-11+2=-9$

25. $6-(-16)=6+16=22$

27. $(-12)-(-6)=-12+6=-6$

29. $8-(-8)=8+8=16$

31. $45-77=45+(-77)=-32$

33. $0+(-15)=-15$

35. $(-21)-(-7)=-21+7=-14$

37. $5-(-6)=5+6=11$

39. $6-(-10)=6+10=16$

41. $7+3-(-3)=7+3+3=13$

43. $$\begin{aligned}-2+(-5)-(-12)&=-2+(-5)+12\\&=-7+12\\&=5\end{aligned}$$

45. $$\begin{aligned}-4+6-9-2&=-4+6+(-9)+(-2)\\&=2+(-9)+(-2)\\&=-7+(-2)\\&=-9\end{aligned}$$

Objective B Exercises

47. $-3\cdot 7=-21$

49. $-5(-7)=35$

51. $4(-9)=-36$

53. $-10(5)=-50$

55. $(-7)(-7)=49$

57. $(-9)(0)=0$

59. $15(4)=60$

61. $-21(6)=-126$

63. $(-3)(-27)=81$

65. $8(-24)=-192$

67. $-5\cdot(17)=-85$

69. $-7(-14)=98$

71. $2\cdot(-8)\cdot 5=-16\cdot 5=-80$

73. $-2(-6)(-3)(4) = 12(-3)(4)$
$= -36(4)$
$= -144$

75. $12 \div (-4) = -3$

77. $(-81) \div (-9) = 9$

79. $0 \div (-4) = 0$

81. $77 \div (-7) = -11$

83. $\frac{36}{-4} = -9$

85. $\frac{-66}{-3} = 22$

87. $-84 \div (-6) = 14$

89. $-48 \div 0$ = undefined

91. $-126 \div 6 = -21$

93. $(-240) \div 6 = -40$

Section T.3

Objective A Exercises

1. $-\frac{3}{4} + \frac{2}{3} = -\frac{9}{12} + \frac{8}{12} = -\frac{1}{12}$

3. $\frac{2}{5} + \left(-\frac{11}{15}\right) = \frac{6}{15} + \left(-\frac{11}{15}\right) = -\frac{5}{15} = -\frac{1}{3}$

5. $-\frac{1}{2} - \frac{3}{8} = -\frac{4}{8} + \left(-\frac{3}{8}\right) = -\frac{7}{8}$

7. $-\frac{3}{10} - \frac{4}{5} = -\frac{3}{10} + \left(-\frac{8}{10}\right) = -\frac{11}{10} = -1\frac{1}{10}$

9. $-\frac{5}{8} - \left(-\frac{7}{12}\right) = -\frac{15}{24} + \frac{14}{24} = -\frac{1}{24}$

11. $-\frac{2}{3} + \left(-\frac{1}{12}\right) = -\frac{8}{12} + \left(-\frac{1}{12}\right) = -\frac{9}{12} = -\frac{3}{4}$

13. $\frac{3}{8} + \left(-\frac{1}{2}\right) + \frac{7}{12} = \frac{9}{24} + \left(-\frac{12}{24}\right) + \frac{14}{24}$
$= -\frac{3}{24} + \frac{14}{24}$
$= \frac{11}{24}$

15. $\frac{2}{3} + \left(-\frac{5}{6}\right) + \frac{1}{4} = \frac{8}{12} + \left(-\frac{10}{12}\right) + \frac{3}{12}$
$= -\frac{2}{12} + \frac{3}{12}$
$= \frac{1}{12}$

17. $-42.1 - 8.6 = -42.1 + (-8.6) = -50.7$

19. $5.73 - 9.042 = 5.73 + (-9.042) = -3.312$

21. $1.09 - (8.3) = 1.09 + 8.3 = 9.39$

23. $-19 - (-2.65) = 19 + 2.65 = -16.35$

25. $-12.3 - 4.07 + 6.82 = -16.37 + 6.82 = -9.55$

27. $-5.6 - (-3.82) - 17.409$
$= -5.6 + 3.82 + (-17.409)$
$= -1.78 + (-17.409)$
$= -19.189$

29. $\frac{11}{12} - \left(-\frac{1}{4}\right)$, because $\frac{11}{12} - \left(-\frac{1}{4}\right) = \frac{11}{12} + \frac{1}{4}$, so the difference is positive, whereas the difference $-\frac{1}{8} - \frac{3}{4}$ is negative.

Objective B Exercises

31. $\frac{3}{8} \cdot \left(-\frac{2}{3}\right) = -\frac{1}{4}$

33. $\left(-\frac{4}{15}\right)\left(-\frac{3}{8}\right) = \frac{1}{10}$

35. $-\frac{3}{4} \cdot \frac{1}{2} = -\frac{3}{8}$

37. $-\frac{7}{12} \cdot \frac{5}{8} \cdot \frac{16}{25} = -\frac{7}{12} \cdot \frac{2}{5} = -\frac{7}{30}$

39. $\left(-\frac{3}{5}\right) \cdot \frac{1}{2} \cdot \left(-\frac{5}{8}\right) = \frac{3}{16}$

41. $12 \cdot \left(-\frac{5}{8}\right) = -\frac{15}{2} = -7\frac{1}{2}$

43. $-9 \cdot \frac{7}{15} = -\frac{21}{5} = -4\frac{1}{5}$

45. $-\frac{5}{2} \cdot 4 = -10$

47. $\left(-\frac{3}{8}\right) \div \frac{7}{8} = -\frac{3}{8} \cdot \frac{8}{7} = -\frac{3}{7}$

49. $\left(-\frac{3}{4}\right) \div \left(-\frac{5}{6}\right) = \left(-\frac{3}{4}\right)\left(-\frac{6}{5}\right) = \frac{9}{10}$

51. $-\frac{2}{3} \div 8 = -\frac{2}{3} \cdot \frac{1}{8} = -\frac{1}{12}$

53. $\frac{3}{8} \div \left(-\frac{5}{12}\right) = \frac{3}{8} \cdot \left(-\frac{12}{5}\right) = -\frac{9}{10}$

55. $(-2.1)(-0.7) = 1.47$

57. $(1.9)(-3.7) = -7.03$

59. $-8.1(-7.5) = 60.75$

61. $-10(0.59) = -5.9$

63. $27.08 \div (-0.4) = -67.7$

65. $(-3.312) \div (-0.8) = 4.14$

67. $-2.501 \div 0.41 = -6.1$

69. $-6.824 \div 0.053 = -128.8$

71. $(-31.792) \div (-0.86) = 37.0$

73. $-\frac{5}{6} \div (-5)$, because the quotient is positive, whereas the product $-\frac{3}{4}\left(\frac{2}{9}\right)$ is negative.

Objective C Exercises

75. $\left(-\frac{2}{7}\right)^3 = \left(-\frac{2}{7}\right)\left(-\frac{2}{7}\right)\left(-\frac{2}{7}\right) = -\frac{8}{343}$

77. $(3.5)^2 = (3.5)(3.5) = 12.25$

79.
$$\begin{aligned}\left(-\frac{9}{11}\right)^2 \cdot \left(\frac{1}{3}\right)^4 &= \left(-\frac{9}{11}\right)\left(-\frac{9}{11}\right)\left(\frac{1}{3}\right)\left(\frac{1}{3}\right)\left(\frac{1}{3}\right)\left(\frac{1}{3}\right)\\ &= \frac{1}{121}\end{aligned}$$

81.
$$\begin{aligned}-3\cdot\left(\frac{2}{5}\right)^2 \cdot \left(-\frac{1}{6}\right)^2 &= -3\left(\frac{2}{5}\right)\left(\frac{2}{5}\right)\left(-\frac{1}{6}\right)\left(-\frac{1}{6}\right)\\ &= -\frac{1}{75}\end{aligned}$$

Objective D Exercises

83.
$$\begin{aligned}0.3(1.7-4.8)+(1.2)^2 &= 0.3(-3.1)+(1.2)(1.2)\\ &= -0.93+1.44\\ &= 0.51\end{aligned}$$

85.
$$\begin{aligned}(1.65-1.05)^2 \div 0.4 + 0.8 &= (0.6)^2 \div 0.4 + 0.8\\ &= 0.36 \div 0.4 + 0.8\\ &= 0.9 + 0.8\\ &= 1.7\end{aligned}$$

87.
$$\begin{aligned}-\frac{3}{4}\left(\frac{11}{12}-\frac{7}{8}\right)+\frac{5}{16} &= -\frac{3}{4}\left(\frac{22}{24}-\frac{21}{24}\right)+\frac{5}{16}\\ &= -\frac{3}{4}\left(\frac{1}{24}\right)+\frac{5}{16}\\ &= -\frac{1}{32}+\frac{5}{16}\\ &= -\frac{1}{32}+\frac{10}{32}\\ &= \frac{9}{32}\end{aligned}$$

89.
$$\begin{aligned}\left(-\frac{2}{3}\right)^2 - \frac{7}{18} + \frac{5}{6} &= \frac{4}{9} - \frac{7}{18} + \frac{5}{6}\\ &= \frac{8}{18} - \frac{7}{18} + \frac{15}{18}\\ &= \frac{1}{18} + \frac{15}{18}\\ &= \frac{16}{18}\\ &= \frac{8}{9}\end{aligned}$$

91.
$$\begin{aligned}\left(-\frac{2}{3}\right)^2 + \left(-\frac{1}{6}\right) \div \frac{3}{8} &= \frac{4}{9} + \left(-\frac{1}{6}\right)\left(\frac{8}{3}\right)\\ &= \frac{4}{9} + \left(-\frac{4}{9}\right)\\ &= 0\end{aligned}$$

93.
$$\begin{aligned}\left(-\frac{1}{4}\right)^2 \div \left(\frac{1}{2}-\frac{3}{4}\right)+\frac{3}{8} &= \frac{1}{16} \div \left(\frac{2}{4}-\frac{3}{4}\right)+\frac{3}{8}\\ &= \frac{1}{16} \div \left(-\frac{1}{4}\right)+\frac{3}{8}\\ &= \frac{1}{16}\left(-\frac{4}{1}\right)+\frac{3}{8}\\ &= -\frac{1}{4}+\frac{3}{8}\\ &= -\frac{2}{8}+\frac{3}{8}\\ &= \frac{1}{8}\end{aligned}$$

95.
$$\begin{aligned}20 \div (6-2^4)+(-5) &= 20 \div (6-16)+(-5)\\ &= 20 \div (-10)+(-5)\\ &= -2+(-5)\\ &= -7\end{aligned}$$

$$\begin{aligned}18 \div |2^3-9|+(-3) &= 18 \div |8-9|+(-3)\\ &= 18 \div |-1|+(-3)\\ &= 18 \div 1+(-3)\\ &= 18+(-3)\\ &= 15\end{aligned}$$

$$\begin{aligned}16+15 \div (-5)-(-4) &= 16+(-3)+4\\ &= 13+4\\ &= 17\end{aligned}$$

$$\begin{aligned}16+15 \div (-5)-(-4) &> 18 \div |2^3-9|+(-3)\\ &> 20 \div (6-2^4)+(-5)\end{aligned}$$

$[17 > 15 > -7]$

Section T.4

Objective A Exercises

1.
$$\begin{aligned}x+7 &= -5\\ x+7-7 &= -5-7\\ x &= -12\end{aligned}$$

3.
$$\begin{aligned}-9 &= z-8\\ -9+8 &= z-8+8\\ -1 &= z\end{aligned}$$

5. $$\begin{aligned} -48 &= 6z \\ \frac{-48}{6} &= \frac{6z}{6} \\ -8 &= z \end{aligned}$$

7. $$\begin{aligned} -\frac{3}{4}x &= 15 \\ \left(-\frac{4}{3}\right)\left(-\frac{3}{4}x\right) &= 15\left(-\frac{4}{3}\right) \\ x &= -20 \end{aligned}$$

9. $$\begin{aligned} -\frac{x}{4} &= -2 \\ (-4)\left(-\frac{x}{4}\right) &= -2(-4) \\ x &= 8 \end{aligned}$$

11. $$\begin{aligned} 3x+8 &= 17 \\ 3x+8-8 &= 17-8 \\ 3x &= 9 \\ \frac{3x}{3} &= \frac{9}{3} \\ x &= 3 \end{aligned}$$

13. $$\begin{aligned} 5 &= 3x-10 \\ 5+10 &= 3x-10+10 \\ 15 &= 3x \\ \frac{15}{3} &= \frac{3x}{3} \\ 5 &= x \end{aligned}$$

15. $$\begin{aligned} \frac{2}{3}x+5 &= 3 \\ \frac{2}{3}x+5-5 &= 3-5 \\ \frac{2}{3}x &= -2 \\ \left(\frac{3}{2}\right)\left(\frac{2}{3}x\right) &= -2\left(\frac{3}{2}\right) \\ x &= -3 \end{aligned}$$

17. $$\begin{aligned} 2b+6-3b &= 4 \\ -b+6 &= 4 \\ -b+6-6 &= 4-6 \\ -b &= -2 \\ \frac{-b}{-1} &= \frac{-2}{-1} \\ b &= 2 \end{aligned}$$

19. $$\begin{aligned} 4-2b &= 2-4b \\ 4-2b+4b &= 2-4b+4b \\ 4+2b &= 2 \\ 4-4+2b &= 2-4 \\ 2b &= -2 \\ \frac{2b}{2} &= \frac{-2}{2} \\ b &= -1 \end{aligned}$$

21. $$\begin{aligned} 5x-3 &= 9x-7 \\ 5x-5x-3 &= 9x-7-5x \\ -3+7 &= 4x-7+7 \\ 4 &= 4x \\ \frac{4}{4} &= \frac{4x}{4} \\ 1 &= x \end{aligned}$$

23. $$\begin{aligned} 2-6y &= 5-7y \\ 2-6y+7y &= 5-7y+7y \\ 2+y &= 5 \\ 2-2+y &= 5-2 \\ y &= 3 \end{aligned}$$

25. $$\begin{aligned} 2(x+1)+5x &= 23 \\ 2x+2+5x &= 23 \\ 7x+2 &= 23 \\ 7x+2-2 &= 23-2 \\ 7x &= 21 \\ \frac{7x}{7} &= \frac{21}{7} \\ x &= 3 \end{aligned}$$

27. $$\begin{aligned} 7a-(3a-4) &= 12 \\ 7a-3a+4 &= 12 \\ 4a+4-4 &= 12-4 \\ 4a &= 8 \\ \frac{4a}{4} &= \frac{8}{4} \\ a &= 2 \end{aligned}$$

29. $$\begin{aligned} 9-7x &= 4(1-3x) \\ 9-7x &= 4-12x \\ 9-7x+12x &= 4-12x+12x \\ 9+5x &= 4 \\ 9-9+5x &= 4-9 \\ 5x &= -5 \\ \frac{5x}{5} &= \frac{-5}{5} \\ x &= -1 \end{aligned}$$

31. $$\begin{aligned} 2z-2 &= 5-(9-6z) \\ 2z-2 &= 5-9+6z \\ 2z-2 &= -4+6z \\ 2z-2z-2 &= -4+6z-2z \\ -2 &= -4+4z \\ -2+4 &= -4+4+4z \\ 2 &= 4z \\ \frac{2}{4} &= \frac{4z}{4} \\ \frac{1}{2} &= z \end{aligned}$$

33.
$$\begin{aligned} 5(6-2x) &= 2(5-3x) \\ 30-10x &= 10-6x \\ 30-10x+10x &= 10-6x+10x \\ 30 &= 10+4x \\ 30-10 &= 10-10+4x \\ 20 &= 4x \\ \frac{20}{4} &= \frac{4x}{4} \\ 5 &= x \end{aligned}$$

35.
$$\begin{aligned} 2(3b-5) &= 4(6b-2) \\ 6b-10 &= 24b-8 \\ 6b-6b-10 &= 24b-6b-8 \\ -10+8 &= 18b-8+8 \\ -2 &= 18b \\ \frac{-2}{18} &= \frac{18b}{18} \\ -\frac{1}{9} &= b \end{aligned}$$

Made in the USA
Middletown, DE
05 November 2026

36552530R10263